Advances in Dryland Farming in the Inland Pacific Northwest

Georgine Yorgey and Chad Kruger, editors

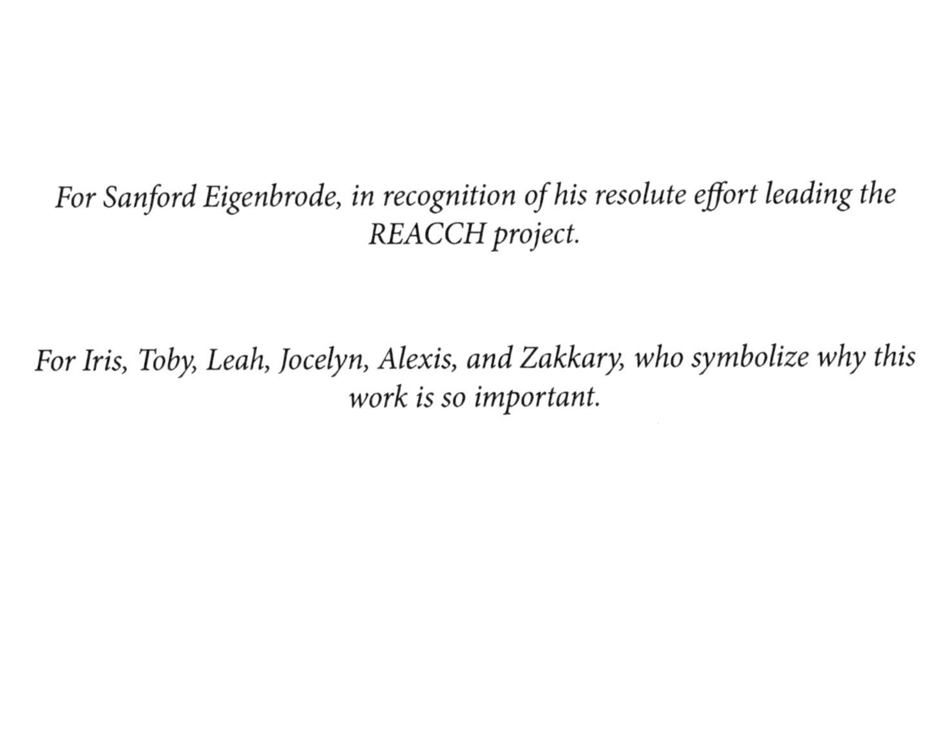

For Sanford Eigenbrode, in recognition of his resolute effort leading the REACCH project.

For Iris, Toby, Leah, Jocelyn, Alexis, and Zakkary, who symbolize why this work is so important.

College of Agricultural, Human, and Natural Resource Sciences

Use pesticides with care. Apply them only to plants, animals, or sites as listed on the label. When mixing and applying pesticides, follow all label precautions to protect yourself and others around you. It is a violation of the law to disregard label directions. If pesticides are spilled on skin or clothing, remove clothing and wash skin thoroughly. Store pesticides in their original containers and keep them out of the reach of children, pets, and livestock.

Advances in Dryland Farming in the Inland Pacific Northwest

Introduction

Georgine Yorgey, Washington State University
Chad Kruger, Washington State University

The Pacific Northwest is an important wheat production region. In 2015, the National Agricultural Statistics Service indicated that Washington, Idaho, and Oregon harvested more than 240 million bushels of wheat, worth an estimated $1.3 billion. The major areas of production in the inland Pacific Northwest are shown below, and incorporate both irrigated and dryland acreage.

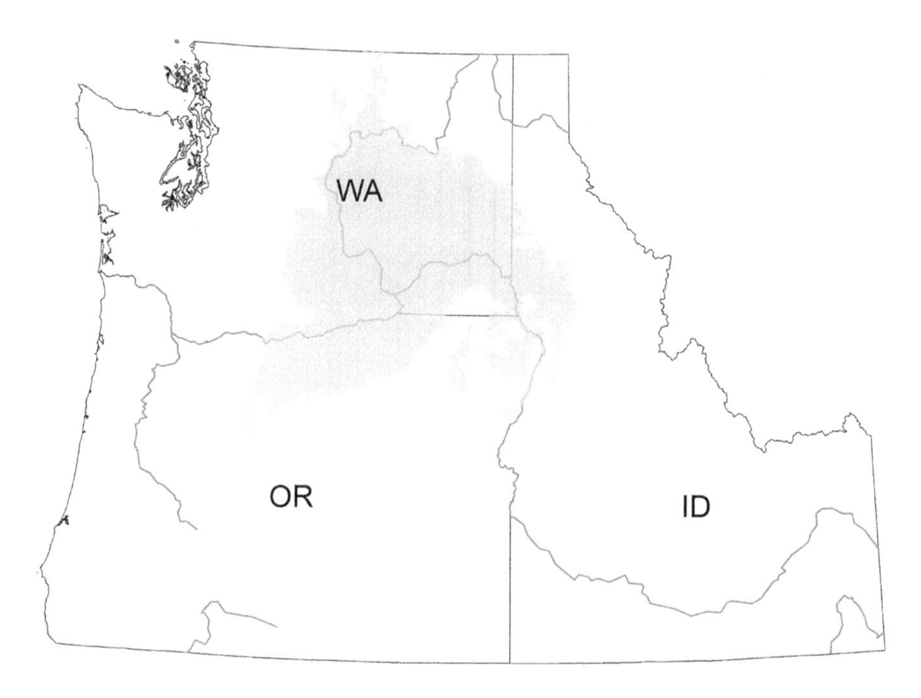

The Columbia Plateau ecoregion, commonly referred to by growers as the inland Pacific Northwest.

The area includes three major land resource areas with distinctive geologic features and soils as defined by the US Department of Agriculture: the Columbia Basin, the Columbia Plateau, and the Palouse and Nez Perce Prairies, all of which are within the Northwestern Wheat and Range Region. It also includes a small portion of dryland cropping in the North Rocky Mountains major land resource area, adjacent to the eastern edge of the Palouse and Nez Perce Prairies. In the dryland areas, which are the focus of this book, wheat is grown in rotation with crop fallow and much smaller acreages of other small grains, legumes, and alternative crops.

This area, identified from here forward by the more familiar term "inland Pacific Northwest," encompasses great diversity, characterized by some common overarching patterns of climate, geography, and agriculture. The inland Pacific Northwest extends eastward from the Cascade Mountain Range in Washington and Oregon into parts of northern Idaho. The landscape includes glacial deposits, coulees, channeled scablands, and rolling terrain with deep, fertile soil. The climate is semi-arid, with cool, wet winters and hot, dry summers.

Across the dryland wheat production areas, there are three major agroecological classes (AECs), with different patterns of cropping:

- Grain-Fallow AEC (defined as areas with greater than 40% fallow)
- Annual Crop-Fallow Transition AEC (with 10–40% fallow)
- Annual Crop AEC (with less than 10% fallow)

There is a considerable precipitation gradient across the region, with drier conditions in the rain shadow immediately east of the Cascades and wetter conditions further inland. The Grain-Fallow AEC is associated with lower precipitation areas, while the Annual Crop AEC is generally, but not always, associated with areas that receive higher levels of precipitation. These AECs, further described in Chapter 1, are dynamic and change as land use and land cover shift over time—with the potential to be influenced by climate, soils, terrain, land and commodity prices, and other factors. Because many recommendations in this book are specific to a farm's AEC characteristics, research results have been coded accordingly: Grain-Fallow ■; Annual Crop-Fallow Transition ▲; and Annual Crop ●. On the first page of each chapter, we have included a legend to help readers easily identify these symbols and the information most pertinent to their AEC.

Climate has always had a dominant influence on dryland production in this region, shaping crop choices, agronomic management systems, and conservation efforts. Though farmers are already highly skilled managers in the context of variable temperature and precipitation patterns, climate change is expected to add uncertainty, stretching the limits of existing management systems. Projected climate change also brings new urgency to questions of sustainability, as changing seasonal climate patterns may exacerbate conditions that have been historically linked to major soil erosion events in the region.

The inland Pacific Northwest has long faced challenges of soil erosion, soil organic matter depletion, and consequent soil fertility loss. One strategy for overcoming these challenges is managing for improved soil health through building and maintaining the continued capacity of soil to function as a vital living ecosystem that sustains plants and animals. Management of other biological, physical, economic, and policy components are also important for achieving a system that can sustain food and fuel production over the long term.

In light of ongoing and new challenges being faced by farmers in the region—and recent significant investments in research to help address these challenges—it is an opportune time to synthesize research-based advances in knowledge to support farmer decision-making and improve the long-term productive capacity of farmland in the region. This book should be viewed as a resource that launches further inquiry rather than an end point. Accordingly, the book includes both citations and links that direct readers to additional resources for more in-depth information. Additional and updated research-based information can also be found through several channels, including:

- The Regional Approaches to Climate Change (REACCH) project website, which houses information resulting from this wide-ranging research effort.
  www.reacchpna.org
- The Extension Libraries for University of Idaho, Washington State University, and Oregon State University.
- The WSU Wheat and Small Grains website.
  http://smallgrains.wsu.edu/

- The WSU Oilseed Cropping Systems website.
  http://css.wsu.edu/biofuels/
- The University of Idaho AgBiz website, which houses Extension economics information.
  www.idahoagbiz.com
- The Agriculture Climate Network, a site providing up-to-date information about research relevant to agriculture and climate change in the Pacific Northwest.
  https://www.agclimate.net/
- The NW Climate Hub, an effort by the USDA to deliver science-based knowledge and practical information to farmers, ranchers, forest landowners, and Native American tribes that will help them adapt to climate change.
  https://www.climatehubs.oce.usda.gov/northwest
- The joint USDA-WSU Long-Term Agroecosystem Research (LTAR) site provides information relating to long-term dryland cropping systems in the region.
  http://ltar.wsu.edu/

This book represents a joint effort by a multi-disciplinary group of research and Extension scientists from across the region. The undertaking was made possible with the support of the USDA National Institute of Food and Agriculture through the REACCH project. This six-year project aimed to enhance the sustainability of Pacific Northwest cereal systems and contribute to climate change mitigation. The REACCH project, led by the University of Idaho, also convened scientists from Washington State University, Oregon State University, the USDA Agricultural Research Service, and Boise State University. In addition to supporting a number of ongoing research and Extension activities and experimental sites across the region, the project provided an opportunity for new projects and collaborations.

Advances in Dryland Farming in the Pacific Northwest is organized into topical chapters that cover major management challenges for dryland farming in the region. Topics are interrelated, and we attempt to make links between chapters clear to the reader throughout.

Chapter 1, *Climate Considerations*, describes the temperature and precipitation patterns that have historically defined crops, yields, and cropping systems across the inland Pacific Northwest. Year-to-year variability in temperature and precipitation have also been an important feature. Human-caused climate change is forecasted to increase the frequency of temperature-induced drought conditions and late summer water deficits, with these changes likely to surpass historic year-to-year climate variability by mid-century. The frequency and severity of extreme weather events will also increase, and may increase production risk. Meanwhile, increased atmospheric carbon dioxide may benefit yields by increasing energy and water use efficiencies. The overall impact of these various factors is likely to vary across the region.

Chapter 2 introduces the concept of *Soil Health* and describes its vital role in sustainable agricultural production. It describes how measurement of the soil's physical, chemical, and biological indicators can be used to assess soil health, though the high degree of spatial variability in the region's soils poses challenges for the selection of indicators sensitive to management in soil quality assessment programs. Nonetheless, adaptive management appropriate to a specific site can benefit soil health, using practices including reduced tillage, cropping intensification, crop diversification, crop residue retention, and application of organic amendments.

Conservation Tillage Systems, Chapter 3, describes the challenges that are presented by conventional tillage-based cropping systems, including soil erosion, soil organic matter depletion, and soil fertility loss. It also describes how conservation tillage systems have been improved and increasingly adopted by growers in the inland Pacific Northwest to address these challenges. Appropriate conservation tillage systems vary across the inland Pacific Northwest, and successful implementation is influenced by crop rotations, equipment choices, residues, soil fertility, economics, and other factors.

The challenges of *Crop Residue Management*, covered in Chapter 4, vary across the region. Heavy residue produced in high-yielding areas can make planting difficult and contribute to unfavorable growing conditions in the early spring. In contrast, in areas with low or intermediate yields, additional residue is desired for enhancing soil and water conservation.

Strategic conservation of crop residue is important for preventing wind and water erosion, enhancing soil water recharge, maintaining soil health, and returning nutrients to the soil. Different residue management strategies, including conservation, harvest, and burning, involve tradeoffs between production, economics, environment, and soil and human health. Calculating immediate economic tradeoffs at the field scale and the within-field scale can support decisions about residue management practices.

Chapter 5 covers *Rotational Diversification and Intensification*. Diversity is low in the wheat-dominated cereal production systems of the inland PNW. Diversifying or intensifying cropping systems can help growers minimize lost production opportunities, improve farm productivity, increase grower income and flexibility, adapt to forecasted climate change, and achieve long-term environmental benefits. These can benefit wheat yields elsewhere in the rotation, an economic benefit that should be accounted for when evaluating potential returns for alternate crop rotations. However, adopting alternative rotations comes with tradeoffs and can also increase risk. Success depends on geographic location, production potential, rotational fit, market opportunity, crop price, and production costs. Specific diversification strategies of interest include inclusion of legumes and oilseeds. Meanwhile practices such as undercutter tillage fallow, no-till fallow, or flex cropping can build soil resiliency and increase opportunities for diversification and intensification in tilled grain-fallow systems.

Chapter 6 discusses *Soil Fertility Management*, which varies depending on precipitation gradients and landscape position that together affect crop fertilizer accessibility, nutrient use efficiencies, and growth. Practices that maximize nitrogen use efficiency include fertilizer placement, source, timing, and rates that match the nitrogen needs of the crop species and variety being grown. Appropriate management strategies and regular soil testing, along with quality recordkeeping, can reduce nutrient loss. This, in turn, can improve farm profitability and reduce the harmful effects of nutrients on air, water, and soil quality. Decreasing soil pH, due mostly to application of nitrogen fertilizers, along with plant nutrient uptake and precipitation, is a growing concern in some areas of the inland Pacific Northwest. Acidification can make nutrients less available to plants and can have a variety of other negative impacts that can decrease yields if not addressed. The chapter also discusses other nutrients such as sulfur and phosphorus that are necessary for wheat crops, but are less susceptible than nitrogen to being depleted annually.

Soil Amendments, the topic of Chapter 7, have historically had limited use in dryland systems of the inland Pacific Northwest, but could improve soil health by benefitting soil carbon, nutrient availability, soil structure, water infiltration and retention, bulk density, and microbial activity. Biosolids can be used by conventional producers, but not certified organic producers, and may be available at relatively low cost, though supply is limited. Manures and composts can be an important resource for building or maintaining soil quality for producers in proximity to concentrations of livestock. Manures with higher nutrient concentrations may be an important nutrient source for some certified organic dryland producers, though cost can be an issue.

Chapter 8, *Precision Agriculture*, describes how technology can be used to manage within-field variability. In cases where variability in the major factors that affect crop yield and quality can be accurately measured at scales relevant to farm management, and when this information can be used to improve the efficiency of crop input use, precision agriculture has the potential to improve yields and crop quality, increase economic returns, and decrease environmental impacts from excessive input use. Precision agriculture technologies can also be used to monitor crop and field variability and help diagnose problems that occur across fields and years.

Integrated Weed Management is discussed in Chapter 9. An integrated approach relies on knowledge and application of ecological principles, including an understanding of weed biology, plant interference, and weed-crop competition. Successful long-term approaches emphasize growing healthy, competitive crops in an effective cropping sequence supported by appropriate use of a variety of preventative, cultural, mechanical, and chemical strategies. Judicious use of herbicides alongside other strategies can ensure that weed management is effective, economical, and prevents the development or spread of herbicide-resistant weed biotypes. Management of selected problematic weeds of inland Pacific Northwest grain production is discussed: downy brome, Russian thistle, jointed goatgrass, and Italian ryegrass.

Disease Management for Wheat and Barley is discussed in Chapter 10. Successful disease management relies on integrated strategies including prevention, avoidance, monitoring, and suppression. The specific implementation of these strategies varies by pathogen and cropping system. This chapter presents information on geographic distribution, disease cycle, diagnostic features, conditions that favor disease, and management practices for selected diseases that affect inland Pacific Northwest cereal production, including stripe rust and those caused by soilborne fungal pathogens and nematodes. For many soilborne pathogens, genetic resistance or chemical controls are not available, and growers rely on cultural practices to favor plant health. System-wide monitoring of crop response is an important tool to determine if changes in cropping practices, or climate effects, reduce the effectiveness of current management strategies.

Chapter 11 discusses *Insect Management Strategies*. Similar to weeds and diseases, effective management of insect pests relies on an integrated approach. That depends on an understanding of each species' distribution, life cycle, crop damage caused, and potential control strategies. The chapter reviews the basic principles of integrated pest management and the implications of technology, a changing climate, and invasion by new pests. It offers summaries of the biology and management for 16 of the most important pests or pest complexes as well as what is known of their responses or potential responses to climate change. These responses are likely to occur through mechanisms including changes in the timing of pest activity, shifts in the geographical range of pests, reductions in the time needed for life cycle completion, and an increased number of generations per year. Finally, this information is placed within the context of the larger agroecosystem and production landscape.

Farm Policies and the Role for Decision Support Tools is the topic of Chapter 12. Policy influence occurs through avenues including the development of risk management options, management recommendations and incentives, and the adoption of agricultural technologies. These influences, in turn, affect management practices in inland Pacific Northwest grain production systems including use of conservation tillage, residue and soil water management, crop rotations, and pest management. The use of spatially explicit data and regional impact models will likely play a larger role in the design and implementation of future farm policies. Decision support tools can also be used to examine the impacts of a targeted conservation policy.

Considering all of the challenges facing dryland agricultural systems in the inland Pacific Northwest, the future can seem more daunting than ever. Fortunately, there is also evidence of significant overlap in the practices that may benefit farmers' bottom lines, enhance long-term productivity and resilience, and reduce dryland farming's contribution to atmospheric greenhouse gases. For example, reducing periods of fallow can potentially increase farm productivity and income, while also increasing the amount of residues that are incorporated into the soil over time. This can enhance soil organic carbon levels, providing a range of benefits in terms of improved soil structure, reduced soil compaction, improved water infiltration, and improved water holding capacity. All of these enhance productivity and resilience over the long run. Meanwhile, increased soil organic carbon allows carbon dioxide to be stored in the soil, reducing atmospheric carbon levels.

These and other practices may create win-win-win strategies: a win for producers by enhancing farm profitability, a win for sustainability and long-term productivity, and a win for climate. The table below summarizes some potential win-win-win strategies for the dryland agricultural systems of the inland Pacific Northwest. While promising, the win-win-win strategies identified here do not fully define a path toward a productive future in the context of a changing climate. There are still many gaps in our knowledge of dryland farming, key barriers that remain to be overcome before some promising strategies can be fully adopted, and uncertain futures in the region.

To be successful, growers need to be highly skilled in evaluating and applying knowledge and experience according to their site-specific context and needs. Research-based recommendations in the book may conflict with each other depending on the specific problem that is a priority for management. Growers will need to exercise caution in the application of information and recommendations contained in this volume based on an understanding of these tradeoffs. Utilizing farmer mentors, knowledgeable technical support providers, and University, USDA, and private-sector researchers will better enable growers to build

Win-win-win strategies can enhance farm profitability, climate, and long-term farm productivity.

Management Strategies	Short-Term Benefits (1-10 years)	Long-Term Benefits (40+ years)
Reduced Tillage/ Direct Seeding	• Decreased soil erosion and nutrient runoff • Increased SOM and improved soil quality • Increased nutrient cycling and storage	• Reduced GHG emissions by storing soil C
Crop Intensification-Reduce Fallow	• Increased food, fuel, and feed production • Increased farm productivity and income	• Removes additional CO_2 from atmosphere by increasing photosynthesis • Increased straw biomass and soil C sequestration
Crop Diversification-Legumes	• Improved control of pests and grass weeds using broadleaf crop in rotation • Reduced N fertilizer costs using biological nitrogen fixation	• Reduced GHG emissions and natural gas use during N fertilizer production • Reduced reactive soil N that leads to GHG emissions
Crop Diversification-Oilseeds	• Improved control of pests and grass weeds using broadleaf crop in rotation • Improved soil structure and water infiltration with canola's strong taproot • Glyphosate-resistant canola is the only Roundup-ready crop that can be grown in PNW rotations	• Increased net productivity, photosynthesis and C fixation • Reduced atmospheric CO_2 through increased soil C sequestration • Reduced GHG emissions and improved N cycling • Avoid summer heat and drought stress with a short season crop

Win-win-win strategies can enhance farm profitability, climate, and long-term farm productivity.

Management Strategies	Short-Term Benefits (1-10 years)	Long-Term Benefits (40+ years)
Customize Wheat Class and Variety to AEC	• Potential to improve protein premiums • Improved overall regional wheat quality and market reputation • Match heat and drought tolerance to AEC • Potential to adapt to pest variability	• Improved resource efficiency and lower loss, as crops are better suited to environment • Tolerant varieties are more adaptable to climate change and associated concerns
Precision N Management	• Reduced N fertilizer costs • Reduced N over-fertilization that can reduce yields • Reduced N runoff and loss	• Reduced GHG emissions and natural gas use during N fertilizer production • Reduced reactive soil N that leads to GHG emissions
Recycle Organic Byproducts as Soil Amendments	• Increased SOM and improved soil quality • Reduced N fertilizer costs • Recycled valuable nutrients • Reduced landfilling of biological wastes	• Reduced GHG emissions and natural gas use during N fertilizer production • Tightened global nutrient cycle reduces GHG emissions

Table by Bill Pan and Kristy Borrelli.

Abbreviations: SOM = soil organic matter; C = carbon; CO_2 = carbon dioxide; N = nitrogen; AEC = agroecological class; GHG = greenhouse gas; PNW = Pacific Northwest.

on the information contained in this volume. Ultimately, the best hedge against the impact of climate change is highly skilled and creative farmers utilizing the best available research and management information to make decisions in an uncertain context.

Acknowledgements

This material is based upon work that is supported by the National Institute of Food and Agriculture, US Department of Agriculture, under award number 2011-68002-30191 (Regional Approaches to Climate Change for Pacific Northwest Agriculture). Chapter 12 is also based upon work supported by the National Institute of Food and Agriculture under award number 2014-51181-22384 (National Needs Graduate and Postgraduate Fellowship Grants Program), Graduate Education in the Economics of Mitigating and Adapting to Climate Change: Evaluating Tradeoffs, Resiliency and Uncertainty using an Interdisciplinary Platform, The Northwest Climate Hub, and Oregon Agricultural Experiment Station.

A project like *Advances* can only be undertaken by a large and committed team to provide the depth and breadth of content that comprises such a comprehensive resource. Literally dozens of researchers, science writers, and support staff contributed to the development of the book over a three-year period. We want to specifically acknowledge Stephen Machado and Kristy Borrelli, who worked with the editors to develop the conceptual plan for this book. All chapters benefitted from the suggestions of convening chapter authors and co-authors, who coordinated the development of chapters, commented on each other's early drafts, and worked with each other to improve the overall quality and integrity of the book. We are also extremely grateful for dozens of high-quality, anonymous peer reviews from numerous scientists throughout the region with relevant expertise—without which this book would not have happened. Sonia Hall reviewed drafts of chapters and provided a number of insightful suggestions to make the book more cohesive, while Karen Hills patiently worked out many details of consistency between chapters. This project also would not have been possible without the WSU editorial and publishing team—with particular recognition to Todd Murray, Therese Harris, Gerald Steffen, Lagene Taylor, Christina Mangiapani, and Melissa Smith.

Chapter 4 authors would like to thank Rakesh Awale for providing a suggestion for a figure that augmented the chapter. Chapter 11 benefitted from a constructive critical review provided by Stephen Clement.

Chapter 1

Climate Considerations

Chad Kruger, Washington State University
Elizabeth Allen, Washington State University
John Abatzoglou, University of Idaho
Kirti Rajagopalan, Washington State University
Elizabeth Kirby, Washington State University

Abstract

Agricultural systems in the inland Pacific Northwest (PNW) have evolved under a Mediterranean-type climate characterized by warm, dry summers and cool, wet winters. Precipitation is the primary limiting factor of production for most of the dryland wheat-growing region. The Cascade Mountain Range creates a rain shadow in its immediate lee that results in a considerable precipitation gradient: drier immediately east of the Cascades with wetter conditions further inland. While year-to-year variability in precipitation is considerable, the climate of the inland PNW has proven sufficiently stable to support dryland small grain production systems that rival the productivity of nearly every other rainfed cereal-producing region of the world. Production practices that enhance the dependability and sustainability of dryland cropping systems in the inland PNW are vitally important in the context of global climate change. Managing under observed and projected climate change impacts will require producers to develop their understanding of changing production uncertainties and risks. This chapter discusses how climatic factors influence regional agricultural systems, considers projected impacts of climate change on inland PNW dryland small grain production, and

Research results are coded by agroecological class, defined in the glossary, as follows:

● Annual Crop ▲ Annual Crop-Fallow Transition ■ Grain-Fallow

explores how producers can learn from, and apply, information from models to a broad range of production management decisions.

Key Points

- Within the Pacific Northwest, local climate patterns and production practices are highly variable from one location to another. The region also has considerable inter-annual variability in precipitation and temperature.
- Regional models project that human-caused climate change will lead to an average annual temperature increase of 3–4°F by the 2050s and 4–6.5°F by 2100, with increased warming during the summer season. Annual precipitation is projected to increase by about 5–15% by 2050. However, summer precipitation is projected to decrease, resulting in reduced soil moisture during the late summer months. These projected shifts in average regional temperature and precipitation surpass current year-to-year climate variability. Climate change will also increase the frequency and severity of extreme weather events.
- Impacts of climate change on dryland wheat production may be variable throughout the region and variable at different time horizons. Increased atmospheric carbon dioxide may benefit yields by increasing energy and water use efficiencies. At the same time, changing temperature and precipitation regimes may also present new risks and challenges for producers.
- Opportunities exist for producers to alter practices to adapt to climate change impacts and reduce greenhouse gas emissions; these adaptation and mitigation strategies often have co-benefits for long-term sustainability.
- Climate models and decision support tools are rapidly evolving. New web-based resources are available to aid producers in obtaining climate information for specific agricultural management decisions.

Introduction

The PNW is characterized by a temperate, Mediterranean-type climate, typically with cool to cold, wet winters and warm to hot, dry summers.

The region's climate is highly variable over the seasons. At the same time, temperature and precipitation patterns in the PNW are also highly variable over space; topography has a strong influence on local climates. Diverse ecosystems within the PNW include old growth evergreen forests, shrub steppe rangelands, and agricultural zones suitable for a diverse array of crops including rainfed (dryland) grain and forage, irrigated grain and forage, wine grapes, vegetables, oilseeds, and tree fruits. This chapter begins with a description the PNW's historical climate, drivers of variability in temperature and precipitation and an overview of projected climate change impacts in the region. Next, climatic characteristics and diversity of the **inland PNW**'s dryland wheat production systems are discussed. We consider how these agricultural systems will be affected by climate change. Climate change impacts may require producers to consider **adaptation** measures in order to support the long-term sustainability of their operations, and there may be emerging incentives to invest in **mitigation** measures that enhance soil carbon storage and reduce greenhouse gas emissions associated with agricultural production. Finally, this chapter presents decision support tools that enable producers to learn about how climate change and variability may affect their operations. This growing array of informational resources for agricultural decision-makers can support identification of locally appropriate management strategies.

Overview of the Climate of the PNW

The Cascade Mountain Range divides the PNW into climatically distinct areas. The high elevation mountain range causes eastward-moving storms to lose their moisture as rain or snow when the storm systems move inland (cooler temperatures at high elevations mean that air is able to hold less moisture). There is a steep downward slope to low valleys east of the Cascades, and as air passes over the Cascades and warms as it descends, condensation and precipitation are even less likely. This process results in a "rain shadow" effect for much of the Columbia River Basin. Leeward of the Cascade Mountains, the maritime influence is limited and mean annual temperature is primarily influenced by elevation. The warmest locations are typically seen at lowest elevations of the Columbia River Basin. Average temperatures are cooler as the topography rises toward the Northern Rockies. Figure 1-1a shows historical (1981–2010) average annual maximum temperature across the PNW. Figure 1-1b shows historical

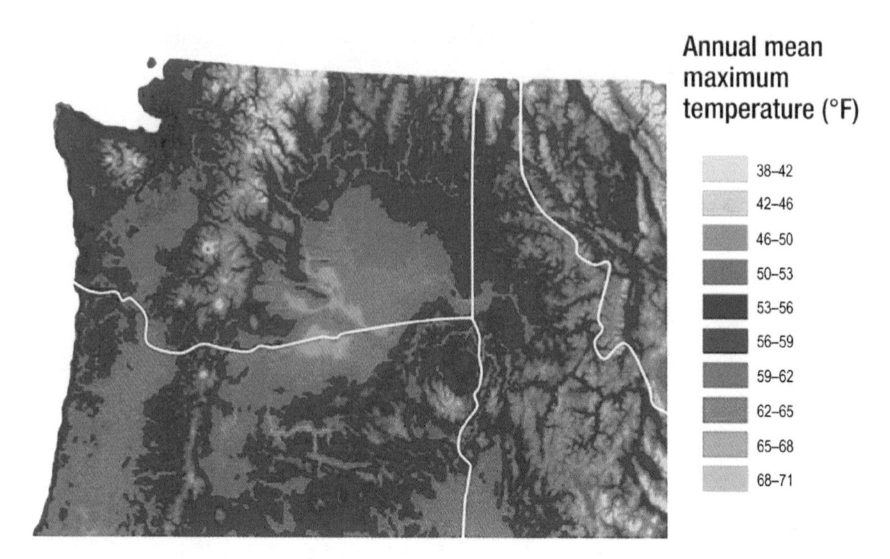

Figure 1-1a. Distribution of annual mean (1981–2010) maximum temperature. (Copyright 2016, PRISM Climate Group, Oregon State University.)

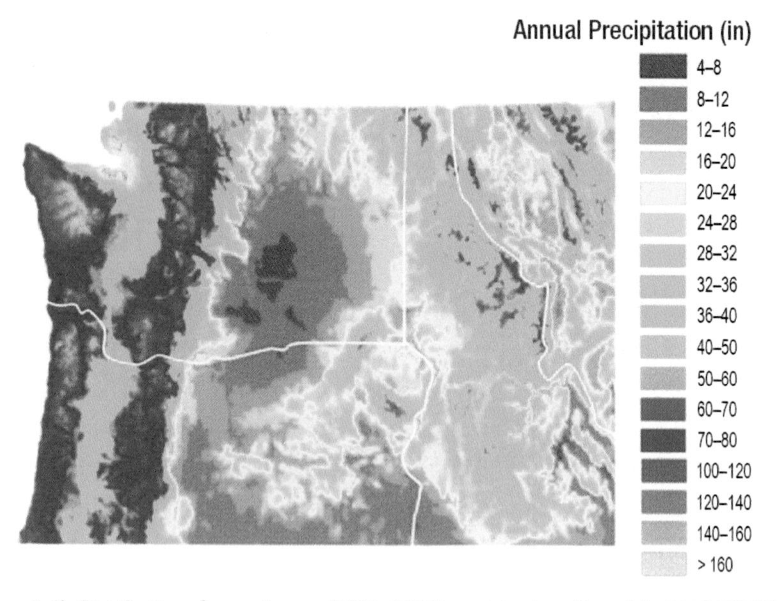

Figure 1-1b. Distribution of annual mean (1981–2010) precipitation. (Copyright 2016, PRISM Climate Group, Oregon State University.)

(1981–2010) average annual precipitation throughout the region.

Substantial year-to-year variability is evident in mean annual temperature (Figure 1-2a) and precipitation (Figure 1-2b) averaged across the PNW.

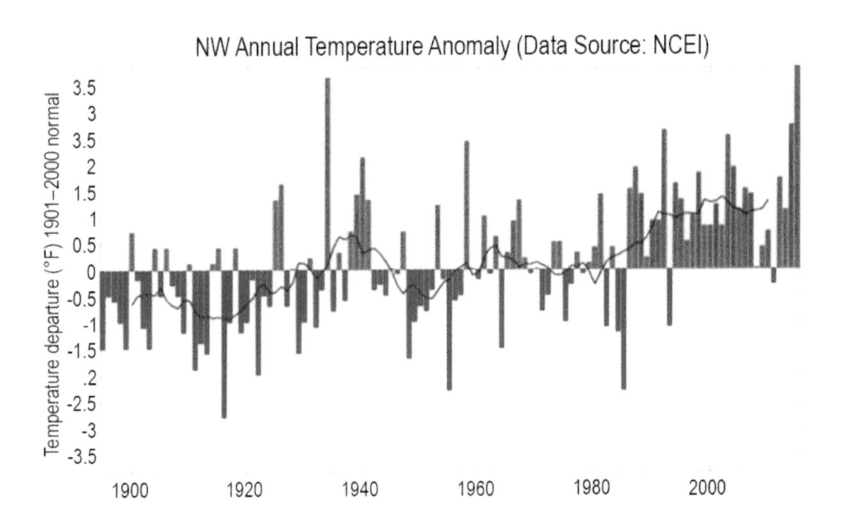

Figure 1-2a: Annual mean temperature anomaly for the National Centers for Environmental Information (NCEI) Northwest region (Washington, Oregon, and Idaho) dataset from 1900 to 2015. Anomalies are taken with respect to the 1900–2000 period. (Figure courtesy of John Abatzoglou.)

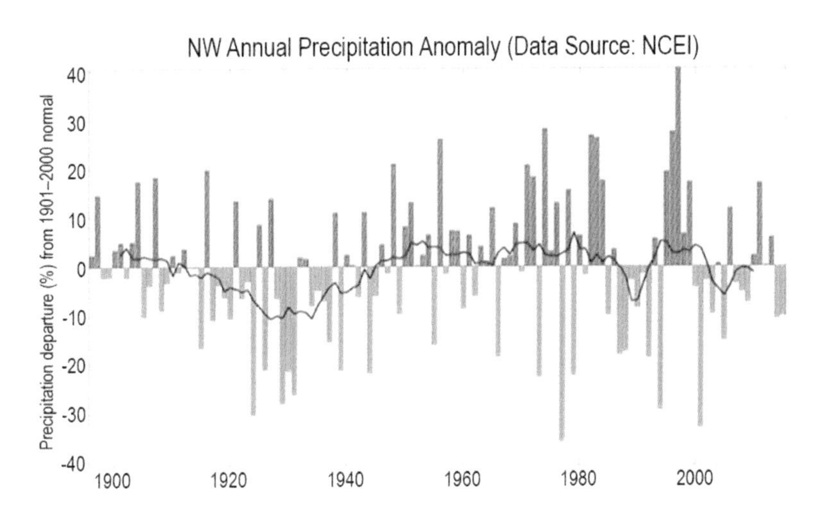

Figure 1-2b: Annual mean precipitation anomaly per water year for the National Centers for Environmental Information (NCEI) Northwest region (Washington, Oregon, and Idaho) dataset from 1900 to 2015. Anomalies are taken with respect to the 1900–2000 period. (Figure courtesy of John Abatzoglou.)

The two primary drivers of natural climate variability in the region are the El Niño Southern Oscillation (ENSO) and the Pacific Decadal Oscillation (PDO). These two cyclical patterns affect the climate of the region on annual to decadal timescales. In their warm phases (El Niño conditions for ENSO or positive phase PDO), the chance of a warmer than average PNW winter and spring increases. Cool phase conditions (La Niña conditions for ENSO and negative phase PDO) increase the odds that PNW winters will be cooler and wetter than an average year. For a detailed discussion of the link between ENSO and PDO cycles and regional temperature and precipitation variability, readers are encouraged to consult "Seasonal Climate Variability and Change in the Pacific Northwest of the United States" by Abatzoglou et al. 2014.

The warmest 10-year period since 1900 in the PNW occurred from 1998 to 2007, and very few years since 1980 have had below average annual mean temperatures. No long-term trends in precipitation are evident across the PNW except for an increase in spring precipitation (Abatzoglou et al. 2014). However, multi-decadal variability in annual precipitation is evident with protracted deficits in precipitation apparent in the 1920s and 1930s and relatively wetter conditions in the 1970s through the mid 1980s.

Earth's physical and biological systems are changing in many ways as a result of anthropogenic, or human-caused, greenhouse gas emissions. Observed and projected changes include rising sea levels, changes in ocean chemistry, warming oceans and air, and changing storm patterns. Consistent with a trend of global and national temperature increase throughout the 20th century, average annual temperatures in the PNW have increased (Abatzoglou et al. 2014). In general for the region, increasing temperatures are linked to longer frost-free seasons, decreased spring mountain snowpack, earlier peak stream flow, and decreased glacial area. Although mean annual temperature is the most frequently cited global indicator of climate change, seasonal temperature and precipitation at regional scales provide more salient links to climate impacts that may be otherwise masked in mean annual temperature (Abatzoglou et al. 2014). Current best estimates of global climate change impacts in the PNW indicate that annual average temperature increases of 3–4°F are expected by the 2050s, with annual average warming of 4–6.5°F projected by 2100 (Abatzoglou et al. 2014). Warming trends are projected to be greatest during the summer months. Annual precipitation is projected to increase 5–15% by the middle of the latter half of the 21st century (Vano et al. 2010). While total annual precipitation is projected to increase, summer season precipitation is expected to decrease and, combined with elevated summer temperatures, this could result in substantial reductions in late summer soil moisture (Vano et al. 2010).

Additional key impacts of forecasted climate change on agriculture in the inland PNW include increased frequency of temperature-induced drought conditions, greater temperature extremes (both minimum and maximum temperatures), and extreme precipitation events due to the fact that warmer air holds more moisture and is linked to more energetic storm events, which could lead to increased soil erosion and flooding (Binder et al. 2010). Warming leads to a longer available growing season, but can result in a shortened actual growth season for most crops due to accelerated growth. Later maturing crops may have reduced quality resulting from heat or inadequate moisture stressors during grain fill (Ortiz et al. 2008). Climate change may also affect the range and severity of agricultural **pests**, plant diseases, and invasive species. Anthropogenic greenhouse gas emissions also increase the concentration of carbon dioxide (CO_2) in the atmosphere, which impacts plant growth and water usage (Stöckle et al. 2010). Over the near-term in the inland PNW, increasing atmospheric CO_2 may offset the negative impacts of climate change on cereal production and may contribute to increased dryland yields because of increased **water use efficiency** in systems that are generally water-limited.

Diversity of Inland PNW Dryland Agricultural Systems

This section of Chapter 1 explores how variations in local climate and topography shape the characteristics of dryland agricultural systems in the inland PNW. The inland PNW's wheat production area corresponds approximately with the Columbia Plateau Ecoregion defined by the US Environmental Protection Agency (2011). This geographic area also parallels the study area for the Regional Approaches to Climate Change (REACCH) research effort, which focuses on dryland farming of the inland PNW where the predominant crops are small grains (wheat/ barley), peas, and canola (Figure 1-3).

Throughout the inland PNW, winter weather is cool to cold with December and January mean daily temperature averaging 30°F and occasionally

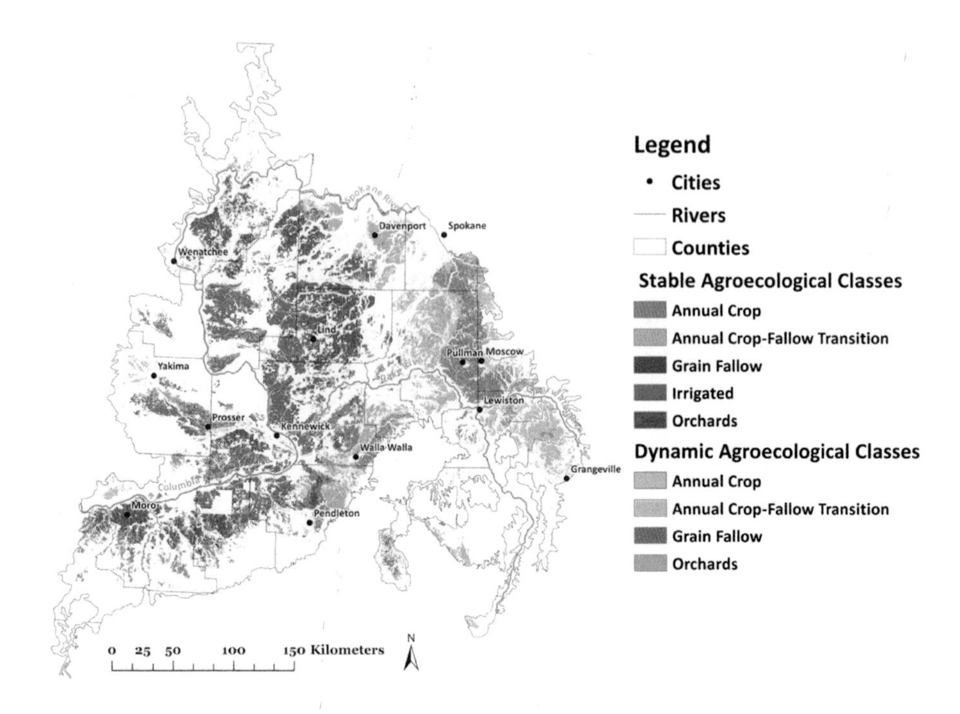

Figure 1-3. Agroecological classes (AECs) of the inland PNW based on 2007–2014 National Agricultural Statistics Service (NASS) Cropland Data Layer data. Much of the heterogeneity in cropping systems reflects climatic constraints, primarily temperature and precipitation regimes, and availability of water for irrigation. Production systems are diverse within each AEC as well. (Huggins et al. in preparation.)

dropping to 14°F or lower (Schillinger et al. 2010). In the summer months, high pressure systems lead to warm and dry conditions with low relative humidity. Across the region, average afternoon temperatures in the summer range between 68 and 95°F (Schillinger et al. 2010). The highest annual mean temperatures occur in the southwest portion of the inland PNW, with cooler annual average temperatures moving to the northeast (Figure 1-1a).

An estimated 70% of the inland PNW's precipitation occurs from October to March and 25% occurs during April to June. July through September are the driest months (Schillinger et al. 2010). As can be seen in Figure 1-1b, there is substantial variation in mean annual precipitation across the inland PNW. A gradient exists from the low precipitation zone in the western portion of the dryland wheat production area (average annual precipitation of 6") to the high precipitation zone in the eastern portion (average annual precipitation on 24") (Schillinger et al. 2010). Often,

Soil Orders of the Inland PNW

Alfisol. Leached basic or slightly acid soil with clay-rich subsurface layer. Alfisols have water available to vegetation for more than half the year or more than 3 consecutive months during a warm season. Alfisols are primarily formed under forest or mixed vegetative cover and are productive for most crops.

Andisol. Soil with high phosphorus retention, available water capacity, and cation exchange capacity. Andisols are most commonly formed from volcanic materials with high proportions of silica, aluminum, and iron-rich compounds. These soils are common in cool areas with moderate to high precipitation and are typically very productive.

Aridisol. Saline or alkaline soil with little organic matter. Aridisols have no available water during most of the time that the soils are warm enough for plant growth and occur in regions with less than 9″ precipitation per year. The vegetation in many areas consists of scattered ephemeral grasses and shrubs. Some Aridisols support limited grazing. If irrigated, many aridisols are suitable for crops.

Entisol. Mineral soil that is not differentiated into distinct soil horizons. The absence of distinct soil horizons may be the result of an inert parent material, insufficient time for horizons to form, occurrence on steep slopes where erosion occurs more rapidly than soil horizon formation, or recent mixing of horizons.

Inceptisol. Freely draining soil that does not have sharply defined soil layers. Inceptisols have water available to plants for more than half the year or more than 3 consecutive months during a warm season and one or more distinct soil horizons. Inceptisols have a wide range of characteristics.

Mollisol. Soil with a dark surface layer rich in organic matter and containing high concentrations of calcium and magnesium. Mollisols characteristically form under grasslands in climates that have a moderate to pronounced seasonal moisture deficit or under forest ecosystems. These soils are very productive for crops.

studies of the region focus on mean annual precipitation as a distinguishing characteristic of different agricultural zones. Generalized precipitation classes for the dryland wheat production region are as follows: high (18"+); intermediate (12–18"); and low (less than 12") mean annual precipitation (Schillinger and Papendick 2008). Roughly 50–60% of the inland PNW's dryland crop production acres occur within the low precipitation zone (Schillinger et al. 2003; Schillinger and Papendick 2008).

Along with precipitation, topography and soil characteristics play a central role in shaping agricultural production. The inland PNW ranges from more gently rolling topography in the west to steep rolling hills in the east. The predominant agricultural soils of the dryland wheat-producing region are silt loams, formed in windblown silt (loess) deposits of depths varying from 3–20 feet in the western portion of the inland PNW and up to 200 feet deep in the east (Douglas et al. 1992). These loess deposits occur over basalt bedrock or flood-deposited sediments (McClellan et al. 2012). Soils are dominated by Mollisol and Aridisol soil orders, but Entisols, Andisols, Alfisols, and Inceptisols are present in localized areas (USDA 1999). (See the Soil Orders of the Inland PNW sidebar for descriptions of these soil orders.) In the drier, warmer western region of the inland PNW, soils developed under steppe and shrub-steppe vegetation, dependent on precipitation (Schillinger et al. 2010). These dry zone soils are relatively low in organic matter, more sandy, and susceptible to wind erosion (Schillinger et al. 2010). Moving to the northeast, in the cooler, high precipitation zone of the inland PNW (including the Palouse Hills), the loess soils developed under native grasslands or forests (Schillinger et al. 2010). These soils are richer in organic matter and clay, and typically susceptible to water erosion when snowmelt or rain runs over recently thawed topsoil (Schillinger et al. 2010; McClellan et al. 2012). In the high precipitation zone, crops are often produced on slopes with an 8% to 30% grade, with some production on slopes as steep as 45% (Schillinger et al. 2003). Across the inland PNW, soil organic matter (SOM) ranges from more than 3% in the high precipitation zone to less than 1% in the low precipitation zone (Schillinger et al. 2006).

Numerous systems have been devised to subdivide the inland PNW into agricultural and ecological zones to communicate recommendations

Table 1-1. Agronomic zones.

Zone	Name	Mean Annual Precipitation	Soil Depth	Cumulative Growing Degree Days (1 Jan-31 May)	Average Winter Wheat Yields (bu/acre)	Soil Organic Matter (%)	Water Holding Capacity (in/ft)
1	Annual crop-wet-cold	Over 16"	—	Under 700; Cold	70–90	4+	2.2–2.6
2	Annual crop-wet-cool	Over 16"	—	700-1000; Cool	80–120	3–4	2.0–2.4
3	Annual crop- fallow - transition	14-16"	Over 40"; deep	700-1000; Cool	60–80	2–3	1.8–2.2
4	Annual crop-dry	10-16"	Under 40"; shallow	Under 1000; Cool	30–40	<1.5	1.6–2.0
5	Grain- fallow-dry	Under 14"	Over 40"; deep	—	40–60	<1.5	1.6–2.0
6	Irrigated	—	—	—	—	—	—

Adapted from Douglas et al. 1999, Table 5.2.

for management, model economic trends, and to study crops, pests and disease vulnerabilities. The low, intermediate, and high precipitation zones defined by Schillinger and Papendick (2008) and discussed earlier in this section are an example of one such classification system that is useful for producers and land managers. Many growers are also familiar with the six agronomic zones defined by Douglas et al. (1992) in which the following criteria are used to differentiate zones: mean annual precipitation, soil depth, and **growing degree days** (Table 1-1). Ranges of average winter wheat yields, soil organic matter, and soil **water holding capacity** have been specified for each of the agronomic zones (Douglas et al. 1999). Agronomic zones 2 through 5 make up the inland PNW's dryland wheat-producing region (Douglas et al. 1992).

Recently, three **agroecological classes** (AECs) have been defined using National Agricultural Statistics Service (NASS) cropland use data to characterize the diversity of agricultural practices in the inland PNW's dryland cereal production region (Kaur et al. 2015). These three classes are: (1) Annual Crop with less than 10% **fallow**, (2) Annual Crop-Fallow Transition with 10–40% fallow, and (3) Grain-Fallow with greater than 40% fallow (Table 1-2). The AEC classification system differs from other classification systems that have been applied in the region because

Table 1-2. Three agroecological classes (AECs) of the inland PNW.

Class	Percent Fallow	Common Grower Practices
Annual Crop ●	<10% fallow	3- or 4-year crop sequence; e.g., winter wheat-spring wheat, or barley-spring broadleaf, or winter wheat-spring grain-winter wheat-spring broadleaf.
Annual Crop-Fallow Transition ▲	10–40% fallow	2- or 3-year crop sequence; e.g., winter wheat-fallow or winter wheat-spring wheat-fallow. Crop choice is more limited by available water.
Grain-Fallow ■	>40% fallow	Typical 2-year crop sequence is winter wheat-fallow. Growers rely on fallow practices to store and retain winter precipitation in the soil profile to establish winter wheat.

distinctions among classes are based on actual land use rather than the biophysical characteristics that play a role in shaping those land-use practices.

The Annual Crop AEC is generally associated with high precipitation zones, while the Grain-Fallow AEC is associated with low precipitation zones. However, AECs are dynamic, changing as land use and land cover shift over time. Figure 1-3 displays the AECs in the inland PNW, including dynamic regions, which have recently transitioned from one use to another (Kaur et al. 2015). Focusing on distinctions among agricultural regions based upon actual land use allows researchers to analyze relationships among biophysical variables (e.g., climate, soils, terrain) and socioeconomic variables (e.g., land prices, commodities grown) (Huggins et al. in preparation). This framework for classification also enables researchers to predict changes in AECs linked to shifts in climate, markets, and management practices. Preliminary analyses of changing grower practices suggest that annual crop regions will decrease, being converted to annual crop-fallow transition systems—this would significantly reduce diversification and increase soil vulnerability to erosion due to increased fallowing (Kaur et al. 2015). Grain-fallow systems are less likely to be affected (Kaur et al. 2015). Figure 1-3 shows the AECs of the inland PNW in 2015, pale colors indicate that a particular area changed from a different prior use to the current AEC over the period of data collection (2007–2014).

Synthesis of Recent Research Developments

Modeling Climate Change

Global climate models (GCMs) simulate oceanic and atmospheric behavior at a coarse spatial resolution (grid cells of 4,000 square miles or more). There are many different GCMs, which differ in how they account for various climatic variables. While essential for understanding global patterns, GCMs have limited direct utility in regional studies because they do not include sufficient detail to capture local-scale spatial variability in climate. To address this limitation, projections of future climate are downscaled to a finer spatial scale. For example, adjustments are made to outputs from GCMs to better capture the rain shadow effect caused by the Cascade Mountains (Abatzoglou and Brown 2012; Mote and Salathe 2010).

It is important to run climate models based on a range of different storylines about future political, economic, technological, and social decisions, all of which affect greenhouse gas emissions. In order to make results from various GCM analyses comparable, the Intergovernmental Panel on Climate Change (IPCC) defined a standard set of possible greenhouse gas emissions pathways. The 5th IPCC Assessment Report compares four different Representative Concentration Pathways (RCPs), described in Table 1-3 (IPCC 2014). The RCPs considered in the IPCC 5th Assessment Report are named according to projected radiative forcing (sunlight absorption) values in the year 2100 relative to pre-industrial values (+2.6, +4.5, +6.0, and +8.5 Watts per square meter). For reference, the additional radiative forcing in 2011 was estimated as +2.2 W/m^2. This simplistic approach gives researchers the ability to place boundaries on the conditions for forecasting while accounting for a range of possible greenhouse gas emissions scenarios and global economic growth trajectories.

Future Climate Projections for the Inland PNW

Understanding the interactions of anthropogenic climate change and natural drivers of climate variability is a key area of research. As described in the introduction to this chapter, climate change is occurring against a background of ongoing year-to-year climate variability. We expect significant regional climate variability to continue in the future even under climate change, with the magnitude of the climate change signal equaling and surpassing the historical climate variability signal for mean annual temperature in the latter half of the 21st century

Table 1-3. Representative concentration pathways used in IPCC Assessment Report 5.

IPCC Representative Concentration Pathways (RCPs) and generalized emissions scenarios	
RCP 2.6	Global annual greenhouse gas emissions (measured in CO_2 equivalents) peak between 2010–2020, with emissions declining substantially thereafter
RCP 4.5	Emissions peak around 2040, then decline
RCP 6	Emissions peak around 2080, then decline
RCP 8.5	Emissions continue to rise throughout the 21st century

(Abatzoglou et al. 2014). The most apparent warming trend in the PNW is for the coldest night of the year, which has warmed significantly in recent decades (Abatzoglou et al. 2014). The only cooling trend observed in the region was for spring temperatures over the last three decades, and this trend is tied to climate variability (Abatzoglou et al. 2014). Future climate projections consistently forecast that the annual maximum temperatures will increase. There will still be cool years, but the frequency at which they occur is expected to dwindle. Figure 1-4 shows projections of future mean annual temperature and precipitation in southeast Washington for two different RCPs. Even under a scenario of moderate global greenhouse gas emissions that peak in 2040 and decline thereafter, substantial temperature and precipitation shifts are projected for the region by mid-century.

Growing degree days are a measure of heat accumulation used by crop models to project plant development rates and determine when crops will reach maturity. Under a scenario of greenhouse gas emissions continuing to increase throughout the 21st century, planting dates, crop maturity dates, feasibility of double cropping systems, and pest and disease pressures may be dramatically affected by warming (Figure 1-5).

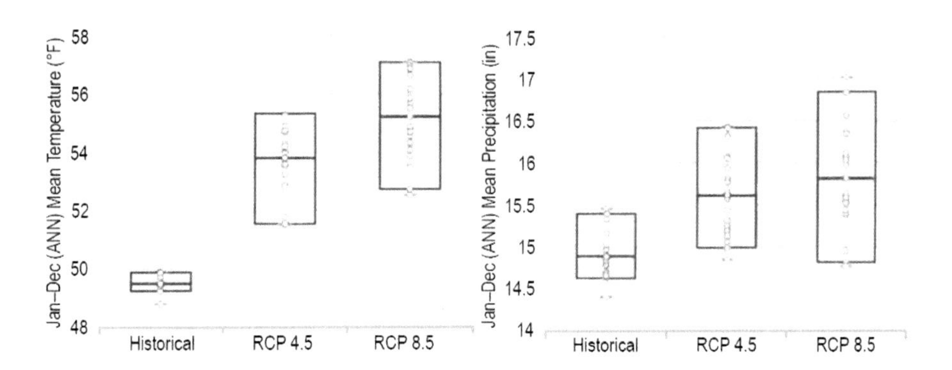

Figure 1-4. Projected changes in mean annual temperature (left) and precipitation (right) for southeastern Washington (45.6-47.8N, 117-120.5W). Historical mean annual temperature and precipitation (1971–2000) are compared with projected annual means for 2040–2060 under two different representative concentration pathways (RCPs). (RCP 4.5 indicates greenhouse gas emissions peaking in 2040 then declining, RCP 8.5 indicates emissions continuing to rise in the 21st century. Figure courtesy of John Abatzoglou.)

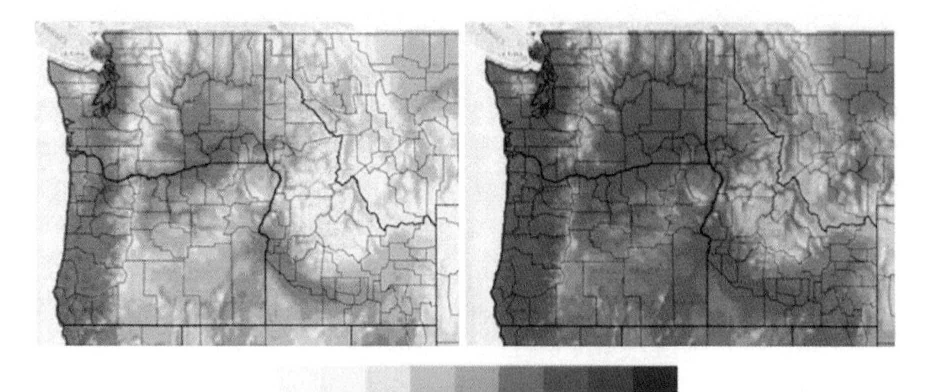

Figure 1-5. Cumulative growing degree days (base 32°F) 1971–2000 (left) and 2040–2069 representative concentration pathway (RCP) 8.5 (right), projections obtained from the AgClimate atlas. See the section about decision support tools at the end of this chapter for more information on how to access and interpret projections such as this. (Figure courtesy of John Abatzoglou.)

Impacts of Climate Change on Small Grains

Dryland small grain production regions are vulnerable to projected reductions in summer precipitation, decreases in summer soil moisture and extended warm weather episodes, which potentially reduce yields or exacerbate production challenges on marginal lands. Wheat and alternative crops are vulnerable to heat stress that can accelerate wheat senescence (the period between maturity and death of a plant) and reduce photosynthesis, which causes shriveling and negatively affects grain quality (Ferris et al. 1998; Ortiz et al. 2008). Warmer, drier conditions are also expected to exacerbate wind-caused soil erosion and reduce early stand establishment of winter wheat on summer fallow (Eigenbrode et al. 2013). Winter wheat could benefit from warmer winters, but may be challenged if drier summers impede late summer and fall planting of these crops, reducing germination and stand establishment. Pest and disease phenology shifts, or changes in the timing of emergence and species ranges, are also projected to impact inland PNW small grain production systems (Eigenbrode et al. 2013). (See Chapter 10: Disease Management for Wheat and Barley and Chapter 11: Insect Management Strategies.)

Projected changes to precipitation patterns in the PNW under climate change are highly variable and more difficult to project than changes in temperature. Cool season (October-March) precipitation is generally projected to increase in the region (Vano et al. 2010). Depending on the timing of precipitation, planting of spring wheat could become more challenging, reducing yields or causing a shift toward more winter cropping (Stöckle et al. 2010). At the same time, increased soil moisture early in the growing season may mitigate the effects of projected reductions in summer precipitation (Stöckle et al. 2010).

Based on crop models, potential yield losses from projected climate change impacts alone would be severe by the end of the 21st century (Stöckle et al. 2010). However, increased atmospheric CO_2 that contributes to more rapid plant development and growth (called **CO_2 fertilization**) and agronomic adaptation are expected to offset the negative effects of climate change on small grain crop yields in the inland PNW. Tubiello et al. (2002) projected US West Coast non-irrigated winter wheat production to increase 10–30% by 2030, relative to baseline climate (1951–1994).

Using the crop model CropSyst, four GCMs, and assuming a scenario of rapid economic growth and technological development with a balanced reliance on fossil fuel-based and renewable energy sources, Stöckle et al. (2010) projected dryland winter wheat yield increases of 13–15% by the 2020s, 13–25% by the 2040s, and 23–35% by the 2080s for a range of locations across Washington state relative to baseline climate (1975–2005) when warming and CO_2 fertilization were included (Stöckle et al. 2010). Dryland spring wheat yields for a range of locations across Washington were projected to change by +7% to +8% by the 2020s, -7% to +2% by the 2040s, and –11% to +0% by the 2080s (Stöckle et al. 2010). The range of values obtained depend upon the production zone and planting date, with lower increases or deficits in lower precipitation zones and better performance occurring if planting is adjusted earlier in the season, avoiding higher temperatures during vulnerable stages (Stöckle et al. 2010). These projections are based on changes in mean temperatures and do not consider the frequency of extreme heat events, which could negatively affect yields, but for which projections are less certain.

Grower Considerations

Recent sociological research suggests that the general public of the inland PNW is increasingly concerned about climate change and that there may be growing public interest in regulation of agricultural practices to support climate change adaptation and mitigation (Wulfhorst et al. 2015). An increasing number of growers expect that future climate change will create conditions that necessitate modifying their agricultural management practices (Bernacchi et al. 2015). This section begins with a discussion of various approaches to climate forecasting at different temporal and spatial scales. Appropriate modeling tools with respect to specific kinds of decisions are described. Next, we outline emerging research about production practices to mitigate and adapt to climate change. Finally, we consider the potential for future economic and regulatory changes linked to climate change impacts.

Climate Forecasting

Producers are accustomed to looking at 7–10 day weather forecasts to make many decisions about planting, fertilizer application, and harvesting. While the limitations of such forecasts are well documented, their utility is indisputable. Making projections about the future climate, however, depends on a different set of modeling tools. Here we'll explore the differences among weather forecasts, seasonal climate forecasting, and longer term climate projections.

Numerical weather forecasts are designed to provide information on what the weather is likely to be like for a specific location and time. These forecasts rely on numerical weather models that are based on the laws of atmospheric physics and thermodynamics. Numerical weather models are initialized and run forward through time starting with initial conditions that are based on measured observations from weather balloons, satellites, and surface weather stations. The simplest forecast to make is for conditions 15-minutes from now, as current observations serve as a guide. Such numerical weather forecasts tend to degrade with time due to errors in initial observations and modeling inaccuracies and typically have limited utility beyond 10 days. Numerical weather models used by the National Weather Service have contributed to an estimated $31.5 billion dollar a year benefit to the US (Lazo et al. 2009).

Seasonal climate forecasts are designed to provide information on climate variability over the next several months. Unlike weather forecasts, they are not intended to provide forecasts for specific dates, but estimates of temperature and precipitation anomalies for the upcoming month of December or December-February, for example. Seasonal climate forecasts take various approaches, but generally rely on linkages between boundary conditions such as ocean temperatures, soil moisture, ice coverage and climate, in addition to the initial conditions that numerical weather forecasts rely on. Seasonal climate forecasts have demonstrated utility for certain seasons and geographic areas; however, they have generally been underutilized in seasonal decision-making. At the same time, the misuse of seasonal forecasts can be detrimental and may limit users' future trust and reliance on seasonal forecasts as guides for decision-making (Hartmann et al. 2002). For example, a water resource manager should not base their decision about how to manage a reservoir exclusively on a seasonal forecast, but information from a seasonal forecast could help them put strategies in place to prepare for a drier than average year (Hartmann et al. 2002).

Climate change projections are designed to provide information on climate statistics over multi-decadal time periods. Unlike weather and seasonal climate forecasts, climate change projections are not designed to yield skillful information about a specific day or month, but rather statistical generalizations about climate over longer term horizons. Climate change projections utilize the same numerical procedures used in numerical weather models and dynamic seasonal forecast models, but also include processes related to integrative aspects of the climate systems including the carbon cycle, vegetation dynamics, and sea-ice dynamics. Individual modeling centers across the world have developed GCMs to conduct such climate experiments. More than 40 models were used in the 5th assessment report of the IPCC, with each modeling group running the same set of experiments with their model to produce a cohesive set of results for comparison. Each modeling group uses different model configurations, including different horizontal resolutions, processes, and feedbacks. The result is a diversity of models that simulate the climate system's response to anthropogenic emissions in various slightly different manners. While this diversity of responses yields "uncertainty" in terms of regional changes in climate, it may be preferable to have a large set of forecasts to draw upon rather than rely on a single or limited set of forecasts (Walker et al. 2003; Hawkins and Sutton 2011).

Selecting the Right Model for the Decision at Hand

Farmers constantly negotiate a wide variety of decisions related to planting and harvesting dates, fertilizer application, financing and insurance, crop varietals and rotations, machinery acquisition, implementing new technology, land use, and more. The range of decisions that dryland small grain producers make require consideration of a variety of different climate-forecasting models. Acquiring and using climate projections has been a challenge for many researchers, let alone agricultural decision-makers. Challenges associated with using outputs from climate model simulations arise for the following reasons: (1) model results are typically stored in formats that require familiarity with computer programming, (2) outputs may be formidably large to download and analyze without high-performance computing, and (3) outputs are often not refined to reflect conditions specific to a location of interest for individual users. These issues create a substantial barrier between data producers and consumers, inhibiting stakeholders' ability to use model outputs to plan for, cope with, or try to optimize climate impacts that may arise in a changing climate (Allen et al. 2015). Fortunately, progress continues to be made in producing, synthesizing, and interpreting climate model projections in a manner that is relevant to agricultural decision-makers.

Researchers are expanding efforts to supply agricultural decision-makers with appropriate tools for specific decisions. It is important to develop an understanding of how to select the right model forecasts for different kinds of decisions. For example, if you were choosing an optimal day for planting within a less-than-10-day range, you would rely on a weather forecast. If, however, you were deciding between planting winter wheat or spring wheat for the upcoming season, you would need a seasonal climate forecast. If you are considering a land acquisition, you would need to review longer term climate change projections to estimate return on investment. There are gaps in this suite of tools: limited information exists for 10–30 day bands and near-term decadal forecasts (next 10–20 years), although these are areas where climate science is rapidly advancing. Emerging research will also support growers in learning from other regions that serve as examples of the future climate of the inland PNW. Such climate analogues can be powerful tools for decision support.

Production Practices for Adaptation and Mitigation

Climate change may exacerbate existing challenges and create new management challenges for inland PNW growers. At the same time, there are likely opportunities associated with a changing climate as other regions increasingly rely on the inland PNW for small grain production, and warmer temperatures and CO_2 fertilization may increase yields for some crops in some locations. Ensuring sustainable small grain production into the future will require growers, policymakers, agriculture industry professionals, and researchers to be familiar with emerging challenges and opportunities. Dryland farmers in the inland PNW have strategies available to them to adapt to changing conditions and, in some cases, work toward climate change mitigation.

Critically important strategies for adapting to increased extreme weather events include improving **soil health** (see Chapter 2: Soil Health), protecting soils from erosion (see Chapter 4: Crop Residue Management), diversifying annual cropping systems (see Chapter 5: Rotational Diversification and Intensification), and monitoring and managing for changing weed, disease, and insect pest pressures (see Chapter 9: Integrated Weed Management, Chapter 10: Disease Management for Wheat and Barley, and Chapter 11: Insect Management Strategies). Ongoing research is investigating the potential benefits of shifting to earlier planting dates and adopting new varieties suited to changing precipitation and temperature regimes.

On-farm climate change mitigation strategies work to reduce greenhouse gas emissions from agriculture while often bringing co-benefits that support long-term sustainability and productivity of farmland. Strategies for mitigation which are applicable to dryland growers include: improving water use efficiency and **nitrogen use efficiency**, reduced tillage or "improved fallow" practices to increase net carbon sequestration (Chapter 3: Conservation Tillage Systems), utilizing organic soil amendments (Chapter 7: Soil Amendments) and using site-specific management to ensure precise application of inputs and reduce greenhouse gas emissions (Chapter 8: Precision Agriculture).

Economic Considerations

Agricultural decision-makers are accustomed to making decisions in the context of multiple uncertainties. Understanding vulnerabilities and risks under future climate conditions is key. For instance, if you are considering land acquisition with a 20–30 year amortization of costs, how should you value—or discount—expected earnings for that period given the range of possible climate futures? Land transactions may be an under-appreciated decision point for farm management because we should no longer assume static climatic conditions (e.g., yield projections or income generation potential) for any given piece of land. A person making a land acquisition decision needs to be able to hedge the investment decision in the context of their relative risk portfolio.

Climate change impacts are likely to change the regulatory context in which producers are operating. Envisioning the future of dryland small grain farming, it is plausible that federal, state, and county programs designed to support farmers and incentivize sustainable farming practices may require increased levels of climate change resilience planning. For more in-depth discussion of changing agricultural policies, see Chapter 12: Farm Policies and the Role for Decision Support Tools.

Previous widely used scenarios describing future social and economic conditions largely neglect to consider agricultural innovations specifically, causing them to be viewed as unrealistic and overly pessimistic by most agricultural analysts. In reality, farming practices are continually developing: practices to manage and monitor the application of fertilizers and pesticides continue to be developed and improved, more energy-efficient machinery is being used on farms, and crops are being developed with new genetic characteristics. Scenarios that account for these various developments in technology, as well as in agricultural trade and policy, enable us to explore specific relationships between global food production and climate (Jones et al. 2016).

Tools and Resources for Growers

The science of climate change impacts on small grain production systems in the inland PNW is rapidly evolving. The authors of this report encourage regional small grain producers to join the community of university-based researchers, Extension professionals, and growers at www.agclimate.net and stay informed about current research and opportunities for continuing education. The informational resources and links to tools listed in this section will be continually updated at on the AgClimate Network website. The following resources will assist growers in accessing: (1) educational material about climate modeling and climate change impacts on inland PNW agricultural systems, (2) locally specific weather and climate data, and (3) decision support tools for agricultural management.

Educational Resources about Climate Modeling and Climate Change Impacts

A Changing Climate for Agriculture: Tools for Kick-Starting Adaptation

https://www.youtube.com/watch?v=7Yc2yPri2hE

Watch this webinar for an up-to-date discussion of how current downscaled climate data can be used to support decision-making in agriculture. This Climate Learning Network webinar presentation by the University of Idaho's John Abatzoglou introduces viewers to several of the climate data resources and decision support tools discussed in this chapter.

What Do We Currently Know about the Impacts of Climate Change on Pacific Northwest Cropland Agriculture?

https://www.surveymonkey.com/r/XFPHB96?sm=RN4RdLEnTR08hZ8awmGxHQ%3d%3d

An accessible webinar presented by Chad Kruger, director of Washington State University's Center for Sustaining Agriculture and Natural Resources. This is a helpful resource for producers looking to get up to speed on the basics of climate change impacts.

Modeling Environmental Change: A Guide to Understanding Results from Models

http://cru.cahe.wsu.edu/CEPublications/FS159E/FS159E.pdf

Written to orient readers from diverse backgrounds to the basics of how environmental models are developed, this Extension publication covers the kinds of input data that go into climate change impact models and considers approaches to defining future scenarios, communicating model outputs, and representing uncertainty.

Assessment of Climate Change Impact on Eastern Washington Agriculture

http://link.springer.com/article/10.1007/s10584-010-9851-4

This is a peer-reviewed analysis of climate change impacts on multiple crops, including dryland wheat, potatoes, and apples in the PNW authored by crop and climate scientists. You can expect to see an updated version of this analysis based on the latest climate model projections in the near future.

Agriculture: Impacts, Adaptation, and Mitigation

http://cses.washington.edu/db/pdf/daltonetal678.pdf

A chapter in the 2013 publication Climate Change in the Northwest, this resource outlines the wide range of projected climate change effects on agriculture in the Northwest US. This is an accessible yet relatively in-depth discussion of the state of current scientific knowledge.

Weather and Climate Data

Climate Engine

http://climateengine.org

A number of spatially and temporally complete historical and current near-surface climate and weather datasets now exist. These data are based on a vast network of place-based observations and overcome some flaws in earlier datasets linked to gaps in observations, sparseness of stations in some regions, and inconsistent variables that were recorded. Abatzoglou (2013) developed a gridded surface meteorological dataset that includes temperature, humidity, winds, solar radiation, and precipitation (***http://metdata.northwestknowledge.net***). To make these data more accessible, the University of Idaho and the Desert Research Institute developed Climate Engine, which lets users analyze maps, examine time series, and download digital data without needing to process the entire data archive.

Office of the Washington State Climatologist

http://www.climate.washington.edu

The Washington State Climatologist collects, shares, and interprets climate data from various sources. The website distributes peer-reviewed climate and weather information for governmental and private decision-makers working on drought, flooding, climate change, and related issues.

National Oceanic and Atmospheric Administration Climate Data Online

https://www.ncdc.noaa.gov/cdo-web/

The Climate Data Online website provides access to the National Climate Data Center's archive of global historical weather and climate data, as well as weather station history information. These data include daily, monthly, seasonal, and yearly measurements of temperature, precipitation, wind, and growing degree days as well as radar data and 30-year climate averages.

AgWeatherNet

http://weather.wsu.edu

Washington State University's AgWeatherNet provides current and historical weather data from WSU's network of 178 automated weather stations. The weather network is managed by the AgWeatherNet team, located at the WSU Irrigated Agriculture Research and Extension Center in Prosser, Washington. WSU's automated weather stations are located primarily in the irrigated regions of eastern Washington. Variables tracked by AgWeatherNet include air temperature, relative humidity, dew point temperature, soil temperature at 8 inches, precipitation, wind speed, wind direction, solar radiation, and leaf wetness. Some stations also measure atmospheric pressure. These variables are recorded every 5 seconds and summarized every 15 minutes by a data logger. The website also connects users with models and decision aid tools.

REACCH Climate and Weather Tools

http://climate.nkn.uidaho.edu/REACCH/climateTools.php

Researchers involved with the Regional Approaches to Climate Change (REACCH) project developed a web tool for agricultural decision makers to get information about location-specific historical climate, current weather, seasonal forecasts and future projections. An example summary table of a number of climate metrics that could be important for agriculture is shown in Figure 1-6. The user can select a location and obtain a summary of contemporary climate normals and projected changes for different time periods.

Decision Support Tools for Agricultural Management

REACCH Decision Support Tools

http://climate.nkn.uidaho.edu/REACCH/decisionTools.php

Related to REACCH's climate and weather tools discussed above, REACCH scientists developed a suite of web-based decision support tools. These tools focus on decision-making issues specific to the crops in the inland PNW. Both short-term decisions, such as scheduling fertilizer application and pest management practices, as well as long-term decisions, such as assessing specific locations for crop suitability, are supported.

AgClimate Atlas

http://climate.nkn.uidaho.edu/NWTOOLBOX/mapping.php

The AgClimate Atlas is a web-based application developed by University of Idaho researchers with support from the USDA NW Climate Hub. The atlas provides downscaled climate data and derived climate metrics including cold hardiness, growing degree days, and reference evapotranspiration that

Climate Dashboard

47.496° N, 120.269° W

2040-2069 vs 1981-2010 under high future emissions scenario

Precipitation		Potential Evapotranspiration	
1981-2010	2040-2069	1981-2010	2040-2069
10.4"	11.1"	37.5"	43.7"
Annual Total	Annual Total	Annual Total	Annual Total

Summer Temperatures		Winter Temperatures	
1981-2010	2040-2069	1981-2010	2040-2069
81.1 °/ 54.8 °	88.5 ° / 60.9 °	36.4 °/ 22.9 °	41.5 ° / 28.8 °
High / Low	High / Low	High / Low	High / Low

Hottest Summer Day		Coldest Winter Day	
1981-2010	2040-2069	1981-2010	2040-2069
97.6 ° / 5	105.5 ° / 7	-0.8 °F / 6b	9 °F / 7b
Average / Heat Zone	Average / Heat Zone	Average / Cold Hardiness Zone	Average / Cold Hardiness Zone

Last/First Freeze		Annual Freeze Free Days	
1981-2010	2040-2069	1981-2010	2040-2069
Apr 21/ Oct 15	Mar 13/ ??	235.2 days	287.1 days
Last Spring / First Fall	Last Spring / First Fall	Annual Average	Annual Average

Figure 1-6. Regional Approaches to Climate Change (REACCH) climate and weather tools sample summary table.

may be useful for agricultural decision making. For example, certain wine grape varietals have growing degree day requirements (base 50°F) that are not currently met across much of the Northwest, providing one limitation on cultivation. However, by the mid-21st century, additional warming is projected to result in a large increase in accumulated heat that may allow for those varietals to reach maturation in the region. Users can select from a variety of metrics, climate scenarios, models, and time horizons (for example, early, mid, or late 21st century). Users can also "mouse" over to extract information for a specific pixel, or download these maps in a format that can be used in ArcGIS or equivalent GIS software.

Integrated Scenario Tools

http://climate.nkn.uidaho.edu/IntegratedScenarios/index.php

The Northwest Climate Science Center funded a project to create a coordinated set of climate, hydrology, and vegetation scenarios called Integrated Scenarios of the Future Northwest Environment. Researchers at the University of Idaho developed tools to visualize data from that project. For example, for irrigated crops in the Northwest, the amount of water storage in the mountain snowpack and the timing of streamflows are key factors. Using the streamflow web tool (***http://climate.nkn.uidaho.edu/IntegratedScenarios/vis_streamflows.php***) users can see graphs that project substantial drops in streamflow during June-August when irrigation demand peaks.

Conclusion

Consistent with global trends, the PNW is projected to experience substantial shifts from historical climate patterns in the coming century. The specific effects of climate change on dryland cropping systems in the inland PNW are expected to be variable over time and variable from one location to another. In the near term, longer growing seasons and increased CO_2 fertilization may lead to increasing yields for dryland cereals in the region. However, later in the 21st century, projections suggest that heat and drought stress and changing pest and disease pressures along with increased incidences of extreme weather events will likely pose new challenges for inland PNW producers. It is hoped that this guide will be informative for regional growers operating under diverse local conditions and utilizing a range of production strategies. This guide, and the tools and resources discussed within, are intended to support producers in identifying appropriate management strategies to ensure the long-term profitability and sustainability of dryland small grain production in the inland PNW.

References

Abatzoglou, J.T., and T.J. Brown. 2012. A Comparison of Statistical Downscaling Methods Suited for Wildfire Applications. *International Journal of Climatology* 32(5): 772–780.

Abatzoglou, J.T. 2013. Development of Gridded Surface Meteorological Data for Ecological Applications and Modelling. *International Journal of Climatology* 33(1): 121–131.

Abatzoglou, J.T., D.E. Rupp, and P.W. Mote. 2014. Seasonal Climate Variability and Change in the Pacific Northwest of the United States. *Journal of Climate* 27(5): 2125–2142.

Allen, E., G. Yorgey, K. Rajagopalan, and C. Kruger. 2015. Modeling Environmental Change: A Guide to Understanding Results from Models that Explore Impacts of Climate Change on Regional Environmental Systems. Washington State University Extension Publication FS159E.

Bernacchi, L.A., J.D. Wulfhorst, L. Nirelli McNamee, and M. Reyna. 2015. Public Perceptions of Climate Change and Pacific Northwest Agriculture. In Regional Approaches to Climate Change for Pacific Northwest Agriculture: Climate Science Northwest Farmers Can Use. REACCH Year 4 Report.

Binder, L.C.W., J.K. Barcelos, D.B. Booth, M. Darzen, M.M. Elsner, R. Fenske, and C. Karr. 2010. Preparing for Climate Change in Washington State. *Climatic Change* 102(1-2): 351–376.

Douglas, Jr., C.L., R.W. Rickman, B.L. Klepper, and J.F. Zuzel. 1992. Agroclimatic Zones for Dryland Winter Wheat Producing Areas of Idaho, Washington, and Oregon. *Northwest Science* 66(1).

Douglas, Jr., C.L., P.M. Chevalier, B. Klepper, A.G. Ogg, and P.E. Rasmussen. 1999. Conservation Cropping Systems and Their Management. In Conservation Farming in the United States, E. Michalson, R. Papendick, and J. Carlson, eds.

Eigenbrode, S.D., S.M. Capalbo, L.L. Houston, J. Johnson-Maynard, C. Kruger, and B. Olen. 2013. Agriculture: Impacts, Adaptation, and Mitigation. In Climate Change in the Northwest, M.M. Dalton and P.W. Mote, eds. Island Press.

Ferris, R., R.H. Ellis, T.R. Wheeler, and P. Hadley. 1998. Effect of High Temperature Stress at Anthesis on Grain Yield and Biomass of Field-Grown Crops of Wheat. *Annals of Botany* 82(5): 631–639.

Hartmann, H., T.C. Pagano, S. Sorooshian, and R. Bales. 2002. Confidence Builders: Evaluating Seasonal Climate Forecasts from User Perspectives. *Bulletin of the American Meteorological Society* 83(5): 683–698

Hawkins, E., and R. Sutton. 2011. The Potential to Narrow Uncertainty in Projections of Regional Precipitation Change. *Climate Dynamics* 37: 407–418.

Huggins, D., R. Rupp, H. Kaur, J. Abatzoglou, and S. Eigenbrode. 2016—in review. Dynamic Agroecological Classes for Assessing Land Use Change in the Inland Pacific Northwest.

Jones, J.W., J.M. Antle, B. Basso, K.J. Boote, R.T. Conant, I. Foster, H.C.J. Godfray, M. Herrero, R.E. Howitt, S. Janssen, B.A. Keating, R. Munoz-Carpena, C.H. Porter, C. Rosenzweig, and B.A. Keating. 2016. Brief History of Agricultural Systems Modeling. *Agricultural Systems.*

Kaur, H., D. Huggins, R. Rupp, J. Abatzoglou, C. Stöckle, and J. Reganold. 2015. Bioclimatic-Driven Future Shifts in Dryland Agroecological Classes. In Regional Approaches to Climate Change for Pacific Northwest Agriculture: Climate Science Northwest Farmers Can Use. REACCH Year 4 Report.

Lazo, J.K., R.E. Morss, and J.L. Demuth. 2009. 300 Billion Served: Sources, Perceptions, Uses, and Values of Weather Forecasts. *Bulletin of the American Meteorological Society* 90(6).

McClellan, R.C., D.K. McCool, and R.W. Rickman. 2012. Grain Yield and Biomass Relationship for Crops in the Inland Pacific Northwest United States. *Journal of Soil and Water Conservation* 67(1): 42–50.

Mote, P.W., and E.P. Salathe, Jr., 2010. Future Climate in the Pacific Northwest. *Climatic Change* 102(1-2): 29–50.

Ortiz, R., K.D. Sayre, B. Govaerts, R. Gupta, G.V. Subbarao, T. Ban, and M. Reynolds. 2008. Climate Change: Can Wheat Beat the Heat? *Agriculture, Ecosystems & Environment* 126(1): 46–58.

Schillinger, W.F., R.I. Papendick, S.O. Guy, P.E. Rasmussen, and C. van Kessel. 2003. Dryland Cropping in the Western United States. In Pacific Northwest Conservation Tillage Handbook Series No. 28, Chapter 2.

Schillinger, W.F., R.I. Papendick, S.O. Guy, P.E. Rasmussen, and C. Van Kessel. 2006. Dryland Cropping in the Western United States. In Dryland Agriculture, G.A. Peterson, P.W. Unger, and W.A. Payne, eds. Agronomy Monograph 23: 365–393. Madison, WI: American Society of Agronomy, Crop Science Society of America, and Soil Science Society of America.

Schillinger, W.F., and R.I. Papendick. 2008. Then and Now: 125 Years of Dryland Wheat Farming in the Inland Pacific Northwest. *Agronomy Journal* 100(3): S-166.

Schillinger, W.F., R.I. Papendick, and D.K. McCool. 2010. Soil and Water Challenges for Pacific Northwest Agriculture. In Soil and Water Conservation Advances in the United States, T.M. Zobeck and W.F. Schillinger, eds. Soil Science Society of America Special Publication 60: 47–79. Madison, WI: Soil Science Society of America.

Stöckle, C.O., R.L. Nelson, S. Higgins, J. Brunner, G. Grove, R. Boydston, and C. Kruger. 2010. Assessment of Climate Change Impact on Eastern Washington Agriculture. *Climatic Change* 102(1-2): 77–102.

Tubiello, F.N., C. Rosenzweig, R.A. Goldberg, S. Jagtap, and J.W. Jones. 2002. Effects of Climate Change on US Crop Production: Simulation Results Using Two Different GCM Scenarios. Part I: Wheat, Potato, Maize, and Citrus. *Climate Research* 20(3): 259–270.

USDA (US Department of Agriculture). 1999. Soil Taxonomy: A Basic System of Soil Classification for Making and Interpreting Soil Surveys, 2nd ed. Natural Resources Conservation Service. USDA Handbook 436. ***https://www.nrcs.usda.gov/Internet/FSE_DOCUMENTS/nrcs142p2_051232.pdf***

US Environmental Protection Agency. 2011. Level III Ecoregions of the Continental United States. National Health and Environmental Effects Research Laboratory. ***https://nctc.fws.gov/courses/csp/csp3200/resources/documents/Eco_Level_III.pdf***

Vano, J.A., M.J. Scott, N. Voisin, C.O. Stöckle, A.F. Hamlet, K.E.B. Mickelson, M.M. Elsner, and D.P. Lettenmaier. 2010. Climate Change Impacts on Water Management and Irrigated Agriculture in the Yakima River Basin, Washington, USA. *Climatic Change* 102: 287–317.

Walker, W.E., P. Harremoës, J. Rotmans, J.P. van der Sluijs, M.B. van Asselt, P. Janssen, and M.P. Krayer von Krauss. 2003. Defining Uncertainty: A Conceptual Basis for Uncertainty Management in Model-Based Decision Support. *Integrated Assessment* 4(1): 5–17.

Wulfhorst, J.D., L.A. Bernacchi, B. Mahler, L. Nirelli McNamee, M. Reyna, and S. Irizarry. 2015. Measuring Producer Trust and Attitudes about Climate Change. In Regional Approaches to Climate Change for Pacific Northwest Agriculture: Climate Science Northwest Farmers Can Use. REACCH Year 4 Report.

Chapter 2
Soil Health

Rakesh Awale, Oregon State University
Stephen Machado, Oregon State University
Rajan Ghimire, New Mexico State University (formerly of Oregon State University)
Prakriti Bista, Oregon State University

Abstract

Soil health refers to a balanced condition of soil physical, chemical, and biological processes conducive to high productivity and environmental quality. Soil health concepts are commonly used to evaluate changes, compare soils, or assess the effectiveness of land-use management. This chapter deals with soil health concepts, soil health indicators and their assessment, current understanding of management effects on soil health, and guidelines for improving soil health and sustainable crop production in the inland Pacific Northwest (PNW). Intensive agricultural systems within the inland PNW have depleted more than 50% of native soil organic matter (SOM), deteriorated soil microbial community, faunal diversity, and soil structure, and increased soil erosion. Crop residue incorporation into soil with intensive tillage, residue burning and removal, little biomass production and return from the dominant cereal-fallow cropping system, lack of crop diversity, farming on steep slopes, continuous application of ammonium-based nitrogen fertilizers, and limited and untimely precipitation are the major factors contributing to the deterioration of soil health and reduced agronomic productivity within the inland PNW. However, through sustained research and extension activities there is evidence that soil health can be improved by adopting practices such

Research results are coded by agroecological class, defined in the glossary, as follows:

● Annual Crop ▲ Annual Crop-Fallow Transition ■ Grain-Fallow

as conservation tillage systems, residue retention, no residue burning, cropping rotation and diversification, annual cropping, balanced and efficient nutrient management, organic soil amendments, reducing acidity, and integrated nutrient, pest, and weed management.

Key Points

- Soil health is vital for sustainable agricultural production and environmental quality.
- Soil health is assessed by evaluating the soil's ability to perform desired ecosystem functions and involves measuring soil physical, chemical, and biological indicators in response to changes in management.
- Soil's high degree of spatial variability poses challenges for the selection of indicators sensitive to management in soil health assessment programs.
- Site-specific adaptive management decisions can lead to long-term sustainability of soil health.
- Reduced tillage practices, cropping intensification and diversification, crop residue retention, and application of organic amendments can enhance soil health.

Soil Health: Concept and Background

Soil is a vital natural resource and its health is fundamental for sustainable agricultural production. **Soil health**, also referred to as soil quality, is defined as "the capacity of soil to function within ecosystem boundaries to sustain biological activity, maintain environmental quality, and promote plant and animal health" (Doran and Zeiss 2000). Soils function to provide ecosystem services that include increased soil water retention and availability, soil aggregation, nutrient cycling and storage, and microbial diversity and function.

Soil health assessment implies the evaluation of fitness of soil to perform desired functions and its capacity to resist and recover from degradation. Land managers, growers, and researchers assign a relative value to soil health using various qualitative and quantitative indicators. For instance, soil **compaction** leads to loss of **soil structure**, limited water and air **infiltration**, and poor root development, rendering soil less productive than a non-compacted, well-structured soil. Here, the suitability of soil for proper root growth (soil function) can be judged by determining soil **bulk density** and penetration resistance (soil health indicators) using appropriate tools. Growing deep-rooted **cover crops** may help reverse soil compaction. A healthy soil should possess the following characteristics:

- High organic matter
- Good soil tilth and structure
- High water infiltration and retention
- Resistance to compaction
- High soil biological activity
- Plant nutrient recycling and availability
- Resistance to erosion
- Devoid of toxic chemicals
- Low in weed and disease pressures

Indicators of Soil Health

Soil health indicators generally describe specific soil properties. Soil properties can be generally categorized as stable or dynamic. Stable soil properties are influenced by soil-forming factors such as climate, organisms, parent material, and topography, which change little with management practices. Examples of stable properties include soil texture, soil type, and soil depth. Dynamic properties can change with land-use and management practices over the course of a short time, generally within a human lifespan, and include SOM, bulk density, and **pH**. Accordingly, soil health assessment programs that include measurement of various physical, chemical, and biological properties of soil that respond to management changes provide clues on soil processes. However, there is no consensus on soil health indicators that are applicable to every agroecosystem. Soil health indicators are site-specific and sometimes temporal in nature. For instance, increased soil water storage would be a desirable indicator in the Palouse region of the **inland PNW** where precipitation is limited, whereas excess water-causing anaerobic conditions are a bad indicator in the west coast areas of the PNW with too much precipitation. Some of the commonly used soil properties to assess soil health are listed in Table 2-1.

Table 2-1. Potential physical, chemical, and biological properties used in soil health assessment.

Physical	Chemical	Biological
✓ Soil color	✓ Organic C and N	✓ Soil respiration
✓ Aggregate stability	✓ Particulate organic matter	✓ Potential mineralizable nitrogen
✓ Water infiltration	✓ Active carbon	✓ Microbial biomass
✓ Bulk density	✓ pH	✓ Soil enzymes
✓ Penetration resistance	✓ Cation exchange capacity and base saturation	✓ Earthworms
✓ Water holding capacity	✓ Electric conductivity	✓ Crop condition, root growth
✓ Runoff and erosion	✓ Heavy metals	✓ Weed and disease pressure
✓ Rooting depth		

Visual Indicators

Farmers often describe soil health based on their perception of its look, feel, smell, and taste. Such visual and morphological observations of soil physical condition and plant growth can track status of soil health. Often, these observations guide subsequent soil health assessments. Changes in soil color, soil crust formation, ephemeral gullies, runoff, physical structure and aggregation, soil depth, root growth, crop emergence, and weed density are some potential visual indicators. For instance, SOM content usually decreases with depth, and is observed as dark-colored soil in top soil horizons that progressively lightens with depth (Figure 2-1). Rills and gullies formed shortly after a rain indicate runoff and slow water infiltration into soil. Soil crusting, reduced aggregation, and surface sealing indicate surface compaction and loss of SOM. Uniform crop emergence and good growth indicate good soil health and management. Deep soils promote SOM, nutrients, water storage, and root growth.

Aggregate Stability

Soil **aggregates** or clods are groups of primary soil particles that are held together by organic (fungal hyphae, bacterial mucilage, root exudates, polysaccharides, lipids, etc.) and inorganic materials (clay, polyvalent

Figure 2-1. A darker color of the surface soil (left) indicates the presence of higher organic matter content compared to the subsoil (right). (Photo by Rakesh Awale.)

metal cations such as calcium and iron, oxides and hydroxides of iron and aluminum, calcium and magnesium carbonates, etc.). **Aggregate stability** refers to the capability of soil aggregates to resist disintegration when exposed to external destructive forces from tillage, water, wind, and freeze-thaw cycles. Soil aggregate stability can be determined by measuring the proportion of aggregates in different size classes, percentage of the stable aggregates (in a specified size class) upon wet or dry sieving, and distribution of stable aggregates into different size ranges.

Generally, high soil aggregate stability and greater amounts of stable aggregates are desirable for sustaining agricultural productivity and protecting environmental quality. Stable aggregates favor high water infiltration rates, provide adequate aeration, resist soil erosion by wind and water, and enhance root growth (Figure 2-2). Stable soil aggregates also provide physical protection of SOM from microbial decomposition, particularly when the proportion of large (>0.25 mm diameter) aggregates increases. Conversely, disintegration of aggregates leads to the formation of surface crusts and result in more runoff, more erosion, and less available water (Figure 2-2). In the Grain-Fallow **agroecological class** (AEC), disking, rodweeding, and packing operations under disk tillage winter wheat-fallow (WW-F) systems reduced the proportion of non-erodible soil aggregates and mean aggregate diameter compared to low disturbance **no-till** spring wheat-chemical fallow (SW-ChF) system (Table 2-2) ■. Consequently, soil water content and **matric potential** (a measure of soil water) were lower, and the susceptibility of soils to wind erosion was higher with disk tillage WW-F than no-till SW-ChF ■. Similarly, in the Annual Crop AEC, no-till increased the proportion of dry aggregates of size above 1 mm, whereas conventional tillage had

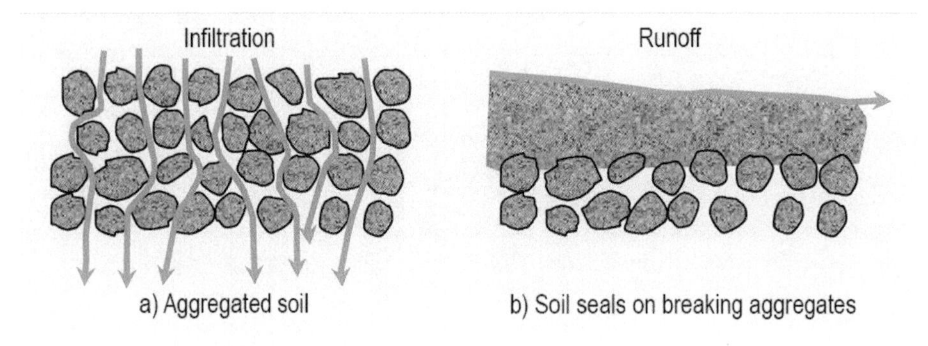

Figure 2-2. Water flow patterns in well-aggregated and weakly aggregated soils. (Reprinted from Magdoff and Van Es 2009.)

Table 2-2. Soil aggregate parameters, water content and susceptibility to wind erosion under no-till spring wheat-chemical fallow (NT SW-ChF) and disk tillage winter wheat-fallow (DT WW-F) cropping systems in east-central Washington. ■

Cropping system	Non-erodible aggregates (%)	Aggregate mean diameter (inch)	Water content (%)	Matric potential (Mpa)	SLR index
NT SW-ChF	50.0	0.09	1.7	−197.8	0.067
DT WW-F	38.8	0.01	1.4	−284.5	0.174

SLR: soil loss ratio, an indicator of wind erosion with 0 = no erosion and 1 = maximum erosion potential.
Adapted from Feng et al. 2011a.

greater proportion of soil aggregates of size below 0.25 mm (Kennedy and Schillinger 2006) ●. See Chapter 1: Climate Considerations for more information on AECs.

The addition of organic materials and retention of crop residues promotes stable soil aggregation by enhancing soil biological activity and the production of various binding agents (such as fungal hyphae, polysaccharides, mucilage, and lipids) and by protecting the aggregates from direct physical impacts of raindrops and wind. A study in the Annual Crop-Fallow Transition AEC revealed that water stability of 1–2 mm soil aggregates as well as that for whole soil was higher with the application of organic manure and pea vines than when synthetic N fertilizers were applied (Table 2-3) ▲. Organic amendments increased the amounts of glomalin (fungal glycoprotein) that act as an insoluble glue to stabilize soil

Table 2-3. Addition of organic materials enhances soil aggregate stability, water movement, and soil biological activity in a winter wheat-fallow (WW-F) rotation under conventional tillage near Pendleton, Oregon. ▲

Soil parameters	Organic amendments		Synthetic N fertilizer	No fertilizer
	Manure	Pea vines		
	100 lb N/ac	30 lb N/ac	80 lb N/ac	0 lb N/ac
Total C (%)	1.590	1.260	1.170	1.090
Total N (%)	0.135	0.103	0.092	0.088
Water stability				
Whole soil (proportion)	0.48	0.41	0.35	0.30
1–2 mm aggregates (proportion)	0.83	0.69	0.65	0.56
Percolation (cubic inch/hr)	1.06	0.95	0.84	0.73
Ponded infiltration (inch/hr)	5.53	4.09	1.49	1.46
Total glomalin (%)	0.259	0.235	0.214	0.213
Earthworm count (per square meter)	128	144	72	80

Adapted from Wuest et al. 2005.

aggregates. Consequently, soil water infiltration rates were higher in soils treated with organic material compared to a synthetic fertilizer addition ▲.

Bulk Density and Compaction

Bulk density is the ratio of the mass of oven-dry soil to its bulk volume. It is a measure of soil compaction. Soil compaction occurs when soil particles are pressed together reducing porosity (pore space) between them. Soil compaction (high bulk density) restricts plant root growth by increasing resistance to root penetration (impedance) (Figure 2-3) resulting in reduced nutrient uptake, nutrient deficiencies, and crop yield. Compaction of soil also reduces air and water permeability resulting in reduced soil water infiltration and increased runoff and soil erosion.

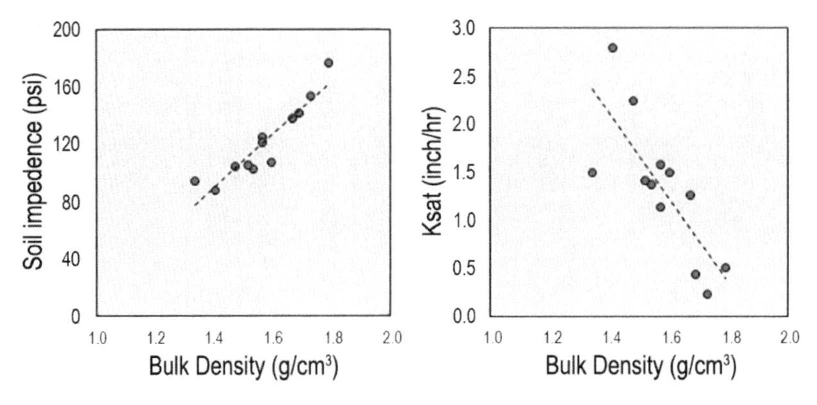

Figure 2-3. Relationships of soil bulk density, impedance (compaction), and saturated hydraulic conductivity (Ksat) within the surface 2 feet of soil after a one-time, high-axle load traffic pass (<5, 10, and 20 ton loads) under winter wheat-spring barley-spring pea (WW-SB-SP) rotation near Moscow, Idaho. (Adapted from Hammel 1994.) ●

The most yield-limiting soil compaction is caused by wheel traffic from heavy equipment. In the Annual AEC, soil compaction under a 20-tons-per-axle load resulted in a 14% yield reduction in spring barley compared to the 5-tons-per-axle load due to water stress and restricted root growth (Figure 2-4) ●. The potential for wheel traffic compaction increases on wet soils because soil water lubricates soil particles, making them easier to move against each other. Therefore, avoiding field operations on wet soils, limiting axle load, and decreasing traffic area can minimize wheel traffic soil compaction.

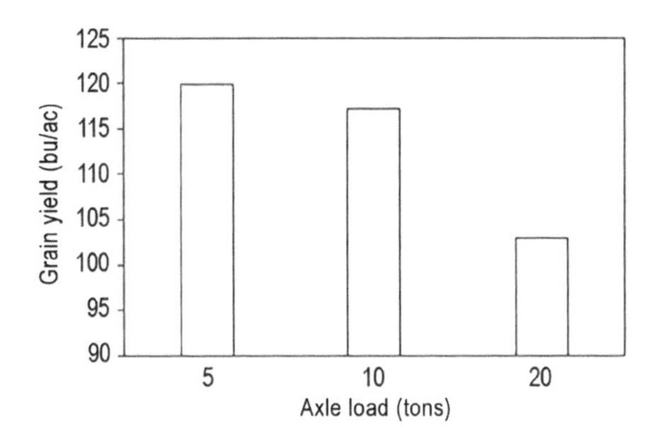

Figure 2-4. Effect of soil compaction from three axle loads on yield of spring barley in a winter wheat-spring barley-spring pea (WW-SB-SP) rotation near Moscow, Idaho. (Adapted from Hammel 1994.) ●

Intensive tillage practices can also increase the susceptibility of a soil to compaction by reducing aggregate stability. For instance, in the Grain-Fallow AEC, greater soil disturbance from disking and packing prior to seeding wheat in late summer increased bulk density under disk tillage WW-F compared to low disturbance systems such as no-till spring wheat-spring barley (SW-SB) and no-till SW-ChF rotations (Table 2-4) ■. Often, a compacted soil layer is formed below the tillage zone due to a continuous smearing action of the tillage implement. In the Grain-Fallow AEC, no soil compaction was seen under either chisel tillage or no-till systems in the top 9 inches of soil, but the plots under chisel tillage were more compacted than no-till plots between 10- to 16-inch depths (Figure 2-5) ■.

Soil bulk density decreases with an increase in SOM content (Figure 2-6) ●. Therefore, management practices which add organic materials such

Table 2-4. Bulk density of no-till spring barley-spring wheat (NT SB-SW), no-till spring wheat-chemical fallow (NT SW-ChF), and disk tillage winter wheat-fallow (DT WW-F) cropping systems in east-central Washington. ■

Cropping system	Bulk density (g/cm^3)	
	Spring	Late summer
NT SB-SW	0.87	0.85
NT SW-ChF	1.03	0.95
DT WW-F	0.98	1.04

Adapted from Feng et al. 2011b.

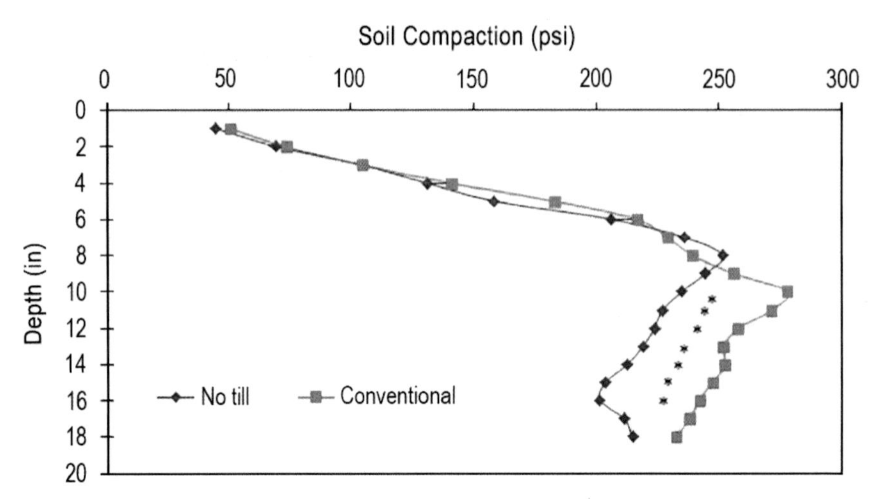

Figure 2-5. Subsoil compaction under chisel tillage in winter wheat-fallow (WW-F) near Wilbur, Washington. (Reprinted from Esser and Jones 2013.) ■

as manure, compost, and crop residues can reduce soil compaction. For instance, annual cropping of SB-SW reduced soil bulk density over grain-fallow rotations (SW-ChF or WW-F) because annual cropping likely increased SOM content from greater crop residue input (Table 2-4) ■. Including cover crops in cropping systems and rotation with deep-rooted crops favors soil aggregation and could offset compaction.

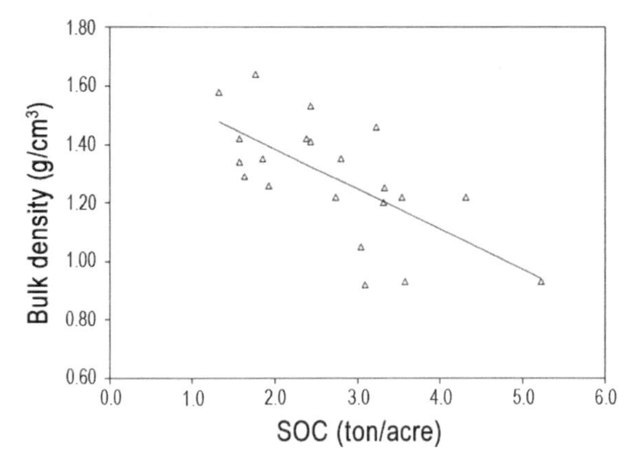

Figure 2-6. Relationship of soil organic carbon (SOC) and bulk density within 0 to 8 inches of Palouse silt loam managed across native prairie, perennial vegetation, no-till, and conventional tillage systems in eastern Washington. (Adapted from Purakayastha et al. 2008.) ●

Soil Water Dynamics: Infiltration, Hydraulic Conductivity, and Water Content

Soil water is the most limiting factor in crop production in the inland PNW. Soil infiltration, **hydraulic conductivity**, water content, and water-holding capacity provide information on soil water movement and plant available water. Soil infiltration is a measure of the rate of water entry into soil and hydraulic conductivity refers to the rate of water movement through soil. Soil hydraulic properties are directly related to soil aggregate stability, compaction, porosity, and pore continuity. Ultimately, these properties regulate soil water storage, nutrient transport, and soil erosion.

Soil type and structure, tillage, SOM content, residue cover, and initial water content influence soil hydraulic properties. Table 2-5 shows that annual cropping increases both water infiltration and saturated hydraulic conductivity in soil compared to a grain-fallow rotation. This is attributed to increased SOM accumulation due to annual residue inputs, increased soil aggregation, and reduced bulk density ■.

Soil disturbances with tillage may temporarily enhance soil water infiltration and conductivity by loosening soils and opening channels. However, tillage also degrades soil structure, breaks pore continuity, and

Table 2-5. Cumulative infiltration and saturated hydraulic conductivity of soil measured under three cropping systems in east-central Washington. ■

Season (time of measurement)	Cropping system†	Cumulative infiltration (in/hr)	Hydraulic conductivity (in/hr)
Spring (April)	NT SB-SW	2.6	21.8
	NT SW-ChF	1.4	5.4
	DT WW-F	2.3	26.6
Late Summer (September)	NT SB-SW	3.9	52.6
	NT SW-ChF	2.4	23.0
	DT WW-F	2.3	4.8

†Cropping systems are no-till spring barley-spring wheat (NT SB-SW), no-till spring wheat-chemical fallow (NT SW-ChF), and disk tillage winter wheat-fallow (DT WW-F). Adapted from Feng et al. 2011b.

creates surface crusts by sealing pores, which eventually reduces hydraulic conductivity and water infiltration. For instance, disk tillage WW-F increased infiltration and conductivity compared to no-till WW-ChF in spring. However, the effect was short-lived and both water infiltration and hydraulic conductivity were higher under no-till than under disk tillage plots in late summer (Table 2-5) ■.

Surface residues enhance soil aggregation, facilitate water infiltration, and prevent surface soil sealing caused by rain impact. No-till annual cropping systems result in increased ground cover that reduces surface runoff and soil erosion by providing greater time for water to infiltrate the soil (Table 2-6) ▲. Surface residues also reduce evaporative loss of soil water and increase plant available water by shielding soil from solar radiation and reducing air movement just above the soil surface (Donk and Klocke 2012). Therefore, evaporation rates are generally lower under residue-covered soil than under bare soil following a precipitation event.

Seasonal variations in soil hydraulic properties occur throughout the growing season in response to root development, earthworm activity, soil disturbances associated with tillage and seeding operations, and changes in precipitation and temperature. For instance, soil infiltration and saturated hydraulic conductivity both increased from spring to late summer, particularly under no-till systems, due to soil fracturing associated with seeding operations under no-till SW-ChF and wheat root growth under no-till SW-SB, as well as due to drier soil in late summer than in spring (Table 2-5) ■. In the Annual AEC, over-winter soil water storage (September to April) and soil water infiltration (measured after

Table 2-6. Mean annual ground cover, runoff, and soil erosion measured for two tillage systems near Pendleton, Oregon. ▲

Tillage	Cropping systems[†]	Ground cover	Runoff	Soil erosion
		%	inches	ton/ac
No-till	WW-ChF-WW-CP/SP	73	1.46	0.10
Conventional	WW-F	44	3.15	4.90

[†]Cropping systems are winter wheat (WW); chemical fallow (ChF); chickpea (CP); spring pea (SP); summer fallow (F).
Adapted from Williams et al. 2014.

crop harvest) remained similar between no-till and conventional tillage sites due to reestablishment of capillary pore continuity from the surface to below the tillage depth by vigorous and extensive growth of wheat roots during the crop-growing season (Kennedy and Schillinger 2006) ●.

Soil pH

Soil pH is a measure of hydrogen ion (H^+) activity in a soil solution and indicates acidity or alkalinity of the soil. The soil pH scale ranges from 0 to 14, where a pH value of 7 is neutral, pH values above 7 are basic or alkaline, and pH values below 7 are acidic. Soil pH influences many physical, chemical, and biological processes in soil that control plant nutrient availability, **cation exchange capacity** (CEC), **element toxicity**, and agronomic yields.

Availability of most macronutrients (nitrogen, potassium, calcium, magnesium, and sulfur) is optimal within a pH range of 6 to 7 and decreases outside this range. Typically, low soil pH (acidity) can lead to significant yield reductions due to nutrient deficiencies in crops. Soil acidity, particularly at pH levels below 5.5, increases solubility of aluminum and manganese in soils (Figure 2-7), causing toxicity to roots and thereby interfering with root growth and plant development (Figure 2-8). Cereal and grass crops are more tolerant to soil acidity compared to legume crops. However, significant yield reductions in wheat and barley occur at pH values below 5.2. For instance, a study in the Midwest showed that aluminum toxicity reduced wheat grain yield by 2.8% for every 1% increase in aluminum saturation (Schroder et al. 2011). Low soil pH also negatively impacts soil microbial communities (particularly bacteria), earthworm populations, rate of SOM decomposition, and efficacy of pesticides, but favors fungal pathogens and certain weed species. Bacterial (*Pseudomonas fluorescens*) inhibition of take-all disease (caused by the fungal pathogen *Gaeumannomyces graminis* var. *tritici*) in wheat reduced with increasing soil acidity due to diminished production of the antibiotic compound phenazine-1-carboxylate upon suppression of bacterial growth at low pH levels (Ownley et al. 2003).

In the inland PNW, continuous use of acid-forming N fertilizer has decreased soil pH. In particular, ammonium-based N fertilizers lower

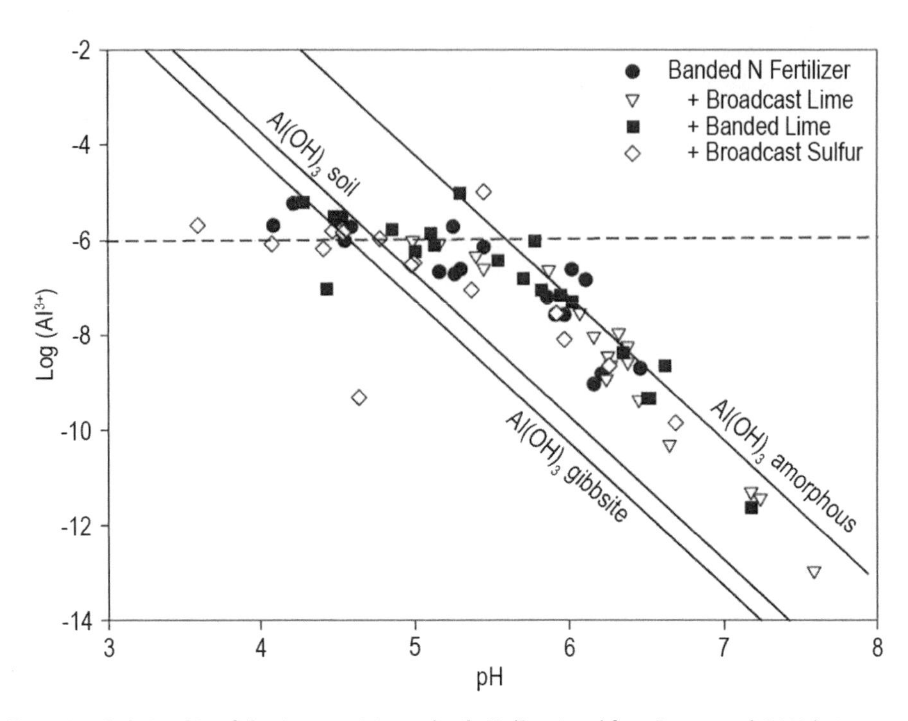

Figure 2-7. Relationship of aluminum activity and soil pH. (Reprinted from Brown et al. 2008.)

Figure 2-8. Retarded wheat root growth (right) due to aluminum toxicity. (Photo credit: CIMMYT 2016.)

soil pH because of the production of acidity (hydrogen ion) during the **nitrification** process [oxidation of ammonium-N (NH_4-N) to nitrate-N (NO_3-N)]. Most native prairie soils in the Annual AEC had neutral to near-neutral pH (6.5 to 7.2) before the onset of cultivation. Three to four decades of N fertilizer application decreased the pH in the surface foot of soil to less than 5.2 (Mahler 2002), which is at or below critical pH levels for optimum production of winter and spring cereals (pH 5.2 to 5.4), and grain legumes (pH 5.4 to 5.6) ●. In the Transition AEC, long-term (47 years) cultivation of winter wheat-spring pea (WW-SP) has considerably decreased the soil pH within the top 2 feet compared to undisturbed grassland pasture. The differences were much pronounced in the top 8-inch soil depth profile (Figure 2-9) ▲. Repeated banding of N fertilizer and limited soil mixing under reduced tillage systems, such as disk tillage and no-till, can further enhance soil acidification, particularly near the soil surface (Figure 2-9) ▲.

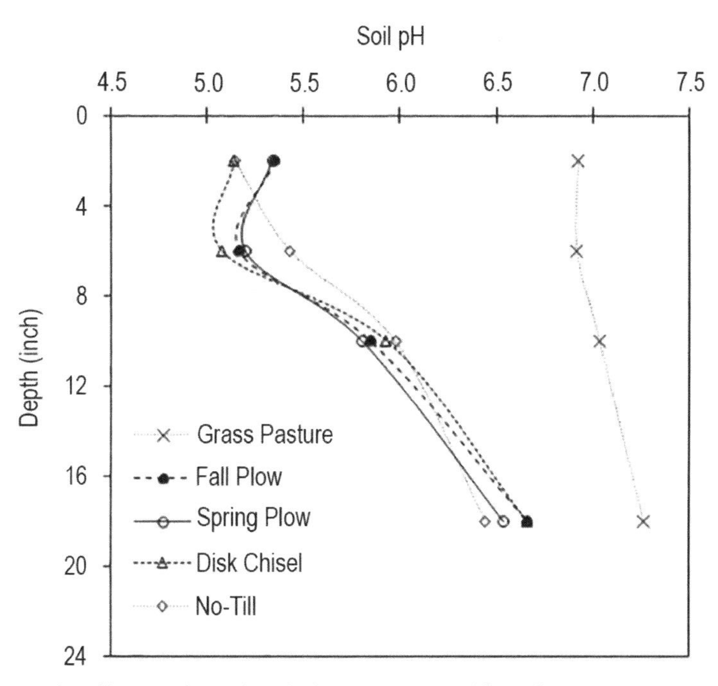

Figure 2-9. Soil profile pH under undisturbed grass pasture and four tillage systems in a long-term winter wheat-spring pea (WW-SP) rotation experiment near Pendleton, Oregon. (Awale et al. unpublished data.) ▲

Application of lime can ameliorate soil acidity (increase pH) and increase soil biological activities and crop yields. Generally, microbial activities tend to decrease with decreasing soil pH due to reduced bioavailability of organic substrates. Acidity suppresses soil enzyme activities by destroying ion and hydrogen bonds in active sites of enzymes and by altering the shape of enzymes. Increased pH caused by lime application enhances the deprotonation of organic substances and decreases bonding between organic compounds and soil particles making organic substances more accessible to microorganisms and soil enzymes. In the Annual AEC, lime application increased soil pH and maintained greater microbial biomass, soil respiration, and acid phosphatase activity; and as a result, improved wheat yield by 3 bushels per acre over non-limed plots (Table 2-7) ●. Another study in the Annual AEC revealed that liming increased soil pH above 5.5, promoted **nitrifier** populations, and increased N availability in soils from enhanced SOM **mineralization** and nitrification (Fuentes et al. 2006) ●. Mixing of a soil profile comprised of a petrocalcic (calcium carbonate) horizon under conventional tillage increased soil pH and, thereby, dehydrogenase enzyme activity over no-till in the Annual AEC due to the suppressive effects of low pH on bacterial growth (Kennedy and Schillinger 2006) ●.

Other alternatives to increase or maintain soil pH include retention of crop residues and application of organic amendments such as manure, compost, and alkaline biochar. Harvesting crop residues removes basic cations (positively charged ions) such as calcium, magnesium, and potassium, which help neutralize soil acidity. Organic materials add negative charges in soils upon their decomposition, and these negative charges can buffer acidity (Brown et al. 2008) ●. Figure 2-10 shows that application of alkaline biochar (pH 10) increased soil pH and wheat yield ▲.

Soil acidification may also be mitigated by using optimum crop N rates, nitrate-based N fertilizers, and including legume crops in cropping systems. Legumes have a natural ability to fix atmospheric N and make it available to plants, thereby reducing synthetic fertilizer N inputs. In the Transition AEC, WW-SP rotation increased soil pH compared to WW-F because of less synthetic N input requirements under a legume-based cropping system (Table 2-8) ▲.

Table 2-7. Soil surface (0–12 inch) pH and microbial parameters between non-limed and lime-applied (1.2 ton/ac) plots measured after two years following lime application under spring barley-winter wheat-spring pea (SP-WW-SP) at two sites in the inland PNW. ●

Soil parameters	Pullman, WA		Genesee, ID	
	No lime	Lime	No lime	Lime
pH	5.19	5.99	5.51	6.36
Microbial biomass carbon (ppm-C)	326	433	289	381
Respiration (ppm-C)	58	60	46.6	71.7
Dehydrogenase (ppm TPF/hr)	6.4	6.1	13.1	15.1
Acid phosphatase (ppm PNP/hr)	211	277	138	145

Adapted from Bezdicek et al. 2003.

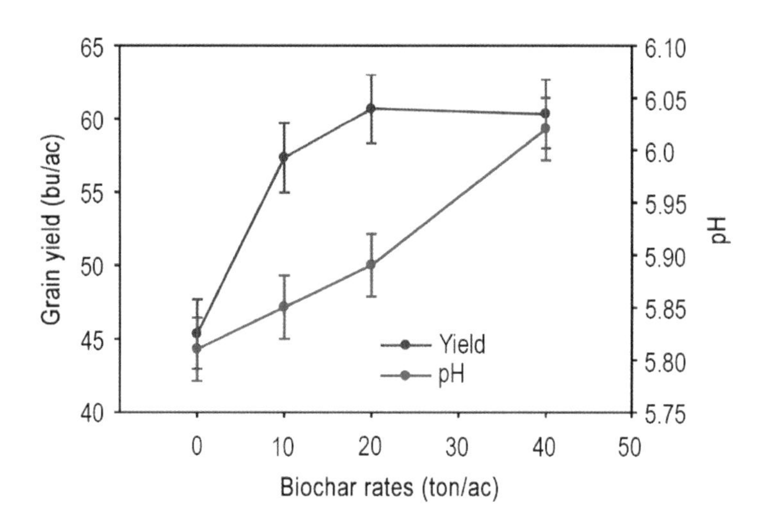

Figure 2-10. Biochar effects on soil pH and winter wheat yield in Athena, Oregon. (Adapted from Machado et al. unpublished.) ▲

Table 2-8. Soil pH measured within a 0 to 8-inch soil profile winter wheat-fallow (WW-F) and winter wheat-spring pea (WW-SP) cropping systems under conventional tillage near Pendleton, Oregon. ▲

Crop rotation	Soil pH	
	0 to 4 inch	4 to 8 inch
WW-F	5.26	5.00
WW-SP	5.35	5.17

Adapted from Awale et al. and Ghimire et al. 2017.

Cation Exchange Capacity

In soils, negative charges develop in clay minerals (permanent charge) and organic matter (pH-dependent charge). The negatively charged soil particles attract and hold positively charged ions or cations such as calcium (Ca^{2+}), magnesium (Mg^{2+}), potassium (K^+), aluminum (Al^{3+}), hydrogen (H^+), zinc (Zn^{2+}), and other molecules much like the opposite poles of a magnet attract each other. The molecules can be nutrients, water, herbicides, and other soil amendments. The adsorbed cations on the clay and organic matter particles are exchangeable with other cations in the soil solution. The sum of exchangeable cations that a soil can hold at a specific pH (or its total negative charge) is the soil's cation exchange capacity (CEC). Table 2-9 lists typical CEC values in some soil types found within the inland PNW.

The CEC of a soil affects soil fertility and plant growth. Calcium, magnesium, potassium, ammonium, and zinc are some important nutrient cations vital for plant growth. Soils with a higher CEC retain nutrient cations and maintain them in soil solution. Conversely, the nutrient cations are susceptible to **leaching** in soils with a low CEC, and as a result, plants are most likely to develop nutrient deficiencies in these soils. To this end, large quantities of fertilizers applied in a single fall application to sandy soils with a low CEC could result in loss of nutrients via leaching below the root zone. In low CEC soils, spring application of nutrient fertilizers during active plant growth stages and in split doses will improve production efficiencies. When applying nutrients to clay soils, it is best to incorporate them to prevent runoff losses.

The CEC can be occupied by either acidic (Al^{3+}, H^+) or basic (Ca^{2+}, Mg^{2+}, Na^+, K^+) cations. Base saturation is the percentage of the total CEC

Table 2-9. Cation exchange capacity (CEC) values and other related soil properties in soils within the inland PNW.

Soil series	CEC	EC	OM	pH	Clay
	meq/100g	dS/m	%		%
Ritzville silt loam	10.2	0.30	0.93	8.29	6.4
Palouse silt loam	18.1	0.22	2.43	6.05	22.4
Larkin silt loam	20.9	0.41	4.42	5.93	19.4
Thatuna silt loam	17.0	0.79	2.54	6.15	18.4
Woodburn silt loam	17.1	0.63	2.95	5.17	28.4
Walla Walla silt loam	16.3	0.33	2.30	6.02	15.4
Puget silt loam	11.4	0.60	2.76	6.14	22.4
Shano silt loam (SSL1)	14.9	3.06	1.13	6.27	16.4
Shano silt loam (SSL2)	10.3	1.01	1.14	5.63	13.4
Quincy fine loamy sand	6.4	0.98	0.85	8.53	7.4

Note: EC = electrical conductivity; OM = organic matter.
Adapted from Ownley et al. 2003.

occupied by basic cations. The higher the base saturation, the higher the soil pH because the basic cations help neutralize acidity. Soil's CEC influences soil's buffering capacity (resistance to rapid pH change) and, therefore, liming requirements. For instance, the lower the CEC of a soil, the faster the soil pH will decrease with time and vice versa. Therefore, low CEC soils generally require frequent liming but at lower rates of application than high CEC soils. In high CEC soils, higher liming rates are needed to reach an optimum pH.

The amounts of a particular basic cation in the CEC is crucial for soil aggregation, water flow, and **water holding capacity**. Generally, high calcium and magnesium and low sodium contents are desirable because Ca^{2+} and Mg^{2+} ions promote soil **flocculation** whereas Na^{+} ions promote soil **dispersion**. Flocculation is important because water moves mainly through large pores between aggregates, and plant roots grow between aggregates (Figure 2-2). Soil dispersion leads to restricted water infiltration and movement through soil by plugging soil pores (Figure 2-2).

Often, soil's CEC is used to determine weed and disease management in wheat production. For example, herbicide application rates may need to be adjusted with regard to soil's CEC. Generally, higher CEC soils require higher rates than low CEC soils because of greater binding ability of soils to the products, rendering them ineffective. In the inland PNW, the biocontrol activity of *Pseudomonas fluorescens* on take-all disease in wheat declined with an increase in soil CEC because high CEC bound and inactivated the antibiotic compound phenazine-1-carboxylate produced by bacteria, or adsorbed nutrients needed by the bacteria to affect biocontrol (Ownley et al. 2003).

Soil CEC is primarily influenced by soil pH, clay type and content, and organic matter content. As pH increases, the number of negative charges in soil increases, thereby increasing CEC. Therefore, adding lime and raising the pH can improve the CEC of soil. Soil CEC increases with clay and organic matter content, and decreases with sand content. Therefore, sandy soils rely heavily on the CEC of organic matter for the retention of nutrients and water. The addition of organic matter will increase the CEC of a soil; however, it requires many years to take effect (Figure 2-11) ▲.

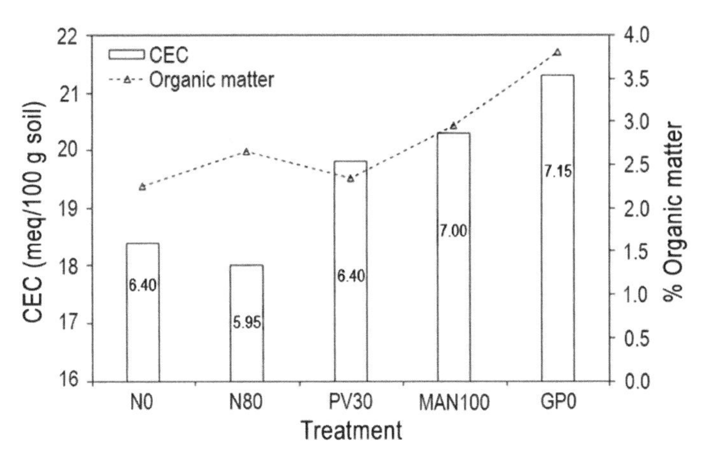

Figure 2-11. Cation exchange capacity (CEC), organic matter, and pH (number within bar) of soils under winter wheat-fallow (WW-F) in long-term (~70 yr) crop residue management experiment and grass pasture near Pendleton, Oregon. Numbers ending in the treatments denote amount of N (lb/ac) applied through inorganic N fertilizer (N), pea vines (PV), beef manure (MAN), and grass pasture (GP). (Adapted from Bandick and Dick 1999.) ▲

Soil Electrical Conductivity

In soils, cations (Ca^{2+}, Mg^{2+}, NH^{4+}, Na^{+}, K^{+}) and anions (Cl^{-}, HCO_3^{-}, $SO4_2^{-}$, NO_3^{-}) from salts dissolved in soil water carry electrical charges and conduct electricity. Soil **electrical conductivity** (EC) is a measure of the ability of soil solution to conduct electric current. The EC increases with ion (salt) concentrations in soil. The higher the concentration of ions in soil, the greater the EC measures and vice versa.

Electrical conductivity has been used to indicate total salt level in soil (salinity), although it does not indicate which salts might be present. In general, most soils contain some levels of salt that are essential for soil structure and plant growth, but excess levels can reduce crop yields by affecting soil-water balance, soil aggregation, soil microorganisms, and by inducing nutrient toxicity. Excess salts increase osmotic potential in the root zone and reduce roots' ability to extract soil water, leading to drought stress.

In non-saline soils, soil EC levels can serve as an indirect indicator of other soil properties, such as soil water content, clay content, organic matter level, soil depth, CEC, and nutrient content (nitrate-N, potassium) (Table 2-9). The conduction of electricity in soils takes place through the moisture-filled pores that occur between individual soil particles, and soil EC increases with the amount of moisture held by soil particles. To this end, EC correlates strongly with soil clay and organic matter contents due to their direct relationships with water-holding capacity as well as with soil CEC. Accordingly, use of soil EC maps has been recently increased in precision farming for delineating soil properties correlated with crop productivity. (For more information on soil EC mapping usage, see Chapter 8: Precision Agriculture.)

In soils, salts primarily develop from the weathering of parent material. However, application of fertilizers, pesticides, and other organic and inorganic amendments can add salts in soils, thereby increasing soil EC. Most microorganisms are sensitive to salt (high EC); and therefore, microbial processes such as soil respiration and nitrification decline as EC increases. Actinomycetes and fungi tend to be less sensitive than bacteria, except for salt-tolerant bacteria.

Soil Organic Matter

Soil organic matter (SOM) comprises organic materials in various states of decomposition, such as tissues of living soil organisms, plant and animal residues, and excretions from plant roots and soil microbes. SOM is a key indicator of soil health due to its influences on soil structure, aggregate stability, water storage and availability, water infiltration, nutrient storage and availability, soil biological activity, CEC, EC, adsorption of metals and agrochemicals, and pH buffering and amelioration. To this end, improvements in such soil physical, chemical, and biological properties with SOM usually correspond to greater agronomic yields (Figure 2-12). In addition, sequestering soil organic carbon (SOC) (58% of SOM) also reduces loss of CO_2, a greenhouse gas, to the atmosphere. Soil organic carbon comes from atmospheric CO_2 that is captured by plants through the process of photosynthesis.

Generally, SOM can be divided into three pools, as determined from their turnover time: active or labile (weeks to years), intermediate (years to decades), and passive or stable (hundreds to thousands of years) (Xiao 2015). The active SOM pool comprises recent plant residues in the early

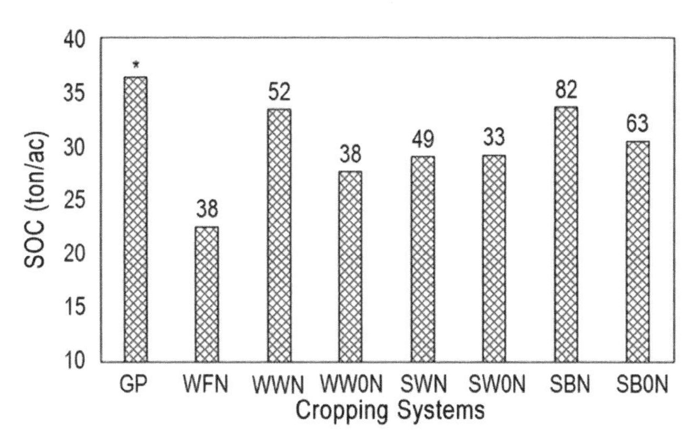

Figure 2-12. Effect of cropping systems (13 to 73 years) on soil organic carbon (SOC; ton/ac) at surface 16 inches of soil and grain yields of cereals under conventional tillage systems (except GP, no-till) near Pendleton, Oregon. Numbers above bars denote 6-year mean grain yields (bu/ac) for winter wheat-fallow (WFN); continuous winter wheat (WWN); continuous spring wheat (SWN); and continuous spring barley (SB) systems. GP represents undisturbed grass pasture. In each treatment, N at the end denotes N applied at 80 to 90 lb/ac, while 0N denotes not fertilized. (Adapted from Machado et al. 2006.) ▲

stages of decomposition and soil organisms. It serves as a food source for the living soil biological community and is responsible for nutrient release. The intermediate or slow SOM pool includes decomposed residues and microbial products that are stabilized through physical and biochemical processes. The slow pool influences soil's physical condition and nutrient buffering capacity. The stable SOM pool is highly recalcitrant (resistant to decomposition) and is important for soil aggregation and soil CEC.

Soil organic matter decomposes over the years and needs to be replaced by fresh biomass for its maintenance. Changes in SOM arise from the imbalance between inputs (from crop residues, manure, and any other organic sources) and outputs (from decay, leaching, and erosion) (Figure 2-13). Soil and crop management practices that increase SOM inputs and optimize the rate of SOM decay play an integral role in the sustainability of cropping systems. Conversely, reduced input of SOM or its rapid decomposition depletes SOM stocks. For instance, conventional tillage incorporates crop residues into soil and facilitates rapid decay of SOM by microbes due to the introduction of oxygen and greater soil residue contact (Table 2-10) ▲. On the other hand, no-till leaves most crop residues on the soil surface and favors the accumulation of SOM ▲.

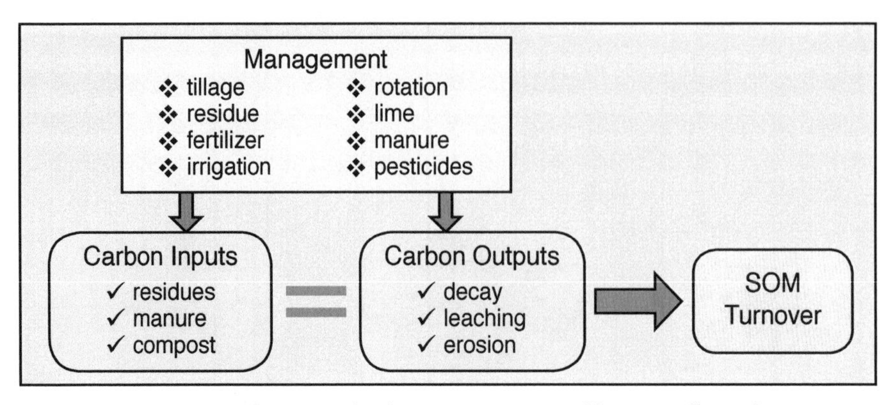

Figure 2-13. Management influences soil carbon input-output equilibrium to affect soil organic matter (SOM) dynamics.

Table 2-10. Tillage effects on soil organic carbon (SOC) dynamics at surface 2 feet in a long-term winter wheat-spring pea (WW-SP) rotation near Pendleton, Oregon.

Tillage	SOC (ton/ac)		ΔSOC
	1995	2005	%
Disk/Chisel	29.2	29.7	1.5
Fall plow	29.3	30.4	3.4
Spring plow	29.3	29.5	0.3
No-till	28.7	32.7	12.0

Adapted from Machado 2011.

SOM dynamics in the inland PNW

In the inland PNW, most native soils in areas where annual precipitation exceeds 9 inches contain about 1.5 to 2.5% SOM compared to about <1% SOM in the drier regions. Conversion of native ecosystems to agriculture has tremendously depleted SOM in the inland PNW. For instance, in the Transition AEC, soil organic carbon (SOC) depleted at the rate of 196 to 428 lb/ac/yr from Walla Walla silt loam soil managed under a long-term conventional WW-F system (Machado 2011). Consequently, SOC and N stocks have depleted by up to 63% and 26%, respectively, in eight decades (Figure 2-14). In the Annual AEC, SOC (surface 4 inches) loss of 55% to 60% with a WW-SW-SP rotation was continuous under conventional tillage for more than 100 years compared to undisturbed natural prairie (Fuentes et al. 2004).

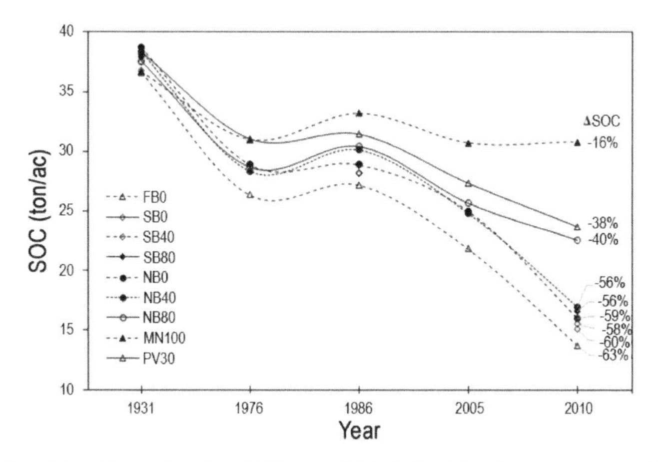

Figure 2-14. Trends in soil organic carbon (SOC) at top 24-inch depth in a long-term crop residue and nitrogen management experiment under conventional tillage winter wheat-fallow (WW-F) system near Pendleton, Oregon. Numbers ending in the treatments denote amount N (lb/ac) applied for fall residue burn (FB); spring residue burn (SB); no burn (NB); manure (MN); and pea vine (PV). Percentages of SOC change (ΔSOC) from 1931 are shown for the year 2010. (Adapted from Machado 2011 and Ghimire et al. 2015.)

The loss of SOM content in the inland PNW is mainly attributed to the following factors:

1. Low quality and quantity of crop residue inputs from the predominant crop-fallow rotation (i.e., WW-F), leaving the soil bare for 14 months between crops.
2. Crop residue removal for biofuel production and for other agricultural practices, such as bedding for animals and mushroom cultivation.
3. Residue burning to eliminate the need for multiple tillage operations for seedbed preparation to ensure better seed germination or plant establishment, as well as weed and disease control.
4. Limited biomass production due to low precipitation (less than 16 inches) and its distribution.
5. Increased biological oxidation of SOM associated with multiple tillage operations (nearly eight or more passes).
6. Farming on steep slopes (up to 45%) with accelerated management-induced soil erosion under cropping systems with high soil disturbance.

The extent to which agricultural management influences changes in SOM depends on several factors, including initial levels of SOM before the management was implemented, duration of management imposed, duration of conservation practices adopted, the degree of system reaching steady-state (SOM saturation), soil and environmental conditions, and crop productivity. Therefore, it takes years to observe significant changes in SOM stocks. Determining SOM fractions sensitive to management practices and that predict SOM changes and future trends should allow early decisions for management changes that lead to SOM build-up and maintenance.

Indicators of SOM dynamics

Management induced changes in SOM are usually evaluated by determining SOC and N contents because SOM consists of approximately 58% carbon (C) and 5% nitrogen (N). However, changes in SOM take time to manifest and are difficult to discern early. Different pools of SOM and soil enzymatic activity are typically more sensitive and provide early signs of the effect of management changes than total SOM alone. Therefore, there is a growing interest in assessing early indicators of SOM dynamics such as particulate organic matter (POM), mineralizable carbon (Cmin) (soil respiration), microbial biomass carbon (MBC), dissolved organic carbon, permanganate oxidizable carbon, and potentially mineralizable nitrogen (PMN) for evaluating soil health changes (Table 2-11). Studies have shown that these indicators are strongly correlated with each other and with SOM (Figure 2-15).

Particulate organic matter

Particulate organic matter (POM) is an intermediate pool of organic matter, and typically makes up a large portion of the light fraction of SOM. It is composed of plant residues as well as microbial and microfaunal debris. Therefore, POM is a transitory or relatively labile fraction of SOM, often of recent origin (weeks to years). POM contributes to soil function in a number of ways including C-cycling and formation of other forms of organic matter such as microbial biomass and soluble C. POM is a food and energy source for soil microbes and other soil fauna, and involved in nutrient cycling, maintenance of soil structure, and stabilization of soil macroaggregates.

Table 2-11. Land-use management effects on soil organic carbon (SOC), particulate organic carbon (POC), microbial biomass carbon (MBC), and C mineralization (Cmin) of 0 to 8 inches of Palouse silt loam in eastern Washington.

†Management	SOC	POC	MBC	26-week Cmin
	ton/ac			
NP	28.4	8.7	1.8	2.8
NTR	26.1	4.6	1.5	2.3
NT28	23.6	3.6	1.0	2.2
NT4	22.3	2.8	0.8	2.0
CRP	16.5	2.8	0.9	2.2
BGNT4	13.5	2.1	0.7	2.3
CT	12.4	1.9	0.9	2.0

†Managements are native prairie (NP); no-till reestablished for 1 year following 10 years no-till and 3 years conventional tillage (NTR); conventional tillage followed by no-till for 28 years (NT28); conventional tillage followed by no-till for 4 years (NT4); conventional tillage followed by 11 years perennial grass under the Conservation Reserve Program (CRP); bluegrass seed production for 9 years followed by no-till for 4 years (BGNT4); >100 conventional tillage (CT).
Reprinted from Purakayastha et al. 2008.

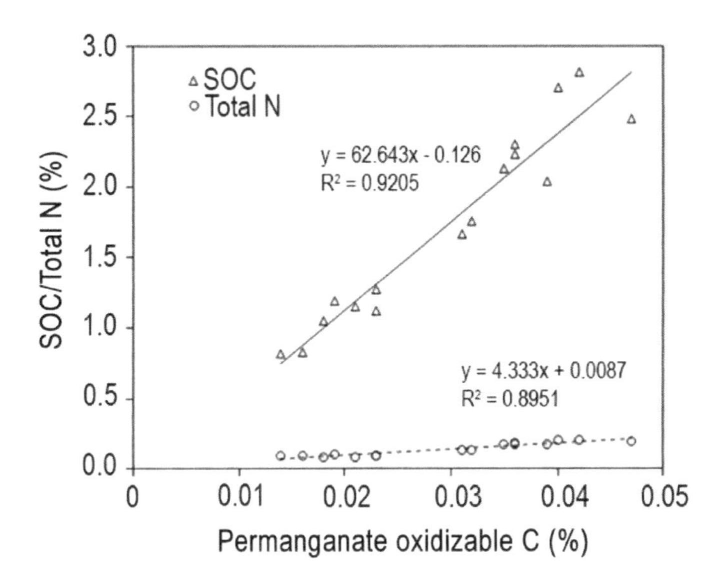

Figure 2-15. Relationship of permanganate oxidizable carbon with soil organic matter (SOM) across dryland cropping regions of the inland PNW. (Adapted from Morrow et al. 2016.)

Permanganate oxidizable carbon

Permanganate oxidizable carbon encompasses all those organic components that can be readily oxidized by potassium permanganate ($KMnO_4$), and hence, is directly related to SOC, total N, POM, **soil microbial biomass**, dissolved organic matter, and enzyme activity.

Dissolved organic matter

Dissolved organic matter is a labile pool of organic matter, present in the dissolved form in soil solution. It is comprised of leachates from plant residues and exudates from soil organisms and plant roots. It serves as energy substrate for soil microbial activity.

Soil microbial biomass and community structure

Soil microbial biomass is the active component of SOM. It is a measure of the total size of the microbial population. Table 2-12 shows typical soil microbial groups along with their relative numbers and biomass. About 80–90% of all the biogeochemical processes carried out in the soil are reactions mediated by microorganisms. Microorganisms play important roles in the decomposition of SOM, nutrient cycling and retention, formation and stability of soil aggregates, degradation of agricultural pollutants, disease suppression, and plant health improvement. Soil microorganisms respond rapidly to changes in soil environment, and microbial communities and functions are excellent indicators of soil health dynamics. Soil microbial biomass dynamics provide early warnings of soil health changes before any detectable changes occur in other soil physical and chemical properties.

Microbes act as agents for biochemical transformations of organic matter, with microbial communities acting on various components of SOM. For instance, bacteria has a low **carbon-to-nitrogen** (C:N) ratio and proliferates using easily available substrates. On the other hand, fungal biomass has a greater C:N ratio and is capable of decomposing more recalcitrant substrates, such as cellulose and lignin. Thus, fungi are typically much more efficient at assimilating and storing nutrients than bacteria. To this end, a shift toward a fungal dominance (high fungal to bacterial ratio) in the microbial community is often thought to enhance

Table 2-12. Relative number and biomass of microorganisms at surface 6-inch soil depth.

Microorganisms	Number/g of soil	Biomass (lb/ac)
Bacteria	10^8 to 10^9	350 to 4460
Actinomycetes	10^7 to 10^8	350 to 4460
Fungi	10^5 to 10^6	890 to 13380
Algae	10^4 to 10^5	10 to 450
Protozoa	10^3 to 10^4	Variable
Nematodes	10^2 to 10^3	Variable

Adapted from Hoorman and Islam 2010.

organic C accumulation and decrease its turnover rate. Accordingly, soil microbial communities differ in their responses to changes in soil management, with fungal communities usually more sensitive to these changes.

Decomposition of SOM by microbes release plant available nutrients such as N, P, and S in soil. For example, heterotrophic bacteria and fungi (requiring an energy source) convert complex forms of organic N into ammonium-N (NH_4-N), which, in turn, are converted into nitrate-N (NO_3-N) by nitrifying bacteria (*Nitrosomonas* spp. and *Nitrobacter* spp.). Certain groups of specialized bacteria such as *Rhizobioum* spp., actinomycetes, and cyanobacteria, with their symbiotic association with plant roots, can fix atmospheric nitrogen (N_2) into plant available form (NH_4). Legume-rhizobium symbiosis is the most important symbiotic relationship in nitrogen fixation. Therefore, cropping systems including legumes usually require less external fertilizer N inputs. Mycorrhizal fungi mobilize and transport nutrients (P, Cu, Zn) and water to plants by increasing the surface area of plant root systems through extensions of fungal hyphae. Microorganisms also hold nutrient elements in their biomass, which are released upon their death. Protozoa and nematodes consume other microbes to mineralize nutrients in the microbial biomass. Retention of nutrients in microbial biomass reduces nutrient losses from soils during periods of slower crop uptake.

Microbes exude sticky binding agents such as glomalin, polysaccharides, and ergosterol as they transform and ingest SOM (Figure 2-16). Fungal hyphae and bacterial filaments also favor the mechanical union of soil particles to enhance soil aggregation (Figure 2-16). Stable soil aggregates are important for soil porosity, water infiltration, and resisting erosion.

Microorganisms have many methods for controlling disease-causing organisms. Protozoa and nematodes help maintain microbial diversity in soils by consuming other soil microbes. Some bacteria and fungi generate antibiotic compounds to suppress other soil microorganisms. For instance, a bacterium *Pseudomonas fluorescens* produces a phenazine compound which inhibits take-all disease-causing fungal pathogen *Gaeumannomyces graminis* in wheat (Ownley et al. 2003). Isolates of *Chryseobacterium soldanellicola* (bacteria) exhibited significant antagonism against

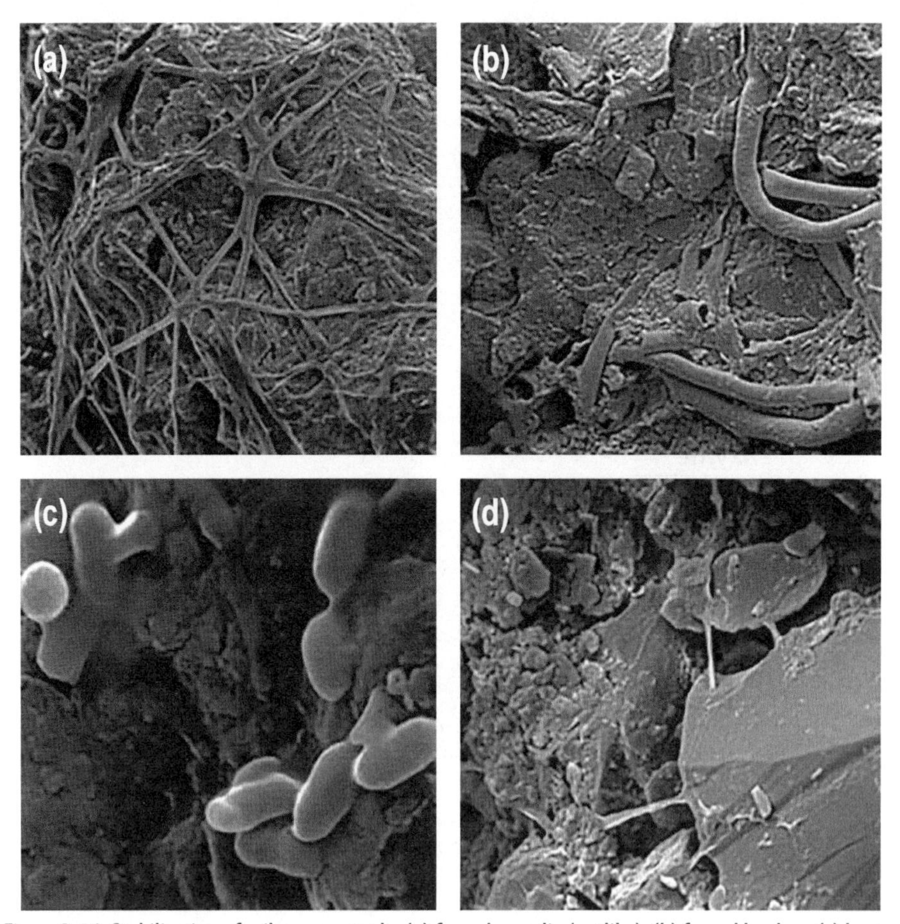

Figure 2-16. Stabilization of soil aggregates by (a) fungal mycelia (netlike), (b) fungal hyphae, (c) bacterial polysaccharides, and (d) actinomycete filaments. (Adapted from Eickhorst and Tippkoetter 2016.)

Rhizoctonia solani (a fungal pathogen that causes bare patch and root rot disease of wheat) in vitro as well as in greenhouse tests (Yin et al. 2013) ■. In addition to suppressing plant diseases, microorganisms can also stimulate plant root growth by producing biochemicals and compounds. For more information on microbial control of diseases, see Chapter 10: Disease Management for Wheat and Barley.

Soil enzymes

Soil enzymes are comprised of living and dead soil microorganisms, plant roots and residues, and animals. These enzymes play key biochemical functions in the overall process of organic matter decomposition and nutrient cycling (Table 2-13). Soil enzymes respond rapidly to changes in soil management and are potential indicators of soil health. Almost all soils contain soil enzymes but their types depend on variable substrates that serve as energy sources for microorganisms.

Soil respiration/mineralizable carbon

Soil respiration involves the oxidation of organic matter to produce CO_2 and water. It is a measure of soil biological activity and usually determined by measuring CO_2 produced by soil microorganisms and plant roots.

Table 2-13. Soil enzymes as indicators of soil health.

Soil enzyme	Role	Significance
Glucosidase	C-cycling	Energy for microbes
Galactosidase	C-cycling	Energy for microbes
Dehydrogenase	C-cycling	Energy for microbes
Invertase	C-cycling	Energy for microbes
Cellulase	C-cycling	Energy for microbes
Phenol oxidase	C-cycling	Energy for microbes
Amidase	N-cycling	Plant available N
Deaminase	N-cycling	Plant available N
Urease	N-cycling	Plant available N
Phosphatase	P-cycling	Plant available P
Arylsulfatase	S-cycling	Plant available S

Adapted from Reardon and Wuest 2016.

Mineralizable carbon refers to CO_2 produced by soil microbes during the decomposition of organic matter.

Potentially mineralizable nitrogen

Potentially mineralizable nitrogen (PMN) is a measure of mineral N released from the mineralizable organic fraction in soil. Microorganisms convert organic forms of nitrogen into plant available inorganic forms such as ammonium-N and nitrate-N (mineralization) and vice versa (**immobilization**). Whether mineral N is released or immobilized depends upon the substrate's C:N ratio. With a low C:N ratio substrate, more N, in excess of microbial requirements, is available than with a high C:N ratio substrate. PMN is an important potential source of N for crop growth and enhances microbial growth and activity. By knowing the soil's inherent ability to supply plant available N, recommended application rates of synthetic N can be determined while minimizing N losses through leaching due to over fertilization, or avoiding yield losses due to under fertilization.

Management effects on SOM indicators

Tillage and crop residue management, crop rotation and cropping intensity, and application of organic and inorganic amendments such as manure, compost, fertilizers, and lime have shown to influence fine indicators of SOM dynamics. Tillage influences soil microbial decomposition rates of SOM by influencing the availability and distribution of SOM in the soil profile, and by regulating other soil properties such as temperature, aeration, water content, and pH. Conservation tillage practices favor SOM accumulation whereas intensive tillage practices break soil aggregates and expose aggregate-protected SOM pools, increase aeration and temperature, increase soil residue contact, and promote SOM decay. Moreover, tillage-induced soil erosion can also transport SOM from eroded areas to other landscape positions (Kennedy and Schillinger 2006) ●. In the Annual AEC, SOC, particulate organic C, microbial biomass C, and 26-week mineralizable C decreased with tillage intensity and duration due to a corresponding increase in residue decomposition (Table 2-11) ●. In a similar experiment in the Annual AEC, no-till increased mineralizable C as well as permanganate oxidizable C over conventional tillage (Figure 2-17) ●. Rotating tillage practices, such as medium duration of no-till (10 years) and short interval of conventional till (3 years) followed by no-till, can lead to greater and more rapid accumulation of SOC and its fractions compared with no-till alone by distributing C accumulated during the no-till phase to deeper depths (Table 2-11) ●.

Tillage also affects soil microbial community structure by destroying fungal networks while favoring bacterial communities by increasing soil-residue contact. Such changes in soil microbial communities can consequently alter SOC storage potential. Usually, fungal dominance (higher fungus-to-bacteria ratio) in the microbial community has been linked to an increase in soil's C storage capacity due to increased carbon use efficiency, whereas bacterial dominance accelerates SOC loss via mineralization. For instance, in the Annual AEC, fungal biomarkers were greater in no-till soils whereas most bacterial biomarkers were greater in soils under conventional tillage (Table 2-14) ●. Fungal dominance under no-till was associated with a dramatic increase in SOC over conventional tillage. For more information on tillage, see Chapter 3: Conservation Tillage Systems.

Cropping systems influence both above- and below-ground residue inputs, and microbial activity. Usually, continuous cropping systems increase SOM from annual residue inputs and promote microbial biomass more than fallow cropping systems. For example, in the Transition AEC, the relative proportion of soil microbial biomass measured under annual

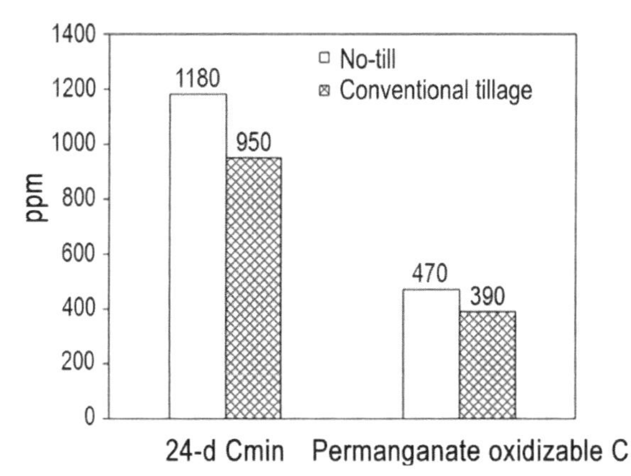

Figure 2-17. Effect of tillage on mineralizable C (24-d Cmin) and permanganate oxidizable C under winter wheat-spring barley-spring (WW-SB-SL) legume cropping system near Genesee, Idaho. (Adapted from Morrow et al. 2016.) ●

Table 2-14. Microbial community (biomarkers), determined as phospholipid fatty acid content, in surface 2-inch soils under no-till and conventional tillage in northeastern Washington. ●

Microbial group		Biomarkers	No-till (%)	Conventional tillage (%)
Bacteria	Gram positive	11:0 iso	0.17	0
		15:0 anteiso	2.48	2.75
		17:0 anteiso	0.75	1.06
	Gram negative	16:1 ω7c	4.62	4.50
		17:0 cyclo	0.98	1.45
		19:0 cyclo	0.27	0.84
Fungi		18:1 ω9c	6.92	6.07
		18:2 ω6c	6.44	4.38
		18:3 ω6c	1.33	0.96

Adapted from Kennedy and Schillinger 2006.

systems (WW-SP and WW-SW) and grain-fallow (WW-F) were about 50% and 25% of undisturbed grass pasture (Figure 2-18) ▲. Diversified cropping systems create a more favorable soil environment for microbial and faunal community due to increased root activity and exudates. For instance, dehydrogenase enzyme activity in a top 2-inch soil depth was higher under a SW-SB rotation than under continuous spring wheat in the Grain-Fallow AEC (Schillinger et al. 2007) ■. Inclusion of cover crops in a cropping sequence, particularly legumes, can increase soil available N along with other improvements in soil properties such as aggregation, infiltration, and weed control. Table 2-15 shows that the winter wheat-winter pea (WW-WP) rotation increased SOC, total N, and mineralizable N compared with other cropping systems without legumes ■. For more information on cropping systems, see Chapter 5: Rotational Diversification and Intensification.

Residue composition determines its decomposition rate, and thus microbial respiration and microbial biomass and community structure. Accordingly, residue composition influences microbial release of nutrients. Microorganisms also require soil nutrients for their own growth, and the C:N ratio of residues generally determines whether nutrients are released (mineralized) and available for plant uptake or

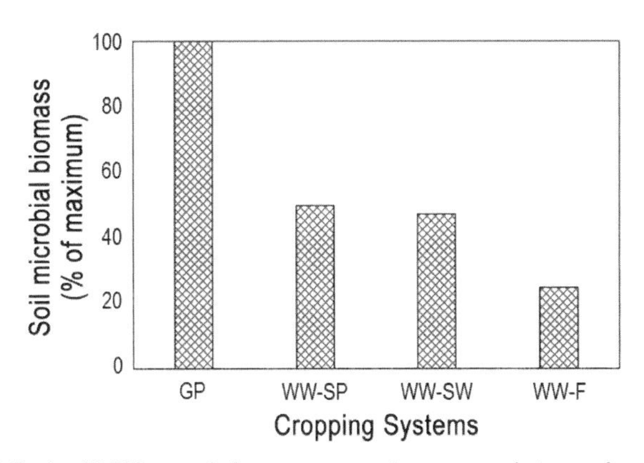

Figure 2-18. Soil microbial biomass in long-term cropping systems (winter wheat-spring pea, WW-SP; winter wheat-spring wheat, WW-SW; and winter wheat-fallow, WW-F) in relation to grass pasture (GP) near Pendleton, Oregon. (Adapted from Collins et al. 1992.)

Table 2-15. Effect of tillage and cropping systems on soil organic carbon (SOC), soil total nitrogen (N), and potentially mineralizable nitrogen (PMN) near Moro, Oregon.

†Management	SOC	Total N	PMN
	%		ppm
NT WW-WP	1.268	0.095	73.9
NT WW-ChF	1.147	0.085	41.1
NT WW-SB-F	1.125	0.093	43.5
CT WW-F	1.053	0.081	41.6

†Managements are no-till winter wheat-winter pea (NT WW-WP), no-till winter wheat-chemical fallow (NT WW-ChF), no-till winter wheat-spring barley-fallow (NT WW-SB-F), and conventional tillage winter wheat-fallow (CT WW-F).
Reprinted from Morrow et al. 2016.

assimilated (immobilized) in the microbial biomass. Therefore, the availability of essential nutrients such as N and P can, in turn, influence soil microbial biomass. Table 2-16 shows N mineralization under wheat-based cropped and undisturbed native soils, with and without wheat straw addition, from five different precipitation zones in northeastern Oregon. In general, soil N mineralization increased with precipitation, undisturbed native soils had higher mineralized N than cropped soils, and wheat straw addition reduced mineralized N (or immobilized) than when no residue was added (Table 2-14). For further information on residue characteristics, see Chapter 4: Crop Residue Management.

Table 2-16. Effect of wheat straw addition on soil nitrogen (N) mineralized from 0 to 8-inch depth of cropped (wheat-based) and native soils across different precipitation zones in northeastern Oregon.

Soil	Precipitation inch	†Soil N mineralized (ppm)		
		No residue added	3 ton/ac residue	6 ton/ac residue
Cropped				
Cowsly	>20	57	26	41
Athena	16 to 20	61	42	48
Walla Walla	14 to 16	41	13	26
Ritzville	10 to 14	29	5	14
Adkins	<10	28	8	23
Native				
Cowsly	>20	114	72	89
Athena	16 to 20	123	60	79
Walla Walla	14 to 16	109	77	70
Ritzville	10 to 14	NA	NA	NA
Adkins	<10	29	−3	−1

†Negative values indicate net immobilization; NA = not available.
Adapted from Douglas et al. 1998.

Addition of organic materials from external sources such as manure, biosolids, and compost represent direct input of organic C as well as other important plant nutrients (e.g. N, P). Similarly, balanced nutrient fertilization and effective management of weeds, **pests**, and diseases would likely increase crop biomass and the amount of residue returned to the soil. Application of lime increases soil pH that creates favorable conditions for soil biological activity. Studies have also shown a strong and positive correlation between additions of organic and inorganic amendments with soil biological activity which, in turn, is responsible for improving soil physical and chemical properties (Tables 2-3 and 2-7). In the Transition AEC, application of organic manure and pea vines increased β-glucosidase, arylsulfatase, and urease activities compared to inorganic N application in a WW-F system (Figure 2-19). See Chapter 7: Soil Amendments for more information on soil organic and inorganic amendments. See Chapter 9: Integrated Weed Management, Chapter 10: Disease Management for Wheat and Barley, and Chapter 11: Insect

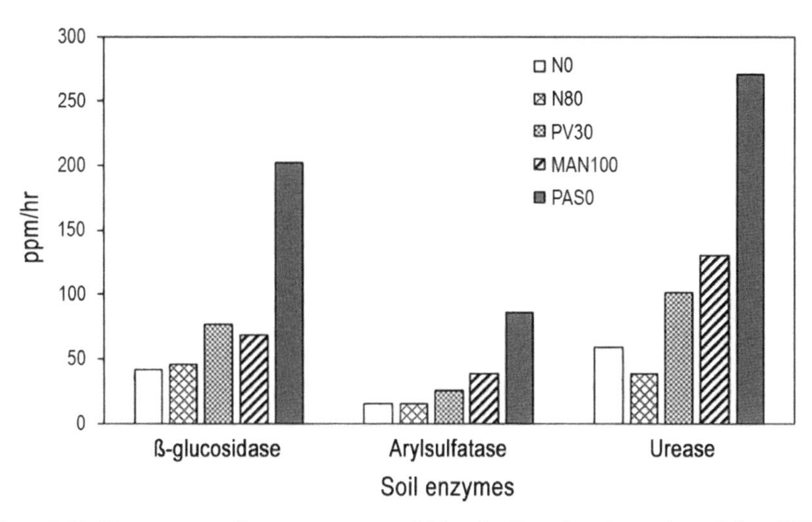

Figure 2-19. Management effects on enzyme activities of soils under winter wheat-fallow (WW-F) in a long-term crop residue management experiment and grass pasture near Pendleton, Oregon. Numbers ending in the treatments denote amount of nitrogen (lb/ac) applied through inorganic nitrogen fertilizer (N), pea vines (PV), beef manure (MAN), and grass pasture (PAS). (Adapted from Bandick and Dick 1999.)

Management Strategies for more information on management of soil fertility, weeds, insects, and pests.

Soil Fauna: Earthworms, Nematodes, and Soil Insects

Soil fauna, including earthworms, nematodes, and insects, are essential in soil structural development; water, air, and nutrient cycling; SOM turnover; suppressing harmful pests; and enhancing beneficial microorganisms in the soil profile. For instance, the burrowing activity of nematodes creates a network of surface-connected tunnels which increase air permeability and water infiltration rates. In addition, burrowing activity also promotes soil mixing and increases soil and plant residue contact that favors organic matter decomposition and nutrient release. The feeding and casting activity of earthworms improve aggregate stability and enhance microbial activity (Figure 2-20).

Soil faunal measurement techniques involve counting the size of their populations. Management practices, such as tillage, residue cover, crop rotation, and liming, that alter soil microenvironments and disturb their

Figure 2-20. Worm excrement (cast) increases soil microbial activity and soil aggregate stabilization. (Photo by Wikimedia Commons.)

habitats and food sources influence the populations of soil fauna. In the Annual AEC, increase in earthworm populations is usually associated with no-till relative to conventional tillage due to lower disturbance, less physical injury, and more food supply (Figure 2-21) ●. Soil fauna is usually concentrated at locations with higher organic matter content because SOM not only serves as a food source to soil fauna but it also retains soil moisture that is necessary for soil faunal survival and reproduction. In the Annual AEC, higher SOM under no-till promoted greater earthworm densities, but the densities remained low under conventional tillage with low SOM content (Umiker et al. 2009) ●. For more information on soil fauna, see Chapter 11: Insect Management Strategies.

Soil Health Assessment

A healthy soil is the foundation of sustainable agricultural production. The concept of soil health is gaining importance due to increasing pressure to sustainably produce food, feed, fiber, and fuel for the world's increasing population. Compared to air and water quality assessments that are well defined, soil health assessments are more difficult in that conditions

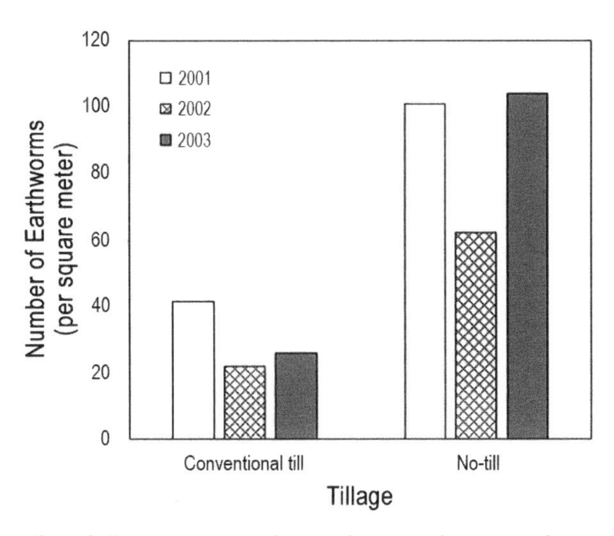

Figure 2-21. Effect of tillage on mean earthworm density under winter wheat-spring barley-spring pea (WW-SB-SP) rotation in northern Idaho. (Adapted from Johnson-Maynard et al. 2007.)

considered healthy for one soil type and crop in a particular environment may not be the same for another soil type and another crop in a different environment. Agricultural landscapes exhibit a high degree of spatial variability and different crops require different conditions. In addition, the slow nature of soil degradation seldom leads to immediate cropping system failures, and therefore subtle deleterious effects of a particular soil management regime on soil health and function are often overlooked by growers until the problem is much worse. The major challenge of soil health research and assessment programs is the lack of a standard set of soil indicators that are sensitive to changes in management and are cheap and accessible. It would be difficult to establish a single health indicator that could effectively judge all soil changes in response to management; hence, comprehensive standards for assessing soil health are needed. In response, the concept of a minimum data set of indicator groups of soil function required for soil health assessment has emerged (Table 2-1; Figure 2-22). However, the components of a minimum data set are not universal because of inherent soil variability. Therefore, any set of indicators may not provide a one-size-fits-all approach to soil health assessment across all soils and all environments. Recently, Morrow et al. (2016) proposed a set of seven criteria that establish a specific framework to judge the effectiveness of soil health indicators for soil health assessments across agroecosystems in the inland PNW:

1. Evidence based: The indicators capture measurable soil properties and processes, supported by scientific research and experimentation.
2. Sensitive to change: The indicators respond to management practices and are applicable across a range of soils and soil conditions.
3. Logistically feasible: The indicators are feasible within time constraints of effective decision making.
4. Accurate and precise: Indicators provide repeatable results with low variability, determined using standardized methods.
5. Cost effective: The cost of implementation is reasonable and within the economic constraints of the land management system.
6. In situ or undisturbed samples: Tests can be performed in situ or on relatively undisturbed samples to capture real field conditions.
7. Valued: The information is useful and interpretable for sound management decisions.

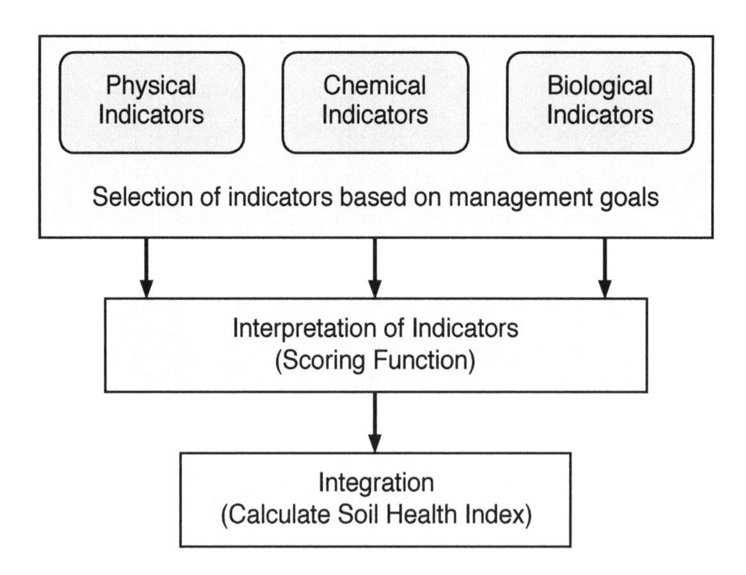

Figure 2-22. Conceptual framework of soil management assessment framework. (Adapted from Andrews et al. 2004.)

Soil Health Assessment Indices

Advances have been made in managing soils by integrating physical, chemical, and biological properties and processes in order to promote long-term soil sustainability. To this end, several indices and tools have been developed to provide a systematic framework for assessing soil health. Examples of such assessment tools include the Cornell assessment of soil health (Moebius-Clune et al. 2016), Haney's soil health test (Haney 2014), the soil management assessment framework (Andrews et al. 2004), and the USDA Natural Resources Conservation Service soil quality test kit (USDA 2001). Soil health assessments are performed by comparing a site against an adjacent, undisturbed site with the same soil type, whenever possible. Under situations where reference sites are not accessible, tracking temporal changes in appropriate indicators can detect management influences on soil health. Accordingly, soil assessment tools are usually framed up with three steps: (1) identification of indicators based on management goals, (2) indicator interpretation or scoring function (e.g., 0 to 10 or low to high, with an indicator score of 10 or high representing the highest potential function for that system), and (3) integration of all indicator scores into a single overall soil health score (Figure 2-22).

Applicability of Soil Health Indices in the Inland PNW

The Haney's soil health testing method is based on soil microbial activity, as determined from the measurement of CO_2 during the first day after rewetting a dried, ground soil sample. The test predicts a soil health score (1 to 50; the higher the better) as well as an estimate of N fertilizer credits from determinations of microbial activity and water soluble organic C and N. However, the reliability of the Haney's test as an indicator of soil health has been questioned because of the difficulty in getting reproducible test results (high random variability associated with test methodology) (Sullivan and Granatstein 2015). Recently, Washington State University researchers (Morrow et al. 2016) reported on their research evaluating the Haney's test using soils under diverse cropping systems and tillage intensities across the inland PNW. It was revealed that 1-day microbial activity (mineralizable C) was highly variable (coefficient of variation from 3% to 50%) such that the water soluble C and N did not correlate with 1-day mineralizable C but correlated with mineralizable C from longer (24-day) incubation time. The results are in sharp contrast from the assumption made in Haney's testing method that 1-day mineralizable C correlates well with water soluble C and N. Overall, the Haney's soil health index was not highly sensitive to tillage and cropping practices in the inland PNW. Therefore, Haney's soil health scoring method would necessitate further validation through extensive field research and calibration prior to its usage in the inland PNW.

Unlike Haney's test, the Cornell soil health assessment, the soil management assessment framework, and the USDA-NRCS soil quality test kit focus on several other physical and chemical properties of soils along with labile SOM pools and biological activity. Accordingly, these soil health assessment methods have been widely used across diverse soils and management practices with effective outcomes (Cherubin et al. 2016; Idowu et al. 2008; Karlen et al. 2008; Seybold et al. 2002; Wienhold et al. 2008). Usage of such indices to assess soil health in the inland PNW has been scarce. However, the indices could potentially be used to manage soil health under dryland farming systems within the inland PNW. For instance, following the approach of the soil management assessment framework, a comprehensive evaluation of soil health assessments across tillage (no-till vs. conventional tillage) and cropping systems (continuous small grains, including wheat, pea, and lentils vs. wheat-fallow) in the inland PNW was conducted (Figure 2-23). The researchers used data sets from Natural Resources Inventory monitoring sites located in Major Land Resource Area 9 (comprising the Palouse and Nez Perce Prairies, southeastern Washington, northwestern Idaho, and northeastern Oregon). Overall, the outcome of the assessment was that a continuous cropping system generally had soil health improvements (higher index values) over a wheat-fallow system (lower index values), with significant benefits observed when coupled with a no-till system (Figure 2-23).

Similarly, a quantitative soil health index was developed to evaluate soil health in fields with three different management systems (organic, conventional or non-organic, and integration of both) under apple orchard production in Washington (Table 2-17). Soil health indicators (physical, chemical, and biological) were scored based upon their effects on four soil functions (water entry, water movement and availability, resistance to

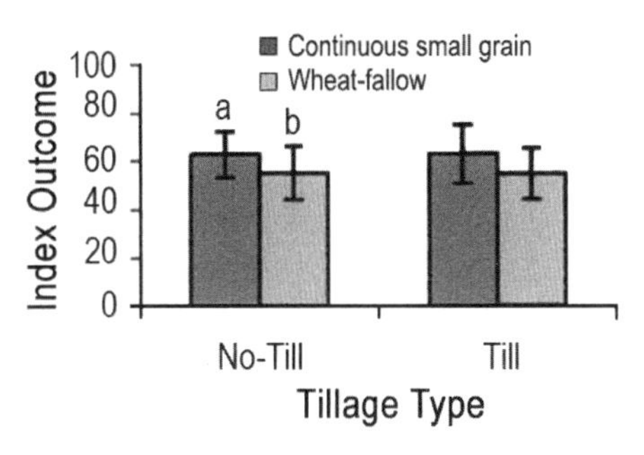

Figure 2-23. Soil management assessment framework index outcomes for cropping system and tillage management across the inland PNW. Treatments or land uses labeled with different letters are significantly different at $\alpha = 0.05$. Error bars represent one standard deviation from the mean. (Reprinted from Andrews et al. 2004.)

Table 2-17. Soil quality rating based on four soil functions for conventional or non-organic, integrated, and organic systems of apple production in Washington.

Treatment	Conventional (non-organic)	Integrated	Organic
Water entry	0.153	0.235	0.213
Water movement and availability	0.208	0.235	0.205
Resist degradation	0.185	0.145	0.225
Sustain productivity	0.255	0.213	0.238
Total soil quality index	0.783	0.923	0.878

Reprinted from Glover et al. 2000.

surface structure degradation, and sustainable production). The total soil quality index for each management was obtained by summing all scores. Overall, the study concluded that the integrated system had soil health advantages over the other two systems and that increasing organic matter and reducing tillage were key aspects.

Several soil health assessment guides, such as the Idaho NRCS soil health assessment card (USDA-NRCS 2014), the Palouse and Nez Perce Prairies soil quality card guides (USDA-NRCS 2004), and the Willamette Valley soil quality card guide (OSU 2009), are also available for use across the inland PNW.

Grower Considerations: Improving Soil Health in the Inland PNW

The main goal of every grower is to develop a farming enterprise that is economically and biologically sustainable. Adhering to soil health concepts to evaluate farm management should assist growers achieve this goal. Soil health assessment indices should be tailored to address the goals of each farming enterprise. Variations in soils, temperature, and environments in the inland PNW AECs necessitate the development of site-specific soil health indices to guide growers in improving soil health on their farms. Working very closely with Extension agents and other government agencies, growers should be able to assess the effect of management practices on soil health. Consequently, the efforts to manage soil health require continuous monitoring of soil health changes and making appropriate adjustments in management practices. Overall, in the inland PNW, soil health can be improved by the adoption of one or a combination of the following management strategies:

- Reduced or minimal soil disturbance through conservation tillage practices such as no-till, strip tillage, subsurface ridge till, and undercutter sweep (Chapter 3: Conservation Tillage Systems).
- Minimization of bare-fallow systems with alternative cropping systems such as chemical-fallow, annual cropping, and cover cropping (Chapter 5: Rotational Diversification and Intensification).
- Increased residue retention and cover (Chapter 4: Crop Residue Management).
- Elimination of crop residue burning (Chapter 4: Crop Residue Management).
- Increased cropping intensification and crop diversity (Chapter 5: Rotational Diversification and Intensification).
- Balanced and efficient fertilization approach to minimize nutrient loss, excessive weed growth, and pest competition and to maximize crop production and biomass return. Inclusion of legume cover crops in a crop rotation and organic amendments reduce N fertilizer needs and slow down soil acidification (Chapter 5: Rotational Diversification and Intensification; Chapter 6: Soil Fertility Management; Chapter 7: Soil Amendments).

- Application of organic amendments such as manure, compost, residues, etc. (Chapter 7: Soil Amendments).
- Integrated management of diseases, pests, and weeds through balanced fertilization, crop rotation, and cover crops (Chapter 5: Rotational Diversification and Intensification); minimum tillage (Chapter 3: Conservation Tillage Systems); and chemical control (Chapter 9: Integrated Weed Management; Chapter 10: Disease Management for Wheat and Barley; Chapter 11: Insect Management Strategies).
- Application of lime and alkaline biochar to ameliorate soil acidity (Chapter 6: Soil Fertility Management).

Additional Resources

Magdoff, F., and H. Van Es. 2009. Building Soils for Better Crops: Sustainable Soil Management, 3rd ed. Handbook Series Book 10. Sustainable Agriculture Research and Education. SARE Outreach Publication: Waldorf, MD.

Moebius-Clune, B.N., D.J. Moebius-Clune, B.K. Gugino, O.J. Idowu, R.R. Schindelbeck, A.J. Ristow, H.M. Van Es, J.E. Thies, H.A. Shayler, M.B. McBride, D.W. Wolfe, and G.S. Abawi. 2016. Comprehensive Assessment of Soil Health - The Cornell Framework Manual, Edition 3.1. Cornell University: Geneva, NY. ***http://www.css.cornell.edu/extension/soil-health/manual.pdf***

OSU. 2009. Willamette Valley Soil Quality Card Guide EM 8710-E. Oregon State University Extension Service. ***https://catalog.extension.oregonstate.edu/em8710***

USDA-NRCS. 2001. Soil Quality Test Kit Guide. USDA-ARS-NRCS Soil Quality Initiative. ***http://www.nrcs.usda.gov/Internet/FSE_DOCUMENTS/nrcs142p2_050956.pdf***

USDA-NRCS. 2004. Palouse and Nezperce Prairies Soil Quality Card Guide. ***http://www.nrcs.usda.gov/wps/portal/nrcs/detail/soils/health/assessment/?cid=nrcs142p2_053871***

USDA-NRCS. 2014. Idaho NRCS Soil Health Assessment Card. ***http://www.nrcs.usda.gov/wps/portal/nrcs/detail/soils/health/assessment/?cid=nrcs142p2_053871***

Xiao, C. 2015. Soil Organic Carbon Storage (Sequestration) Principles and Management: Potential Role for Recycled Organic Materials in Agricultural Soils of Washington State. Washington State Department of Ecology Pub. No. 15-07-005. ***https://fortress.wa.gov/ecy/publications/SummaryPages/1507005.html***

References

Andrews, S.S., D.L. Karlen, and C.A. Camberdella. 2004. The Soil Management Assessment Framework: A Quantitative Soil Evaluation Method. *Soil Science Society of America Journal* 68: 1945–1962.

Awale, R., et al. Unpublished. Long-Term Tillage Effects on Soil Organic Carbon, Nitrogen, and Grain Yields of Winter Wheat-Spring Pea Rotation in the Pacific Northwest.

Bandick, A.K., and R.P. Dick. 1999. Field Management Effects on Soil Enzyme Activities. *Soil Biology and Biochemistry* 31: 1471–1479.

Bezdicek, D.F., T. Beaver, and D. Granatstein. 2003. Subsoil Ridge Tillage and Lime Effects on Soil Microbial Activity, Soil pH, Erosion, and Wheat and Pea Yield in the Pacific Northwest, USA. *Soil and Tillage Research* 74: 55–63.

Brown, T.T., R.T. Koenig, D.R. Huggins, J.B. Harsh, and R.E. Rossi. 2008. Lime Effects on Soil Acidity, Crop Yield, and Aluminum Chemistry in Direct-Seeded Cropping Systems. *Soil Science Society of America Journal* 72: 634–640.

Cherubin, M.R., D.L. Karlen, A.L.C. Franco, and C.E.P. Cerri. 2016. A Soil Management Assessment Framework (SMAF) Evaluation of Brazilian Sugarcane Expansion on Soil Quality. *Soil Science Society of America Journal* 80: 215–226.

CIMMYT (International Maize and Wheat Improvement Center). 2016. Mineral and Environmental Stresses. ***http://wheatdoctor.org/image-galleries/mineral-and-environmental-stresses/aluminum-toxicity***

Collins, H.P., P.E. Rasmussen, and C.J. Douglas, Jr. 1992. Crop Rotation and Residue Management Effects on Soil Carbon and Microbial Dynamics. *Soil Science Society of America Journal* 56: 783–788.

Donk, S.V., and N.L. Klocke. 2012. Tillage and Crop Residue Removal Effects on Evaporation, Irrigation Requirement, and Yield. West Central Research and Extension Center, North Platte. Paper 62. ***http://digitalcommons.unl.edu/westcentresext/62/***

Doran, J.W., and M.R. Zeiss. 2000. Soil Health and Sustainability: Managing the Biotic Component of Soil Quality. *Applied Soil Ecology* 15: 3–11.

Douglas, Jr., C.L., P.E. Rasmussen, H.P. Collins, and S.L. Albrecht. 1998. Nitrogen Mineralization Across a Climosequence in the Pacific Northwest. *Soil Biology and Biochemistry* 30: 1765–1772.

Eickhorst, T., and R. Tippkoetter. 2016. The Hidden World of Soils. University of Bremen, Germany. ***http://www.microped.uni-bremen.de***

Esser, A.D., and R. Jones. 2013. No-Till and Conventional Tillage Fallow Winter Wheat Production Comparison in the Dryland Cropping Region of Eastern Washington. *Journal of the NACAA*. ***http://www.nacaa.com/journal/index.php?jid=227***

Feng, G., B. Sharratt, and F. Young. 2011a. Soil Properties Governing Soil Erosion Affected by Cropping Systems in the US Pacific Northwest. *Soil and Tillage Research* 111: 168–174.

Feng, G., B. Sharratt, and F. Young. 2011b. Influence of Long-Term Tillage and Crop Rotations on Soil Hydraulic Properties in the US Pacific Northwest. *Journal of Soil and Water Conservation* 66: 233–241.

Fuentes, J., D.F. Bezdicek, M. Flury, S. Albrect, and J.L. Smith. 2006. Microbial Activity Affected by Lime in a Long-Term No-Till Soil. *Soil and Tillage Research* 88: 123–131.

Fuentes, J., M. Flury, and D.F. Bezdicek. 2004. Hydraulic Properties in a Silt Loam Soil under Natural Prairie, Conventional Till, and No-Till. *Soil Science Society of America Journal* 68: 1679–1688.

Ghimire, R., S. Machado, and P. Bista. 2017. Soil pH, Soil Organic Carbon, Soil Nitrogen, and Grain Yield in Winter Wheat-Summer Fallow Systems. *Agronomy Journal* 109 (in press). doi:10.2134/agronj2016.08.0462.

Ghimire, R., S. Machado, and K. Rhinhart. 2015. Long-Term Crop Residue and Nitrogen Management Effects on Soil Profile Carbon and Nitrogen in Wheat-Fallow Systems. *Agronomy Journal* 107: 2230–2240.

Glover, J.D., J.P. Reganold, and P.K. Andrews. 2000. Systematic Method for Rating Soil Quality of Conventional, Organic, and Integrated Apple Orchards in Washington State. *Agriculture, Ecosystems & Environment* 80: 29–45.

Hammel, J.E. 1994. Effect of High-Axle Load Traffic on Soil Physical Properties and Crop Yields in the Pacific Northwest USA. *Soil and Tillage Research* 29: 195–203.

Haney, R.L. 2014. Soil Health. USDA-ARS. ***http://www.nrcs.usda.gov/Internet/FSE_DOCUMENTS/nrcs144p2_043902.pdf***

Hoorman, J.J., and R. Islam. 2010. Understanding Soil Microbes and Nutrient Cycling. Ohio State University Extension. ***http://ohioline.osu.edu/factsheet/SAG-16***

Idowu, O.J., H.M. van Es, G.S. Abawi, D.W. Wolfe, J.I. Ball, B.K. Gugino, B.N. Moebius, R.R. Schindelbeck, and A.V. Bilgili. 2008. Farmer-Oriented Assessment of Soil Quality Using Field, Laboratory, and VNIR Spectroscopy Methods. *Plant and Soil* 307: 243–253.

Johnson-Maynard, J.L., K.J. Umiker, and S.O. Guy. 2007. Earthworm Dynamics and Soil Physical Properties in the First Three Years of No-Till Management. *Soil and Tillage Research* 94: 338–345.

Karlen, D.L., M.D. Tomer, J. Neppel, and C.A Camberdella. 2008. A Preliminary Watershed Scale Soil Quality Assessment in North Central Iowa, USA. *Soil and Tillage Research* 99: 291–299.

Kennedy, A.C., and W.F. Schillinger. 2006. Soil Quality and Water Intake in Traditional-Till vs. No-Till Paired Farms in Washington's Palouse Region. *Soil Science Society of America Journal* 70: 940–949.

Machado, S., et al. Unpublished. Biochar Rate Effects on Soil pH and Grain Yield of Winter Wheat in the Pacific Northwest.

Machado, S. 2011. Soil Organic Carbon Dynamics in the Pendleton Long-Term Experiments: Implications for Biofuel Production in Pacific Northwest. *Agronomy Journal* 103: 253–260.

Machado, S., K. Rhinhart, and S. Petrie. 2006. Long-Term Cropping System Effects on Carbon Sequestration in Eastern Oregon. *Journal of Environmental Quality* 35: 1548–1553.

Mahler, R.L. 2002. Impacts and Management of Soil Acidity Under Direct-Seed Systems - Status and Effects on Crop Production. In Pacific Northwest Conservation Tillage Systems Information Source - Direct Seed Conference: January 16-18, 2002. ***http://pnwsteep.wsu.edu/directseed/conf2k2/dscmahler.htm***

Morrow, J.G., D.R. Huggins, L.A. Carpenter-Boggs, and J.P. Reganold. 2016. Evaluating Measures to Assess Soil Health in Long-Term Agroecosystem Trials. *Soil Science Society of America Journal* 80: 450–462.

Ownley, B.H., B.K. Duffy, and D.M. Weller. 2003. Identification and Manipulation of Soil Properties to Improve the Biocontrol of Performance of Phenazine-Producing *Pseudomonas fluorescence. Applied and Environmental Microbiology* 69: 3333–3343.

Purakayastha, T.J., D.R. Huggins, and J.L. Smith. 2008. Carbon Sequestration in Native Prairie, Perennial Grass, No-Till and Cultivated Palouse Silt Loam. *Soil Science Society of America Journal* 72: 534–540.

Reardon, C., and S. Wuest. 2016. Soil Amendments Yield Persisting Effects on the Microbial Communities: A 7-year Study. *Applied Soil Ecology* 101: 107–116.

Schillinger, W.F., A.C. Kennedy, and D.L. Young. 2007. Eight Years of Annual No-Till Cropping in Washington's Winter Wheat-Summer Fallow Region. *Agriculture, Ecosystems & Environment* 120: 345–3589.

Schroder, J.L., H. Zhang, K. Girma, W.R. Raun, C.J. Penn, and M.E. Payton. 2011. Soil Acidification from Long-Term Use of Nitrogen Fertilizers on Winter Wheat. *Soil Science Society of America Journal* 75: 957–964.

Seybold, C.A., M.D. Hubbs, and D.D. Tyler. 2002. On-Farm Tests Indicate Effects of Long-Term Tillage Systems on Soil Quality. *Journal of Sustainable Agriculture* 19: 61–73.

Sullivan, D.M., and D. Granatstein. 2015. Are "Haney Tests" Meaningful Indicators of Soil Health and Estimators of Nitrogen Fertilizer Credits? *Nutrient Digest* 7: 1–2 ***http://landresources.montana.edu/soilfertility/documents/PDF/reports/NutDigSu2015.pdf***

Umiker, K.J., J.L. Johnson-Maynard, T.D. Hatten, S.D. Eigenbrode, and N.A. Bosque-Perez. 2009. Soil Carbon, Nitrogen, pH, and Earthworm Density as Influenced by Cropping Practices in the Inland Pacific Northwest. *Soil and Tillage Research* 105: 184–191.

Wienhold, B.J., S.S. Andrews, H. Kuykendall, and D.L. Karlen. 2008. Recent Advances in Soil Quality Assessment in the United States. *Journal of the Indian Society of Soil Science* 56: 1–10.

Williams, J.D., S.B. Wuest, and D.S. Long. 2014. Soil and Water Conservation in the Pacific Northwest through No-Tillage and Intensified Crop Rotations. *Journal of Soil and Water Conservation* 69: 495–504.

Wuest, S.B., T.C. Caesar-TonThat, S.F. Wright, and J.D. Williams. 2005. Organic Matter Addition, N, and Residue Burning Effects on Infiltration, Biological, and Physical Properties of an Insensitively Tilled Silt-Loam Soil. *Soil and Tillage Research* 84: 154–167.

Yin, C., S.H. Hulbert, K.L. Schoroeder, O. Mavrodi, D. Mavrodi, A. Dhingra, W.F. Schillinger, and T.C. Paulitz. 2013. Role of Bacterial Communities in the Natural Suppression of *Rhizoctonia solani* Bare Patch Disease of Wheat (*Triticum aestivum* L.). *Applied and Environmental Microbiology* 79: 7428–7438.

Chapter 3

Conservation Tillage Systems

Prakriti Bista, Oregon State University
Stephen Machado, Oregon State University
Rajan Ghimire, New Mexico State University (formerly of Oregon State University)
Georgine Yorgey, Washington State University
Donald Wysocki, Oregon State University

Abstract

Conservation tillage may improve the sustainability of winter wheat-based crop rotations in the dryland areas of the inland Pacific Northwest (PNW). Intensive tillage systems often bury most surface crop residues, pulverize soil, and reduce surface roughness. The tilled systems also have the potential to accelerate soil fertility loss and soil erosion, reducing the long-term sustainability of dryland agriculture. This chapter reviews the sustainability challenges posed by conventional tillage, including soil erosion, soil organic matter (SOM) depletion, soil fertility loss, and soil acidification. It also synthesizes recent studies in the region and evaluates agronomic and environmental benefits of direct seeding, undercutter tillage fallow, and other forms of reduced tillage. Conservation tillage systems are contributing to enhanced sustainability of dryland agriculture in the region by reducing erosion, and improving soil health and ecosystem services.

Key Points

- Conventional tillage-based cropping systems deplete SOM, enhance soil erosion, and threaten sustainable crop production.

Research results are coded by agroecological class, defined in the glossary, as follows:

● Annual Crop ▲ Annual Crop-Fallow Transition ■ Grain-Fallow

- Conservation tillage systems have been increasingly adopted by growers in the inland PNW to conserve soil fertility and SOM, reduce soil erosion, and improve sustainability of dryland cropping systems in the region.
- Adoption of conservation tillage systems is dependent on considerations such as agroecological class, crop rotations, equipment, residue management, soil fertility management, support systems, and economics.

Introduction

Sustainable agricultural systems produce sufficient yields of farm products at profitable levels while conserving natural resources over the long-term (Wysocki 1990). For a system to be sustainable, it must be biologically productive, economically viable, environmentally sound, and socially beneficial. Soil erosion and SOM depletion are among the biggest sustainability challenges for conventional tillage dryland agriculture that is predominant in the **inland PNW**. The adoption of **conservation tillage** practices can address these issues and therefore contribute to sustainable farming systems in the region.

The Natural Resource Conservation Service (NRCS) and the Conservation Technology Information Center (CTIC) define conservation systems as crop management systems that leave at least 30% of crop residue on the soil surface after planting, to reduce soil erosion by water. In areas where wind erosion is a concern, any system that maintains at least the equivalent of 1,000 pounds per acre of crop residue from small grains on the surface throughout the critical erosion period is known as conservation tillage (CTIC 2016). In the inland PNW, the primary rationale for adopting conservation systems was to mitigate soil erosion by water and wind (Papendick 2004). This chapter provides an overview of tillage systems drawing on sources including the conservation tillage handbook and reports from Solution to Environmental and Economic Problems (STEEP) and the Columbia Plateau Project (PM10), reviews past and present literature related to conservation systems, and provides grower considerations for enhancing the sustainability of dryland agriculture in the inland PNW region.

Conventional System

Conventional tillage practices require four or more intensive tillage operations a year for seedbed preparation, weed control during **fallow**, and fertilization prior to planting. Conventional tillage has many variations and depends on cropping intensity and rotation, but a typical system in the region for managing summer fallow as described by Schillinger (2001) is (1) sweep tillage in August following winter wheat harvest (for weed control), (2) chiseling in November with straight point shanks (to prevent runoff from frozen ground), (3) glyphosate herbicide application in late winter (to control late fall and winter germinating weeds), (4) primary tillage in March with a cultivator equipped with sweeps and tine harrows, (5) a shank anhydrous ammonia application in April, and (6) rodweeding in May, June, and July. In total, a typical conventional system can have up to eight tillage operations during a 14-month fallow period, not including sowing. The repeated tillage often buries up to 90% of crop residue, pulverizes soil clods, and reduces surface roughness (Feng et al. 2011; Schillinger and Papendick 2008).

Because the dryland area of the inland PNW has diverse tillage, challenges and solutions also vary across the region. This section provides a brief overview of the major cropping systems of the inland PNW. Additional details on crop rotations can be found in Chapter 5: Rotational Diversification and Intensification.

Winter Wheat-Summer Fallow

Under a winter wheat-summer fallow rotation, only one crop is produced in two years. About 2.52 million acres of crop land are part of the Grain-Fallow **agroecological class** (AEC) (Huggins et al. 2015) that receives less than 12 inches of precipitation annually (Huggins et al. 2015; Schillinger et al. 2006a). In this system, winter wheat planted in fall or late summer is harvested the following summer (July). After crop harvest, the land is left fallow until the following September/October, a fallow period of about 14 months. The main purpose of the fallow is to store winter precipitation to enable the successful establishment of winter wheat planted in the fall. Fallow also helps to control weeds, reduce the risk of crop failure, and lessen the effects of drought.

Three-Year Winter Wheat-Based Rotation

In the Annual Crop-Fallow Transition AEC, covering 1.85 million acres of cropped land, crops are grown in two out of every three years. Rotations generally incorporate winter wheat, a spring cereal or legume, and fallow. This AEC covers areas receiving 12–18 inches of precipitation annually. More intensive cropping reduces the potential for soil erosion compared to the Grain-Fallow AEC. The enhanced diversity of the three-year rotation, especially when a non-cereal crop is included, also reduces weed and disease pressure (Ogg et al. 1999; Schillinger et al. 2006b; Smiley et al. 2013). The spring crops usually grown in rotation with winter wheat are spring barley, spring wheat, pea, lentil, chickpea, canola, and condiment mustard. ▲

Annual Cropping

In the Annual Crop AEC, about 1.44 million acres are annually cropped. This AEC generally receives more than 18 inches of precipitation per year. In addition to the spring crops, rotations often include winter triticale, winter canola, winter barley, and winter peas, with no fallow. Between 2007 and 2013, diversification and cropping intensity were found to be higher in the Annual AEC than in the Grain-Fallow AEC (Huggins et al. 2015). The Annual AEC presents more opportunities to vary crops making this AEC less vulnerable to weather or potential climate change than Grain-Fallow AEC (Huggins et al. 2015). ●

Conservation Systems

Conservation tillage practices are useful for erosion control, **soil health**, crop productivity, farm efficiency, and profitability. The three types of conservation tillage systems and one other tillage system defined by CTIC are described below.

Ridge Tillage

Ridge tillage eliminates full-width tillage. The soil is left undisturbed from harvest to planting except for strips up to one-third of the row width. Planting is completed on the ridge and usually involves removal of the top of the ridge. Equipment for such tillage often includes sweeps, disk openers, coulters, or row cleaners. Ridges are rebuilt during row cultivation and residue is left on the surface between ridges. Weed control is accomplished with crop protection products.

Mulch Tillage

Mulch tillage is designated as full-width tillage that disturbs the entire soil surface, and it is done prior to and/or during planting. Equipment used for this type of tillage includes chisel, disks, field cultivator, sweeps, or blades and harrows.

No-Till/Chemical Fallow

No-till/chemical fallow leaves the soil undisturbed from harvesting to planting. In the inland PNW, no-till is commonly described as **direct seeding**. Direct seeding eliminates full-width tillage for seedbed preparation. However, there are some variations within this system (Veseth 1999). Planting, seeding, or drilling is done using hoe drills. Weeds are controlled with crop protection products.

Low-disturbance direct seeding

Low-disturbance direct seeding involves the use of narrow knives, single discs, or double discs (standard or offset with one leading edge) that typically disturb less than 40% of the row width and retain nearly all residues on the surface.

High-disturbance direct seeding

Under high-disturbance direct seeding, hoe or sweep openers may disturb up to about 65% of the row width, but still retain much of the crop residue on the soil surface. With some flatter sweep blades, the surface soil and residue disturbance can be minimal even though much of the surface layer is undercut with the opener. Obviously, the furrow size, soil disturbance, and residue retention will vary with opener designs, speed, soil moisture, and other factors.

One-pass and two-pass direct seed systems

Growers can choose between one-pass direct fertilize and seed systems, and two-pass systems with direct fertilizing and direct seeding in separate operations. In both cases, there are no other tillage operations for seedbed preparation before seeding. The choice depends on the precipitation zone and seasonal distribution, length of planting windows, equipment availability, cost, crop choices, available labor, and other considerations.

The high-disturbance direct seed implement with wider hoe or sweep openers may not fit the classic no-till definition, but rather fall in the "mulch till" category because of full-width tillage between harvest and planting.

Reduced Tillage

Reduced tillage is designated as full-width tillage that disturbs the entire soil surface, leaving 15% to 30% of residue cover after planting.

Other conservation tillage practices in the inland PNW include minimum tillage, delayed minimum tillage, undercutter fallow, chisel, discs, and sweep tillage systems.

The undercutter method of fallow management uses wide V blade sweeps that slice beneath the soil surface and simultaneously deliver nitrogen during primary spring tillage followed by one or two non-inversion rodweeding operations during the summer to control weeds (Schillinger et al. 2010; Schillinger and Young 2014).

Both minimum tillage and delayed minimum tillage use undercutter V-sweep as a primary tillage. Herbicides may be used to control weeds following primary tillage, but secondary tillage such as rodweeding is used more commonly. Delayed minimum tillage is similar to minimum tillage except primary spring tillage with undercutter V-sweep is delayed until at least mid-May (Schillinger 2001).

Adoption of Conservation Systems in the Inland PNW

In general, there are considerable variations among conservation tillage practices. Certain conservation practices in the inland PNW are unique to the region. For example, farmers prefer hoe-type drills for cereal planting in narrow rows. As per the CTIC definition for no-till/direct seed, the threshold limit for soil disturbance is less than one-third the row width, which is difficult to achieve with the hoe drills used in this region. Hence, many acres of wheat or barley planted using direct seed are categorized as mulch tillage rather than no-till systems (Smiley et al. 2005). Other conservation tillage practices followed in the inland PNW besides direct seed are undercutter, chisel, discs, and sweep tillage systems. All forms of conservation practices, however, are aimed at protecting soil and water resources. Effects of different tillage implements on residue cover, SOM, and erosion in the inland PNW are summarized in Table 3-1.

As farmers in different AECs gain experience and confidence in conservation tillage systems and are motivated by fuel savings and government programs to promote such practices, the number of conservation farmers in the inland PNW is growing (Schillinger et al. 2010). Advances in no-till grain drill technology have allowed precise seed and fertilizer placement in one pass, saving growers the cost of multiple tillage operations needed under conventional systems. No-till acreage in Oregon for winter wheat has increased from less than 1% in 1996 to 16% (102,000 acres) in 2004, whereas no-till spring wheat acreage increased to 19% (434,000 acres) in 2004 from less than 2% in 1996. Similarly, in Washington no-till planting increased for both winter and spring wheat. Acreage under no-till winter wheat increased from 3% in 1990 to 11% (182,900 acres) in 2004, and no-till spring wheat acreage increased from 2% in 1990 to 18% in 2000 and remained steady throughout 2004 (Smiley et al. 2005). The increase in direct-seeded acres was attributed partly to the Pacific Northwest Direct Seed Association (***http://www.directseed.org/***), a grower-based organization formed in 2000 to promote conservation tillage and no-till farming in the region (Kok et al. 2009). A survey conducted in Columbia County, Washington, showed that 94% of winter crop land and 40% of spring crop land were direct seeded in 2007–2008 (***http://www.nacaa.com/presentations/presentation_list.php?app_id=407***). A recent representative survey of wheat growers from 33 different counties in Washington, Idaho, and Oregon showed that nearly 70% of the growers were using no-till or another form of conservation tillage in 2012–2013 (Figure 3-1).

Sustainability Challenges and Benefits of Conservation Systems

Dryland farming in this region faces three major sustainability challenges: erosion, loss of SOM, and soil acidification. This section describes each,

Table 3-1. Summary on the effects of crop rotations, tillage equipment, and tillage depths on residue cover, soil organic matter (SOM), and soil erosion.

Tillage system[†]	Crop rotation	Equipment	Years under current management	Tillage depth (cm)	Residue cover/ Ground cover (%)	SOM (ton/ac/yr) Loss (–) Gain (+)	Soil erosion (lb/ac)
Machado 2011 study site: Pendleton, OR ■							
Conventional	WW-F	MB plow	>65	23	22	–0.14 to –0.26	NR*
Conventional	WW-Pea	MB plow	42	23	21–27	+ 0.014 (SP) +0.07	NR
Reduced	WW-F	Disc Sweep	>65	15	63–66	–0.14 to –0.26	NR
Reduced	WW-Pea	Disc Chisel	42	15	49	+0.07	NR
Direct seed	WW-Pea	No-till drill	23	–	97	+0.56	NR
Machado et al. 2015 study site: Moro, OR ■							
Conventional	WW-F	Chisel Sweep	6	15	<15	NR	NR
Direct Seed	WW-WW SW-SW SB-SB WW-ChF WW-WP WW-SB-ChF	None	6	–	25 to >65	NR	NR
Williams and Wuest 2011 study site: Pendleton, OR ■							
Reduced	WW-SP-WW-F	Disc Chisel	4	31	59	NR	19

Tillage system[†]	Crop rotation	Equipment	Years under current management	Tillage depth (cm)	Residue cover/ Ground cover (%)	SOM (ton/ac/yr) Loss (–) Gain (+)	Soil erosion (lb/ac)
Direct seed	WW-SP-WW-F	None	4	–	81	NR	9
Williams et al. 2009 study site: Wildhorse Creek Watershed, OR ▲							
Conventional	WW-F WW-ChF	MB plow	4	–	5	NR	375
Direct seed	WW-F-WW-CP	None	4	–	67	NR	9
Riar et al. 2010 study sites: Davenport, WA and Helix, OR ▲							
Conventional	WW-F	Tandem disk	2	12	33.4 (OR) 69.3 (WA)	NR	NR
Reduced	WW-F	Sweep undercutter	2	12	37-42 (OR) 77-82 (WA)		
Direct seed	WW-F	None	2		43-48 (OR) 88-93 (WA)		
Stockle et al. 2012 study sites: multiple in eastern Washington ●▲■							
Conventional to Direct seed	WW-F WW-SB-F WW-SB-SW SW-SB-SP SC-SC-P	NR	–	NR	NR	+0.13 to +0.24 CO_2 ton/ac/yr	NR

Tillage system[†]	Crop rotation	Equipment	Years under current management	Tillage depth (cm)	Residue cover/ Ground cover (%)	SOM (ton/ac/yr) Loss (–) Gain (+)	Soil erosion (lb/ac)
Brown and Huggins 2012 study sites: multiple (non-irrigated inland PNW region) ● ▲ ■							
Conventional to Direct seed	Multiple	–	10-12	–	NR	+0.8 to +0.13	NR
Umiker et al. 2009 study sites: Palouse and Nez Perce, ID							
Conventional	WW-SW-SP	MB plow Chisel	2	NR	NR	1.79% SOM at 0-10 cm	NR
Direct seed	WW-SW-SP	None	2	–	>30	2.05% SOM at 0-10 cm	NR
Kok et al. 2009 study sites: multiple, inland PNW ● ▲ ■							
Conventional to Reduced/Direct seed	Multiple	–	30	–	–	–	Reduced erosion from 40,000 to 10,000 lb/ac/yr

*NR: Not Reported

[†]Tillage system: Conventional tillage included moldboard plowing or offset heavy disk or tandem disk followed by tillage with secondary tillage – straight point chisel, harrow, cultivators, rodweeder. Reduced-tillage included combination two or more – sweep, chisel, cultivator, harrow, undercutter, rodweeder – minimum soil disturbance and direct seed included no-till hoe drill, no-till drill. SP = spring plow, FP = fall plow.

[‡]Crops in rotations: F = fallow, CP = chick pea, SW = spring wheat, SB = spring barley, WP = winter pea, SP = spring pea, ChF= chemical fallow.

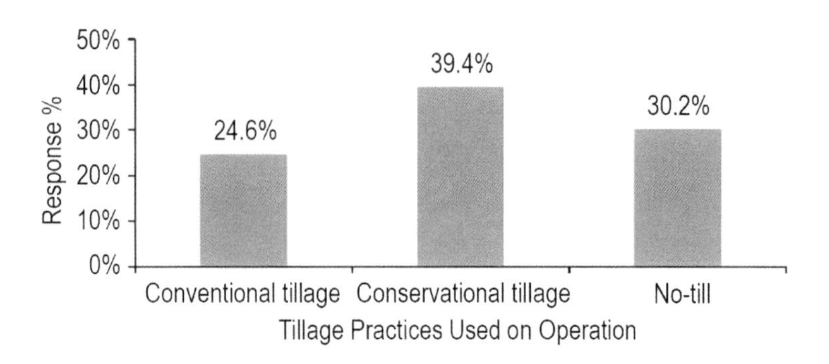

Figure 3-1: Distribution of tillage practices used by growers in inland PNW. Final response rate was 46.2% (900 surveys) with a sampling margin of error of +/- 3% at the 95% confidence interval. (Source: REACCH 2015)

explores the impacts of tillage on these issues, and provides insight on the effect of conservation systems on enhancing various aspects of sustainability including crop yield and farm economy.

Soil Erosion

Wind and water erosion are major factors affecting the sustainability of the cereal-producing regions of the inland PNW influencing both crop productivity and soil health. Historically, conventional farming practices had annual erosion rates of 10 to 30 ton/acre/year, resulting in topsoil loss equivalent to 0.75 ton of soil per bushel of wheat (Kok et al. 2009). ● ▲ ■

Wind erosion and dust emissions mostly occur in low precipitation areas with sandy silt loam soils that are poorly **aggregated** and dominated by particulates <100 µm in diameter, which are vulnerable to wind erosion by direct suspension and have a great potential to emit particulate matter (PM10) (Feng et al. 2011). ■ Short duration, high-velocity winds affect nearly 6 million acres of crop land, posing an especially severe threat during fall and spring when soil is dry and soil cover is very limited (McCool et al. 2001; Papendick 2004). The dominance of winter wheat-fallow in this area often means that residue is produced in only one out of every two years, and, even in cropped years, water limitations constrain residue production (Papendick 2004). Excessive tillage during summer fallow pulverizes soil clods and buries residue (Young and Schillinger 2012).

Meanwhile, water erosion is a significant issue in wetter areas of the region, resulting in the loss of millions of tons of topsoil annually (Kok et al. 2009). Water erosion is caused by a combination of factors including winter precipitation with high potential for frozen soil runoff, steep and irregular topography (35% to 45% slope), and crop management systems that leave the soil with inadequate protection during the winter rainy season (Kok et al. 2009; Michalson 1999).

Planting winter wheat in early September in bare soil following intensive tillage causes up to two-thirds of annual soil erosion across the inland PNW (Papendick 2004).

Erosion is problematic for a number of reasons. Approximately one-third of eroded soil is deposited on surface water and can adversely affect water quality. Erosion of topsoil also results in the loss of nutrients resulting in declines in crop productivity and increased input costs (to replace lost nutrients) to sustain yield (Schillinger et al. 2010). In addition, loss of soil removes SOM and reduces water storage potential, negatively influencing root zone and seedbed environments, and the nutrient-supplying capacity of soil. Eastern Oregon fields with residue burning and no fertilizer application had higher soil erosion rates (1.47 ton/acre/year vs. 0.04 ton/acre/year) and lower SOM content compared to fields with standing stubble (Williams 2008). ■

While tillage is not the only factor that causes erosion, it is a major contributing factor. The tillage-intensive conventional systems create a dry, loose zone of fine soil particles which are susceptible to erosion by strong winds prevalent in the spring, late summer, and early autumn. Intensive inversion tillage increases total runoff to as high as 0.2 inches and soil erosion to 0.20 ton/acre compared to 0.03 inches and 0.005 ton/acre, respectively, in a no-till system at a similar slope position (Williams et al. 2009). ■ Remarkable improvements in erosion control have been achieved over the last 30 years mostly through the reduction in tillage (Kok et al. 2009). However, erosion remains an ongoing threat to the resources, environment, and agricultural economy of the region (Schillinger et al. 2010), emphasizing the need for conservation tillage practices.

Conservation tillage practices have been effective in minimizing soil erosion. The undercutter method of summer fallow management left sufficient surface residue to reduce soil loss from wind erosion by 65% and PM10 (an air quality indicator) by 70% compared to conventional (disk) tillage in winter wheat-fallow in a low precipitation zone of the Columbia Plateau (Sharrat and Feng 2009). Similarly, spring-sown cereal and chemical fallow or direct seed systems were reported to increase stored water, residue cover, soil aggregation, and soil strength, reducing the risk of wind erosion when compared to the conventional winter wheat-fallow rotation in east-central Washington (Feng et al. 2011). ■ Greater soil aggregation in a conservation tillage rather than in a conventional tillage system is shown in Figure 3-2.

Direct seed was also found highly effective in controlling runoff and soil erosion compared with inversion tillage systems in northeastern Oregon. The runoff and erosion totaled 0.2 inches and 0.19 ton/acre, respectively, under moldboard plowing, versus 0.03 inches and 0.005 ton/acre, under direct seed (Williams et al. 2009). ▲ In a four-year rotation of winter wheat-spring pea-winter wheat-fallow, direct seed had increased ground cover and **infiltration** rates, and decreased runoff and soil erosion when compared to tilled conservation practices such as mulch tillage, chisel plow, and undercutter (Williams and Wuest 2011). ■

Soil Organic Matter

Soil organic matter is essential for long-term sustainability of agricultural systems. It promotes soil aggregation, increases soil water and nutrient

Figure 3-2: Soil aggregation in conventional and reduced tillage. (Photo credit: Rajan Ghimire.)

holding capacity, serves as a sink for sequestration of atmospheric carbon, and facilitates **mitigation** of greenhouse gas emissions. Further discussion of the soil health impacts of SOM is presented in Chapter 2: Soil Health. Repeated tillage during field operations loosens the soils, makes it susceptible to erosion, and facilitates SOM loss through **mineralization** and oxidation.

Winter wheat-summer fallow systems in low precipitation zones of the inland PNW have lost more than 60% of SOM from topsoil (Brown and Huggins 2012; Ghimire et al. 2015; Machado 2011; Rasmussen and Smiley 1997). Similarly, a study on high precipitation regions of the Palouse has shown that conversion of native prairie to wheat cropping systems using intensive inversion tillage has caused substantial loss of organic matter pools including 56% of soil organic carbon, 79% of particulate organic carbon, 50% of microbial biomass carbon, and 28% of mineralizable carbon (Purakayastha et al. 2008). In addition to loss of SOM under intensive tillage, disease and weed incidence and weather variability negatively affected winter wheat production in the region (Camara et al. 2003; Sharma-Poudyal and Chen 2011; Smiley et al. 2009). ●▲■ The continued decline in SOM content and higher yield variability under intensive tillage threatens long-term agronomic and environmental sustainability of the winter wheat-fallow systems in the inland PNW.

Increasing SOM is a prerequisite for sustainable agricultural production. Conservation tillage systems are recognized for their ability to sequester carbon, minimize greenhouse gas emissions, and reduce the threat of climate change (Stöckle et al. 2012). Restoring SOM that has been lost due to years of intensive tillage in cold and dry regions of the inland PNW, however, is a great challenge because of low biomass production in dryland areas (Brown and Huggins 2012). Conservation tillage practices, along with intensifying cropping rotations, are recommended for increasing soil organic carbon sequestration in dryland areas where biomass production limits SOM accumulation (Machado et al. 2006). In a study evaluating the effects of different tillage and cropping systems on soil carbon sequestration, continuous cropping under direct seed was able to increase SOM in the top 10 cm of soil within a short period of six years compared to 73 years in a conventional winter wheat-fallow system in long-term experiments in Pendleton, Oregon (Machado et al. 2006). ■ Reduced tillage practices minimize SOM loss by eliminating or reducing tillage operations for field preparation, which benefits dryland cropping systems through soil water conservation (Lenssen et al. 2007) and protects biomass carbon from decomposition. In the low precipitation regions of eastern Washington, greater total SOM was observed with continuous direct seed spring cropping than with tillage fallow and direct seed chemical fallow (Gollany et al. 2013).■ The increased SOM in continuous direct seed spring cropping was mainly due to the accumulation of undecomposed crop residues that increased readily useable SOM for soil microbes (Gollany et al. 2013). Direct seed also increased near surface SOM content in the high precipitation region under spring wheat and pea compared to the conventional tillage system (Umiker et al. 2009).● Long-term use of direct seed systems has the potential to recover lost SOM compared with intensive tillage systems (Bista et al. 2016; Brown and Huggins 2012). ●▲■ Converting intensive tillage winter wheat-fallow to direct seed can reduce SOM loss by 17% to 47% depending on the residue and nutrient management practices (Bista et al. 2016). (Figure 3-3). ■

Soil pH and Soil Fertility

Acidification of soils is a major concern in the inland PNW. Soils with **pH** below 5 have been reported across all three AECs in the region (McFarland and Huggins 2015). Low soil pH affects many chemical and biological reactions in soil that influence nutrient availability and crop productivity. Agricultural management practices accelerate the rate of soil acidification mainly due to the continuous application of ammonium-based nitrogen fertilizers, continuous depletion of basic cations by crop removal, and accelerated rate of SOM decomposition (Mahler 2002). (See Chapter 2: Soil Health and Chapter 6: Soil Fertility Management for further detail.)

Conventional and conservation systems influence soil profile acidification differently. Stratified layers of acid soil at the depth of fertilizer placement has been observed in both direct seed fields in Palouse, Washington, and tilled long-term fertility trials in Pendleton, Oregon (Koenig et al. 2013). In a direct seed system, soil acidity develops more rapidly at the depth of fertilizer placement when compared to conventional tillage systems

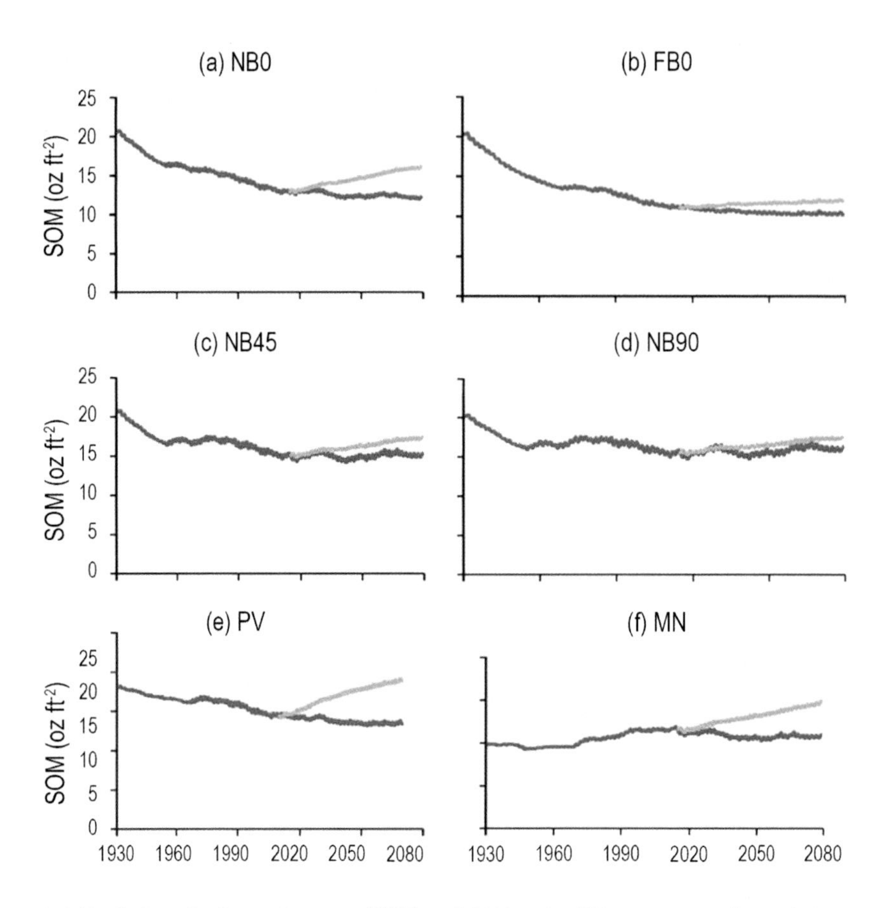

Figure 3-3: Prediction of soil organic matter (SOM) until 2080 under different crop residue and nitrogen (N) management treatments (a) to (f) with the baseline management (conventional tillage: red line) and alternative management (direct seed: orange line) in a moldboard plowed winter wheat-fallow system at Pendleton, Oregon, long-term experiments. (FB = fall stubble burn, NB = no burn, MN = manure application at primary tillage, PV = pea vine application at primary tillage. Accompanying numbers 0, 45, and 90 indicate amount of N applied from chemical fertilizer.) (Modified from Bista et al., 2016).

due to the absence of mechanical mixing (McCool et al. 2001). Therefore, there is some concern that direct seeding may exacerbate soil acidity.

Tillage management also influences nutrient availability and crop performance (Pan et al. 1997). Intensive tillage facilitates SOM decomposition and nutrient release. Over the long-term, this can deplete the nutrient bank in soil. (Fertility management strategies are discussed in Chapter 6: Soil Fertility Management.) For example, continuous use of conventional tillage in a winter wheat-fallow system for 84 years depleted nearly 30% of the soil N reservoir from the 0–60 cm soil profile in eastern Oregon (Ghimire et al. 2015). ■

Conservation tillage can also create challenges for managing fertility (Veseth 1999). Greater residue cover on the soil surface under a conservation tillage system sometimes immobilizes nutrients and makes them unavailable for the following crop. Reduced availability of nutrients, particularly nitrogen, is one of the many impacts of high concentrations of residue in conservation systems. (See Chapter 4: Crop Residue Management.)

Yield and Economics

The effects of direct seed, which is widely accepted for efficient erosion control and SOM sequestration, on crop yield and farm economy need to be further explored. Recent studies suggest that sufficiently high yield and greater farm profitability from conservation tillage compared to conventional tillage can be achieved. Similar wheat yield and grain quality as in a conventional system (disc/chisel) was obtained with a conservation tillage system (sweep tillage) in an intermediate precipitation region of Washington (Riar et al. 2016). However, the surface residue cover was greater with the conservation tillage system. ▲ Although sweep tillage systems had similar yields as conventional systems, they were more profitable because of reduced tillage operations and associated production costs. In Moro, Oregon (11-inch precipitation) comparable yields were observed under the conventional (chisel) winter wheat-fallow, direct seed winter wheat-chemical fallow, and winter wheat-spring barley-chemical fallow (Machado et al. 2015). ■ Given the conservation benefits from a direct seed system, such as greater residue cover and ecosystems services, direct seeding was recommended as an alternative system to conventional tillage. In eastern Washington, conservation tillage practices such as minimum tillage and delayed minimum tillage were found to be more profitable as they reduced fuel and farm labor expenses compared to conventional tillage winter wheat-fallow (Nail et al. 2007).

Similarly, a survey of 47 farmers in the inland PNW showed equivalent winter wheat grain yields and profitability in undercutter systems and conventional tillage fallow systems (Young and Schillinger 2016). In the low precipitation region of Washington, greater profitability from undercutter fallow systems than conventional dust mulch fallow systems were due to reduced costs of production (Zaikin et al. 2007). ■ Moreover, the undercutter system is eligible for conservation payments, but the traditional system is not. Such benefits further strengthen the profitability advantage of the undercutter system over the conventional system.

Additional Grower Considerations

Adoption of conservation farming systems vary based on factors such as climatic conditions, available equipment, crop rotations, soil type, topography, cash flow, information resources, federal farm programs, and other factors. When used in combination with other sustainable management practices, reduced tillage practices (e.g., direct seeding, undercutter tillage fallow, delayed planting, and minimum tillage) can help achieve favorable yields, attain farm profitability, and maintain environmental integrity. Details of alternative crop management practices such as legume incorporation are discussed in Chapter 5: Rotational Diversification and Intensification. The impacts of tillage on weed and disease control are described in Chapter 9: Integrated Weed Management and Chapter 10: Disease Management for Wheat and Barley. Differences among conventional and conservation tillage systems are given in Table 3-2.

Resources and Further Reading

Conservation Tillage Handbook

http://pnwsteep.wsu.edu/tillagehandbook/chapter1/index.htm

Columbia Plateau PM10 Project

http://pnw-winderosion.wsu.edu/

Regional Approaches to Climate Change – Pacific Northwest Agriculture

https://www.reacchpna.org/

Table 3-2. Differences between conservation and conventional tillage.

Parameters	Conservation tillage	Conventional tillage
Tillage operation	Minimum soil disturbance	Requires intensive tillage (more than four per year)
Crop residue	Leaves more than 30% (≈1,000 lb/ac) on surface	Crop residues are incorporated in soil
Soil organic matter (SOM)	Increase SOM sequestration in surface soil	Increase SOM loss from surface soil
Greenhouse gas emission	Reduce greenhouse gas emission such as CO_2	Increases greenhouse gas emissions
Erosion	Reduce soil loss from wind and water erosion	High risk of soil loss from wind and water erosion
Soil water storage	Increase infiltration and reduce evaporation	More soil water loss from evaporation and poor infiltration
Water body pollution	Minimum water body pollution with sediment load and field-applied chemicals	High risk of water body pollution
Aggregate stability	Increase soil aggregate stability	Lower soil aggregate stability
Labor and fuel	Low fuel use and labor cost	High fuel use and labor costs due to more trips over the field
Tillage equipment	Direct seed drills costlier than conventional drills	Machinery is widely available
Weed control	Reliance on herbicide during fallow	Tillage used to control weeds
Crop management	Information on new crop management strategies evolving	Relatively more information on crop management strategies
Germination	Potential slower germination	Well-tilled and clean seeding facilitates germination and plant establishment
Fertilizer	May initially require more nitrogen	Initial nitrogen requirement does not increase

Costs of Owning and Operating Farm Machinery in the Pacific Northwest. Pacific Northwest Extension Publication PNW346.

http://whatcom.wsu.edu/ag/documents/enterbudgets/CostOwnOperFarmMachPNW.pdf

References

Bista, P., S. Machado, R. Ghimire. S.J. Delgrosso, and M. Reyes-Fox. 2016. Simulating Organic Carbon in a Wheat-Fallow System Using Daycent Model. *Agronomy Journal* 108: 2554–2565.

Brown, T.T., and D.R. Huggins. 2012. Soil Carbon Sequestration in the Dryland Cropping Region of the Pacific Northwest. *Journal of Soil Water Conservation* 67: 406–415.

Camara, K.M., W.A. Payne, and P.E. Rasmussen. 2003. Long-Term Effects of Tillage, Nitrogen, and Rainfall on Winter Wheat Yields in the Pacific Northwest. *Agronomy Journal* 95: 828–835.

CTIC (Conservation Technology Information Center). 2016. West Lafayette, Indiana. ***http://www.ctic.org/resourcedisplay/322/***

Feng, G., B. Sharratt, and F. Young. 2011. Soil Properties Governing Soil Erosion Affected by Cropping Systems in the U.S. Pacific Northwest. *Soil and Tillage Research* 111: 168–174.

Ghimire, R., S. Machado, and K. Rhinhart. 2015. Long-Term Crop Residue and Nitrogen Management Effects on Soil Profile Carbon and Nitrogen in Wheat-Fallow Systems. *Agronomy Journal* 107: 2230–2240.

Gollany, H.T., A.M. Fortuna, M.K. Samuel, F.L. Young, W.L. Pan, and M. Pecharko. 2013. Soil Organic Carbon Accretion vs. Sequestration Using Physicochemical Fractionation and CQESTR Simulation. *Soil Science Society of America Journal* 77: 618–629.

Huggins, D.R., R. Rupp, P. Gessler, W. Pan, D. Brown, S. Machado, J. Abatzoglou, V. Walden, and S. Eigenbrode. 2012. Dynamic Agroecological Zones for the Inland Pacific Northwest, USA. ASA Annual Meeting, Oct. 21–24, Cincinnati, OH.

Huggins, D., Pan, W., Schillinger, W., Young, F., Machado, S. and K. Painter. 2015. Crop Diversity and Intensity in the Pacific Northwest Dryland Cropping Systems in Regional Approaches to Climate Change for Pacific Northwest Agriculture: Climate Science Northwest Farmers Can Use. REACCH Annual Report Year 4: 38-41. University of Idaho, Washington State University, and Oregon State University.

Koenig, R., K. Schroeder, A. Carter, M. Pumphrey, T. Paulitz, K. Campbell, and D. Huggins. 2013. Soil Acidity and Aluminum Toxicity in the Palouse Region of the Pacific Northwest. Washington State University Extension Publication FS050E. ***http://pubs.wpdev.cahnrs.wsu.edu/pubs/fs050e/?pub-pdf=true***

Kok, H., R.I. Papendick, and K.E. Saxton. 2009. STEEP: Impact of Long-Term Conservation Farming Research and Education in Pacific Northwest Wheatlands. *Journal of Soil and Water Conservation* 64: 253–264.

Lenssen, A.W., G.D. Johnson, and G.R. Carlson. 2007. Cropping Sequence and Tillage System Influences Annual Crop Production and Water Use in Semiarid Montana, USA. *Field Crop Research* 100: 32–43.

Machado, S. 2011. Soil Organic Carbon Dynamics in the Pendleton Long-Term Experiments: Implications for Biofuel Production in Pacific Northwest. *Agronomy Journal* 103: 253–260.

Machado, S.M., K. Rhinhart, and S.E. Petrie. 2006. Long-Term Cropping System Effects on Carbon Sequestration in Eastern Oregon. *Journal of Environmental Quality* 35: 1548–1553

Machado, S.M., L. Pritchett, and S.E. Petrie 2015. No-Tillage Cropping Systems Can Replace Traditional Summer Fallow in North-Central Oregon. *Agronomy Journal* 107: 1863–1887.

Mahler, R.L. 2002. Impacts and Management of Soil Acidity Under Direct-Seed Systems: Status and Effects on Crop Production. Proceedings of Northwest Direct-Seed Cropping Systems Conference, Jan. 16–18, Spokane, WA. ***http://pnwsteep.wsu.edu/directseed/conf2k2/dscmahler.htm***

McCool, D.K., D.R. Huggins, K.E. Saxton, and A.C. Kennedy. 2001. Factors Affecting Agricultural Sustainability in the Pacific Northwest, USA: An Overview. 10th Sustaining the Global Farm: International Soil Conservation Organization Meeting, West Lafayette, IN. 24–29 May 1999. National Soil Erosion Research Laboratory, West Lafayette, IN. p. 255–260.

McFarland, C., and D.R. Huggins. 2015. Acidification in the Inland Pacific Northwest. *Crop and Soils Magazine* March-April: 4–12.

Michalson, E.L. 1999. A History of Conservation Research in the Pacific Northwest. In Conservation Farming in the United States—The Methods and Accomplishments of the STEEP program, E.L. Michalson, R.I. Papendick, and J.E. Carlson, eds. CRC Press, Boca Raton 1–10.

Nail, E.L., D.L. Young, and W.F. Schillinger, 2007. Diesel and Glyphosate Price Changes Benefit the Economics of Conservation Tillage versus Traditional Tillage. *Soil and Tillage Research* 94: 321–327.

Ogg, A.G., R.W. Smiley, K.S. Pike, J.P. McCaffrey, D.C. Thill, and S.S. Quisenberry. 1999. Integrated Pest Management for Conservation Systems. In Conservation Farming in the United States: The Methods and Accomplishments of the STEEP program, E.L. Michalson, R.I. Papendick, and J.E. Carlson, eds. CRC Press, Boca Raton 97–127.

Pan, W.L., D.R. Huggins, G.L. Malzer, C.L. Douglas, Jr., and J.L. Smith. 1997. Field Heterogeneity in Soil-Plant Nitrogen Relationships: Implications for Site-Specific Management. In the Site-Specific Management for Agriculture Systems. ASA-CSSA-SSSA, Madison, Wisconsin.

Papendick, R.I. 2004. Cropping Systems Research to Control Wind Erosion and Dust Emissions on Dryland Farms. In Farming with the Wind II. Wind Erosion and Air Quality Control of the Columbia Plateau and Columbia Basin. University Publishing, WSU, Pullman, Washington.

Purakayastha, T.J., D.R. Huggins, and J.L. Smith. 2008. Carbon Sequestration in Native Prairie, Perennial Grass, No-Till, and Cultivated Palouse Silt Loam. *Soil Science Society of America Journal* 72: 534–540.

Rasmussen, P.E., and R.W. Smiley. 1997. Soil Carbon and Nitrogen Change in Long-Term Agricultural Experiments at Pendleton, Oregon. In Soil Organic Matter in Temperate Agroecosystems: Long-Term Experiments in North America, E.A. Paul et al., eds. CRC Press, Boca Raton, FL.

REACCH (Regional Approaches to Climate Change). 2015. Variation in Tillage Practices among Inland Northwest Producers. Regional Approach to Climate Change – Pacific Northwest Agriculture report by J. Gray, S. Gantla, L. McNamee, L. Bernacchi, K. Borrelli, B. Mahler, M. Reyna, B. Foltz, S. Kane, and J.D. Wulfhorst.

Riar, D.D., D.A. Ball, J.P. Yenish, S.B. Wuest, and M.K. Corp. 2010. Comparison of Fallow Tillage Methods in the Intermediate Rainfall Inland Pacific Northwest. *Agronomy Journal* 102: 1664–1673.

Schillinger, W.F., and R.I. Papendick. 2008. Then and Now: 125 Years of Dryland Wheat Farming in the Inland Pacific Northwest. *Agronomy Journal* 100: 166–182.

Schillinger, W.F., R.I. Papendick, S.O. Guy, P.E. Rasmussen, and C. van Kessel. 2006a. Dryland Cropping in the Western United States. In Dryland Agriculture, 2nd ed., G.A. Peterson et al., eds. *Agronomy Monograph* 23. ASA, CSSA, and SSSA, Madison, WI.

Schillinger, W., T. Paulitz, R. Jirava, H. Schafer, and S.E. Schofstoll. 2006b. Reduction of Rhizoctonia Bare Patch in Wheat with Barley Rotations. Washington State University Publication XB1045E. ***http://www.pnw-winderosion.wsu.edu/Docs/Publications/06%20 pubs/RedRhizocXB.pdf***

Schillinger, W.F. 2001. Minimum and Delayed Conservation Tillage for Wheat-Fallow Farming. *Soil Science Society of America Journal* 65: 1203–1209.

Schillinger, W.F., and D.L. Young. 2014. Best Management Practices for Summer Fallow in the World's Driest Rainfed Wheat Region. *Soil Science Society of America Journal* 78: 1707–1715.

Schillinger, W.F., R.I. Papendick, and D.K. McCool. 2010. Soil and Water Challenges for Pacific Northwest Agriculture. In Soil and Water Conservation Advances in the United States, T.M. Zobeck and W.F. Schillinger, eds. Soil Science Society of America Special Publication 60, 47–79. Soil Science Society of America, Madison, WI.

Sharma-Poudyal, D., and X.M. Chen. 2011. Models for Predicting Potential Yield Loss of Wheat Caused by Stripe Rust in U.S. Pacific Northwest. *Disease Control and Pest Management* 101: 544–554.

Sharratt, B.S., and G. Feng. 2009. Windblown Dust Influenced by Conventional and Undercutter Tillage within the Columbia Plateau, USA. Earth Surface Processes and Landforms 34: 1323–1332.

Smiley, R., W.S. Machado, J.A. Gourlie, L.C. Pritchett, G.P. Yan, and E.E. Jacobsen. 2013. Influence of Semi-Arid Cropping Systems on Root Diseases and Inoculum Density of Soil Borne Pathogens. *Plant Disease* 97: 547–555.

Smiley, R.W., D. Backhouse, P. Lucas, and T.C. Paulitz. 2009. Diseases which Challenge Global Wheat Production—Root, Crown, and Culm Rots. In Wheat: Science and Trade. Blackwell Publishing, Ames, IA. 125–153.

Smiley, R.W., M.C. Siemens, T.M. Gohlke, and J.K. Poore. 2005. Small Grain Acreage and Management Trends for Eastern Oregon and Washington. Dryland Agricultural Research Annual Report (Special Report 1061): 30–50

Stöckle, C., S. Higgins, A. Kemanian, R. Nelson, D. Huggins, J. Marcos, and H. Collins. 2012. Carbon Storage and Nitrous Oxide Emissions of Cropping Systems in Eastern Washington: A Simulation Study. *Journal of Soil Water Conservation* 67: 365–377.

Umiker, K.J., J.L. Johnson-Maynard, T.D. Hatten, S.D. Eigenborode, and N.A. Bosque-Pérez. 2009. Soil Carbon, Nitrogen, pH, and Earthworm Density as Influenced by Cropping Practices in the Inland Pacific Northwest. *Soil and Tillage Research* 105: 184–191.

Veseth, R.J. 1999. PNW Direct Seeding Status and What's Driving it. PNW Conservation Tillage Handbook Series No. 25, Chapter 2.

Williams, J.D. 2008. Soil Loss from Long-Term Winter-Wheat/Summer Fallow Residue and Nutrient Management Experiment at Columbia Basin Agricultural Research Center, Pendleton, Oregon. In 2008 Dryland Agricultural Research Annual Report. Agricultural Experiment Station, Oregon State University.

Williams, J.D., and S.B. Wuest. 2011. Tillage and No-Tillage Conservation Effectiveness in the Intermediate Precipitation Zone of the Inland Pacific Northwest, United States. *Journal of Soil and Water Conservation* 66: 242–249.

Williams, H.D., H.T. Gollany, M.C. Siemens, S.B. Wuest, and D.S. Long. 2009. Comparison of Runoff, Soil Erosion, and Winter Wheat Yields from No-Till and Inversion Tillage Production Systems in Northeastern Oregon. *Journal of Soil and Water Conservation* 1: 43–52.

Wysocki, D. 1990. Conservation Farming and Sustainability. In Pacific Northwest Conservation Tillage Handbook Series No. 12, Chapter 1: Soil Erosion Impacts on Productivity.

Young, D.L., and W.F. Schillinger. 2012. Wheat Farmers Adopt the Undercutter Fallow Method to Reduce Wind Erosion and Sustain Profitability. *Soil and Tillage Research* 124: 240–244

Young, D.L., and W.F. Schillinger. 2016. Wheat Farmers Adopt the Undercutter Fallow Method to Reduce Wind Erosion and Sustain Profitability. Washington State University Extension Bulletin.

Zaikin, A.A., D.L. Young, and W.F. Schillinger. 2007. Economic Comparison of the Undercutter and Traditional Tillage Systems for Winter Wheat-Summer Fallow Farming. Washington State University Working Paper Series 2007–15. ***http://faculty.ses.wsu.edu/WorkingPapers/WP_2007-15_Undercutter.pdf***

Chapter 4

Crop Residue Management

Haiying Tao, Washington State University
Georgine Yorgey, Washington State University
David Huggins, USDA-ARS and Washington State University
Donald Wysocki, Oregon State University

Abstract

Crop residue, a byproduct of harvested food and fiber, makes up a substantial amount of crop production biomass. Although traditionally considered an agricultural waste, residue is now recognized for its value in reducing soil susceptibility to wind and water erosion and contributing to soil health and soil fertility. This chapter reviews the benefits of residue retention and the methods for estimating residue coverage and biomass. Additionally, recent and emerging residue management tools are described, including a stripper header that leaves nearly all standing residue, improved information about managing decomposition rates through utilization of different crops and varieties, and new budgeting tools for balancing the tradeoffs related to harvesting or burning residues. This chapter provides growers with general principles about sustainable residue management practices for inland Pacific Northwest (PNW) cropping systems.

Research results are coded by agroecological class, defined in the glossary, as follows:

● Annual Crop ▲ Annual Crop-Fallow Transition ■ Grain-Fallow

Key Points

- Soil erosion is a major contributor to soil degradation and air pollution in the inland PNW. Erosion can be mitigated by strategic management of crop residue.
- The challenges of residue management vary across the inland PNW. Heavy residue produced in high-yield areas can make planting difficult and contribute to unfavorable growing conditions in the early spring. In areas with low or intermediate yields, additional residue is desired for enhancing soil and water conservation benefits.
- Appropriate residue management strategies depend on agroecological class, tillage, cropping system, and varieties.
- Different residue management strategies have tradeoffs between production, economics, environment, and soil health. The estimate of immediate economic tradeoff of harvest residue and burning can help support decisions about residue management practices.

Introduction

Crop residue, a byproduct of harvested food and fiber, makes up a substantial amount of crop production biomass. Residues are traditionally considered an agricultural waste. However, they are increasingly recognized for their value in reducing soil's susceptibility to wind and water erosion, improving soil water conservation, and contributing to **soil health**. When harvested, crop residues are also valuable as livestock feed and bedding, and as feedstock for mushroom, fiberboard, and paper production. More recently, the prospect of using crop residue for energy production has also emerged.

The challenges of residue management vary across the **inland PNW**. One important factor is that residue production varies widely across the region, with estimated residue production for winter wheat ranging from roughly 0.9 ton/acre in the Grain-Fallow **agroecological class** (AEC) ■ to 8.5 ton/acre in the Annual Crop AEC ● (see Chapter 5: Rotational Diversification and Intensification). Too much residue can be problematic in the Annual Crop AEC ●. High residue levels can lead to colder, wetter soils in the early spring, complicate planting, and create conditions that benefit soilborne pathogens. However, residue production is generally lower than desired for soil health in areas with low or intermediate yields in the Grain-Fallow AEC ■. Meanwhile, residue decomposition rates vary across the inland PNW and in different seasons of a year. Residue decomposition proceeds rapidly during spring, summer, and fall when soil moisture is adequate and air temperature is optimal, conditions that occur more frequently in wetter areas of the inland PNW. This is because the soil microbes responsible for decomposition are most active in warm (77–95°F) and moist (50–70% water-filled pore space) conditions (Havlin et al. 2005). Decomposition is very slow when soil temperatures are below 50°F or above 105°F, or when soil moisture is <40% water-filled pore space.

The purpose of this chapter is to provide growers with general principles about sustainable residue management practices that balance the agronomic, environmental, soil health, and economic tradeoffs of residue use. Specific objectives include: (1) discuss the benefits of crop residue in a sustainable agricultural system, (2) provide a critical evaluation of methods for estimating residue production, and (3) review current residue management practices in the inland PNW, including emerging residue management strategies and tools to help evaluate their benefits and tradeoffs.

Benefits of Residue Retention

In the inland PNW, under **conventional tillage** systems such as plowing, disking, or chiseling, residues are recycled by incorporating them into the soil. Alternatively, under **conservation tillage** systems, residues are recycled by leaving them to decay on the field surface. In combination with limiting the frequency and intensity of soil disturbance by tillage, residue return plays the following important roles: (1) protecting soil from erosion, (2) improving soil health, (3) increasing soil water retention and availability, (4) moderating soil temperature, (5) providing wildlife habitat, and (6) building soil organic matter (SOM). These factors, in turn, support long-term crop productivity. Tillage is covered in more detail in Chapter 3: Conservation Tillage Systems.

Protecting Soil Against Erosion

As discussed in Chapter 3: Conservation Tillage Systems, soil erosion is one of the biggest challenges for sustainable agricultural production in the inland PNW. Water erosion is the major concern in the Annual Crop AEC ●; wind erosion is the major concern in the Grain-Fallow AEC ■. A surface cover of crop residue can effectively reduce both water and wind erosion. The optimal ground coverage for erosion control is linked to soil topography and slope, evenness of residue distribution, tillage, type of residue, and residue decomposition rate.

Residue cover and conservation tillage reduce water erosion primarily by protecting the soil from the impact of raindrops that disperse soil **aggregates** and cause soil surface crusting. Thus, the surface residue slows rain or melting snow movement across the soil surface, allowing more time for **infiltration** and reducing the extent of soil freezing under snow cover (Dickey et al. 1986; Smil 1999; Hatfield et al. 2001).

Crop residue on the soil surface reduces wind erosion by reducing wind speeds near the soil surface to below the threshold level for lifting soil particulates. The most important factors influencing the effectiveness of residue management for controlling wind erosion include: (1) mass of residue, (2) percentage of soil covered by residue (Figure 4-1), (3) degree of residue contact with soil, which ensures residue remains in place and does not blow away (Papendick and Moldenhauer 1995), and (4) the height, diameter, and population of standing stems, because these characteristics determine the silhouette area through which the wind passes (McMaster et al. 2000).

Mass of residue is determined not only by residue production, but also by tillage practices (see Chapter 3: Conservation Tillage Systems). In general, residue level and ground cover decrease in order of **no-till** > conservation tillage > conventional tillage (Table 4-1). For the same erosion control effectiveness, more residue is needed in a conventional tillage system than in a no-till system (Figure 4-2). Moreover, a greater amount of residue is needed in a more intensive tillage system, such as moldboard plow tillage, than in a conservation tillage system, such as undercutter tillage (Figure 4-3) (Elbert et al. 1981). The Agronomy Guide from Purdue University summarizes the effects of tillage operations on the amount of post-tillage residue cover (Eck et al. 2001).

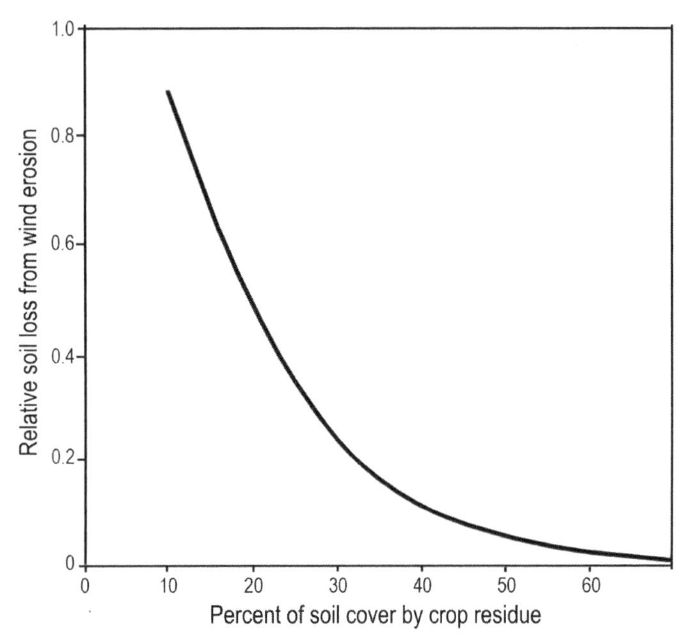

Figure 4-1. Relationship between relative soil loss from wind erosion and percentage of soil covered by residue. (Adapted from Fryrear 1985.)

Table 4-1. Mean annual ground cover measured during winter crop growth in the Grain-Fallow agroecological class in Umatilla County, Oregon ■.

Site†	Tillage	% ground cover	Tillage	% ground cover
Drainage	no-till	73a‡	traditional/ moldboard plow	44b
Hillslope	no-till	81a	minimum tillage	64b
Draw	no-till	81a	minimum tillage	59b

†The maximum slopes were 30%, 20%, 23%, and 4% in no-till, traditional/moldboard plow, hillslope, and draw sites, respectively.
‡Values in rows are significantly different at $P \leq 0.05$ with different letters.
Adapted from Williams et al. 2014.

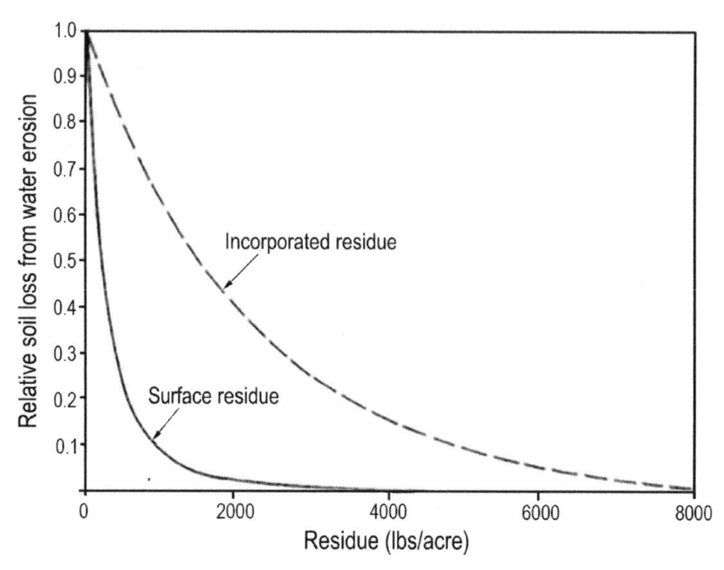

Figure 4-2. Relative soil loss from water erosion on land with surface and incorporated residues for northwest Washington where rill erosion is the dominant type of erosion on crop land. (Adapted from Papendick and Moldenhauer 1995.)

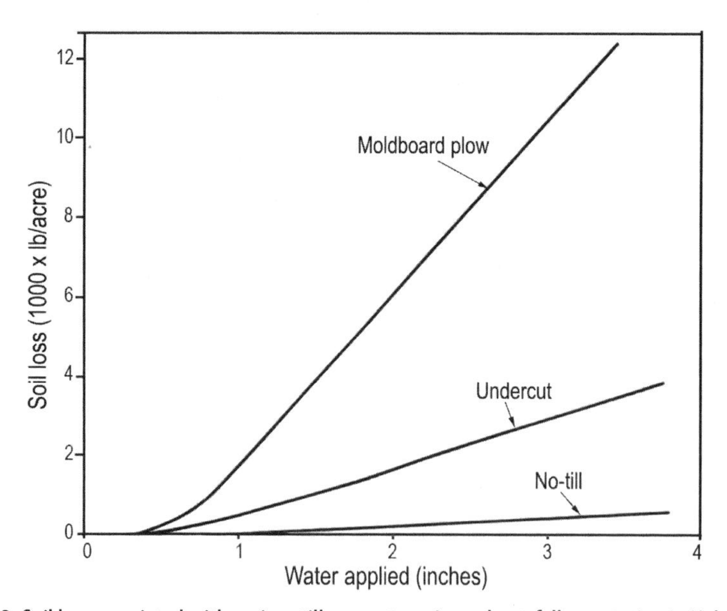

Figure 4-3. Soil loss associated with various tillage systems in a wheat-fallow rotation in Nebraska. (Adapted from Dickey et al. 1981.) Note: lab simulation study on 4% slope hill after planting wheat in a wheat-fallow rotation.

Adequate soil surface protection with crop residue cover is particularly important in the Grain-Fallow AEC during the critical periods for wind erosion, including in the fall and in April and May, when high winds and vulnerable soil conditions occur (Papendick and Moldenhauer 1995). Residue cover is generally <15% soil surface cover (<500 lb/acre) by November to March in areas where conventional tillage is practiced during summer **fallow** (Thorne et al. 2003; Williams et al. 2014). Practices that increase residue biomass production or no-till is especially important in these areas to preserve surface cover during vulnerable times. No-till with tall standing stubble is especially effective. For example, a study found that doubling the mass of 10-inch-high wheat residue (from 450 to 906 lb/acre) has been shown to cut wind erosion by more than 95% (Smil 1999). ■

In contrast, in the high-yield areas in the Annual Crop AEC, residue production significantly exceeds the amount required for erosion control. The concern is to retain enough residue for effective erosion control in fields where conventional tillage, or harvesting or burning residue, is practiced (see Chapter 3: Conservation Tillage Systems). The amount of residue that can be harvested or the frequency of burning should be carefully estimated so that an adequate amount of residue is retained for erosion control and soil health improvement. ●

Researchers generally agree that 30% residue coverage (approximately 1,000 lb/acre residue) is adequate to control both wind and water erosion in flat fields; but, coverage requirements increase to as much as 60% in sloped fields under a conservation tillage system (USDA-NRCS 2005; 2008). The Natural Resources Conservation Service (NRCS) Conservation Plan currently requires at least 30% of last year's crop residue on the soil at planting for a conservation tillage system. In addition, for best water and wind erosion management, surface residue should be spread as uniformly as possible. At harvest, straw chopped into smaller-sized pieces is more likely to be spread uniformly. However, the smaller size is also more likely to be redistributed by wind or water and will decompose faster due to greater surface area contact with soil and water.

Improving Water Conservation

Water availability is a major limiting factor for dryland crop production. Winter wheat requires an estimated 2.32 inches of available water for vegetative growth prior to reproductive development. Each additional 0.39 inch of available stored soil water and spring rainfall (April-June) produces an average of 134 and 155 lb/acre grain, respectively, in eastern Washington (Schillinger et al. 2010).

Soil water recharge and storage is especially important in the inland PNW because an estimated 70% of the region's precipitation occurs between October and March (as discussed in Chapter 1: Climate Considerations). Because daily potential evaporation during the rainy season is low, soil water can percolate beyond the surface soil layers if water runoff can be effectively controlled (Ramig et al. 1983; Ramig and Ekin 1984). Residue and tillage management strategies can be used to increase infiltration, reduce evaporation, enhance snowfall catch, and improve **water holding capacity**, therefore increasing soil water storage.

Surface residue cover can increase infiltration and suppress evaporation. Though, the extent of this effect depends largely on the AEC, the amount of residue, and the tillage system. For no-till producers in the Grain-Fallow AEC, large amounts of surface residue cover are required to effectively reduce evaporation due to the extended dry, hot summers (Wuest and Schillinger 2011). A 6-year field study conducted in this AEC concluded that, even with 4 or 7 times the regional residue average (1.4 ton/acre), evaporation reduction in no-till surface cover remained limited compared to tilled fallow ■. This limited benefit of no-till summer fallow on water storage efficiency has also been documented in other parts of the US: the percentage of precipitation that generally can be stored in soils is only 10% in Texas, 22% in eastern Colorado, and 25–30% in western Kansas for the 14-month winter wheat-summer fallow rotation (Peterson et al. 1996).

In the Annual Crop-Fallow Transition AEC, no-till that leaves residue on the soil surface provides significant benefits in soil water storage over conventional tillage. Research conducted in Pendleton, Oregon, during a dry year suggested that conserving residues results in higher water infiltration and greater soil water storage than when residues were incorporated (Figure 4-4). The average soil water storage in the 41-inch profile was 7.36, 6.61, and 6.10 inches under no-till (94% ground cover), residue returned after tillage (83% ground cover), and residue incorporated with tillage (23% ground cover), respectively (Williams and Wuest 2014) ▲.

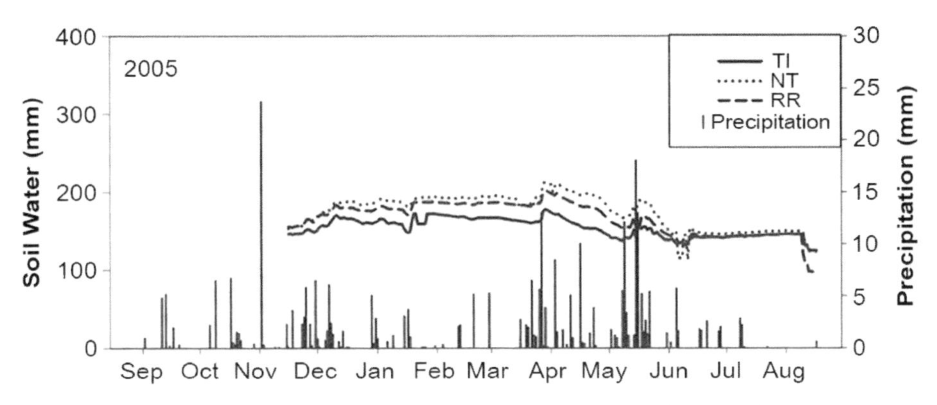

Figure 4-4. Soil water in a 41-inch profile during a dry crop year (2005) under TI (residue incorporated into soil), NT (no-till), and RR (residue returned on soil surface after tillage) treatments of annual winter wheat near Pendleton, Oregon. (Adapted from Williams and Wuest 2014.)

Increases in soil water storage in no-till that leaves residue on the soil surface have also been seen in the Annual Crop AEC. Near Troy, Idaho, soil moisture in the surface 6 inches of winter wheat, managed with no-till, was found to be significantly higher in the fall between precipitation events than winter wheat managed with no-till plus stubble reductions or with conventional tillage (Huggins and Pan 1991).

Understanding the impacts of residue on snow capture is important to understanding water storage impacts of residue. Across the inland PNW, roughly a third of precipitation is in the form of snow during the primary soil-water recharge period. Trapping more snow can increase soil water storage, and can also provide insulation that protects plants from winterkill. In the unique land topography of the Palouse, redistribution of snow by wind and snowmelt runoff can also cause substantial spatial variation in soil water availability. Ridge tops and south-facing slopes generally retain the least amount of snow, and valley areas retain the thickest snowpack regardless of tillage system (Figure 4-5).

In this topography of the Palouse, no-till retains more soil water with less spatial variation of snow depth at all topographic locations compared with conventional tillage. In Pullman, Washington, during the 2007–2008 season, no-till ridge tops and south-facing slopes (with 3.54 to 13.00 inches standing stubble) retained 3.94 to 4.72 inches and 3.94 to 5.51 inches more snow, respectively, during two separate snow events,

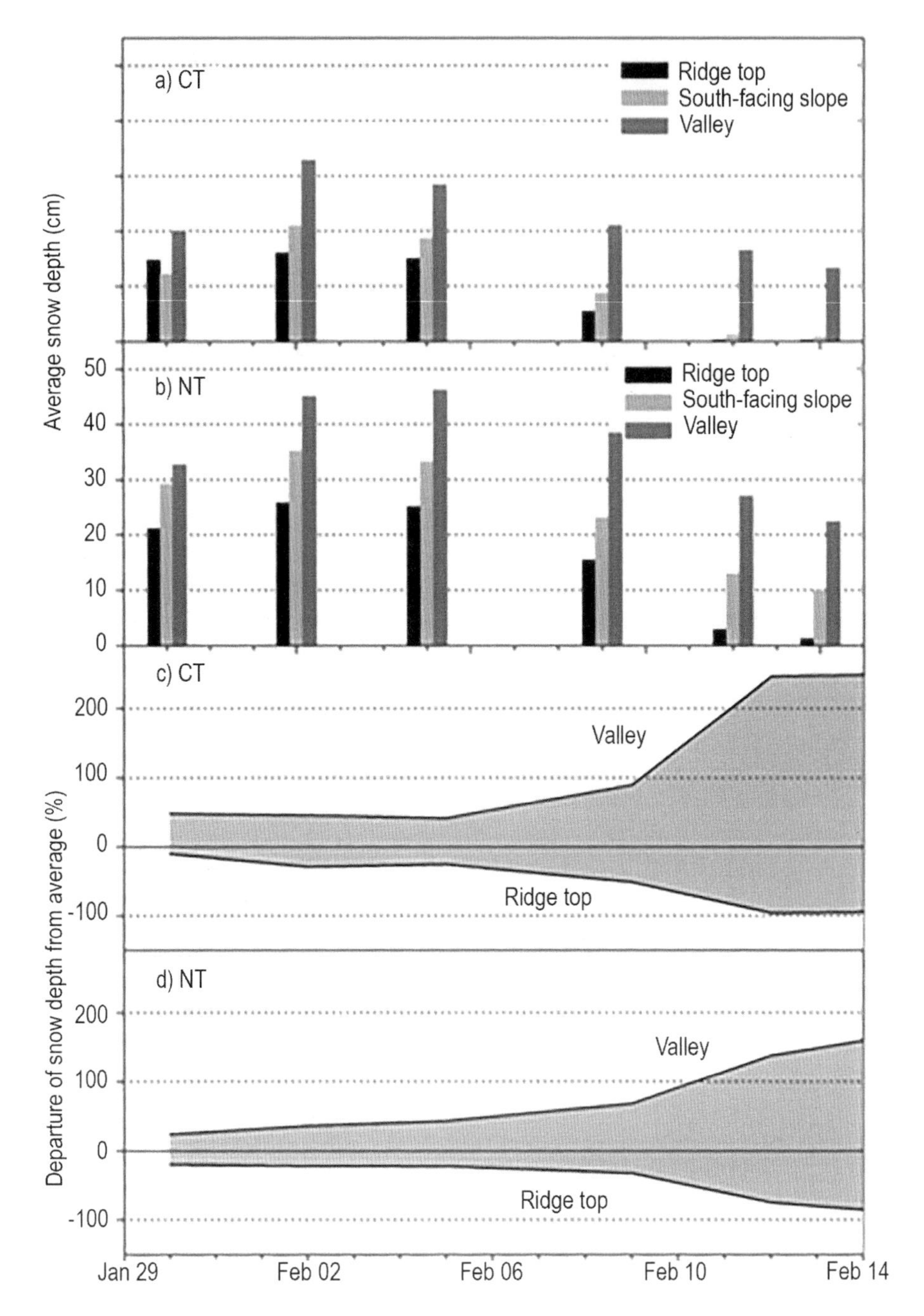

Figure 4-5. Average snow depth observed during January 29 through February 14, 2008, on two adjacent fields in Pullman, Washington, for (a) conventional tillage (CT) on a private farm and (b) no-till (NT) on Washington State University Cook Agronomy Farm, and departure of snow depth at ridge top and valley from the average for (c) CT and (d) NT. (Adapted from Qiu et al. 2011.)

compared with conventionally tilled fields (Qiu et al. 2011). By spring, no-till stored 2.36, 1.14, and 0.51 inches more water in the 5-foot soil profile at ridge tops, south-facing slopes, and valley locations, respectively, than conventional till.

Standing crop residue, such as wheat and sunflower stubble, is more effective not only in reducing wind speed and evaporation but also in increasing snow catch than chopped residue left on the soil surface (Nielsen 2013). Snow catch generally increases as stubble height increases in no-till (Figure 4-6). A long-term study in Saskatchewan concluded that leaving 35–47 inches wide standing stubble strips of residue about 16–24 inches tall every 19.5 feet trapped 1.6 times as much snow as shorter stubble at 12 inches tall (Campbell et al. 1992).

Improving Soil Health

In addition to conserving soil and water, residue remaining in the field positively affects soil physical, chemical, and biological properties and productivity, mostly via increasing SOM. A more complete discussion of soil health and the benefits of SOM are presented in Chapter 2: Soil

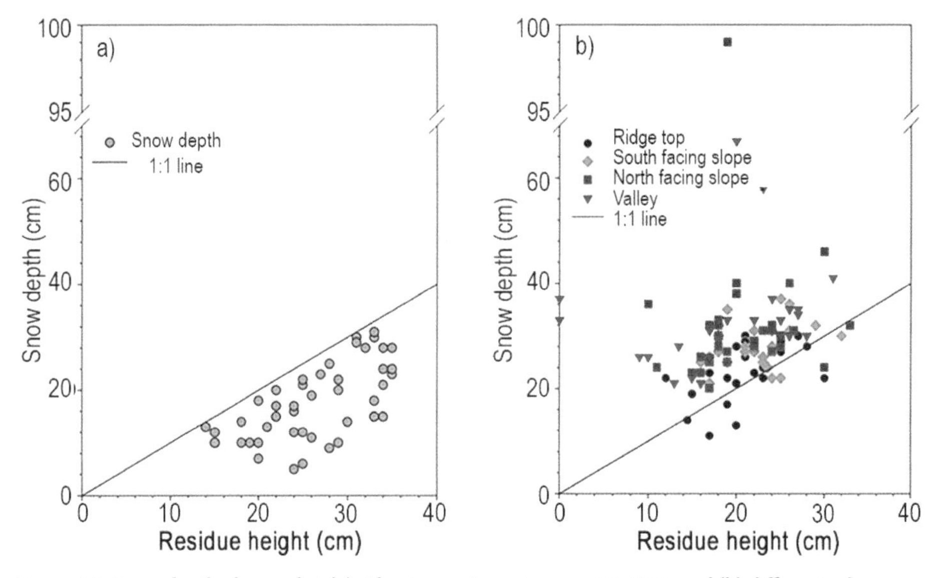

Figure 4-6. Snow depth observed at (a) ridge top route on January 15, 2008, and (b) different sub-sections on January 10, 2008, at the Washington State University Cook Agronomy Farm in Pullman, Washington (under no-till). (Adapted from Qiu et al. 2011.)

Health. Residue return and reduced tillage can be a cost-effective way to maintain soil health. Crop residue, including roots, is the primary source of organic matter for most dryland cropping systems in the inland PNW. Although, surface residue retention in no-till, arid cereal systems only has limited impact on soil organic carbon (SOC) accumulation (Gollany et al. 2011).

Adding crop residue to soil can also increase total soil porosity and reduce soil **bulk density**, surface sealing, and crust strength, which benefits crop emergence and water infiltration. Microbial decomposition of crop residue produces polysaccharides and other compounds that help bind soil particles together into stable soil aggregates, which is one of the major mechanisms of aggregate stabilization in soils (Turmel et al. 2015). Soil aggregates, in turn, can protect SOM from decomposition by making it less accessible to microorganisms.

Clearly, surface residue retention improves **aggregate stability** of the surface soils (Campbell and Souster 1982; Baker et al. 2007). Yet, the effects of residue management practices on subsoil physical quality remains unclear. Li et al. (2012) found decreased macroaggregate proportions in the 2–12 inches of subsoil under no-till. Other research found no differences in subsoil macroaggregate proportions under different tillage systems when residue was retained in the fields (Jacobs et al. 2009).

Crop residues also provide nutrients (see Chapter 6: Soil Fertility Management). Nutrient availability from decomposition of crop residue depends on residue type and quality. Pulse and canola residues contain higher concentrations of nitrogen (N) and phosphorus (P), and therefore return more of these nutrients than cereal crop residues. Research conducted in Alberta, Canada, found that canola straw returned 45 lb N per acre and pea returned 20 lb N per acre, whereas wheat only returned 14 lb N per acre (Soon and Arshad 2002). In a New Zealand study, only an estimated 7% of N in lentil straw was mineralized during the following growing season. The remaining N can become a long-term source of N (Bremer and Kessel 1992).

Estimating Residue Ground Coverage and Biomass

Estimating Residue Ground Coverage

The percentage of ground coverage by residue after planting is an important benchmark generally used to determine effectiveness of erosion control (Figure 4-7). Several methods can be used for measuring crop residue in fields, including weight per unit area (Figure 4-8), the line-transect method, the meter stick method, or the photo comparison method. Detailed descriptions of each measurement method, along with guidance for interpreting results can be found on the USDA-NRCS website: ***http://www.nrcs.usda.gov/Internet/FSE_DOCUMENTS/nrcs144p2_042684.pdf.***

Estimating Residue Biomass

Crop residue biomass is typically estimated indirectly. Historically, these estimates have been made on the basis of **harvest index** or **residue-to-grain** (R:G) ratio. Harvest index is the ratio of crop yield to the crop's total aboveground biomass (Donaldson et al. 2001; Smil 1999). R:G ratio is simply the ratio of dry residue yield to grain yield. The equation below

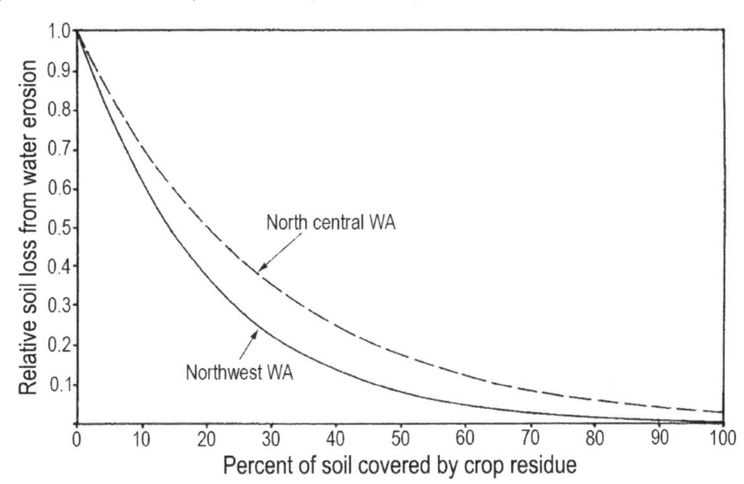

Figure 4-7. Relationship between relative soil loss from water erosion and percentage of soil covered by small grain residue. (North central WA: where mixed interill/rill erosion is dominant type water erosion; Northwest WA: where rill erosion is the dominant type of water erosion on crop land.) (Adapted from Papendick and Moldenhauer 1995.)

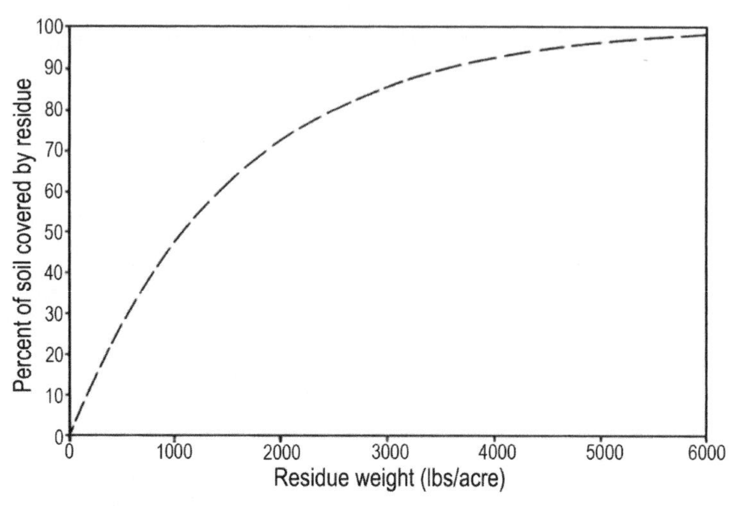

Figure 4-8. A general relationship between percentage ground cover and residue weight per acre for common small grains and legumes in the non-irrigated areas of the inland PNW. (Adapted from Papendick and Moldenhauer 1995.)

is typically used to calculate the amount of residue production for wheat based on grain dry yield and R:G ratio.

Existing literature reports a wide range of R:G ratios. The ranges of R:G ratios commonly used are 1.10–1.70 for wheat, 0.82–2.50 for barley, 1.20–1.70 for rye, 1.08–1.32 for grain triticale, and 1.86–4.00 for canola (Koenig et al. 2011; McClellan et al. 1987; Behl and Singh 1998; Lal 2005). Using wheat production in 2012 (USDA-NASS) as an example, the ranges of estimated total wheat residue production in the inland PNW states of Washington, Idaho, and Oregon (assuming the mean test weight was 60 lb/bu) are listed in Table 4-2.

The estimation method using the R:G ratio results in a wide range of estimated residue quantities (Table 4-2). Environmental factors, nitrogen fertility, and genotype (especially crop height), greatly impact the R:G ratio.

In response to this difficulty, McClellan et al. (2012) developed improved residue-to-grain yield relationships for inland PNW dryland cereal and legume production to estimate crop residue production. These relationships were described using linear models fitted to data from a large number of research sites across eastern Washington and north-central

Table 4-2. Estimated total wheat residue production by state in the inland PNW.

State	Total grain production	Calculated total residue production		
		R:G ratio = 1.1	R:G ratio = 1.7	Range
	million bushels	---------- million metric tons ----------		
WA	141.02	4.22	6.52	4.22-6.52
ID	96.84	2.90	4.48	2.90-4.48
OR	57.51	1.72	2.66	1.72-2.66

Note: total grain production was published by USDA-NASS 2012.

Oregon (Table 4-3; Figure 4-9). These linear models could explain 31% of the variation in spring wheat residue yield, and a much higher percentage (55 to 69%) of variation in winter wheat, winter barley, and spring barley residue yield. Although the model found a positive linear relationship between lentil grain yield and residue yield, it could only explain a small amount of the variation (9%) for lentils—much lower than for the cereal crops (McClellan et al. 2012). Their results suggest that the fixed R:G ratio overestimates residue production of high-yielding winter wheat by as much as 35%, and underestimates residue production of low-yielding spring wheat by as much as 66%. These differences are large enough to have implications for decisions about residue management.

Table 4-3. Linear regression analysis at a three standard deviation rejection limit for an eastern Washington dataset.

Crop	Number of Samples	Regression Equation For Calculating Residue[†]	Standard Deviation	R^2
			lb/acre	
Winter wheat	1135	$y = 1.1274x + 1,175.3$	2787	0.69
Spring wheat	112	$y = 0.8613x + 2068.1$	2159	0.31
Winter barley	53	$y = 0.8310x + 1747.7$	2089	0.65
Spring barley	737	$y = 0.7013x + 1302.9$	1771	0.55
Lentils	144	$y = 0.4684x + 1843.8$	904	0.09
Peas	117	$y = 0.7187x + 940.9$	975	0.29
Austrian winter peas	12	$y = 1.3427x + 1327.2$	1236	0.63

[†]x = grain yield (lb/acre) and y = residue dry yield (lb/acre).
Adapted from McClellan et al. 2012.

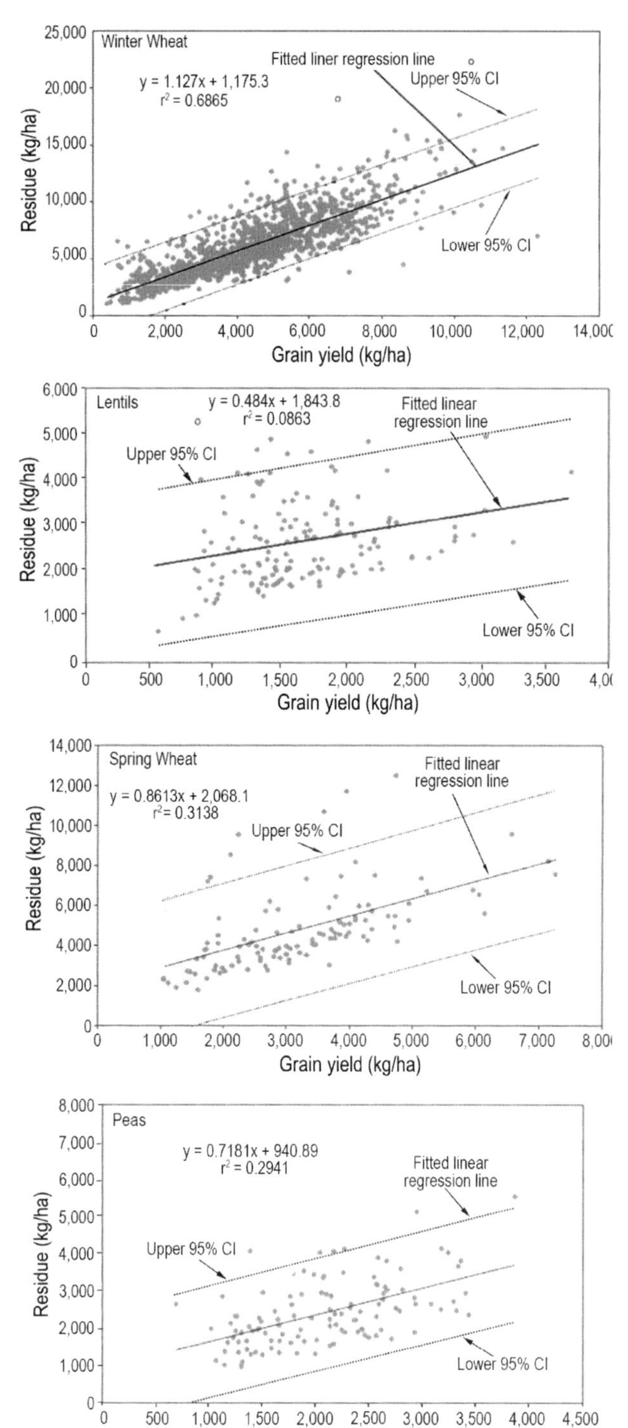

Figure 4-9. Regression line, fitted equation, and 95% confidence interval (CI) for winter wheat, spring wheat, lentils, and peas in eastern Washington. (Adapted from McClellan et al. 2012.)

Including factors such as plant height and N status, in addition to grain yield, can significantly improve the accuracy of straw yield predictions over the use of R:G ratio alone (Engel et al. 2003; Long and McCallum 2013). Research conducted on three neighboring commercial production fields in Oregon suggested that wheat height was a better predictor of residue yield for irrigated hard red spring wheat than grain yield or grain protein concentration (Long and McCallum 2013). A strong linear relationship was found between straw yield and wheat height for this cultivar and within this environment. When relationships between residue production and factors such as plant height, N status, and yield are confirmed, a GIS, on-combine lidar sensor, yield monitor, and protein monitor-equipped combine can simultaneously collect wheat height, yield, and protein data as appropriate (Long and McCallum 2013). These tools can generate maps that can be used to make site-specific decisions about sustainable residue harvest.

Remote sensing spectral indices have also been evaluated to predict crop residue cover and density (Aguilar et al. 2012). These indices include: (1) broadband spectral normalized difference indices derived using Landsat Thematic Mapper (TM) bands, such as the normalized difference tillage index, the normalized difference index 5 and 7, and the normalized differential senescent vegetation index; (2) reflectance-band height indices such as the lignin-cellulose absorption index and the cellulose absorption index; and (3) spectral angle methods. These methods work well for distinguishing crop residue from background soils, and therefore percentage of ground coverage of laying residues. However, more work needs to be done to use these technologies to measure residue density and quantify the amount of crop residue for both laying residues and tall standing stubble of stripper header harvest (Aguilar et al. 2012).

Managing Residues in Inland PNW Dryland Cropping Systems

As described previously, the challenges for residue management vary across the region, necessitating different management strategies. In wetter areas, where too much residue can be an issue, tillage can effectively decrease residue levels by accelerating decomposition, but this strategy can diminish the benefits provided by crop residue. Mowing to cut residue into shorter pieces is one common strategy for coping with high residues without tillage. Crop rotation, for example, adding canola, pea, lentil, rapeseed, or wheat cultivars with faster decomposition genotypes into rotation can also be used to reduce overall residue levels throughout the rotation, as these crops produce less residue with a higher decomposition rate (Brown 2015). Burning or harvesting can also be used to reduce the amount of residue, though with tradeoffs for soil health and nutrients.

On the other hand, in areas with low or intermediate yields, where residue production is generally lower than desired, different management strategies are needed. When feasible, crop intensification and diversification to reduce fallow can be used to increase biomass and soil carbon (C) sequestration while increasing soil and water conservation (Gollany et al. 2013; Schillinger et al. 1999; and Young et al. 2015). Examples of crop intensification include replacing winter wheat-summer fallow with summer fallow-winter pea-winter wheat, or by replacing summer fallow by short-season, spring-planted crops such as spring wheat, barley, canola, sunflower, and others to make a winter wheat-spring crop-summer fallow rotation in the Transition AEC ▲. Research has shown that annual no-till spring cereal cropping systems can provide greater wind erosion protection and reduce SOM loss in the Grain-Fallow AEC, but the economic returns were less than winter wheat-summer fallow cropping systems (Rasmussen et al. 1998; Young et al. 2015) ■. In addition, other management strategies such as early seeding, higher seeding rates, planting tall varieties or crops with high residues (such as barley or winter triticale), and optimizing fertility can increase residue density.

The sections below discuss several relevant management strategies in more detail: harvest with a stripper header, managing decomposition rates through utilization of different crops and varieties, burning, and harvesting residues for other uses. Intensification strategies are discussed in more detail in Chapter 5: Rotational Diversification and Intensification.

Harvest with a Stripper Header

Use of a stripper header to conserve tall standing stubble is a promising residue management strategy that is being explored in the inland PNW for erosion control and water conservation by research and innovative growers (Port 2016; Yorgey et al. in preparation).

Tall standing stubble in a no-till system, achieved by harvesting with a stripper header and leaving standing stubble at full-crop height, can reduce residue decomposition rates and conserve water (McMaster et al. 2000; Port 2016). Tall standing stubble is especially important for sparse stands. A 4-year study conducted in Ralston, Washington, found that stripper header winter triticale stubble in no-till chemical fallow influences the microclimate at the soil surface. The tall standing stubble can reduce soil temperatures and reduce average wind speed at the soil surface to less than one half of average wind speed. The stripper header winter triticale stubble also preserved greater amounts of soil moisture and more uniform soil moisture in the 0 to 3-inch seed zone. This allows for timely planting and establishment of fall-seeded canola (Port 2016). Another 5-year study in Fort Collins, Colorado, concluded that tall standing residue provides numerous benefits such as increased snow trapping, decreased decomposition rates, wind speed, weed pressure, soil temperature during the fallow period, and within-field variation in snow cover and water storage (McMaster et al. 2000).

Managing Residue Decomposition Rates through Crop, Variety, and Fertility

Residue decomposition rates are influenced by a variety of management factors. An awareness of these influences is helpful and, in some cases, may offer opportunities to growers in heavy residue systems for reducing residue buildup or to those in low residue systems for preserving residue cover.

Residue structural components and chemical composition help determine residue decomposition rates in soils. Loss of simple sugars and amino acids occurs rapidly, whereas polysaccharides, proteins, and lipids take much more time to decompose. Lignin is even more resistant to decomposition, and is a major contributor to humus in soils.

Different crops have different residue structure components and chemical composition. Pulse crops and canola residues usually contain higher N concentrations and have lower **carbon-to-nitrogen** (C:N) ratios than cereal residues, and thus decompose more rapidly. Thus, including pulses and canola in rotation with cereals can reduce residue levels across the rotation. Different environmental conditions can also result in different residue chemical composition and thus decomposition, with faster decomposition for residues grown in drier years or areas (Stubbs 2009).

Residue structural components and chemical composition can also differ significantly among varieties for a single crop, and this can impact decomposition rates. A study conducted in eastern Washington on the chemical composition of residue from different cultivars of spring barley, spring wheat, and winter wheat suggested that the percentage of acid detergent fiber (ADF), acid detergent lignin (ADL), C:N ratio, and N content all correlated with residue decomposition. The percentage of ADF and total N were both found to be best correlated with decomposition after 8 weeks of incubation. Total N was also a good indicator of decomposition at 16 weeks (Figure 4-10). Foot rot-resistant cultivars had higher ADF, ADL, and C:N ratio than foot rot-susceptible cultivars, and were therefore more resistant to decomposition.

The residue C:N ratio also determines whether net N **mineralization** or **immobilization** occurs from freshly added residue. Thus, N management should take into consideration the previous crop residue. Although soil microorganisms have a C:N ratio of 8:1, they require a crop residue C:N ratio of 24:1 for decomposition activity (of the 24 parts, 8 parts remain in microorganism biomass and 16 parts are lost as CO_2 during respiration). If a crop residue has a C:N ratio of ≤24, net N mineralization occurs. If the C:N ratio is >24, a temporary net N immobilization occurs until microorganisms die and subsequently release N from their biomass. If high C:N ratio residue is repeatedly added to the soil, soil microorganisms can tie up N, increasing the likelihood of N deficiency during decomposition of high C:N crop residue. The USDA-NRCS provides typical C:N ratios of different crop residues at: ***http://www.nrcs.usda.gov/wps/PA_NRCS-Consumption/download?cid=nrcs142p2_052823&ext=pdf***. For more discussion on managing soil fertility in light of mineralization and immobilization, see Chapter 6: Soil Fertility Management.

The composition of roots can be different than the above-ground portions of crops, and this also impacts decomposition rates. In canola, pea, and wheat, C:N ratios in straw are found to be higher than in roots, but lignin

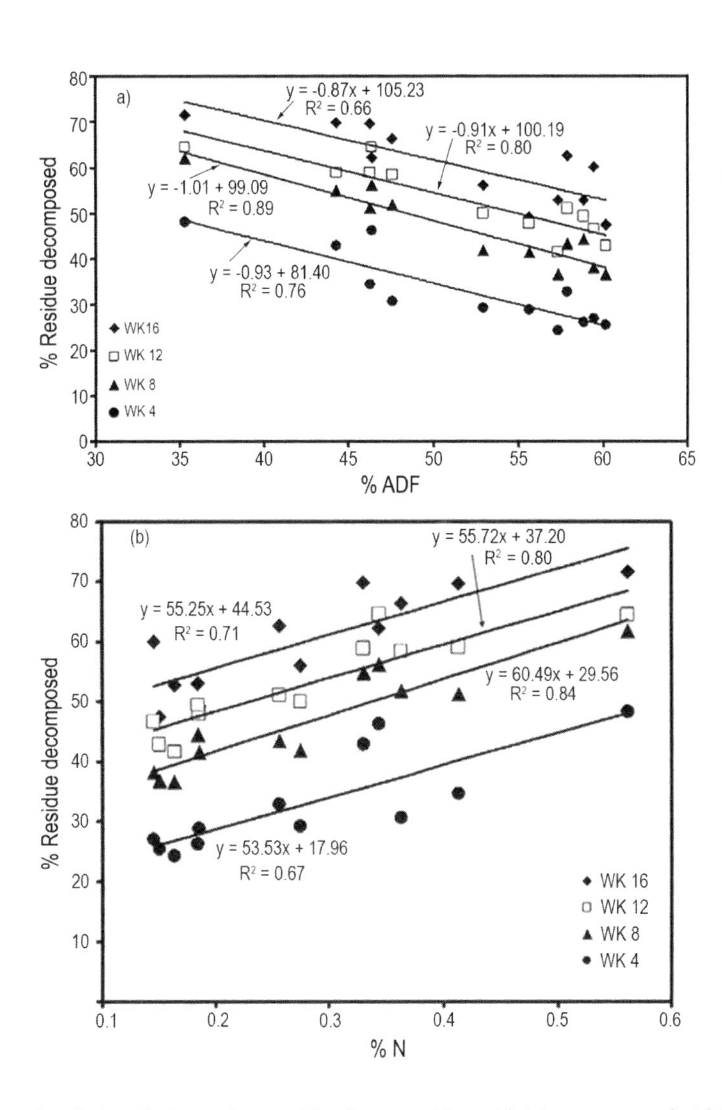

Figure 4-10. Correlation of winter wheat residue decomposition with (a) percentage of acid detergent fiber (ADF) and (b) percentage of total N after 16-week incubation of six cultivars from Pullman and Dusty, Washington. (Adapted from Stubbs et al. 2009.)

content is higher in roots than in straw (Soon and Arshad 2002). Unlike above-ground residue, the decomposition of roots was neither correlated with N concentration nor C:N ratio.

Residue Harvest

Particularly in areas where residue production is plentiful, grain residue can be harvested for livestock feed, bedding, mushroom production, as feedstock for fiberboard or paper production, and **biochar** production. Crop residue with high cellulose content, including residues of corn, wheat, sorghum, rice, and barley, was also identified by the USDA as a potential future feedstock for second generation biofuel production because of its large quantity, easy availability, and renewability (Perlack et al. 2005). Unlike first generation biofuel feedstocks such as corn grain and soybeans, using crop residues can avoid displacement of food production by allowing grain to be harvested for food or feed while the residues are harvested for ethanol production.

However, although large quantities of residue are produced, not all the residue produced can be or should be removed. Harvesting crop residues involves tradeoffs between other uses and the agroecosystem services described previously in this chapter (Laird and Chang 2013; Huggins et al. 2014). Specific considerations that can help determine how much straw should be harvested include a determination of the amount of residue needed for effective soil erosion control, maintenance of SOM and soil health, and economics.

Determining sustainable residue harvest

Calculating the amount of residues needed for erosion control and maintenance of soil health can provide a minimum amount of residues that need to be conserved for long-term agroecosystem sustainability. Residues produced beyond that level may thus be considered for harvesting.

Changes in SOC are highly correlated with residue input in the Grain-Fallow AEC of the inland PNW (Gollany et al. 2011; Rasmussen and Parton 1994), implying that SOC sequestration is particularly sensitive to crop residue removal in this system ■. Long-term studies established in 1931 in Pendleton, Oregon (average 15.75 inches of precipitation), estimate that 3.57 to 4.46 ton/acre/year of crop residue should be retained in this system for SOC maintenance (Rasmussen et al. 1980; 1998). However, wheat residue produced in winter wheat-summer fallow in this region was not able to maintain initial SOC levels regardless of crop rotation, fertilizer rates, and tillage practices (Machado 2011). The main reason was that growing one crop in two years did not produce sufficient biomass to maintain SOC. Harvesting wheat residue from this system would accelerate SOC depletion (Table 4-4). However, other long-term studies at the same site suggested that SOC could be maintained or increased in a continuous annual cropping system even if residues were the only SOC input (Table 4-5) (Machado 2011).

In the Annual AEC, where rotations are more diversified, a budget for sustainable cereal straw harvest should be based on the crop rotation instead of cereal residue alone (Papendick and Moldenhauer 1995; Huggins et al. 2014). For example, a comparison of three crop rotations including a 2-year rotation of winter wheat-spring pea (WW-SP), a 3-year rotation of winter wheat-spring pea-spring wheat (WW-SP-SW), and a 3-year rotation of winter wheat-spring barley-spring wheat (WW-SB-SW) indicated that harvestable residue was greatest for the WW-SB-SW rotation and least for the WW-SP rotation (Huggins et al. 2014). Tillage was also important, with smaller adverse effects of residue harvest in the 3-year rotations for no-till compared with conventional tillage.

Economic tradeoffs for cereal residue harvesting

Once needs for erosion control and soil health maintenance are established, residue beyond these amounts may be available for baling, if economically feasible. The economic tradeoff calculated using partial budgeting (Table 4-6) is the most direct calculation that can be used to support decisions on residue baling in a given year. Using this approach, costs of residue harvest can be calculated based on the value of the nutrients removed and the cost of the baling process, and this cost should be offset by the sale of the straw for economic sustainability (Huggins et al. 2014).

The value of nutrients in crop residue is a function of the nutrient content and replacement fertilizer prices (Table 4-7). The costs of residue harvest and residue swathing, baling, and stacking vary greatly depending on

Table 4-4. Tillage effects on biomass, grain yield, and soil organic carbon (SOC) in the tillage fertility long-term experiment in a winter wheat-summer fallow rotation conducted in Pendleton, Oregon.

Tillage	TDM†	GYD†	SDM†	SOC‡			Significance level§	Residue cover at seeding %
	1989-2003			1984	1995	2005		2004-2005
	------ lb/acre/year ------			------ Metric ton/acre ------				
Plow	6,067a	2,498a	4,015a	25.8a	24.9b	22.5a	***	22b
Disc	5,621b	2,320b	3,747b	27.1a	26.4a	23.0a	***	63a
Sweep	5,532b	2,231b	3,747b	26.7a	25.1b	22.6a	***	66a

Note: *** indicates that SOC means in 1984 and 2005 are significantly different at $\alpha = 0.001$; †TDM = total bundle dry matter, GYD = combine grain yield, SDM = bundle straw yield, SOC = soil organic carbon; ‡Within columns, means followed by the same letter are not significantly different at $\alpha = 0.05$; §This column compares SOC across years. Adapted from Machado 2011.

Table 4-5. Tillage effects on biomass, grain yield, and soil organic carbon (SOC) in the wheat-pea rotation of a continuous annual cropping system in a long-term experiment conducted in Pendleton, Oregon.

Tillage	TDM†	GYD†	SDM†	SOC‡		Significance level§	Residue cover at seeding %
	1995-2004			1995	2005		2005
	------ lb/acre/year ------			Metric ton/acre			
Maximum Tillage	8,030a	3,569ab	4,550a	26.6a	27.0b	ns	49b
Fall Plow	8,119a	3,569ab	4,639a	26.7a	27.7b	ns	27c
Spring Plow	8,208a	3,658a	4,639a	26.7a	26.8b	ns	21c
No-till	8,030a	3,390b	4,639a	26.1a	29.7a	***	97a

Note: *** indicates that SOC means in 1995 and 2005 are significantly different at $\alpha = 0.001$; ns = not significant. †TDM = total bundle dry matter, GYD = combine grain yield, SDM = bundle straw yield, SOC = soil organic carbon; ‡Within columns, means followed by the same letter are not significantly different at $\alpha = 0.05$; §This column compares SOC across years; Adapted from Machado 2011.

Table 4-6. Partial budgeting comparing residue removal and residue return to fields for growers.

Partial Budgeting Alternative: residue removal	
Increased cost Residue harvest Residue swathing, baling, and stacking Sensors, imageries	**Increased revenue** Sale of the straw
Reduced revenue Value of fertilizer replacement for N, P, K, S, and Cl	**Reduced costs** none
A. Total increased costs and reduced revenue	B. Total increased revenue and reduced costs
Expected change in net revenue (B – A)	

the density of stubble, header width, baling method and size of bales, and field conditions. If farmers own the equipment, the cost may be lower than hiring custom operators (Duft and Pray 2002). The custom operators' rates for Idaho agricultural operations is updated by University of Idaho Extension in the BUL729 publication. If the expected change in net revenue calculated by partial budgeting is positive, then it is economically feasible to bale.

Site-specific cereal residue harvesting

Crop residue production varies both between and within fields. In general, between-field variability can be explained by the type of crop, harvest methods, plant height and other genotype characteristics, environmental factors such as plant-available water and nitrogen fertility, and field management practices. Within-field variability can result from differences in soils, slopes, water, nutrients, **pests**, and interactions of these factors.

Because of this variability, site-specific harvesting of cereal residues may help ensure long-term economic and environmental sustainability. Over three years, within-field variability in cereal straw production in a field near Pullman, Washington, was more than twofold, and the variability in nutrient removal ranged over fivefold if all the residue was removed (Huggins et al. 2014). Site-specific harvesting could be made feasible by

Table 4-7. Estimated average nutrients export and their fertilizer replacement value for cereal for a winter wheat-spring barley-spring wheat (WW-SB-SW) rotation conducted at the WSU Cook Agronomy Farm in Pullman, Washington in 2009–2011.

nutrients	Nutrient			Nutrient value		
	WW	SW	SB	WW	SW	SB
	-------- lb/acre --------			-------- $/acre --------		
C	1,698	1,305	1,411	na	na	na
N	17.0	13.4	14.3	10.20	8.05	8.59
P_2O_5	3.0	2.3	2.5	1.89	1.44	1.55
K_2O	22.3	17.8	18.7	11.84	9.47	9.95
S	2.3	1.7	1.9	0.77	0.59	0.64
Total nutrient value				24.70	19.57	20.73
Cost of harvest, $/ton				30.73	31.46	31.22
Cost of swathing, baling, and stacking, $/acre				14.50	11.41	12.24

Note: WW = winter wheat; SW = spring wheat; SB = spring barley. Nutrient value varies based on prices for fertilizers N, P, K, and S. Adapted from Huggins et al. 2014.

the use of **precision agriculture** technology, such as lidar sensors, to measure real-time wheat height to generate residue yield maps. For more information on these technologies, see Chapter 8: Precision Agriculture.

Opportunities for harvest with biomass return after processing

One intriguing potential for reducing the downsides of harvest is to return residuals to fields after processing, along with the retained carbon and nutrients. For example, when residues from cereal systems are used for paper-making, the process results in "black liquor," an organic waste effluent that contains lignin, nutrients, and other organics. If returned to soils, black liquor can be used as a soil amendment. In an alternative example, thermochemical energy production in an oxygen-limited environment can result in bio-oil and biochar, with emphasis on one or the other products depending on conditions. Bio-oil provides energy, while biochar is a solid, carbon-rich, porous material that can be returned to soils, sequestering carbon and providing a mild liming effect. Black liquor and biochar are described in Chapter 7: Soil Amendments.

Residue Burning

Grain producers in the inland PNW have burned wheat residue for a number of reasons. First, burning can eliminate seedbed tillage operations and enable producers to use existing machinery to plant winter wheat in no-till systems (McCool et al. 2008). Second, burning is sometimes perceived to positively impact crop growth and yields although this effect has not always been seen; for example, research conducted in the UK found no yield advantage from burning winter wheat straw over incorporation (Smil 1999). Third, burning can reduce the incidences of disease, weeds, and insects. Research has shown that burning can reduce seeds located on the soil surface by 97% for brome, 50% for wild oat, 61–94% for blackgrass, and 43–65% for goatgrass spikelets (Young et al. 1990), though efficacy can sometimes be impacted by non-uniform high temperatures over the soil surface during burning (McCool et al. 2008; Smil 1999; Young et al. 1990). Ongoing work has also indicated a potential role for limited burning, such as windrow burning, as one part of an integrated weed management strategy (Lyon et al. 2016; Young et al. 2010). This is discussed in Chapter 9: Integrated Weed Management.

Although there are advantages to burning residues in some cases, there are also tradeoffs, including the adverse impacts on soil health and productivity. Long-term burning can reduce total C and N pools, SOM, net N mineralization rates, C:N ratio, microbial biomass, extractable C and polysaccharides (readily available carbon sources for microbes), ammonium, and available P (Fasching 2001). Long-term crop residue burning has negative impacts on soil physical characteristics including decreased water stability of soil aggregates due to reduction of soil biological activities (glomalin, basidiomycetes, and earthworm counts) (Wuest et al. 2005), increased erodibility and soil density, and decreased water and nutrient retention (Papendick and Moldenhauer 1995; Holmgren et al. 2014). A literature review indicates that there is no measurable negative effect from occasional and short-term burning (Holmgren et al. 2014), although the practice can cause ground hardening and reduce water infiltration as a result of temporary soil surface sealing after burning.

Beyond the farm, residue burning also has negative environmental impacts. It contributes to greenhouse gas emissions including CO_2, nitrous oxide (N_2O), methane (CH_4), and carbonyl sulphide (COS, which has a greenhouse gas potential 724 times that of CO_2) (Smil 1999; Jain 2014). It also releases air pollutants including carbon monoxide (CO), ammonia (NH_3), nitric oxide and nitrogen dioxide (NO_x), sulfur dioxide (SO_2), non-methane hydrocarbon, volatile organic compounds, particulate matter less than 10 μm (PM_{10}) and 2.5 μm ($PM_{2.5}$) in size, and smoke. These air pollutants exacerbate respiratory and lung disease with public health impacts (***http://community.seattletimes.nwsource.com/archive/?date=19981001&slug=2775191***). Research has found that there were significant increases in hospital admissions related to the increase in pollutants during agricultural crop residue burning (Agarwal et al. 2012).

Economic tradeoffs of residue burning

The economic benefit of residue burning can best be estimated as a function of change in subsequent crop yield. However, this change is hard to identify because there are differing results of burning dependent on other field management practices (Smil 1999; Huggins et al. 2011; Young et al. 2010). Meanwhile, the cost of residue burning can be estimated using the cost of the burning permit and the fertilizer replacement value of nutrients lost from burning, generally including N, P, K, and S. For example, based on average fertilizer prices from 2008 to 2010, the cost for fall field burning was estimated at $28.62/acre and $9.64/acre for spring field burning in eastern Washington (Table 4-8). The burning permit can be a substantial percentage of the total cost. In 2016, the Washington Department of Ecology charged a minimum fee of $37.50 for the first 10 acres and $3.75/acre for each additional acre.

In calculating the nutrient value, C content varied only a little, but other nutrients in residue varied significantly depending on time of burning, crop, water and nutrient supply, management practices, and other factors (Table 4-9). A field burning study on surface winter wheat residue conducted in Pullman, Washington, found that fall burning reduced residue by 62% whereas spring burning reduced residue by 55% (Huggins et al. 2011). The difference between fall and spring burning is a result of winter wheat residue reduction (by 36%) from decomposition and/or mixing with soil by biota between fall and spring. Spring burn resulted in greater residue N loss (40%) compared with fall burn (33%). Losses of K, P, and S from fall burning averaged 70%, 37%, 57%, respectively, and from spring burning averaged 56%, 39%, 45%, respectively (Table 4-10).

Table 4-8. Fertilizer replacement cost for nutrient loss during fall and spring burning of winter wheat residue; research was conducted in Pullman, Washington, within the Annual Crop agroecological class. ●

Residue nutrient	Fertilizer cost†	Fall burn nutrient loss	Spring burn nutrient loss
	---- $/lb ----	------------- $/acre -------------	
N	0.45 (82-0-0)	5.27	4.95
K_2O	0.62 (0-0-60)	20.40	2.83
P_2O_5	0.73 (10-34-0)	1.72	1.20
S	0.47 (12-0-0-26)	1.23	0.66
Nutrient replacement cost		28.62	9.64

†Average fertilizer prices for 2008–2010 from Idaho input cost publication series (***http://www.uidaho.edu/cals/idaho-agbiz***).
Source: Huggins et al. 2011.

Table 4-9. Nutrient content in harvested straw and nutrient content in one ton of harvested straw and ash from spring wheat, oats, and flax sampled in western Canada.

Nutrient	Nutrient content in straw			Nutrient content in one ton straw before/after burn					
	Spring wheat	Oats	Flax	Spring wheat	Oats	Flax	Spring wheat	Oats	Flax
	(%)			Pre-burn (lb)			Post-burn (lb)		
C	41	42	46	826	832	910	77	31	28
N	0.97	0.64	0.86	22	10	28	0.4	0.1	0.05
K	1.44	2.34	0.24	29	47	4.7	24	30	2.6
P	0.14	0.08	0.07	2.7	1.5	1.4	2.4	1.3	0.9
S	0.11	0.22	0.06	2.2	4.4	1.1	0.7	2.2	0.14

Adapted from Heard et al. 2006.

Table 4-10. Fall and spring burning effects on winter wheat residue loads and residue nutrient contents trials conducted in Pullman, Washington.

	% of nutrient in residue				Nutrient content			
	Fall burn		Spring burn		Fall burn		Spring burn	
	Pre-burn	Post-burn	Pre-burn	Post-burn	Pre-burn	Post-burn	Pre-burn	Post-burn
	(%)				lb/acre			
Residue					8093	3059	5168	2354
C	39.9	39.9	43.0	40.5	3228	1218	2226	955
N	0.44	0.78	0.52	0.69	35.9	24.2	27.3	16.3
K	0.46	0.39	0.13	0.13	38.8	11.47	6.90	3.07
P	0.03	0.06	0.04	0.05	2.82	1.79	1.85	1.13
S	0.06	0.07	0.05	0.06	4.56	1.95	2.39	1.32

Source: Huggins et al. 2011.

Conclusion

Crop residue is a nutrient and SOC source for soil health when returned to soil, a valuable feedstock when harvested, and a ground cover for protecting from soil erosion. Making decisions for best residue management practices is complicated because it requires balancing tradeoffs between agronomic productivity, soil health improvement, soil erosion and environmental protection, and economics. The review of benefits and tradeoffs of residue management practices and tools discussed in this chapter can be used to help growers make decisions that are best for their specific situations.

References

Agarwal, R., A. Awasthi, N. Singh, P.K. Gupta, and S.K. Mittal. 2012. Effects of Exposure to Rice-Crop Residue Burning Smoke on Pulmonary Functions and Oxygen Saturation Level of Human Beings in Patiala (India). *Science of the Total Environment* 429: 161–166.

Aguilar, J., R. Evans, M. Vigil, and C.S.T. Daughtry. 2012. Spectral Estimates of Crop Residue Cover and Density for Standing and Flat Wheat Stubble. *Agronomy Journal* 104: 271–279.

Baker, J.M., T.E., Ochsner, R.T., Venterea, and T.J. Griffis. 2007. Tillage and Soil Carbon Sequestration—What Do We Really Know? *Agriculture, Ecosystems & Environment* 118: 1–5.

Behl, R.K., and V.P. Singh. 1998. Combining Ability Analysis for Harvest Index and Related Characters in Triticale. *Crop Improvement* 15: 110–112.

Bremer, E., and C. van Kessel. 1992. Plant-Available Nitrogen from Lentil and Wheat Residues during a Subsequent Growing Season. *Soil Science Society American Journal* 56: 1155–1160.

Brown, T.T. 2015. Variable Rate Nitrogen and Seeding to Improve Nitrogen Use Efficiency. Ph.D. diss., Washington State University, Pullman, WA.

Campbell, C.A., B.G. McConkey, R.P. Zentner, F. Selles, and F.B. Dyck. 1992. Benefits of Wheat Stubble Strips for Conserving Snow in Southwestern Saskatchewan. *Journal of Soil and Water Conservation* 47: 112–115.

Campbell, C., and W. Souster. 1982. Loss of Organic Matter and Potentially Mineralization Nitrogen from Saskachewan Soils due to Cropping. *Canadian Journal of Soil Science* 62: 651–656.

Dickey, E.C., D.P. Shelton, and P.J. Jasa. 1986. Residue Management for Soil Erosion Control. Historical Materials from University of Nebraska-Lincoln Extension. University of Nebraska-Lincoln.

Donaldson, E., W.F. Schillinger, and S.M. Dofing. 2001, Straw Production and Grain Yield Relationships in Winter Wheat. *Crop Science* 41: 100–106.

Duft, K.D., and J. Pray. 2002. The Prospects for an Electrical Generation and Transmission Cooperative Fueled by Straw Produced in Eastern Washington. Farm Business Management Reports. EB1946E. Washington State University Cooperative Extension.

Eck, K.J., D.E. Brown, and A.B. Brown. 2001. Managing Crop Residue with Farm Machinery. Agronomy Guide. Purdue University Cooperative Extension Service. ***https://www.agry.purdue.edu/ext/pubs/AY-280-W.pdf.***

Elbert, C.D., P.W. Harlan, and D. Vokal. 1981. Crop Residue Management for Water Erosion Control. Biological Systems Engineering: Papers and Publications. Paper 244.

Engel, R.E., D.S. Long, and G.R. Carlson. 2003. Predicting Straw Yield of Hard Red Spring Wheat. *Agronomy Journal* 95: 1454–1460.

Fasching, R.A. 2001. Burning-Effects on Soil Quality. Agronomy Technical Note No. 150.16. USDA-NRCS, October.

Fryrear, D.W. 1985. Soil Cover and Wind Erosion. Transactions of the ASAE 28: 781–784.

Gollany, H.T., A.M. Fortuna, M.K. Samuel, F.L. Young, W.L. Pan, and M. Pecharko. 2013. Soil Organic Carbon Accretion vs. Sequestration Using Physicochemical Fractionation and CQESTR Simulation. *Soil Science Society of America Journal* 77(2): 618.

Gollany H. T., R.W. Rickman, Y. Liang, S.L. Albrecht, S. Machado, and S. Kang. 2011. Predicting Agricultural Management Influence on Long-Term Soil Organic Carbon Dynamics: Implications for Biofuel Production. *Agronomy Journal* 103: 234–246.

Havlin, J.L., J.D. Beaton, S.L. Tisdale, and W.L. Nelson. 2005. Soil Fertility and Fertilizers: An Introduction to Nutrient Management. Pearson Education, Inc.: New Jersey.

Hatfield, J.L., T.J. Sauer, and J.H. Prueger. 2001. Managing Soils to Achieve Greater Water Use Efficiency: A Review. *Agronomy Journal* 93: 271–280.

Holmgren, L., G. Cardon, and C. Hill. 2014. Economic and Soil Quality Impacts from Crop/Rangeland Residue Burning. ***http://extension.usu.edu/files/publications/publication/AG_Forages_2014-02pr.pdf***

Huggins, D.R., C.E. Kruger, K.M. Painter, and D.P. Uberuaga. 2014. Site-Specific Trade-Offs of Harvesting Cereal Residues as Biofuel Feedstocks in Dryland Annual Cropping Systems of the Pacific Northwest, USA. *BioEnergy Research* 7(2): 598–608.

Huggins, D.R., and W.L. Pan. 1991. Wheat Stubble Management Affects Growth, Survival, and Yield of Winter Grain Legumes. *Soil Science Society of America Journal* 55(3): 823–829.

Huggins, D., T. Paulitz, K. Painter, G. Birkhauser, R. Davis, and D. Uberuaga. 2011. Straw Management and Crop Rotation Alternatives to Burning Wheat Stubble: Assessing Economic and Environmental Trade-Offs. Final Project Report to Agricultural Burning Practices and Research Task Force.

Jacobs, A., R. Rauber, and B. Ludwig. 2009. Impact of Reduced Tillage on Carbon and Nitrogen Storage of Two Haplic Luvisols after 40 Years. *Soil & Tillage Research* 102: 158–164.

Jain, N., A. Bhatia, and H. Pathak. 2014. Emission of Air Pollutants from Crop Residue Burning in India. *Aerosol and Air Quality Research* 14: 422–430.

Koenig, R.T., W.A. Hammac, and W.L. Pan. 2011. Canola Growth, Development, and Fertility. Washington State University Extension Publication FS045E.

Laird, D.A., and C.W. Chang. 2013. Long-Term Impacts of Residue Harvesting on Soil Quality. *Soil & Tillage Research* 134: 33–40.

Lal, R. 2005. World Crop Residues Production and Implications of Its Use as a Biofuel. *Environmental International* 31: 575–584.

Li, C., L. Yue, Z. Kou, Z. Zhang, J. Wang, and C. Cao. 2012. Short-Term Effects of Conservation Management Practices on Soil Labile Organic Carbon Fractions Under a Rape-Rice Rotation in Central China. *Soil & Tillage Research* 119: 31–37.

Long, D.S., and J.D. McCallum. 2013. Mapping Straw Yield Using On-Combine Light Detection and Ranging (Lidar). *International Journal of Remote Sensing* 34(17): 6121–6134.

Lyon, D.J., D.R. Huggins, and J.F. Spring. 2016. Windrow Burning Eliminates Intalian Ryegrass (*Lolium perenne* ssp. *Multiflorum*) Seed Viability. *Weed Technology* 30: 279-283.

Machado, S. 2011. Soil Organic Carbon Dynamics in the Pendleton Long-Term Experiments: Implications for Biofuel Production in Pacific Northwest. *Agronomy Journal* 103: 253–260.

McClellan, R.C., T.L. Nelson, and M.A. Sporcic. 1987. Measurements of Residue to Grain Ratio and Relative Amounts of Straw, Chaff, Awns and Grain Yield of Wheat and Barley Varieties Common to Eastern Washington. In Proceedings of the 10th Annual STEEP Conference, ed. L.F. Elliott. Spokane, Washington, May 20-21, 1987. *STEEP-Conservation Concepts and Accomplishments* 617–624. Pullman, WA: Washington State University Press.

McClellan, R.C., D.K. McCool, and R.W. Rickman. 2012. Grain Yield and Biomass Relationship for Crops in the Inland Pacific Northwest United States. *Journal of Soil and Water Conservation* 67: 42–50.

McCool, D.K., C.D. Pannkuk, A.C. Kennedy, and P.S. Fletcher. 2008. Effects of Burn/Low-Till on Erosion and Soil Quality. *Soil & Tillage Research* 101: 2–9.

McMaster, G.S., R.M. Aiken, and D.C. Nielsen. 2000. Optimizing Wheat Harvest Cutting Height for Harvest Efficiency and Soil and Water Conservation. *Agronomy Journal* 92 (6): 1104–1108.

Nielsen, D.C. 2013. Snow Catch and Soil Water Recharge in Standing Sunflower Residue. *Journal of Production Agriculture* 11(4): 476–480.

Papendick, R.I., and W.C. Moldenhauer. 1995. Crop Residue Management to Reduce Erosion and Improve Soil Quality-Northwest. Conservation Research Report Number 40. USDA-ARS.

Perlack, R.D., L.L. Wright, A.F. Turhollow, R.L. Graham, B.J. Stokes, and D.C. Erbach. 2005. Biomass as Feedstock for a Bioenergy and Bioproducts Industry: The Technical Feasibility of a Billion-Ton Annual Supply. US Department of Energy and USDA.

Peterson, G.A., A.J. Schiegel, D.L. Tanaka, and O.R. Jones. 1996. Precipitation Use Efficiency as Affected by Cropping and Tillage Systems. *Journal of Production Agriculture* 9: 180–186.

Port, L.E. 2016. A Diversified High-Residue No-Till Cropping System for the Low-Rainfall Zone. M.S. Thesis. Washington State University.

Qiu, H., D.R. Huggins, J.Q. Wu, M.E. Barber, D.K. McCool, and S. Dun. 2011. Residue Management Impacts on Field-Scale Snow Distribution and Soil Water Storage. *American Society of Agricultural and Biological Engineers* 54: 1639–1647.

Ramig, R.E., R.R. Allmaras, and R.I. Papendick. 1983. Water Conservation: Pacific Northwest. Dryland Agriculture-Agronomy Monograph No. 23. ASA-CSSA-SSSA, Madison, WI.

Ramig, R.E., and L.C. Ekin. 1984. Effect of Stubble Management in Wheat-Fallow Rotation on Water Conservation and Storage in Eastern Oregon. Columbia Basin Agricultural Research, Special Rep. 713. Oregon State University, Corvallis, OR.

Rasmussen, P.E., R.R. Allmaras, C.R. Rohde, and N.C. Roager, Jr. 1980. Crop Residue Influences on Soil Carbon and Nitrogen in a Wheat-Fallow System. *Soil Science Society America Journal* 44: 596–600.

Rasmussen, P.E., S.L. Albrecht, and R.W. Smiley. 1998. Soil C and N Change under Tillage and Cropping Systems in Semi-Arid Pacific Northwest Agriculture. *Soil & Tillage Research* 47: 197–205.

Rasmussen, P.E., and W.J. Parton. 1994. Long-Term Effects of Residue Management in Wheat-Fallow: I. Inputs, Yield, and Soil Organic Matter. *Soil Science Society America Journal* 58: 523–530.

Schillinger, W.F., R.J. Cook, and R.I. Papendick. 1999. Increased Dryland Cropping Intensity with No-Till Barley. *Agronomy Journal* 91: 744–752.

Schillinger, W.F., R.I. Papendick, and D.K. McCool. 2010. Soil and Water Challenges for Pacific Northwest Agriculture. Soil and Water Conservation Advances in the United States. SSSA Special Publication 60. 47–79.

Smil, V. 1999. Crop Residues: Agriculture's Largest Harvest. *BioScience* 49(4): 299–308.

Soon, Y.K., and M.A. Arshad. 2002. Comparison of the Decomposition and N and P Mineralization of Canola, Pea and Wheat Residues. *Biology and Fertility of Soils* 36: 10–17.

Stubbs, T.L., A.C. Kennedy, P.E. Reisenauer, and J.W. Burns. 2009. Chemical Composition of Residue from Cereal Crops and Cultivars in Dryland Ecosystems. *Agronomy Journal* 101: 538–545.

Thorne, M.E., F.L. Young, W.L. Pan, R. Bafus, and J.R. Alldredge. 2003. No-Till Spring Cereal Cropping Systems Reduce Wind Erosion Susceptibility in the Wheat/Fallow Region of the Pacific Northwest. *Journal of Soil and Water Conservation* 58(5): 250–257.

Turmel, M., A. Speratti, F. Baudron, N. Verhulst, and B. Govaerts. 2015. Crop Residue Management and Soil Health: A Systems Analysis. *Agricultural Systems* 134: 6–16.

Williams, J.D., and S.B. Wuest. 2014. Soil and Water Dynamics in Continuous Winter Wheat in the Semiarid Pacific Northwest, USA. *Soil Science Society of America Journal* 78: 571–578.

Williams, J.D., S.B. Wuest, and D.S. Long. 2014. Soil and Water Conservation in the Pacific Northwest through No-Tillage and Intensified Crop Rotations. *Journal of Soil and Water Conservation* 69: 495–504.

USDA-NRCS. 2005. Conservation Practices that Save: Crop Residue Management. ***http://www.nrcs.usda.gov/wps/portal/nrcs/detailfull/national/energy/conservation/?cid=nrcs143_023637***.

USDA-NRCS. 2008. CORE4 Conservation Practices Training Guide: The Common Sense Approach to Natural Resource Conservation. ***http://www.nrcs.usda.gov/Internet/FSE_DOCUMENTS/nrcs143_025540.pdf***

Wuest, S.B., T.C. Caesar-TonThat, S.F. Wright, and J.D. Williams. 2005. Organic Matter Addition, N, and Residue Burning Effects on Infiltration, Biological, and Physical Properties of an Insensitively Tilled Silt-Loam Soil. *Soil and Tillage Research* 84: 154–167.

Wuest, S.B., and W.F. Schillinger. 2011. Evaporation from High Residue No-Till versus Tilled Fallow in a Dry Summer Climate. *Soil Science Society of America Journal* 75: 1513–1519.

Yorgey, G.G., K. Borrelli, K.M. Painter, and H. Davis. In press. Stripper Header and Direct Seeding, Ron and Andy Juris: Ron and Andy Juris (Farmer to Farmer Case Study Series). Pacific Northwest Extension Publication. Washington State University.

Young, F.L., D.A. Ball, D.C. Thill, J.R. Alldredge, A.G. Ogg, Jr., and S.S. Seefeldt. 2010. Integrated Weed Management Systems Identified for Jointed Goatgrass (*Aegilops cylindrica*) in the Pacific Northwest. *Weed Technology* 24(4): 430–439.

Young, F.L., A.G. Ogg, Jr., and P.A. Dotray. 1990. Effect of Postharvest Field Burning on Jointed Goatgrass (*Aegilops cylindrica*) Germination. *Weed Technology* 4(1): 123–127.

Young, F.L., J.R. Alldredge, W.L. Pan, and C. Hennings. 2015. Comparisons of Annual No-Till Spring Cereal Cropping Systems in the Pacific Northwest. *Crop, Forage & Turfgrass Management*.

Chapter 5

Rotational Diversification and Intensification

Elizabeth Kirby, Washington State University
William Pan, Washington State University
David Huggins, USDA-ARS and Washington State University
Kathleen Painter, University of Idaho
Prakriti Bista, Oregon State University

Abstract

Diversification and intensification of inland Pacific Northwest (PNW) dryland cereal cropping systems can present win-win scenarios that deliver short and long-term benefits for producers and the environment, stabilizing profit and increasing adaptability to and mitigation of climate change. Improving diversity, or reducing fallow, can enhance current farm productivity and income levels, pest management, soil structure, and water infiltration. Alternating oilseeds and grain legumes, in rotation with cereals, can reduce greenhouse gas emissions and improve nitrogen cycling; replacing fallow with crops can increase straw residues and the potential for carbon sequestration. Growers seek reliable, site-specific information on the management and potential of alternative cash crops and cover crops. Recent studies help to interpret the agronomic and economic feasibility of alternative cropping systems as well as understanding their role in potential climate change adaptation and mitigation.

Research results are coded by agroecological class, defined in the glossary, as follows:

● Annual Crop ▲ Annual Crop-Fallow Transition ■ Grain-Fallow

Key Points

- Rotational diversity is low in the wheat-dominated cereal production systems of the inland PNW. Diversifying or intensifying cropping systems helps producers minimize lost production opportunities, improve farm productivity, increase grower income and flexibility, adapt to predicted climate change, and achieve long-term environmental benefits.
- Adopting alternative rotations comes with tradeoffs and can increase risk. Success is dependent on geographic location, production potential, rotational fit, market opportunity, crop price, and production costs.
- Broadleaf crop sequences can improve cereal pest management, nutrient cycling, or soil structure. For example, legumes can reduce nitrogen fertilizer costs and greenhouse gas emissions and canola can increase water infiltration, break up hardpans, disrupt weed and disease cycles, and access water and nutrients deep in the soil.
- Adopting improved fallow practices can be an important step toward building soil resiliency and increasing future opportunity for diversification and intensification in tilled grain-fallow systems.
- Rotational benefit to wheat yield should be accounted for when evaluating potential returns for alternate crop rotations.

Introduction

Wheat has been the dominant crop in the **inland PNW** dryland region since land was first broken out of native bunchgrass and sagebrush. Cool season small grain cereals are well-suited to the region and the development and adoption of locally adapted, semi-dwarf varieties along with access to chemical fertilizers and pesticides have made it possible to grow wheat profitably for long periods. However, intensive tillage and fallow-based production have contributed to degraded **soil health** and declining productivity. Growers are increasingly interested in rotational diversification with alternate crops and intensification strategies such as fallow replacement, increased cropping with alternate winter crops, and **cover cropping**. These strategies target improved long-term productivity and more flexible adaptation to ongoing and predicted climate change.

Acronyms Used in Crop Rotations

AWP – Austrian winter pea
ChF – chemical fallow
CP – chickpea
F – fallow
FB – facultative barley
FW – facultative wheat
HRSW – hard red spring wheat
HRWW – hard red winter wheat
L – spring lentil
P – spring pea
SAF – safflower
SB – spring barley
SC – spring canola
SW – spring wheat
SWSW – soft white spring wheat
SWWW – soft white winter wheat
UTF – undercutter tillage fallow
WC – winter canola
WL – winter lentil
WP – winter pea
WT – winter triticale
WW – winter wheat

The region's climate, topography, and soils are highly diverse, yet crop diversity remains low, particularly in the driest areas where winter wheat-fallow has been the most profitable rotation. Cropping systems that lack diversity are more vulnerable to changes in commodity prices, production costs, and weather and climate. Strategies that increase diversity and intensification can provide growers greater flexibility, productivity, and income stability. Diversification is useful to break pest cycles, broaden pest management options, and manage herbicide resistance; crop sequences can be managed to increase carbon sequestration, reduce petroleum use, and mitigate greenhouse gas emissions. A broader selection of economically viable crops could advance the adoption of **no-till** cropping and decrease production costs, soil erosion, and degradation (Huggins and Reganold 2008; Kirkegaard et al. 2008a; Long et al. 2016; Pan et al. 2016).

Alternative crops have been evaluated in the inland PNW for more than 100 years (Figure 5-A1), but adoption has been limited by agronomic and economic challenges and government policy (Guy and Karow 2009; Kephart et al. 1990; Machado et al. 2004). Regional climate change, evolving markets, and more supportive government policies are motivating producers to further explore alternate crop options and maximize production opportunities. Heavy reliance on the volatile wheat market jeopardizes short-term profitability when wheat prices decline. Economic vulnerability is particularly evident in the drier regions where monocrop wheat systems dominate. Climate models predict warmer, drier summers, highlighting the need for flexibility and adaptation to increased temperature and drought stressors. Recent changes in federal farm support programs encourage more crop diversification; Farm Bill provisions encourage increased consumption of pulse crops and reduction of the farm energy footprint, providing incentives to produce grain legumes. Domestic consumption of dry pea, lentil, and chickpea (hummus) has increased from less than 0.5 pound to more than one pound per person since the early 1980s. Revisions to the US Standards for Whole Dry Peas and Split Peas and development of food quality varieties are broadening opportunities for growers to adopt winter peas. New federal insurance policies also help limit risk to producers as they develop alternative systems. In addition, the Energy Independence and Security Act of 2007 aimed to increase biofuel use and reduce petroleum consumption and greenhouse gas emissions has supported infrastructure and expanded opportunities for oilseed production in the inland PNW.

Recent research and grower efforts have focused on developing viable diversification strategies to further integrate oilseeds, grain legumes, alternate cereals, and cover crops into inland PNW cropping systems to improve agronomic, economic, and environmental performance, and to better understand their roles in adaptation to, or **mitigation** of, regional climate change. Producers face complex management decisions and assessment of potential benefits and tradeoffs. Greater crop choice would allow growers to better respond to commodity-driven opportunities and to plan crop choice and sequence in order to benefit wheat yield, **water use efficiency**, and nutrient cycling; enhance soil quality and residue management; or spread field workload. Successful diversification strategies will have a good agronomic fit, and meet short- and long-term economic and environmental goals. Crops must be adapted to local conditions, able to perform consistently, and not require extensive equipment modifications. In order to optimize these new crop rotations, alterations in other agronomic practices such as planting, soil and nutrient management, and harvesting techniques require re-evaluation.

Agroecological Class Characteristics, Production Challenges, and Adaptive Strategies

Three **agroecological classes** (AECs) are defined for the inland PNW dryland cereal production region based on the dominant cropping system and percentage of area in **fallow**: (1) Annual Crop (<10% fallow), (2) Annual Crop-Fallow Transition (10–40% fallow), and (3) Grain-Fallow (>40% fallow). Yield potential, limited by biogeographical factors and crop markets, determine the dominant cropping system and relative opportunities for diversification and intensification within each AEC (Huggins et al. 2015; Kaur et al. 2015). Biogeographical factors include climate (e.g., precipitation, potential evapotranspiration, temperature), soil characteristics (e.g., depth, texture, organic matter, soil water recharge, and **water holding capacity**), and topography. Cold, wet winters and warm to hot, dry summers are typical across the inland PNW. Areas with greater than 16" mean annual precipitation (MAP) can typically support annual cropping, whereas producers in drier areas rely on summer fallow practices for adequate recharge of soil water to support a subsequent crop. A map of the AECs and a description of the regional diversity, climate patterns, and precipitation and agronomic zone classifications are found in Chapter 1: Climate Considerations.

USDA National Agricultural Statistics Service (NASS) cropland layer data illustrate low crop diversity across the dryland AECs, discussed individually in the following chapter sections. The area of the Annual Crop AEC averaged 1.44 million acres from 2007 to 2014 compared to the Annual Crop-Fallow Transition and Grain-Fallow AECs with 1.85 and 2.52 million acres, respectively (Table 5-1). Predictably, crop diversity was greatest in the Annual Crop AEC and lowest in the Grain-Fallow AEC. Whereas opportunity for diversification varies by AEC, the fractions of crop area in winter wheat (40% to 45%) were fairly stable across the classes, indicating that growers make crop choices based on commodity opportunity rather than following set rotations. Crop diversity increased in all three AECs from 2007 to 2014. Spring pea and chickpea acreage increased in the Annual Crop and Transition AECs; lentil acreage declined from 2010 to 2014; the 2011 acreage drop-off represents lost cropping opportunities from excessively wet conditions that prevented

Table 5-1. Average percentage of crop and fallow area by agroecological class (2007–2014).

	Annual Crop ●	Transition ▲	Grain-Fallow ■	Total
Average million acres	1.44	1.85	2.52	5.81
Average % crop and fallow area				**Crop acres**
Fallow	3.2	27.6	50.1	1,822,796
Winter wheat	40.8	40.7	45.1	2,478,807
Spring wheat	16.6	15.8	2.8	599,657
Spring barley	5.5	4.4	0.4	169,884
Chickpea	6.8	0.6	0.03	109,115
Lentil	6.1	0.3	0	92,031
Pea	5.4	1.7	0.07	110,752
Canola	0.9	0.4	0.2	26,966
Alfalfa	5.4	5.3	0.8	197,614

Source: Unpublished values (Huggins pers. comm.) were compiled using NASS cropland layer data (2007–2015).

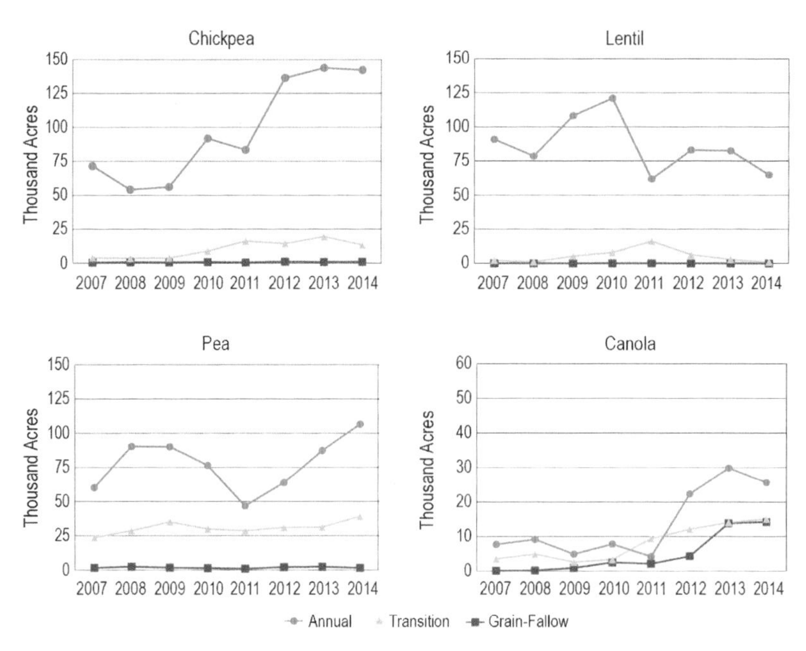

Figure 5-1. Annual grain legume and canola acreage trends by AEC (2007-2014). Unpublished values (Huggins pers. comm.) were compiled using NASS cropland layer data (2007-2015)

Table 5-2. Alternative crop sequences to improve rotational diversity by agroecological class.

	Traditional Sequences	Alternate Sequences
Annual Crop ●	WW-(SW or SB)- (P, L, or CP); WW-SW-WW-(P, L, or CP);	WW-SW-SC; WW-SW- (WP or WL); WW-SC-(P, L, or CP) back-to-back broadleaf
Transition ▲	WW-SW-F; WW-F	WW-SW-(WP or WL); WW-F-WC; WW-F-SC; WW-SW-SC; WW-SB-F
Grain-Fallow ■	WW-F	WW-F-WP; WW-F-WC; WW-F-WC-F; WW-F-WT-F

See Acronyms Used in Crop Rotations sidebar for abbreviation definitions.

Table 5-3. Diversification strategies by agroecological class.

	Traditional Sequences	Alternate Sequences
Annual Crop ●	• High productivity; heavy residue load • Steep slopes and erosion (water) • Reduced tillage • Persistent winter annual grass weeds • Cold wet springs (delayed or prevented planting)	• No-till • Winter legumes • Spring canola and other oilseeds • Herbicide-resistant canola • Cover crop • Perennial crops
Transition ▲	• Moderate productivity • Erosion (wind and water) • Deficient seed zone moisture • Reduced tillage • Persistent winter annual grass weeds • Areas of shallow soils	• No-till or improved fallow practices (e.g., tall cereals, stripper header, undercutter method) • Diversify • Flex crop (intensification) • Cover crop

Grain-Fallow
■
	• Poor soil health; low productivity and residue
	• Reliance on fallow
	• Erosion (wind); fine, poorly aggregated soils
	• Deficient seed zone moisture
	• Intensive tillage
	• Persistent winter annual grass weeds
	• Marginal profitability
		• Diversify winter wheat phase
		  - Winter triticale or barley, pea or lentil, canola
		  - Facultative wheat or barley
		• Flex crop with adequate moisture
		  - Cereal or broadleaf
		• Improved fallow practices

planting spring crops. Canola acreage increased in all the dryland AECs (Figure 5-1).

Traditional and alternative rotations by AEC are listed in Table 5-2, and Table 5-3 summarizes production issues and adaptive strategies.

Annual Crop AEC ●

From 2007–2014, an average 63% of the Annual Crop AEC acreage was planted in small grain cereals including winter wheat (41%), spring wheat (17%), and spring barley (5%). Grain legumes (pea, lentil, and chickpea) accounted for 18% of the area, canola had nearly 1%, and just 3% of the area was in fallow (Table 5-1). Acreage of annual broadleaf crops increased nearly 50% during this period; chickpea acreage doubled to more than 140,000 acres, whereas canola acreage tripled, ranging from 22,000 to nearly 35,000 acres in 2012, 2013, and 2014 (Figure 5-1).

The Annual Crop AEC generally has sufficient available water to support continuous cropping and is characterized by high productivity and heavy post-harvest residue. Deep, silt loam soils can have up to 3–4% soil organic matter and 2.2–2.4 in/ft soil water holding capacity. Steep topography and winter precipitation make this region vulnerable to runoff and high rates of erosion. Exposed subsoil is common on hilltops and bare knobs, where productivity has been degraded. Improved wheat varieties and increased chemical inputs have helped maintain high yields in the region, and adoption of reduced tillage or no-till, including **direct seeding**, has helped to slow erosion and loss of soil organic matter (Douglas et al. 1999; Douglas et al. 1992; Schillinger et al. 2003; Schillinger and Papendick 2009). Wet, cold spring conditions can delay or even prevent planting; excessive residue keeps soils cool and wet, can hinder direct-seed practices, and favors soilborne pathogens. Annual grass weeds are a severe problem and can reduce yields by nearly half.

Growers commonly use 3- or 4-year crop sequences such as winter wheat-spring grain (wheat or barley)-spring legume and winter wheat-spring grain-winter wheat-spring broadleaf (legume or oilseed), shown in Table 5-2. Potential adaptive strategies include rotational diversification with no-till or reduced-till spring canola, winter peas or lentils, cover crops, and increased perennial plantings (Tables 5-2 and 5-3). Some growers are trying 4-year rotations including two consecutive years of broadleaf crops (e.g., WW-SC-P, L, or CP) to enhance weed management options. More specific information on integrating adaptive strategies is found in later sections of this chapter. Information on **conservation tillage** is presented in Chapter 3: Conservation Tillage Systems, and weed management is discussed in Chapter 9: Integrated Weed Management.

Annual Crop-Fallow Transition AEC ▲

The average fractions of winter wheat (41%), spring wheat (16%), and spring barley (4%) in the Transition AEC were very similar to the Annual Crop AEC, whereas nearly 28% of the transition acreage was in fallow and just 3% in broadleaf crops from 2007 to 2014 (Table 5-1). Crop diversity noticeably increased during this period. Total broadleaf crop acreage doubled; peas increased 65%, to more than 39,000 acres in 2014. Beginning in 2011, chickpea and canola acreage grew rapidly, to more than 13,500 and 15,000 acres, respectively, in 2014 (Figure 5-1).

Much of the Transition AEC is cropped in the 2-year WW-F sequence, shifting to an intensified 3-year rotation, typically winter wheat-spring grain–fallow (WW-SW-F or WW-SB-F) or, less commonly, winter wheat-spring broadleaf-fallow in areas with sufficient moisture (Table 5-2). The transition region has more variable moisture conditions and lower overall productivity compared to the Annual Crop AEC. Crop choice is limited by available water; winter wheat has been the most reliable crop. Soils generally have lower soil water holding capacity (1.8–2.2 in/ft) and soil organic matter (2–3%). The topography includes steep to gentle slopes that are susceptible to erosion by water or wind. Pockets of shallow soils (<40") that do not benefit from precipitation storage during a fallow period are generally cropped annually. These areas have low annual yield potential (Douglas et al. 1999). Tilled fallow is common, especially in the drier areas, to maintain seed zone moisture and control weeds, but most growers in the Transition AEC have adopted some form of reduced tillage, or no-till (Douglas et al. 1992; Schillinger et al. 2003; Schillinger and Papendick 2009).

Both diversification and intensification strategies have potential in the Transition AEC (Table 5-3). No-till, **flex cropping**, and practices such as integrating tall cereals, harvesting with a stripper header, or undercutter tillage fallow increase the potential for intensification of traditional rotations and limit missed production opportunities. Flex cropping when conditions are favorable (good market price, low weed pressure, adequate moisture), can increase carbon sequestration and enhance soil organic matter (Lutcher et al. 2013). A few highly innovative producers have had success with 4- and 5-year crop sequences integrating no-till winter and spring cereals, spring pea, winter pea, canola, and camelina with 12–14" precipitation, and with direct seed flex cropping systems on shallow soils (2–3 feet deep) in a traditional winter wheat-fallow area with 12" average precipitation (Yorgey et al. 2016a; 2016b). Warm season crops (corn, safflower, sunflower, and proso millet) have had limited success due to high water demand, inadequate heat units, highly variable yields, and limited market access (Schillinger et al. 2003).

Grain-Fallow AEC ■

The Grain-Fallow AEC is the largest dryland production area in the inland PNW. This region is characterized by poor soil health, drought, high pest pressure, and low grain and residue productivity. Diversity is very low and opportunities to diversify or to intensify production are limited. The traditional crop sequence is the 2-year WW-F sequence with an average 50% of the acreage in fallow, annually. From 2007 to 2014, 45% of acreage was planted to winter wheat. Just over 3% of the area was in high-yielding spring wheats and barley with small areas (less than 1%) of canola, dry pea, and chickpea (Tables 5-1, 5-2, and 5-3). Both chickpea and canola acreage grew rapidly. Chickpea acreage doubled to more than 1,000 acres from 2007 to 2014, and canola acreage expanded from less than 200 acres in 2007 to around 14,000 in 2013 and 2014 (Figure 5-1). However, these values still represent a tiny fraction of the Grain-Fallow AEC.

Producers are seeking more sustainable alternatives to the intensively tilled fallow system, which exposes soil to erosion, degrades soil organic matter, and represents missed production opportunities. Moisture is generally insufficient to support profitable annual cropping, and growers rely on fallow practices to store and retain winter precipitation in the soil profile, maintain seed zone moisture to establish winter wheat, mineralize soil nitrogen (N), and stabilize yield and profitability. Blowing dust continues to be a severe environmental issue. Poorly **aggregated** soils with relatively low soil water holding capacity (1.6–2 in/ft) and low organic matter (<1.5%) are extremely susceptible to high rates of wind erosion.

Annual cropping of no-till spring grains has not proven economical to date, but there are opportunities to diversify the winter wheat phase of the rotation with winter triticale, winter peas, and winter canola, and to intensify rotations using flex cropping, depending on yield potential and commodity prices. The adoption of no-till fallow is limited by excess soil water evaporation from the seed zone compared to **conventional tillage**; inadequate seed zone moisture at optimal planting dates can delay seeding and reduce yields. Reduced-till fallow using undercutter tillage shows promise to successfully control weeds, reduce erosion potential, and retain seed zone moisture (Huggins et al. 2015; Schillinger et al. 2003; Schillinger et al. 2010; Young et al. 2015).

Integrating Diversification Strategies: Grower Considerations and Supporting Research

The following section presents considerations for integrating alternate crops, cover crops, or flex cropping, such as rotational fit, stand establishment, weed and N management, and the effect of an alternate crop on subsequent wheat yield. Potential for alternate crop adaptation to the dryland AECs, typical grain and residue yields, N requirements, and water use are compared in Appendix Tables 5-A1, 5-A2, and 5-A3.

Integrating Grain Legumes

Rotational fit

Diversifying with grain legumes has an important role in cereal production systems, providing short- and long-term benefits. Short-term benefits include (1) biological N fixation which improves soil fertility and reduces reliance on N inputs, (2) options for controlling grass weeds that are persistent in annual cereal systems, (3) reduced disease and pest pressure, (4) moderate water use conserves soil water reserves for subsequent crops, and (5) ability to flex crop or plant an opportunity crop (Chen et al. 2006; McPhee and Muehlbauer 2005). Improved soil tilth and reduced greenhouse gas emissions associated with N fertilizer production and reactive soil N are examples of long-term benefits. European studies found that arable cropping systems with legumes reduced nitrous oxide emissions and N fertilizer use 18% and 24%, respectively, compared to systems with no legumes, mitigating climate change and saving growers money (Reckling et al. 2016). Grain legumes efficiently utilize residual soil nitrates, reducing the potential for N loss by **leaching** (Mahler 2005a; Muehlbauer and Rhoades 2016) and wheat yields following pea or lentil can be 10–20% greater than in a wheat-fallow rotation (Guy and Gareau 1998; Guy and Karow 2009). ● ▲

Dry peas and lentils have been produced in the inland PNW since the 1920s, primarily in the Annual Crop AEC, replacing a year of cereal or fallow. Low pea and lentil prices in the 1970s spurred interest in chickpea production, and the area is now the leading chickpea production region in the US.

Cool season peas, lentils, and chickpeas are well-adapted to the inland PNW cereal production system, yet performance and yield of spring-planted legumes can be limited by late-season drought and heat stress. Early planting can offset risk, but busy spring workloads or excessive wet and cool soil conditions can delay or even prevent spring field operations. Predicted climate change may intensify these limitations; increased late-winter and early-spring precipitation may make early-spring planting more difficult, and hotter, drier summers may increase drought and heat stress. Researchers found a strong relationship between dry pea yield and available soil moisture during the June-August period; pea yields were reduced 20% in years with below-average moisture (Abatzoglou and Eigenbrode 2016). The deeper rooting habit, drought tolerance, and lower susceptibility to high temperatures of chickpeas during flowering may be better adapted to future conditions than peas. Planting winter-hardy legumes can be useful for adapting to climate change with the advantages of increased yield and improved water use efficiency compared to spring-sown legumes or fallow; maximum crop growth of fall-seeded legumes occurs in early spring when evapotranspiration is low (Gan et al. 2015; Muehlbauer and McPhee 2007).

Successful integration of grain legumes depends on defining the best rotational fit, within each AEC, to enhance overall productivity of the cropping system. Direct seed or reduced tillage systems enhance stand establishment and help protect soils when integrating legume crops which produce less residue that decomposes more rapidly than cereals. Spring grain legumes are suited to the Annual Crop AEC where sufficient precipitation supports continuous cropping and higher productivity. In the Transition and Grain-Fallow AECs, spring legumes can be used to replant failed fall-sown crops such as winter canola, winter pea, or lentil, or planted as an opportunity or flex crop option to replace fallow. In low production areas where shallow soils have less total available water than deep soils, spring dry pea or lentils can provide an alternate crop with less water uptake and similar water use efficiency as spring wheat. In eastern Oregon studies, lentils outperformed other legumes in locations with less than 14" precipitation and have potential as an alternate crop; chickpea has potential to replace dry pea in a traditional WW-P rotation, but typically uses too much soil water to replace fallow in the WW-F rotation without reducing wheat yield (Machado et al. 2006a; 2006b). ▲ ■

Recent releases of high-yielding winter cultivars provide producers a viable alternative to integrate or increase legumes in their rotations and

Why Winter Pea Can Work in Conservation Systems

1. Excellent rotation crop for winter wheat
2. Viable economic potential
    a. High yield potential
    b. Reliable market
    c. Diversified farm income
3. Planting and emergence flexibility
4. Improved winter survival
5. N fixation – low fertilizer input
6. Wide adaptation across precipitation zones
7. Low soil acidification during winter pea sequence
8. Good water use efficiency
9. Residue is easily managed

Source: Guy 2016

avoid many of the challenges associated with spring direct-seed planting. Fall-planted winter peas can be adapted across the region and provide needed diversity in the Transition and Grain-Fallow AECs (see the Why Winter Pea Can Work in Conservation Systems sidebar). Fall-sown winter cultivars offer many advantages over spring pulses: (1) 30–50% or more greater yield, (2) improved water use efficiency, (3) better weed competitiveness, (4) earlier maturity, and (5) better protection against soil erosion with over-winter surface cover and higher biomass production (Chen et al. 2006; Kephart et al. 1990; McPhee and Muehlbauer 2005).

Austrian winter peas were first produced in the PNW in the 1930s, and grown for feed and green manures. Turkish red 'Morton' winter lentils were developed specifically for use in direct seed or reduced tillage systems and released in 2004. As interest in winter legumes has grown, PNW breeding and variety trial programs have broadened efforts to develop improved winter hardiness in food quality winter peas; winter peas for forage, feed, and cover cropping; and winter lentils. Chickpea and Austrian winter feed pea studies also continue (McGee et al. 2014; McPhee and Muehlbauer 2005). As locally adapted, food quality (non-pigmented), winter-hardy pea cultivars become more available, market opportunities will expand. Recent grade standard revisions allow producers to market food quality winter peas as smooth, dry, yellow, or green peas, similar to spring types, when size is adequate (Table 5-4). Food quality winter peas have a clear seed coat and hilum, white flowers, and high palatability compared to Austrian winter peas that have pigmented seed coats, purple flowers, slightly lower palatability due to tannin content, smaller seed size and, typically, longer vines. 'Lynx' peas have improved winter hardiness to –5°F and a clear seed coat with potential for the food market.

Winter pea yields are variable with location, variety, and crop year. McGee and McPhee (2012) reported winter pea yields at four locations from 2008–2011, averaged across four varieties ('Lynx,' 'Whistler,' 'Windham,' and 'Specter') and ranging from 817 lb/acre near Wilbur, Washington to just under 3,100 lb/acre near Pullman, Washington (data not shown), and Guy (2016) reported 1,810–3,840 lb/acre grain yield of 'Windham' peas at locations in Transition and Grain-Fallow AECs (Table 5-5; Figure 5-2). These values represent yields that a grower may expect to achieve. Austrian winter pea acreage grew rapidly during the 2011–2015 period, most of which is likely located in the Annual

Table 5-4. Winter pea characteristics.

Variety	Class	Vine Type	Winter Hardiness	Height, Max (in)	Grams/ 100 Seeds	Seed Coat	Flower Color	Tannin-Free
'Lynx'	Green	Short-semi leafless	–5°F	35	15.4	Clear	White	Yes
'Windham'	Yellow	Short-semi leafless	0°F	29	14.5	Subtly Mottled	White	Yes
'Whistler'	Yellow	Short-semi leafless	+5°F	32	16.60	Mottled	White	Yes
'Specter'	Yellow	Tall-semi leafless	+5°F	40	13.5	Subtly Mottled	White	Yes
Austrian	Purple	Tall-normal	+10°F	65	12–14	Mottled	Purple	No

Sources: McGee et al. 2012; McPhee and Muehlbauer 2007.

Table 5-5. 'Windham' winter pea yield and returns in eastern Washington (2009). ▲ ■

Location	MAP[1] (in)	Acres	Yield (lb/ac)	$/lb	Gross $/ac
Ritzville ■	12	47	1,810	0.18	325
Waterville ▲	15	217	3,660	0.18	660
Sprague ▲	16	146	3,840	0.18	690

[1]Mean annual precipitation
Source: Guy 2016.

Figure 5-2. 'Windham' winter peas near Ritzville, Washington, in 2009. (Photo by Stephen Guy.)

Crop AEC (Table 5-6). Yields were similar across years; high yields in 2011 were likely a result of plentiful winter precipitation. Currently, NASS does not track yellow and green winter peas.

Small, red lentils have potential to substitute for fallow sequences in traditional WW-F rotations in the Transition AEC. Small reds have good yield potential and marketability, do not require N inputs, are adaptable to no-till systems, use less water than larger lentils, and have potential for **recrop** wheat. Early studies (1987–1988) near Davenport, Washington

(14" MAP), showed spring-planted yields over 1,200 lb/acre (data not shown). Small red spring lentils are less adapted to the Annual Crop AEC where large red or yellow lentils have higher yield potential (Veseth 1989).

High-value, food-quality peas may earn a 50% price premium over feed peas (McGee and McPhee 2012; McGee et al. 2014; McPhee and Muehlbauer 2007). However, at Wilbur, 2015 revenues for fall-planted peas for food were less than revenues for peas for cover crop seed market and for winter wheat; cover crop peas had a $7 per hundredweight premium over food market winter peas. Revenues were between $350–$400/acre for 'Windham' and 'Lynx' peas for cover, higher than for most other winter crops and spring grains (Nelson 2016; data not shown). Guy (2016) reported gross revenues for 'Windham' peas from $325 to $690/acre based on yield and location (Table 5-6).

Small, red 'Morton' winter lentils are well-adapted to the Annual AEC and flex crop conservation tillage systems. ● ▲ Muehlbauer and McPhee (2007) reported that fall-planted 'Morton' lentils had 108% greater yield than spring lentils (73% more than highest yielding spring lentils). Recent 'Morton' winter lentil average yields ranged from 2,065 to 5,195 lb/acre over multiple years (2009–2014), similar to 'Windham' winter pea (2,439 to 5,642 lb/acre) in USDA-ARS variety trials (Table 5-7).

Winter peas offer advantages over spring legumes: overwinter soil cover, greater yield potential, and earlier maturation than spring-planted peas, avoiding heat and water deficits that occur later in the growing season. Winter pea yields can more than double spring pea yields (Table 5-8). While there are many advantages to integrating winter legumes in crop rotations, significant agronomic challenges need to be addressed including (1) optimal sowing dates and rates, (2) improved winter survival, (3) control of late-emerging broadleaf weeds, and (4) sensitivity to sulfonylurea (SU) herbicide carryover.

Plant establishment

Early seeding of cool season spring grain legumes enables plants to flower and set pods prior to droughty, hot conditions. These crops benefit from seeding as soon as field work can be done, typically mid-March to mid-April, once soil temperatures reach 40°F. Pea and lentil yield potential

Table 5-6. PNW dryland Austrian winter pea production for Idaho, Washington, and Oregon.

Year	AWP	
	Acres	lb/ac
2015	17,000	1,441
2014	7,638	1,604
2013	10,707	1,598
2012	6,950	1,577
2011	5,800	1,723
5-year average	9,619	1,588

Source: Todd Scholz, USA Dry Pea and Lentil Council.

Table 5-7. Average yields of 'Morton' winter lentil and 'Windham' winter pea in the Annual Crop agroecological class. ●

Year	'Morton' Lentil ●	'Windham' Pea ●
	lb/ac	
2014	2,065	2,995
2013	3,362	1,784
2012	4,248	4,231
2011	5,195	5,642
2010	—	2,435
2009	2,592	2,521

Compiled from USDA-ARS winter legume breeding variety trial reports (2012–2015).

Table 5-8. Average yield of winter and spring peas by location, in eastern Washington (2014). ● ▲

Location	Winter Pea	Spring Pea
	lb/ac	
Garfield ▲	5,135	1,589
Pullman ●	4,159	2,116
Dayton ●	3,464	1,961

Source: McGee 2016

declines when planted in May. Chickpea is somewhat less sensitive to later planting dates and has improved germination with 45°F soil temperatures for some cultivars. However, in eastern Oregon studies, chickpea yield and bean quality were favored by planting as early as possible in March, and lentil performance was affected by annual precipitation, seeding rate, date, and location; seed zone moisture and weed pressure affected stand establishment and yield potential. Low seed zone moisture at planting (after winter wheat) resulted in poor winter lentil establishment and low yield (Machado et al. 2006a; 2006b).

Seeding rates and depth vary with seed size as determined by crop and cultivar. Seeding rates are targeted to 3–4.5 plant/ft^2 for chickpea, 8–10 plant/ft^2 for pea, and 10–12 plant/ft^2 for lentil. Row spacing of 6–7" with seeding depths of 1–3" for lentil, pea, and chickpea are routinely recommended. Relatively large seed size allows for deep planting peas, lentils, and chickpeas into moisture; peas and kabuli chickpeas can be planted 4" or more to moisture when needed. Guy and Lauver (2015) conducted spring seeding rate trials at Annual Crop and Transition AEC sites using 6–11 (pea), 7–12 (lentil), and 2–8 (chickpea) seed/ft^2 and found varied results. The lowest seeding rate for pea resulted in significantly lower yields; yields at higher rates were similar. The lentil seeding rate did not affect yield, and results were consistent enough to support a 10 seed/ft^2 planting recommendation. Chickpea yields indicated that seeding rates of 3–4 seed/ft^2 yielded better than 2 seed/ft^2 and there were incremental increases in yield with increased seeding rates of 3–8 seed/ft^2. At Moro (11" MAP) and Pendleton (16" MAP), Oregon, Machado et al. (2006a) compared lentil seeding rates of 20 and 10 seed/ft^2. The higher seeding rate resulted in more plants (6–7 plant/ft^2) and higher yields than the lower seeding rate. Narrow row spacing of legumes (6") helps control weeds while wider row spacing (12") can achieve the same results at low precipitation sites. Corp et al. (2004) found no yield differences for 6" or 12" row spacing in chickpea.

Fall-sown, winter-hardy pulse crops are well-adapted for direct seeding into standing cereal stubble and yield more than conventionally seeded pulses. Studies from the northern Plains and the PNW show that stubble enhances early growth and winter survival, reduces erosion and evaporation, and improves soil water recharge, storage, and water use efficiency. Stubble height does not appear to affect yield, and improves harvest of legume crops

(Chen et al. 2006; Cutforth et al. 2002; Huggins and Pan 1991; Muehlbauer and Rhoades 2016; Papendick and Miller 1977).

Recent studies indicate that there is a reasonably wide planting window for fall-seeded legumes. Optimal seeding dates are similar to those for winter wheat seeding (late August through October, depending on location). Timing should allow for adequate fall growth to support winter survival and early spring vigor; plants should reach the 2–3" tall rosette stage before winter dormancy (Guy pers. comm.; McPhee and Muehlbauer 2005). Chen et al. (2006) documented that winter pea and lentil cultivars, seeded into stubble, have greater yield potential than spring cultivars when planted at both 'early' and 'late' seeding dates in central Montana, although the earlier seeding can increase winter survivability and yield.

Establishing winter peas is easier than small-seeded oilseeds, which may be helpful if future hotter, drier, summers create more challenging conditions. Larger seed size allows for deeper planting in order to access seed zone moisture; winter pea cultivars such as 'Windham' can be deep-planted into moisture and emerge through 6" or more of soil (Guy 2016; Nelson 2016). Also, an extended planting window may allow fall-seeded legumes to benefit from early fall precipitation. Seeding rates vary from 30 lb seed/acre for winter lentil and up to 120 lb seed/acre for winter pea.

Weed management

Incorporating broadleaf crop sequences into cereal production systems provides opportunities for chemical control of grassy weeds such as downy brome, jointed goatgrass, wild oats, and feral cereal rye that persist in annual crop cereal systems and reduce yields in subsequent wheat crops. For more information on weeds and alternative rotations for management, see Chapter 9: Integrated Weed Management.

The PNW Weed Management Handbook and other sources provide guidelines for chemical weed control in grain legumes. Post-emergent herbicides are labeled for use in dry pea, chickpea, and lentil to control annual grasses. However, Italian ryegrass herbicide resistance has developed to the Group 1 post-emergent herbicides, which are the only post-emergent grass weed options registered for use in lentil, chickpea, and winter pea; control is useful only against non-resistant biotypes.

In-crop weed control is one of the most significant production challenges for grain legumes, particularly spring-sown crops. Few broadleaf herbicides are registered for use in lentil and chickpea. Fields with high broadleaf-weed seed infestation are not suitable for lentil and chickpea production. Pre-emergent herbicides have limited effectiveness, and there are no post-emergent chemical options for control of annual broadleaf weeds registered for use in chickpeas; metribuzin is registered for use in lentils, along with imazamox (only on Clearfield lentils). Additional materials and modes of action are available for dry pea, for both grass and broadleaf weeds.

Producers should avoid planting legumes in fields with heavy weed pressure; weeds can reduce grain legume yields 67% and complicate harvest operations (Campbell 2016). Short-statured lentils and chickpeas have slow initial growth and open canopy habits, and lack competitiveness against both early- and late-emerging weeds. Pea stands establish and close the canopy more quickly than lentils and chickpea; leafy pea cultivars are more competitive than semi-leafless types if they do not have a strong branching habit. When peas lodge, weeds can grow above the canopy. A wider number of herbicides are registered for use in peas compared to lentil and chickpea.

Cultural practices are not highly effective for weed control in pulses. Shallow seeding and earlier emergence can give seedlings a head start on weeds that emerge earlier than deep-planted pulses; however, this practice increases the potential for herbicide damage. Increased seeding rates have been shown to be only slightly beneficial for controlling weeds. Delayed seeding, to allow for mechanical or chemical control of early weeds, reduces yield potential resulting from increased temperatures and drought before maturity. Weedy stands may require application of a pre-harvest dessicant, increasing production costs. Inadequate weed control in grain legume sequences can impact subsequent crops from higher weed seed populations (Campbell 2016).

Nitrogen management

Grain legumes can improve soil N status, reduce N leaching, and improve the carbon footprint of cropping systems by replacing N fertilizer inputs; production of commercial N fertilizer accounts for a third of the total energy input to crop production. Pulse crops are able to utilize residual soil N, supplement their N needs by symbiotic fixation of atmospheric N, and leave surplus available soil N for use by a subsequent crop such as winter wheat (Bezdicek et al. 1989; Mahler 2005a; 2005b; Muehlbauer and Rhoades 2016; NRCS 2014). Whereas both the fallow system and legumes improve the soil N balance, fallowing releases N from the mineralization of soil organic matter and depletes the organic matter in the process. Canadian dryland studies showed that the benefits of summer fallow (stored precipitation and mineralized N) could be successfully replaced by diversifying with grain legumes. Also, an L-SW rotation had a 127% lower per-area carbon footprint than continuous SW and was 153% lower than F-SW-SW rotations; similarly, the L-SW rotation had the lowest per-grain carbon footprint (Gan et al. 2014; 2015).

Legume N management

Applying N fertilizer to legume crops is generally not economical (Mahler 2015; Muehlbauer et al. 1981). Peas, lentils, and chickpeas planted in soils with less than 20 lb N/acre may benefit from low rates (20–30 lb N/acre) to sustain seedlings until nodulation occurs and biological N fixation begins. High soil N may reduce biological N fixation, and fall-planted legumes with elevated N content and excess vegetative growth are more prone to winter injury (Corp et al. 2004; Murray et al. 1987).

Peas and lentils benefit from inoculation with a specific *Rhizobium* strain when they are planted in new fields or where they have not been grown in the past 20 years. *Rhizobium* specific to chickpea are not common in PNW fields and may not persist between crops; seeds should be inoculated where chickpea has not been grown in the past two years (Mahler 2005a; 2005b; 2015; Muehlbauer et al. 1981).

N management for crops following grain legumes

Nitrogen from legume residues increases the soil N balance and should be credited against total N requirements for subsequent crops. The N credits to wheat following grain legumes are shown in Table 5-9 and are based on legume grain production. Estimated N credits from lentils producing 1,000 lb/acre seed and dry peas producing 2,500 lb/acre seed are 10 and 20 lb N/acre, respectively. Using a rough conversion factor of grain yield × 0.008 helps estimate potential N credit from chickpea, winter lentil, or pea.

Mahler and Guy (2005) illustrate a slightly different method to estimate credits to N requirements of spring canola following legumes (Table 5-10). At planting, an estimated 60% of the previous legume residue will have decomposed and should be accounted for in a pre-plant soil test as plant-available N. To account for N from the fraction of the residue that has not yet broken down, 1 ton of residue is equivalent to 6 lb N/acre credit. See Chapter 6: Soil Fertility Management for more information on N management.

Rotation effect

Peas, lentil, and chickpeas are short season, shallow-rooted crops and generally use 15–35% less water than cereals or oilseeds, leaving more water in the profile for subsequent crops. Dry pea has the highest productivity (grain and residue) and water use efficiency compared to

Table 5-9. Estimated nitrogen (N) credit to wheat from preceding grain legume.

Crop	Grain yield lb/ac	N credit lb N/ac
Lentil[1]	>1,000	10
Chickpea	>2,000	15
Dry pea[1]	>2,500	20
Winter lentil	>2,500	20
Winter pea	>3,750	30

[1]Koenig 2005.

Table 5-10. Estimated nitrogen (N) credit to spring canola from legume residue breakdown.

Tons Residue	Grain yield lb/ac[1]	N credit lb N/ac
0.5	500	3
1	1,000	6
2	2,000	12
3	3,000	18
4	4,000	24

Source: Mahler and Guy 2005.
[1]residue-to-grain ratio used = 1 ton residue per 1,000/lb pea (or lentil) seed.

lentil and chickpea. In general, water use is lowest for lentil and highest for chickpea (Table 5-A3). Peas and lentils extract most of their water from the upper 2 feet of the soil profile; chickpea has a deeper effective rooting depth of 3–4 feet and a longer growing season, leading to the higher water use (Corp et al. 2004; Gan et al. 2009; Gan et al. 2015).

Recent studies have aimed to define the rotation effect of integrating a winter pea sequence. Results have been mixed. Ritzville, Washington (12" MAP), studies at the Jirava farm (2010–2015) showed a benefit to spring wheat yield after 'Windham' winter pea in a 3-year rotation (WP-SW-F) compared to spring wheat yield after winter wheat (WW-SW-F). Winter pea and winter wheat yields averaged 2,094 lb/acre and 72 bu/acre, respectively, over 4 years. Subsequent spring wheat yields averaged 30 bu/acre following winter pea compared to 28 bu/acre after winter wheat. Winter pea used significantly less water than winter wheat. Likewise, available soil water levels for planting spring wheat were higher after winter pea than after winter wheat, even though winter precipitation storage efficiency was higher following winter wheat, a result of drier soil and greater residue production (Schillinger et al. 2016). ▲ ■

At a traditional WW-F site in Montana (14" MAP), Chen et al. (2012) found that winter wheat yield and N recovery in grain were higher in a WP(hay)-WW sequence compared to SW-WW and similar to WW-F, and benefitted from income for hay. Smiley and Machado (2009) reported that replacing summer fallow with winter pea reduced subsequent winter wheat yields at Moro, but had little effect on winter wheat yields at Pendleton (2000–2005). However, nematode populations were greater under winter pea than spring cereals.

Integrating Canola

Rotational fit

Interest in regional energy crops and rotational diversification spurred feasibility research on canola for food, feed, and fuel production beginning in the 1970s; however, commercial adaptation of canola in the inland PNW has lagged behind other dryland production regions. Successful adaptation depends on defining wheat-canola sequences, specific to each AEC, that improve weed control and stand establishment under hot, dry conditions during optimal planting date windows, maximize winter survival, and enhance soil water and N recharge and uptake throughout the growing season (Long et al. 2016; Pan et al. 2016).

Spring canola is the better-adapted option for annual crop systems and can replace either spring legumes or cereals. Pan et al. (2016) found that spring canola yields are correlated with total available water and had water use efficiency of 182 lb grain yield per inch of water used. Winter canola is less suited to the Annual Crop AEC; soil moisture and growing degree days after cereal harvest are often insufficient to establish and grow winter canola to an adequate size (3–4 leaf rosette stage) before freezing winter conditions. Winter oilseeds have a role in annual crop systems when seeded after fallow or in 'delayed planting' in years with unfavorable spring seeding conditions, when soil water is sufficient for fall planting. ● ▲

Both winter and spring canola can be adapted to the Transition AEC with positive benefits, and grain-fallow producers are integrating winter canola to improve pest management strategies, diversify markets, and increase sustainability. Spring canola is less commonly grown in the Grain-Fallow AEC because yields are only 50–60% those of winter canola, limiting profitability (Hulbert et al. 2012; Karow 2014; Long 2016; Pakish et al. 2015; Pan et al. 2016). According to Young et al. (2012), spring canola could fit into grain-fallow systems "as a replant crop in instances of winterkilled, fall-planted canola, or as an opportunity crop during cycles of above-normal precipitation." As an opportunity crop, or flex crop, growers might include spring canola when prices are strong and moisture conditions are favorable (based on historical in-season precipitation data), or when available soil moisture is sufficient to a 48" rooting depth. New information from research on canola production in the Grain-Fallow AEC is becoming available which will give growers tools to reduce risk and support adoption. ▲ ■

The greatest agronomic challenges to adoption of oilseeds across the region include:

- Inconsistent stand establishment and yields
- Environmental limitations during optimal planting windows
- Winter survival of winter crop types

- Sensitivity to imidazolinone (IMI) and sulfonylurea (SU) herbicide carryover

Canola can be direct seeded into stubble, or into reduced-till or conventionally tilled soils. Cold, wet conditions may delay spring planting operations past optimal yield windows or early plantings may suffer frost damage. Fall operations are hindered by lack of seed zone moisture and high ambient and soil temperatures. Direct seeding may slow seedling development but conserves moisture for the crop and reduces erosion; increased seeding rates can enhance stand establishment.

Notable agronomic benefits growers have experienced include:

- 10–30% yield benefits to subsequent wheat crops
- Integrating herbicide-resistant canola has improved grassy weed control
- Improved **soil structure** and **infiltration**, and reduced runoff

Several eastern Washington growers have shared their experiences with integrating canola and other oilseeds in a series of case studies (Sowers et al. 2011; 2012).

Rotation effect

Improved winter wheat yields following grain legumes are well-documented. Guy and Karow (2009) compared winter wheat yields following several rotational crops, expressed as a percentage of yield following pea (100%). Relative winter wheat yields after cereals were 74–86% of wheat after pea at Moscow and Genesee, Idaho (Annual Crop AEC); yields after brassica crops were 85–99% of yields after pea, indicating similarly strong potential rotation benefits (Figure 5-3). ●▲ Calculations derived from recent biofuel cropping systems studies (2011–2013) showed similar results based on three years of yields of winter wheat following spring crops: relative wheat yields were highest following pea (100%) and lentil (99%); wheat yields after spring canola and camelina were 90% of yield after pea. Earlier studies showed winter wheat grown after five different broadleaf crops averaged 29% greater yield than winter wheat following winter wheat, while the rotation benefit of winter wheat after spring cereals averaged just 9% (Guy 2014; data not shown).

Wheat yields following canola are generally expected to be greater than yields after wheat. However, yield impacts are not always positive, or clearly understood. Select north-central Washington growers have achieved 30% yield increases of winter wheat in rotation with canola (Sowers et al. 2012), and on-farm studies near Ritzville, from 2006 to 2009, showed winter wheat following canola had 39% greater yields than winter wheat following wheat (Esser and Hennings 2012). ▲ ■ In contrast, rotation studies near Reardan, Washington (14" MAP), showed spring wheat yields following winter wheat (58 bu/acre) were similar to spring wheat yields following winter canola (49 bu/acre) over 5 years (2009–2010; 2012–2014). Soil water use by winter wheat and canola were similar, as was soil water content when spring wheat was seeded, whereas suppressed mycorrhizal fungi populations under spring wheat after canola may have limited the potential rotation benefit (Schillinger et al. 2013; 2014a; Hansen et al. 2016).

Recent studies at the Washington State University Wilke Research and Extension Farm (Davenport), showed variable spring wheat yield

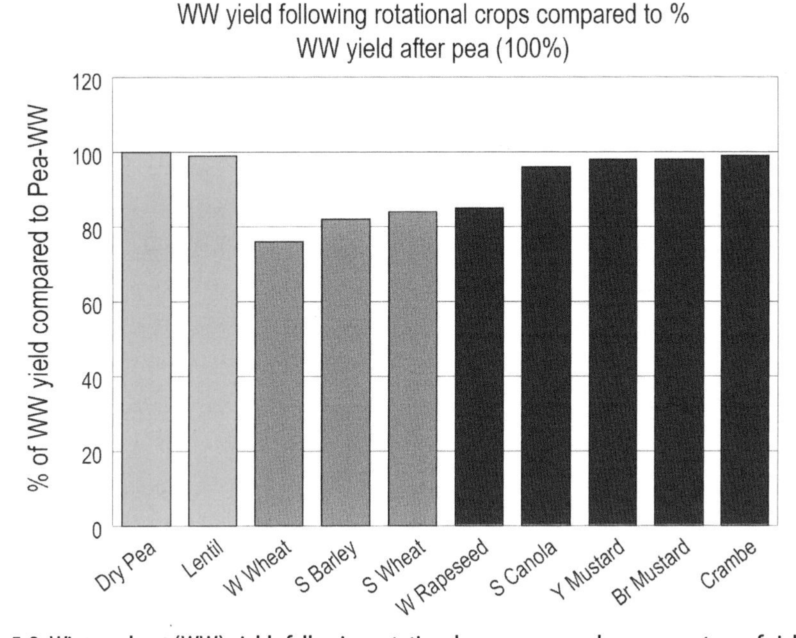

Figure 5-3. Winter wheat (WW) yields following rotational crops expressed as a percentage of yield following pea (5-year average at Genesee and Moscow, Idaho). Legumes are in blue, cereals in gray, and oilseed crops in red. (Adapted from Guy and Karow 2009.) ●

responses (2013–2015): two of three years showed yield benefits (8% and 16%) for spring wheat following canola compared to winter wheat. However, yields of spring wheat following canola were 6% less than following winter wheat in 2015 (Esser and Appel 2016). Yield benefits to wheat after canola have not yet been documented from long-term research sites in the PNW (Long et al. 2016). Integrating canola offers other positive rotation effects; canola stubble can trap snow, reduce runoff, and improve soil water recharge. Additional soil moisture can increase potential biomass production, soil organic matter, and water-holding capacity, leading to more opportunities for intensification.

Spring canola establishment

Canola emergence is impacted by wet soils, soil crusting, and sensitivity to herbicide carryover, such as Pursuit. For optimal germination and yield, spring seeding should be done as soon as soil temperatures reach 49°F and fields are suitable for machinery. Seeding spring canola into heavy winter wheat residue is challenging; establishment is typically just 50–60% of the seeding rate. Cold soil temperatures slow seedling growth in early plantings, while later seeding reduces yield potential. Typical canola seeding rates are 4.5–6 lb/acre for the Annual Crop AEC, targeted to establish a stand count of 4–10 plant/ft^2 at harvest. Canola seed size and weight are highly variable; seeding rates should be adjusted based on seed lot information. Direct seeding or broadcast methods benefit from higher rates than for conventionally tilled fields (Brown et al. 2009). Spring canola stands are more consistent when planted before seed zone moisture declines and temperatures warm; later plantings result in increased heat and drought stress during flowering and grain fill (Brown et al. 2009; Gan et al. 2004; Pan et al. 2016). Spring canola direct-seeded into heavy wheat residue over 9 years showed an inverse relationship between seeding date and grain yield (Huggins and Painter 2011). A 2-year study showed yields were not affected by row spacing (10” or 20”) with a 5 lb/acre seeding rate. Advantages of wider row spacing include lower machine and fuel costs and less drill plugging (Pan et al. 2016; Young et al. 2012). Karow (2014) recommends 5–8 lb/acre seeding rates and 12–16” row spacing for spring canola in eastern Oregon; narrower row spacing helps control weeds. ●▲

Winter canola establishment

The greatest challenge growers face with winter canola is consistent stand establishment. Insufficient seed zone moisture and excessive temperatures during germination can lead to inconsistent stands or even crop failure; poor establishment increases risk of winterkill. Conditions are most favorable with adequate seed zone moisture within an inch of the soil surface and cool air temperatures (<84°F) for a week after planting. Small seeded canola (0.2–0.6 g/100 seed) cannot be planted deeply into moisture like wheat or peas. Canola emerges best when planted just 0.8–1.2" below the soil surface, but can emerge from twice that depth when needed to access moisture (Karow 2014; Pan et al. 2016). In the Grain-Fallow AEC, soil moisture is often 4–6" below the surface in summer fallow in August. Planting with deep furrow drills can be successful when seeds need to emerge just to the bottom of furrow. Young et al. (2014a) found 22% improved plant density, more uniform distribution, better growth, and improved weed suppression growth when canola was planted with deep furrow drills modified with 10–15" shovels to move hot, dry soil away from the seed zone. ▲ ■

Guidelines for optimal winter canola planting dates and rates have been lacking for the low precipitation grain-fallow systems. Information from recent studies helps growers reduce risk and achieve successful stands and yields. Mid-August to early-September planting dates are optimal for achieving adequate growth for winter survival (rosette stage) and yield potential (Brown et al. 2009; Karow 2014); Young et al. (2014a) found early August to about August 25th to be the optimal planting window with soil moisture less than 4" below the soil. Late seeding of winter canola, to take advantage of fall rains, significantly reduces both survival and yield. September plantings showed reduced yields of nearly 40% compared with an August planting (Pan et al. 2016).

Integration of biennial, dual-purpose canola cultivars, provides growers the intensification option to produce both forage and grain from a single planting (Kirkegaard et al. 2008b; Neely 2010). This allows earlier seeding of winter canola, enabling better seed germination and plant growth while the soil moisture during fallow is still close to the soil surface, and improves yield consistency (Karow 2014). Several growers in the Transition AEC have had success with early-July seeding. The first season's forage can

Figure 5-4. Cattle grazing a biennial, dual purpose canola stand near Ritzville, WA. (Photo by Karen Sowers.)

be grazed (Figure 5-4) or used to produce silage (Kincaid et al. 2011). Recovering plants are capable of overwintering, and then proceed into stem elongation and reproductive phases in producing competitive grain yields in the following season. Continued development of varieties with improved tolerance to cold temperatures and open winters will reduce production risks for this region.

Canola seed size is highly variable; having accurate seed lot weights helps set appropriate seed rates to achieve target plant populations. Brown et al. (2009) guidelines recommend that growers determine a seed rate targeted to 10–16 seedling/ft^2 at establishment to give 5–10 plant/ft^2 at maturity. Karow (2014) recommends a seeding rate of 4–7 lb/acre for eastern Oregon; Young et al. (2014a) found 4 lb/acre to be optimal for seeding in grain-fallow locations with 10" MAP. A Ritzville area grower reduces risk by increasing his seed rate to 7 lb/acre when the soil moisture line is lower than 4" below the surface. Canola yields generally decrease significantly when the mature stand population drops below 4 plant/ft^2. However, canola growth is indeterminate, and less dense stands can compensate in growth and yields; stands do not need to be uniform to achieve economic yields. Brown et al. (2009) reported that canola's compensatory growth habit makes it possible for winter canola stands with just 1–2 plant/ft^2 in spring to produce 70–80% of yields achieved with the higher density. Similarly, Young et al. (2014a) found that spring stands with populations of 2–4 plant/ft^2 achieved excellent yields (>1,500 lb/acre) in the Grain-Fallow AEC.

In recent planting rate and date studies at Okanogan and Bridgeport, Washington (10–10.5" MAP), Young (2012; 2014a) found optimum yields with a 4 lb/acre rate; trials had 56–83% winter survival rates, resulting in a spring stand of 2–4 plant/ft^2. For seeding spring canola following a failed winter crop, results were better with drilled seed; broadcast spring canola was more vulnerable to frost damage and yielded just 30–67% compared to drilled canola.

Winter canola is less suited to the Annual Crop AEC; moisture reserves are too low following harvest of a previous crop to plant canola in mid-August. Dry conditions extend into October and delayed seeding allows seedling growth to only a 2–3 leaf stage prior to winter conditions, leaving seedlings more vulnerable to winterkill. Canola is less winter hardy than wheat because of canola's 'epigeal' type emergence, where cotyledons and the shoot growing point emerge above the soil surface, increasing the plant's exposure and sensitivity to harsh conditions. In contrast, cereals exhibit 'hypogeal emergence,' where the shoot growing point is below-ground and protected from severe cold and other environmental stressors (Karow 2014; Klepper et al. 1984; Koenig et al 2011; Long et al. 2016; Pan et al. 2016).

Winter canola establishment considerations:

- Depth to moisture
- Soil and air temperature
- Seed rate, date, and N effect on winter survival

Weed management

Integrated weed management practices are critical to successful weed control in cereal-based cropping systems, reducing reliance on in-crop herbicides and preventing herbicide resistance. Using diverse crop sequences, reduced tillage or chemical fallow, competitive crop varieties, and seeding practices to establish optimal stands help reduce weed pressure. Well-established canola stands maximize competitiveness; canola seedlings grow and close canopy rapidly, competing well with annual weeds, while late plantings or poor stands are less competitive. (Brown et al. 2009; Karow 2014; Long 2016).

Herbicide-resistant spring canola varieties provide opportunities for better control of the grassy weeds that persist in cereal-dominated annual crop systems, such as downy brome, jointed goatgrass, feral rye, and Italian ryegrass (Brown et al. 2009; Young et al. 2016b). Replacing spring legumes with glyphosate-resistant spring canola in rotation with winter and spring cereals, over several years, was effective in reducing Italian ryegrass in studies near Pullman (Huggins and Painter 2011), and also improved control of broadleaf weeds, including mayweed chamomile and common lambsquarters that are difficult to control in legume sequences. Spring canola cultivars are available with resistance to glyphosate, glufosinate, or imazamox; volunteer herbicide-resistant canola may need to be controlled during fallow or subsequent crop sequences. ● ▲

Wilke Farm studies at Davenport, found that integrating herbicide-resistant spring canola in a 4-year rotation improved feral cereal rye control and improved economic returns resulting from improved canola and wheat yields (Hulbert et al. 2013); canola yields and wheat yields following canola benefitted from the improved weed control. ▲ ■

Similarly, growers in the Grain-Fallow AEC may choose to grow winter canola to improve grass weed control. Downy brome, Italian ryegrass, and jointed goatgrass persist in winter wheat along with feral rye. Typical weed control options include mechanical rod weeding in conventional or reduced tillage systems, or chemical fallow where direct seeding is utilized to reduce wind erosion (Pan et al. 2016; Young and Thorne 2004). Recent research in north-central Washington found that spring and split applications of quizalofop and glyphosate in canola controlled 90% of feral rye, eliminated seed production, and increased canola yield more than 40% (Young et al. 2016b). Use of glyphosate-resistant winter canola, plus glyphosate application, can provide additional opportunities to control feral rye. Adopting herbicide-resistant cultivars also allows for easier chemical rotation management, circumventing plant-back restrictions for IMI and SU herbicides.

Nitrogen management

Canola nutrient requirements, timing, and placement vary from traditional wheat fertilization programs. Karow (2014), Koenig et al. (2011), Mahler and Guy (2005), Pan et al. (2016), Wysocki et al. (2007), and others describe several important N management factors to consider:

- Canola N, phosphorus (P), sulfur (S), and potassium (K) uptake per unit yield can be 50–100% greater than soft white and hard red wheat uptake indicating that higher levels of available N are required for regional canola production compared with wheat nutrient management.
- The percentage of N, P, and K removed in canola grain is lower than that removed in wheat; nutrients left in oilseed residues contribute to subsequent crop sequences.
- Canola's strong taproot and extensive root hairs enhance utilization of soil N to its full rooting depth, reducing reliance on applied N.
- Fertilizer N = (yield goal × base N recommendation of 6–8 lb N/100 lb seed yield) – soil N credits.
- Fertilizer application at or near planting date, placed to the side of the seed row, will improve germination and stand establishment, enhance early growth, improve **nitrogen use efficiency**, and limit root injury from ammonia toxicity.
- High N going into winter reduces survival rates.
- Hybrid cultivars with higher yield potential may need higher levels of N for optimal yield compared to open-pollinated varieties.

Canola response to N is influenced by climate, available soil N, cultivar, water availability and management practices; further research is needed to determine optimal timing and rates for the different AECs. Spring canola yield responds well to applied N when residual soil N levels are low, and shows minimal fertilizer N response when soil N supply is high and yield potential is low (Maaz et al. 2016; Pan et al. 2016).

Unit nitrogen requirements (UNRs) for wheat are estimated at 4.5 to 6 lb N/100 lb grain, for soft and hard wheat, respectively. Recommended UNRs for canola range from 6 to 10.7 lb N/100 lb seed for dryland production (Karow 2014; Koenig 2011). Observations over 6 years in the Annual Crop AEC showed variable UNRs of 7–13 lb N/100 lb grain for spring canola, affected by water availability. Requirements at the Wilke Farm site in the Transition AEC ranged from 9–17 lb N/100 lb grain. High UNR values resulted from lower yields and nitrogen use efficiency in this region and more complete accounting of root zone residual N and crediting of soil N mineralization contributions to the total N supply (Pan et al. 2016). Spring canola yielding up to 916 lb/acre had an average optimal N rate of 20 lb N/acre; zero N was required following fallow, compared to 5–58 lb N/acre following wheat despite lower available soil water and canola yield potential. The modest optimal N rates were influenced by high residual soil N carryover and mineralization.

Optimal fertilizer rate, timing, and placement need to be defined to maximize winter canola yield and winter hardiness. Winter canola N requirements are based on two growth phases: (1) fall growth from planting to winter dieback and dormancy, and (2) spring growth through maturity (Long et al. 2016; Pan et al. 2016). High N status of vegetative winter oilseeds decreases cold hardiness and survivability, supporting use of conservative N rates at winter canola planting, allowing winter canola to use up residual soil N during early establishment. Karow (2014) recommends that if fall N is needed for winter canola, apply 30 to 50 lb N/acre prior to planting and apply the remainder in the spring.

Integrating Other Oilseeds

Camelina

Camelina requires few cultural inputs, is more drought and stress tolerant than canola, and performs well in fields with marginal productivity, thus has the potential to help mitigate climate change and improve sustainability of dryland cereal production systems. More commonly grown as a summer annual, camelina is also adapted as a winter annual with hardiness similar to winter wheat. Camelina's short season (85–100 days) could offer more resilience to hotter, drier summers (Ehrensing and Guy 2008; Hulbert et al. 2012).

Similar to canola, camelina establishment is challenging. Small, variable-sized seeds require shallow planting (0.25") for successful emergence; seeds may have difficulty emerging if soil crusting occurs after rain showers. Seedlings have good frost tolerance. Camelina is highly sensitive to IMI and SU herbicides and has similar plant-back restrictions as canola. Efforts are ongoing to develop herbicide-resistant camelina varieties (Hulbert et al. 2012).

Planting date studies have shown highest yields with late-winter planting (February 15-March 1) at Lind, Washington and Pendleton, Oregon, compared to late-fall and mid-winter plantings when inadequate precipitation and control of fall-emerging weeds likely reduced stands. Establishment was similar with direct seed or broadcast, into standing stubble, with seed rates of 3–5 lb/acre; broadcast seeding required less time and expense. Late-planted camelina had greater Russian thistle populations at Lind (Hulbert et al. 2012; Schillinger et al. 2014b).

No broadleaf herbicides have been registered for weed control in camelina; dense, early planting in clean fields reduces weeds and increases competitiveness. Camelina's short season allows for harvest prior to seed set of many weeds.

Researchers are evaluating the potential for a 3-year WW-Camelina-F rotation to replace traditional 2-year WW-F rotations; good yields have been achieved by replacing fallow with camelina following winter wheat. Yield potential is dependent on annual precipitation. The taproot of camelina can efficiently extract subsoil water and N; UNRs are 5–6 lb N/100 lb seed. Expected yields run 60–70 lb seed per inch of precipitation in the PNW. Studies showed a yield range of 1,610 to 3,070 lb/acre in the annual crop region (Moscow-Pullman), 1,549–2,000 lb/acre at transition sites (Lacrosse, Washington and Pendleton), and lower, inconsistent yields of 115–1,030 lb/acre at Lind (Hulbert et al. 2012). Schillinger et al. (2014b) found similar water use efficiency among camelina trials across dryland areas (65 lb seed per acre inch water used) indicating that camelina yield can be predicted by annual precipitation. Camelina produces relatively low amounts of residue compared to cereals, a major disincentive to adopting camelina in the Grain-Fallow AEC. Sharratt and Schillinger (2014; 2016) found 57–212% increased wind erosion potential in summer fallow following camelina and safflower, compared to fallow after winter wheat, due to differences in crop residue characteristics. To protect against increased erosion, potential growers should either replace fallow with a spring crop or exclusively use no-till fallow practices.

Yellow mustard

Yellow mustard has good agronomic and economic feasibility in the dryland PNW, performs well in no-till systems, and requires few chemical inputs. Well-adapted to hot, dry conditions, yellow mustard can provide an adaptive strategy for predicted climate change, although the crop is sensitive at flowering. Mustard has relatively high water use, and wheat and mustard prices determine economic feasibility. Grain yields are similar to spring canola in the higher precipitation areas (2,000–2,500 lb/acre) and out-yields spring canola by 55% in regions with less than 12" annual precipitation. Predicted grain yield is 95 lb grain per inch annual precipitation, but will vary by cultivar and practices (J. Brown et al. 2005). Growers have greater flexibility with planting dates compared to other spring crops, and planting can be delayed for better weed control. Yellow mustard establishes quickly, can close canopy in 30 days, and is highly competitive with weeds compared to canola and safflower; some growers have had success using no herbicides. Growers should avoid planting mustard in fields with potential for catchweed bedstraw infestation that can impact seed quality and price. Crops generally mature in 80–85 days. Planting depth (0.5–1") and spacing (6–8") is similar to other oilseeds; 12" row spacing can be used in direct seed conditions. Recommended seeding rates are 7–8 lb/acre for conventional tillage and 8–9 lb/acre for direct seeding. Seeds can be planted after soil temperature reaches 40°F, but mustard is highly susceptible to frost damage. In eastern Oregon, planting may begin in mid-March; mid-April to early May is more suitable in the higher precipitation areas. Estimated required N (lb N/100 lb seed) range from 12.8 in high precipitation areas to 8 in low precipitation areas. Similar to other oilseeds, mustard is highly sensitive to carryover of imazamox herbicides (J. Brown et al. 2005; Wysocki and Corp 2002).

University of Idaho and Oregon State University Columbia Basin Agricultural Research Center (CBARC) variety yield trials between 2007 and 2015 showed average yields of just under 2,000 lb/acre (Moscow), 1,170 lb/acre (Davenport), 800 lb/acre (Pendleton), and 1,056 lb/acre

(Moro). Variety yield data can be found at ***http://www.cals.uidaho.edu/brassica/growers.asp***.

The rotation effect of yellow mustard on subsequent wheat crops has been variable. Guy and Karow (2009) found improved yields of wheat following yellow mustard compared to following cereal crops, while a grower near Ione, Oregon, found poor wheat yields following spring mustard, attributed to a notable increase of root-lesion nematodes following mustard (Yorgey et al. 2016b).

Safflower

Both spring and fall-seeded **facultative** safflower cultivars have potential in no-till dryland cropping systems. Low residue production increases risk of erosion; safflower should not be followed by fallow (Sharrat and Schillinger 2016). Safflower is relatively drought- and heat-tolerant due to its long taproot; however, it also has higher water use and lower water-use efficiency than wheat or other alternative crops in the region, which can reduce yield of succeeding crops. General seeding recommendations include an optimal planting window in April and May, seed depth of 1–1.5", and a seed rate target of 3 plant/ft^2. Soil crusting can hinder emergence. Safflower lacks competitiveness with weeds due to slow emergence and initial rosette growth; no broadleaf herbicides have been registered for weed control in safflower (Armah-Agyeman et al. 2002; Petrie et al. 2010).

Preliminary evaluations at Moro and Pendleton showed spring safflower grain yields ranging from 400 to 1,400 lb/acre; fall-sown, winter-hardy facultative safflower with earlier flowering and maturity showed increased yields up to 1,900 lb/acre compared to spring safflower. Yields in higher precipitation areas ranged from 2,575 to 3,135 lb/acre (Petrie et al. 2010). Safflower yields ranged from 125–1,130 lb/acre in Ritzville trials with an average 483 lb/acre over 6 years (2010–2015). ▲ ■

Relatively high water use by safflower depletes soil moisture at higher rates than other crops in rotation and can carry over through a year of fallow, reducing subsequent wheat yields. At Ritzville, wheat yield was lower in a WW-SAF-UTF sequence compared to WW-SW-UTF and WW-UTF over 4 years, but was significantly lower in only one year

(2012), indicating that wheat may benefit from some rotation effect that partially offsets lower soil water availability (Schillinger et al. 2016).

Integrating Alternate Spring Cereals for Reduced- and No-Till Late Planting

From 2007 to 2014, 20–22% of the Transition and Annual Crop AECs was planted to spring wheat and spring barley compared to 3% of the Grain-Fallow AEC (Table 5-1). Spring grains typically follow winter wheat in 3- or 4-year cropping sequences where moisture permits. Spring plantings broaden opportunities for control of winter annual grass weeds, and adequate seed zone moisture helps stands establish. However, wet spring conditions can delay or prevent planting, especially in the Annual Crop AEC.

Replacing or supplementing summer fallow with spring grains can enhance soil quality; no-till annual spring cropping could reduce susceptibility to wind erosion an estimated 95% in grain-fallow systems (Thorne et al. 2003). Annual cropping reduces the time soil is left bare between crops and increases crop residue and surface roughness, providing year-round protection from erosion. However, spring wheat yields are typically just 50–70% of winter wheat yields, and soil moisture deficits during flowering or grain fill can further reduce profitability. Annual cropping systems have had greater income risk and resulted in lower annual net returns than WW-F (Juergens et al. 2004; Schillinger and Young 2004; Young et al. 2015). Cropping sequences that improve water and nitrogen use efficiency and reduce erosion can help mitigate effects of climate change and past soil degradation.

Conventional and reduced tillage WW-F remains the most profitable crop sequence in the Grain-Fallow AEC. As an alternative to annual spring cropping, improved winter wheat harvest (e.g., stripper header), fallow (e.g., undercutter tillage), and flex cropping practices can improve residue cover and soil health, increasing potential for rotational diversification and intensification in the future. ▲ ■

Hard red spring wheat

Low prices for soft white wheat and favorable hard red wheat prices are incentives for growers to increase hard red spring wheat acreage. Whereas yields are typically lower and more variable than for soft white spring wheat, differences may be offset by potential price premiums earned for high percentage protein. However, the higher UNRs of 3–3.7 lb N/bu grain to achieve 14% protein can impact profitability and reduce nitrogen use efficiency, particularly in areas with high yield potential. Optimal N rates are dependent on yield potential, fertilizer costs and premium or discount price values related to grain protein concentration (Baker et al. 2004; B. Brown et al. 2005). More information on N management is described in Chapter 6: Soil Fertility Management.

Hard red spring wheat is well-adapted to dry areas with shallow soils, low yield potential and lower N requirements. Annual cropping with hard red spring wheat to replace WW-F can help reduce erosion, but has not been profitable in the short-term. Young et al. (2015) evaluated annual no-till hard red spring wheat cropping in a 6-year (1996–2001) study in the WW-F region near Ralston, Washington (11.5" MAP). Results showed that continuous HRSW, HRSW-SB, and SWSW-ChF no-till crop systems reduced wind erosion but were generally not profitable. This study also provided the first evaluation of a no-till SWSW-ChF system in the region, which benefitted from an 18-month window for control of winter annual grasses and cereal rye. Yields were generally greater than for continuous spring cereal sequences, but less than in a reduced-till WW-F system.

Similarly, annual no-till cropping has not been economically competitive in the Horse Heaven Hills (6" MAP) where growers are seeking alternatives to the WW-F system. Continuous HRSW had a 6-year average annual yield of 473 lb/acre compared to 1,062 lb/acre winter wheat, every other year. No differences in precipitation storage efficiency were found, and straw production was similar for both crop sequences (Schillinger and Young 2004). ▲ ■

White spring wheat

Schillinger et al. (2007) evaluated annual no-till cropping as an alternative to WW-F in an 8-year study near Ritzville. SWSW-SB, HWSW-SB, and continuous SWSW and HWSW sequences generally had lower average profitability and higher economic variability compared to values reported by traditional WW-F producers. Continuous SWSW had 4-year average yields similar to spring barley and out-yielded hard white spring wheat. ▲ ■

Spring barley

Historically, spring barley was a preferred alternate crop in the inland PNW. However, Washington production declined significantly from 500,000 acres in 2000 to just 110,000 in 2015. NASS (2015) cropland data showed an average 169,000 barley acres across the dryland AECs during the 2007 to 2014 period, accounting for 5% of the cropped ground in the Annual and Transition AECs, and less than 1% in the Grain-Fallow AEC (Table 5-1). Several factors have contributed to the decline of barley acreage: low barley feed prices ($1.93–$5.10/bu), lack of herbicide-resistant varieties, susceptibility to root and crown pathogens, low grain lysine content (reducing feed quality), and condensed tannins ill-suited for food uses.

Barley end uses include feed, malt, and food; straw and grain are also potential feedstocks for ethanol production. Improved prices and recent interest in food quality barley may offer growers additional market incentives, especially in the drier regions. Development of food quality winter barley cultivars with improved hardiness could provide additional options, and value, to growers in the future (Petrie 2008; Rustgi et al. 2015).

Barley is well-adapted across the region and provides several rotational benefits. For example, barley has a shorter growing season than wheat and may prove more able to avoid the late-season stressors predicted with climate change. Barley can suppress select soilborne pathogens; Oregon studies found that cereal rotations including a spring barley sequence had the lowest root-lesion nematode infection rate (Smiley and Machado 2009). Feed protein requirements are low (10% or less) and the 2 lb N/bu UNR for feed barley is much lower than for wheat, whereas high-quality malt barleys require 11–12% protein and have a slightly higher UNR than soft white wheat. Barley typically has a higher straw-to-grain ratio than wheat, which can enhance soil health and carbon sequestration.

Currently there are no price premiums for high protein barley. However, in response to market interest in the food quality barley niche, Rey et al. (2009) looked into the feasibility of producing high beta-glucan, no-till barley at Moro and Pendleton. Results showed that the high beta-glucan, waxy, hulled varieties 'Salute' and BZ 502-563 performed competitively with the commonly grown feed barleys 'Baronesse' and 'Camas' at both sites, and are a good alternative for dryland cereal producers in the inland PNW. ▲ ■

Integrating Alternate Winter Cereals and Improved Fallow Practices

Alternate winter cereals, such as winter triticale, hard red winter wheat, and facultative wheat and barley, add diversity to traditional soft white winter wheat acreage and are adaptable to the warmer, drier summers predicted for the PNW. Winter crops benefit from a longer growing season, deeper rooting and more efficient utilization of winter precipitation, and earlier grain fill (Pakish et al. 2015; Schillinger et al. 2010). Autumn-sown grains protect soils during winter precipitation and have a higher grain and straw yield potential than their spring counterparts; earlier maturity avoids drought and heat later in the growing season.

Winter triticale

Winter triticale is highly promising as an alternative crop to diversify the Grain-Fallow AEC and is adapted to late-planted, no-till systems. Production may increase the opportunity for adoption of no-till systems in the WW-F region where inadequate seed zone moisture in early fall limits success. Low prices and lack of insurance have limited grower adoption; triticale crop insurance is expected to be available beginning in 2017.

Ritzville, studies found that early-planted winter triticale out-yielded early-planted winter wheat 22% over 6 years (2011–2016); average yields were 5,005 lb/acre and 4,085 lb/acre, respectively (Figure 5-5). Late-planted winter triticale yield (3,735 lb/acre) was similar to early-planted winter wheat (Schillinger, unpublished data, with permission). ▲ ■ Late-planted winter triticale has better yield potential compared to winter wheat, which can suffer 36% yield reduction over early-planted wheat. The crop can be grown with the same equipment and inputs as wheat, and in-crop grass weed herbicides are available for use. Winter triticale produces larger amounts of residue than wheat, has a root mass that is double that of most other cereals, can enhance soil quality and

Figure 5-5. Early-seeded winter triticale (right) out-yields early-seeded winter wheat (left) at Ritzville, WA. (Photo by Bill Schillinger.)

carbon sequestration, and provide erosion protection. Winter triticale has low susceptibility to insect pest and disease problems and good weed competition due to vigorous growth habits, leafiness, and height. Low feed grain prices have limited interest in triticale production in this area in the past, but improved prices support economic opportunity. Triticale can outperform wheat in marginal conditions, produce more biomass, potentially sequestering additional carbon, and is more tolerant of low soil pH and several soilborne pathogens and nematodes. Winter triticale's extensive root mass is effective against erosion (Schillinger et al. 2012; 2015).

Hard red winter wheat

Hard red winter wheat has been an attractive alternative in the Grain-Fallow AEC where yield potential is just marginally less than for soft white winter wheat and growers can more economically achieve the high protein percentage required to garner market premiums. Additional needed N inputs are low due to water stress and lower yield potential in this region. Initially, hard red winter wheat was best adapted to the driest production regions (<9" MAP). However, improved hard red winter wheat varieties with greater yield potential, disease or herbicide resistance may lead to increased production in higher precipitation areas. Depending on relative prices, returns on lower yielding hard red winter wheat can be similar to soft white winter wheat returns (Esser et al. 2008). Adoption of no-till hard red winter wheat is limited in the Grain-Fallow AEC as late planting reduces yield potential. However, no-till studies in Morrow Co. Oregon (4–6' soil depth and 7.2–9.4" MAP) found that late-planted AgriPro 'Paladin' and 'Norwest 553' hard red winter wheat cultivars performed well in late-planted situations and yielded similarly to 'Tubbs' soft white winter wheat, ranging from 33.6 to 35.4 bu/acre. Recently released cultivars such as 'Farnum' have not yet been evaluated for performance in low precipitation, no-till systems. Fallow area producers will benefit from continued development of varieties that perform well under late-planted, no-till systems (Lutcher et al. 2012). ▲ ■

Facultative wheat and barley

Facultative wheat or barley show potential as a replacement for winter wheat in no-till cropping systems where late-planting of winter wheat and annual spring cropping are not feasible. No-till fallow practices can reduce seed zone moisture, delaying winter wheat planting and reducing yield potential compared to conventional till WW-F systems. Facultative cereals can be planted later than winter wheat, thus are better adapted to no-till chemical fallow systems such as FW-ChF or FW-SW-ChF, reducing erosion and enhancing soil health in the Grain-Fallow AEC. Late-planted facultative wheat generally out-yields spring wheat and provides winter cover; facultative wheat is more competitive with summer annual weeds, and appears to be less susceptible to stripe rust and root disease pathogens than spring wheat.

Results from long-term studies at Ralston, showed that a no-till FW-ChF sequence had less yield variability than reduced-till WW-F, but had lower yields and net returns. Facultative wheat is more susceptible to winter damage and provides less winter cover than winter wheat, but begins spring growth and flowering earlier than true winter wheat, potentially avoiding late-season heat and drought stressors. Researchers concluded that FW-ChF shows potential for no-till late planting purposes, but that conservation cost-share incentives would likely be needed for growers to adopt this system, and that further research would be useful (Bewick et al. 2008; Sullivan et al. 2013). ▲ ■

Tall stubble no-till and undercutter tillage systems

No-till fallow systems generally reduce yield potential in the Grain-Fallow AEC due to delayed fall planting in response to excess evaporation from the seed zone compared to conventional tillage. However, studies at Ralston show that replacing semi-dwarf cultivars with tall, high residue winter wheat or triticale and harvesting with a stripper header can support timely fall planting in no-till systems. Tall, standing stubble, and heavy residue, protects the soil surface, reducing wind speed and surface temperature, and conserving seed zone moisture. These improved seeding conditions lead to better establishment of fall-seeded crops; Young et al. (2016a) found that this no-till system improved winter canola establishment 35–40% over reduced-till fallow. ▲ ■

Studies at the Jirava farm (Ritzville) found that seed zone moisture is generally better with undercutter tillage fallow than for standard no-till fallow systems. At Lind (9" MAP), winter wheat yields in a WW-UTF system were 35% greater than yields in the late-planted, no-till winter wheat system. In addition, the undercutter method can reduce blowing dust 70% (Schillinger 2016; Schillinger et al. 2016). ▲ ■

Integrating Flex Cropping

Flex cropping practices provide producers with options to reduce fallow, gain production opportunities, and increase crop biomass, carbon sequestration, and soil surface cover. Adequate moisture, favorable crop prices, and low weed and disease pressures help determine profitability of flex crop options. Growers can assess yield potential for a spring or fall flex crop using soil water content prior to planting, historic precipitation values, and site-specific yield history. Growers may be able to take advantage of late-summer rains to support recrop winter wheat or alternate crops such as winter canola or peas following wheat harvest, or to plant a spring crop during a traditional fallow sequence.

Available soil water and expected crop season precipitation can be useful in determining profitability prior to planting a crop. Schillinger et al. (2012) calculated that wheat requires 2.3" of available water for vegetative growth with an average 5.8 bu/acre production with each additional inch of available water, including stored soil water and spring precipitation. Winter wheat yields following summer fallow increased 7.3 bu/acre with each additional inch of water compared to 5.4 bu/acre for recrop spring wheat. Growers can use this tool to predict wheat yields after summer fallow or recrop spring wheat using the following equations and site or region-specific real-time and historical moisture values:

$$\text{WW after SF: Yield} = 6.7\ \text{SFW} + 7.9\ \text{OWG} + 4.4\ \text{A} + 7.6\ \text{M} + 12.2\ \text{J} - 16.4$$

$$\text{Recrop SW: Yield} = 5.4\ \text{OWG} + 1.4\ \text{A} + 6.4\ \text{M} + 5.7\ \text{J} - 10.6$$

Where Yield is grain yield in bu/acre, SFW is summer fallow available soil water in inches, OWG is net over-winter soil water gain in inches, A is April rain, M is May rain, and J is June rain in inches.

Lutcher et al. (2013) provide excellent guidelines for optional fall or spring flex planting decisions based on soil depth, crop choice, MAP, effective rooting depth of crop, and total plant-available soil water content at planting time. Table 5-11 illustrates the minimum plant-available soil water content needed for successful cropping.

Table 5-11. Recommended minimum plant-available soil water content needed for fall and spring planting.

Average annual precipitation (in)	Minimum plant-available soil water content (in)*	
	Fall planting	Spring planting
<10	3.5	4.5
10 to 12	3.0	4.0
12 to 14	2.5	3.5
14 to 16	2.0	3.0
16 to 18	1.5	2.5
>18	1.0	2.0

Source: Lutcher et al. 2013 Note: Values listed in this table are guidelines only. *Effective rooting depth. Decisions to plant may be based solely on the anticipated quantity and timing of precipitation later in the growing season.

Integrating Cover Crops

Cover crops are "close growing crops such as grasses or legumes that are used primarily to provide seasonal protection against soil erosion and for soil improvement" (Unger et al. 2006). Historically, growers in the inland PNW used green manure cover crops to supplement N before the introduction of chemical fertilizer, as well as to control erosion and provide forage and hay for livestock. Cover crop benefits include maintaining soil organic matter; fixing N, reducing soil evaporation; increasing infiltration; suppressing weed, disease, and pest pressure; improving soil structure; providing soil erosion protection; and promoting cash crop productivity (Snapp et al. 2005).

There is renewed interest in cover cropping to enhance crop diversification and improve soil quality in PNW dryland cereal production systems. Researchers and growers are evaluating cover crop plant biomass production, soil fertility, soil moisture dynamics, and other factors that affect production and profitability of the following cash crop (e.g., winter wheat) and soil quality indicators. Cover crops can offer positive on-farm benefits, but there are major challenges to successful integration in dryland cropping systems including establishment, weed competition, water demand, and effect on yield of subsequent cash crops. Preliminary research results have shown that establishing mixed cover crops after cash crop harvest may be impractical due to soil moisture deficits (Thompson and Carter 2014). Similarly, Roberts et al. (2016) indicated that cover crop mixtures pose a high risk; cover crops may extract excessive water, limiting the available water for the following season's cash crop. Growers may also be reluctant to take on additional labor and operation costs such as seeds, tillage, weed control, and cover crop termination. With no immediate cash return to producers, cover crops need to be further evaluated for intermediate- and long-term economic and cropping system benefits. Research with single and multiple cover crop species are ongoing in various precipitation regions of southeastern Washington and in Pendleton, Oregon. Cover crop mixes that contain cruciferous crop seeds (e.g., mustard, canola, radish) are a potential source of black leg disease caused by the fungal pathogen *Phoma lingam*. Growers should only plant seed that has been tested and certified to be black leg-free.

Research efforts are focused on finding beneficial ways to include cover crops in dryland systems that will complement cash crop production. Some potential cover cropping options in the PNW follow:

1. Companion cover crop grown for a short period with a cash crop.
2. Added biomass can reduce erosion, provide protection from winter freeze, and improve soil organic matter (e.g., seeding low rates of faba bean, radish, and buckwheat with the standard rate of winter wheat; Roberts et al. 2016).
3. Cover crop in reduced or no-till systems. Minimizing tillage intensity can improve water infiltration and reduce evaporative losses, which can counterbalance the moisture utilized by a cover crop.
4. Cover crop as forage.
5. Cover crop mixes and cattle grazing were integrated into small grain and oilseed rotation in a high precipitation region in north-central Idaho (Finkelnburg et al. 2016). The three-year study demonstrated a gain in heifer body weight and winter wheat yield.
6. Plant cover crop in "prevented planting" acreage.
7. Climate change is expected to increase spring precipitation. Excess soil moisture can lead to more frequent delayed or prevented planting acreage; planting a cover crop in lieu of leaving the ground fallow can reduce erosion and improve soil quality without negative financial impact (Steury 2014).

The USDA Natural Resource Conservation Service (NRCS) has recently developed the PNW Cover Crop Selection Tool to help growers and conservation planners select cover crop species adapted to their climate, soils, and intended purposes. More information on the tool and its use can be found in ***http://www.nrcs.usda.gov/wps/portal/nrcs/detail/plantmaterials/technical/toolsdata/plant/?cid=nrcseprd894840***.

Measuring Economic Impacts of Diversification on a Rotational Basis

There are both short- and long-term economic impacts of rotational diversification, and measuring long-term impacts may be difficult. However, longer term impacts, such as increased soil organic matter and therefore increased water holding capacity, may provide a stronger incentive for changing farming practices. This section discusses short- and long-term costs and benefits of diversifying your cropping system.

Short run costs of changing cropping systems will typically involve some management challenges:

- Growers will need to evaluate potential alternative crops and varieties, along with accompanying changes in pest control, marketing, and other factors.
- A new crop or management practice may require either adapting current machinery or purchasing new machinery.
- A new crop or management practice will involve some increase in risk as well as additional management.

Short run benefits of changing your cropping system might include:

- Reduced weed, disease, or insect problems
- Modified spring or fall workload
- Ability to take advantage of strong market prices, or avoid weak market prices

Long run benefits of changing your typical cropping system may include:

- Reduced erosion
- Improved soil health
- Reduced risk associated with a more diverse portfolio of potential crops
- Increased returns due to increased flexibility with respect to timing of fallow operations

Several farmers in the region who converted to no-till many years ago are finding that their soils are able to support a more diverse crop mix today due to improved soil health. The rotational diversity they use today would not have been possible without earlier efforts to improve their soil. In grain-fallow areas where no-till may result in excessive evaporation of seed zone moisture, improved fallow practices and flex cropping can support diversification. Grower case studies are available online at ***http://pnwsteep.wsu.edu/dscases*** and ***https://www.reacchpna.org/case_studies***.

Enterprise budgets and worksheets for each rainfall zone are available to compare annual profitability by cropping system and are referenced below. Measuring average annual profitability for diverse rotations allows economic comparisons across rotations with varying lengths. Adjusting budgets to reflect rotational impacts can be done by changing crop yields, prices, herbicide use and other assumptions in specific worksheets. Many different rotational scenarios can be created, and the resulting comparisons will be calculated automatically in the summary tab of the worksheet. The following sections discuss budget scenarios comparing profitability for each dryland AEC (Painter 2016a; 2016b; 2016c). Since the relative economics of crop choices vary each year, all examples use 5-year average prices (2011–2015) received by PNW farmers.

In a detailed economic analysis of net returns by crop under conventional tillage for the Annual Crop AEC, (Painter 2016a; 2016d), assuming typical yields and prices as stated in Table 5-12, soft white winter wheat, chickpea, and hard red spring wheat were the most profitable crops, averaging $60/acre or more net returns over total costs. Austrian winter pea, soft white spring wheat, and lentil averaged $32, $27, and $18/acre, respectively, whereas net returns were negative for pea and spring barley at –$7 and –$24/acre, respectively, and lowest for spring canola at –$35/acre, based on the assumptions and underlying budget values in the Painter (2016a) worksheet. ●

However, on any one piece of land, average net returns over time need to be calculated on a rotational basis. Farmers rotate crops for many reasons, including reducing disease and pest issues, and to improve overall soil health and crop yields. Average net returns by rotation are calculated as a simple average of net returns by crop (Figures 5-6, 5-7, and 5-8) and may reflect rotation benefits to subsequent crops. In the Annual AEC, net returns were highest for a rotation of soft white winter wheat, hard red spring wheat, and chickpea, averaging $62 per rotational acre per year, and lowest for a rotation of winter wheat, spring barley, and spring canola, at $2/acre/year (Painter 2016a).

On a crop-by-crop basis, choices such as peas, barley, or canola appear unprofitable, but from a rotational or longer term standpoint, these crops may improve overall profitability. For example, research shows

Table 5-12. Crop yield and price assumptions and net returns over total costs by crop for the Annual Crop agroecological class, 2011–2015 average farmgate prices. ●

Annual Crop ●	Unit	Yield unit/ac	Price $/unit	Net Return $/ac/yr[1]
Soft White Winter Wheat	bu	80	$6.44	$64
Soft White Spring Wheat	bu	58	$6.44	$27
Hard Red Spring Wheat	bu	58	$8.41	$60
Spring Barley	ton	1.5	$188.00	–$18
Pea	lb	1700	$0.19	–$7
Lentil	lb	1100	$0.30	$18
Chickpea	lb	1200	$0.34	$63
Spring Canola	lb	1500	$0.21	–$35
Austrian Winter Pea	lb	2000	$0.22	$32

[1]Net returns over total costs using 2013 input costs.

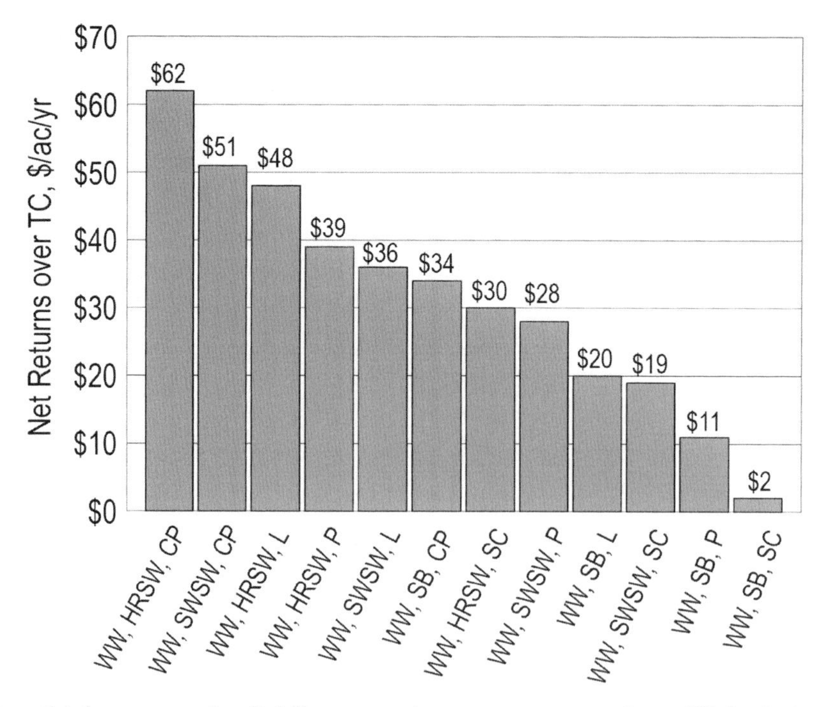

Figure 5-6. Average annual profitability, expressed as net returns over total costs (TC), for the Annual Crop agroecological class, 2011–2015 average farmgate crop prices. ●

a yield increase in wheat following peas and canola of 19% and 15%, respectively (Guy and Karow 2009); or, use of an herbicide-tolerant canola crop can reduce persistent annual grass weeds, increasing subsequent crop yields and reducing herbicide costs.

Detailed enterprise budgets and worksheets for low rainfall (Connolly et al. 2015a; 2015b) and intermediate rainfall (Connolly et al. 2016a; 2016b) regions, include an oilseed rotation and a grain rotation and separate wheat budgets for each. These budgets were adapted to reflect 2011–2015 average farmgate crop prices in the following profitability scenarios (see Painter 2016b; 2016c).

In the Transition AEC example, net returns over total costs were highest for a 45-bushel hard red spring wheat crop, averaging $47/acre (Table 5-13). Net returns for an 86-bushel winter wheat crop in the oilseed rotation (F-SWWW-SC) were $58/acre, but this included the costs of the preceding fallow year, so it is a 2-year return (Painter 2016b). ▲

The predominant crop sequence in the Transition AEC is a 3-year F-SWWW-SWSW rotation. Diversifying from this rotation to include an oilseed such as spring canola (F-SWWW-SC) can provide many rotational benefits, from reducing disease and weed pressure to breaking up hardpan layers and improving nutrient cycling. Assuming a 10% yield advantage for winter wheat in the oilseed rotation, and replacing the spring wheat crop with a spring canola crop, net returns for the 3-year period average $19/acre/year, compared to $16/acre/year in the F-SWWW-SWSW rotation, or $25/acre for F-SWWW-HRSW (Figure 5-7). Thus, the oilseed rotation is competitive with the grain rotation under the assumption of a 10% yield advantage for soft white winter wheat.

In the Grain-Fallow AEC example, net returns over total costs for 2011–2015 using five-year average prices for PNW farmers were greatest ($16/acre over a 2-year period) for soft white winter wheat preceded by summer fallow in a 4-year oilseed rotation (F-WC-F-SWWW), which assumes a 10% yield advantage for winter wheat (Table 5-14; Painter 2016c). The standard 2-year (F-SWWW) cropping system showed net returns over total costs of –$6/acre over a 2-year period (–$3/acre/year), which is not an economically sustainable system. For winter canola

Table 5-13. Crop yield and price assumptions and net return over total costs by crop for the Transition agroecological class, 2011–2015 average farmgate prices.

Transition	Rotation[1]	Unit	Yield unit/ac	Price $/unit	Net Return $/ac[2]
Soft White Winter Wheat	OR	bu	86	$6.44	$58
Spring Canola	OR	lb	1500	$0.21	–$2
Soft White Winter Wheat	GR	bu	78	$6.44	$30
Soft White Spring Wheat	GR	bu	50	$6.44	$18
Hard Red Spring Wheat	GR	bu	45	$8.41	$47
Spring Barley	GR	ton	1.5	$188.00	$4

[1]OR = oilseed rotation with canola (F-WW-SC); GR = grain rotation (F-WW-SW or SB)
[2]Net returns over costs using 2013 input costs.

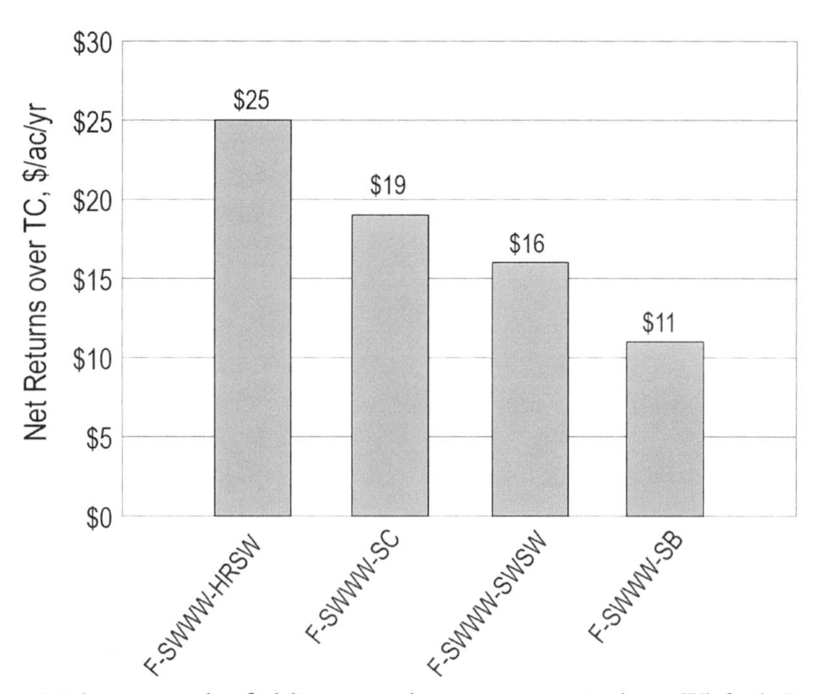

Figure 5-7. Average annual profitability, expressed as net returns over total costs (TC), for the Transition agroecological class, 2011–2015 average farmgate prices.

preceded by summer fallow (F-WC), average returns over total costs were even less profitable at –$22/acre over a 2-year period. ■

For the 4-year oilseed rotation (F-SWWW-F-WC), average net returns over total costs were –$1/acre/year (Figure 5-8), assuming the 10% average yield advantage for soft white winter wheat, compared to –$3/acre/year for the standard cropping system of F-SWWW. The 10% yield advantage for winter wheat in the oilseed rotation is attributed to a more diversified rotation, particularly beneficial in the presence of problems such as persistent grass weeds or wheat disease. Lack of profitability in both of these systems highlights the production challenges in the Grain-Fallow AEC.

Obviously, economic feasibility is critical to sustainability. Growers are not motivated to plant spring and winter canola if estimated net returns over total costs are negative. When spring canola prices were rising between 2008 and 2012 (Figure 5-9), planted acreage of this crop responded, just as chickpea acreage expanded in response to the high relative expected returns from this crop. However, annual production of winter wheat across all AECs typically occurs on more than 40% of the total acreage. Continuous cropping of small grains results in yield decline and decreased returns. Relatively small gains in yields or cost savings can make diversification into alternative crops economically advantageous. Producers may be willing to grow a less profitable crop in the current year to increase resiliency and economic returns in subsequent years, particularly if they can use tools such as these budget worksheets to estimate impacts under different assumptions. Quantifying the potential risks and benefits associated with new crops or rotations may be an important step in convincing growers, bankers, landlords, and other farming partners to try new practices for enhancing overall sustainability.

Table 5-14. Crop yield and price assumptions and net returns (2-year) over total costs for the Grain-Fallow agroecological class, 2011–2015 average farmgate prices. ■

Grain-Fallow ■	Rotation[1]	Unit	Yield unit/ac	Price $/unit	Net Return $/ac[2]
Winter Canola	OR	lb	1500	$0.21	-$22
Soft White Winter Wheat	OR	bu	50	$6.44	$16
Soft White Winter Wheat	GR	bu	45	$6.44	-$6

[1]OR = oilseed rotation with canola; GR = grain rotation
[2]Net returns over costs ($/acre, 2-year crop-fallow cycle) using 2013 input costs.

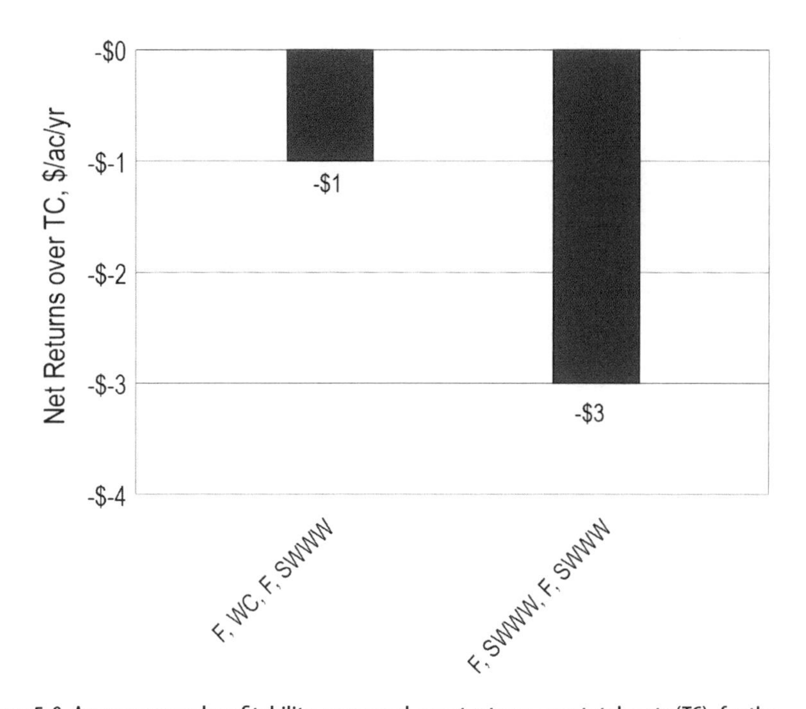

Figure 5-8. Average annual profitability, expressed as net returns over total costs (TC), for the Grain-Fallow agroecological class, 2011–2015 average farmgate prices. ■

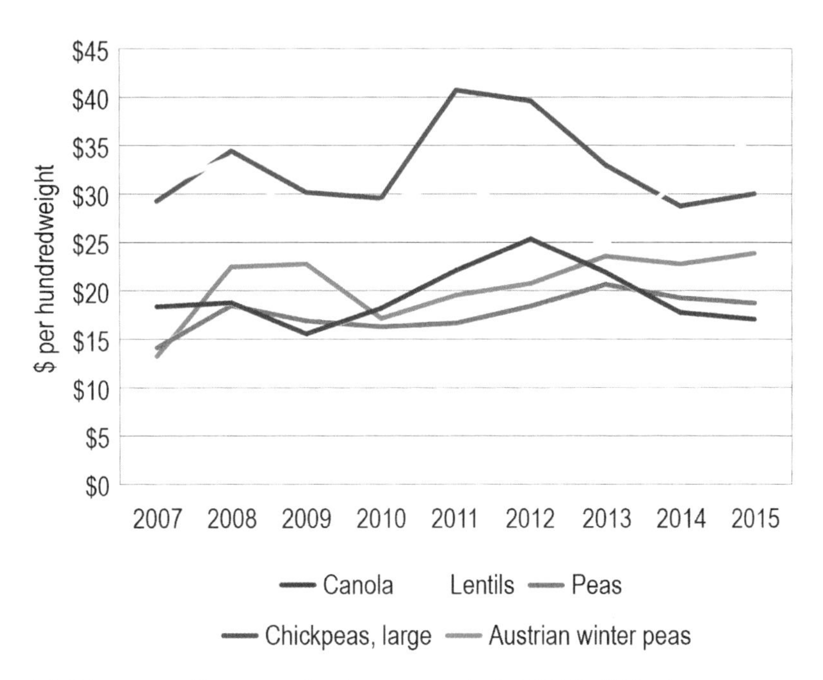

Figure 5-9. Prices received by growers for alternate crops in the PNW (NASS data).

Grower Resources

Oregon State University AgBiz Logic Website

http://www.agbizlogic.com/

Oregon State University Wheat Research Website

http://cropandsoil.oregonstate.edu/group/wheat

REACCH Farm Enterprise Budgets

https://www.reacchpna.org/farm-enterprise-budgets

REACCH Grower Case Studies

https://www.reacchpna.org/case_studies

STEEP Grower Case Studies

http://pnwsteep.wsu.edu/dscases/

University of Idaho Brassica Breeding and Research Website

http://www.cals.uidaho.edu/brassica/index.asp

University of Idaho AgBiz Website

http://www.uidaho.edu/cals/idaho-agbiz

STEEP Grower Case Studies

http://pnwsteep.wsu.edu/dscases/

Washington State University Washington Oilseed Cropping Systems Website

http://css.wsu.edu/biofuels/

Washington State University Small Grains Website

http://smallgrains.wsu.edu/

References

Abatzoglou, J., and S. Eigenbrode. 2016. Climate Impacts on Palouse Pea Yields. In 2016 Dryland Field Day Abstracts: Highlights of Research Progress. Washington State University, Oregon State University, and University of Idaho.

Armah-Agyeman, G., J. Loiland, R. Karow, and A.N. Hang. 2002. Safflower. Oregon State University Extension Publication EM8792. *http://extension.oregonstate.edu/gilliam/sites/default/files/Safflower.pdf*

Baker, D.A., D.L. Young, D.R. Huggins, and W.L. Pan. 2004. Economically Optimal Nitrogen Fertilization for Yield and Protein in Hard Red Spring Wheat. *Agronomy Journal* 96: 116–123.

Bewick, L.S., F.L. Young, J.R. Alldredge, and D.L. Young. 2008. Agronomics and Economics of No-Till Facultative Wheat in the Pacific Northwest, USA. *Crop Protection* 27: 932–942.

Bezdicek, D.F., and D.M. Granatstein. 1989. Crop Rotation Efficiencies and Biological Diversity in Farming Systems. *American Journal of Alternative Agriculture* 4(3-4).

Brown, B., M. Westcott, N. Christensen, W. Pan, and J. Stark. 2005. Nitrogen Management for Hard Wheat Protein Enhancement. Pacific Northwest Extension Publication PNW578. ***http://plantbreeding.wsu.edu/pnw0578.pdf***

Brown, J., J.B. Davis, and A. Esser. 2005. Pacific Northwest Condiment Mustard (*Sinapis alba* L.) Grower Guide 2000-2002. University of Idaho. NREL Subcontract report NREL/SR-501-36307.

Brown, J., J.B. Davis, M. Lauver, and D. Wysocki. 2009. USCA Canola Growers Manual. University of Idaho and Oregon State University Extension Publication. ***http://www.uscanola.com/site/files/956/102387/363729/502632/Canola_Grower_Manual_FINAL_reduce.pdf***

Campbell, J. 2016. Weed Control in Pulse Crops. In 2016 Dryland Field Day Abstracts: Highlights of Research Progress. Washington State University, Oregon State University, and University of Idaho.

Chen, C., P. Miller, F. Muehlbauer, K. Neill, D. Wichman, and K. McPhee. 2006. Winter Pea and Lentil Response to Seeding Date and Micro- and Macroenvironments. *Agronomy Journal* 98: 1655–1663.

Chen, C., K. Neill, M. Burgess, and A. Bekkerman. 2012. Agronomic Benefit and Economic Potential of Introducing Fall-Seeded Pea and Lentil into Conventional Wheat-Based Crop Rotations. *Agronomy Journal* 104: 215–224.

Connolly, J.R., V.A. McCracken, and K.M. Painter. 2016a. Enterprise Budgets: Wheat & Canola Rotations in Eastern WA Intermediate Rainfall (12-16") zone. Oilseed Series. Washington State University Extension Publication TB10E. ***http://cru.cahe.wsu.edu/CEPublications/TB10E/TB10.pdf***

Connolly, J.R., V.A. McCracken, and K.M. Painter. 2016b. Worksheet for TB10E. ***https://boundaryagblog.files.wordpress.com/2016/10/intermediaterainfallenterprisebudget_tb10spreadsheet.xlsx***

Connolly, J.R., V.A. McCracken, and K.M. Painter. 2015a. Enterprise Budgets: Wheat and Canola Rotations in Eastern Washington Low Rainfall (<12") Region. Oilseed Series. Washington State University Extension Publication TB09. ***http://cru.cahe.wsu.edu/CEPublications/TB09E/TB09.pdf***

Connolly, J.R., V.A. McCracken, and K.M. Painter. 2015b. Worksheet for TB09. ***https://boundaryagblog.files.wordpress.com/2016/12/lowrainfallenterprisebudget_tb09spreadsheet.xlsx***

Corp, M., S. Machado, R. Smiley, D. Ball, S. Petrie, M. Siemens, and S. Guy. 2004. Dryland Cropping Systems: Chickpea Production Guide. Oregon State University Extension Publication EM8791. ***https://catalog.extension.oregonstate.edu/sites/catalog/files/project/pdf/em8791.pdf***

Cutforth, H.W., B.G. McConkey, D. Ulrich, P.R. Miller, and S.V. Angadi. 2002. Yield and Water Use Efficiency of Pulses Seeded Directly into Standing Stubble in the Semiarid Canadian Prairie. *Canadian Journal of Plant Science* 82: 681–686.

Douglas, Jr., C.L., P.M. Chevalier, B. Klepper, A.G. Ogg, and P.E. Rasmussen. 1999. Conservation Cropping Systems and Their Management. In Conservation Farming in the United States, E. Michalson, R. Papendick, and J. Carlson, eds.

Douglas, Jr., C.L., R.W. Rickman, B.L. Klepper, and J.F. Zuzel. 1992. Agroclimatic Zones for Dryland Winter Wheat Producing Areas of Idaho, Washington, and Oregon. *Northwest Science* 66: 1.

Ehrensing, D.T., and S.O. Guy. 2008. Camelina. Oregon State University Extension Publication EM8953E. ***http://extension.oregonstate.edu/gilliam/sites/default/files/Camelina_em8953-e.pdf***

Esser, A.D., J. Knodel, and J. Knodel. 2008. Hard Red Winter Wheat Feasibility in Comparison to Soft White Winter Wheat. Washington State University Extension Publication OFT 08-2.

Esser, A., and R. Hennings. 2012. Winter Canola Feasibility in Rotation with Winter Wheat. Washington State University Extension Publication FS068E. ***http://www.pnw-winderosion.wsu.edu/Docs/Publications/08/Esser-Knodel.pdf***

Esser, A.D., and D. Appel. 2016. Spring Canola in Rotation at WSU Wilke Farm. Presentation given at the WSU Oilseed Meeting, Pullman WA. ***http://css.wsu.edu/biofuels/files/2016/03/EsserMarch2016WOCS.pdf***

Finkelnburg, D., K. Hart, and J. Church. 2016. Cover Crop Demonstration Project in North-Central Idaho. In 2016 Dryland Field Day Abstracts. University of Idaho, Oregon State University, and Washington State University.

Gan Y., S.V. Angadi, H. Cutforth, D. Potts, V.V. Angadi, and C.L. McDonald. 2004. Canola and Mustard Response to Short Periods of Temperature and Water Stress at Different Developmental Stages. *Canadian Journal of Plant Science* 84: 697–704.

Gan Y.T., C.A. Campbell, H.H. Janzen, R. Lemke, L.P. Liu, P. Basnyat, and C.L. McDonald. 2009. Root Mass for Oilseed and Pulse Crops: Growth and Distribution in the Soil Profile. *Canadian Journal of Plant Science* 89: 883–893.

Gan, Y., C. Hamel, J.T. O'Donovan, H. Cutforth, R.P. Zentner, C.A. Campbell, Y. Niu, and L. Poppy. 2015. Diversifying Crop Rotations with Pulses Enhances System Productivity. *Scientific Reports* 5: 4625.

Gan, Y., C. Liang, Q. Chai, R.L. Lemke, C.A. Campbell, and R.P. Zentner. 2014. Improving Farming Practices Reduces the Carbon of Spring Wheat Production. *Nature Communications* 5: 5012.

Guy, S. 2014. Rotational Influence of Brassica Biofuel and Other Crops on Winter Wheat. In Washington Oilseed Cropping Systems Project 2013 Annual Progress Report. K.E. Sowers and W.L. Pan, eds.

Guy, S. 2016. Pulse Crop Production and Management for Successful Conservation Tillage Cropping Systems. Pacific Northwest Direct Seed Association Annual Conference, January 2016. ***http://www.directseed.org/files/7714/5335/1277/Pulse_Crop_Production_for_Successful_Conservation_Cropping_System_Stephen_Guy.pdf***

Guy, S.O., and R.M. Gareau. 1998. Crop Rotation, Residue Durability and N Fertilizer Effects on Winter Wheat Production. *Journal of Production Agriculture* 11(4): 457–461.

Guy, S., and R. Karow. 2009. Alternate Crops for Direct Seeding in the Dryland Inland Northwest. ***http://pnwsteep.wsu.edu/directseed/conf98/alternat2.htm***

Guy, S.O., and M.A. Lauver. 2015. Grain Legume Variety and Agronomic Performance Trials 2015. Washington State University Extension Variety Testing Program. ***http://smallgrains.wsu.edu/wp-content/uploads/2014/09/2015-Legume-Book.pdf***

Hansen, J., W.F. Schillinger, A. Kennedy, and T. Sullivan. 2016. Rotational Effects of Winter Canola on Subsequent Spring Wheat as Related to the Soil Microbial Community. In 2016 Field Day Abstracts. University of Idaho, Oregon State University, and Washington State University.

Huggins, D., and K. Painter. 2011. Spring and Winter Canola Research at the WSU Cook Agronomy Farm. In Washington Oilseed Cropping Systems Project 2011 Annual Progress Report. K.E. Sowers and W.L. Pan, eds.

Huggins, D.R., and W.L. Pan. 1991. Wheat Stubble Management Affects Growth, Survival, and Yield of Winter Grain Legumes. *Soil Science Society of America Journal* 55: 823–829.

Huggins, D., W. Pan, W. Schillinger, F. Young, S. Machado, and K. Painter. 2015. Crop Diversity and Intensity in the Pacific Northwest Dryland Cropping Systems. In Regional Approaches to Climate Change for Pacific Northwest Agriculture: Climate Science Northwest Farmers Can Use. REACCH Annual Report Year 4: 38-41. University of Idaho, Washington State University, and Oregon State University.

Huggins, D.R., and J.P. Reganold. 2008. No-Till: The Quiet Revolution. *Scientific American* July: 71–77.

Hulbert, S., A.D. Esser, and D. Appel. 2013. Oilseed Production and Outreach Report: Region 2. In Washington Oilseed Cropping Systems Project 2013 Annual Progress Report. K.E. Sowers and W.L. Pan, eds.

Hulbert, S., S. Guy, W. Pan, T. Paulitz, W. Schillinger, D. Wysocki, and K. Sowers. 2012. Camelina Production in the Dryland Pacific Northwest. Washington State University Extension Publication FS073E. ***http://cru.cahe.wsu.edu/CEPublications/FS073E/FS073E.pdf***

Juergens, L.A., D.L. Young, W.F. Schillinger, and H.R. Hinman. 2004. Economics of Alternative No-Till Spring Crop Rotations in Washington's Wheat-Fallow Region. *Agronomy Journal* 96: 154–158.

Karow, R. 2014. Canola. Oregon State University Extension Publication EM8955. ***https://catalog.extension.oregonstate.edu/sites/catalog/files/project/pdf/em8955.pdf***

Kaur, H., D. Huggins, R. Rupp, J. Abatzoglou, C. Stockle, and J. Reganold. 2015. Bioclimatic-Driven Future Shifts in Dryland Agroecological Classes. In Regional Approaches to Climate Change for Pacific Northwest Agriculture: Climate Science Northwest Farmers Can Use. REACCH Annual Report Year 4: 10–11. University of Idaho, Washington State University, and Oregon State University.

Kephart, K.D., G.A. Murray, and D.L. Auld. 1990. Alternate Crops for Dryland Production Systems in Northern Idaho. In Advances in New Crops, J. Janick and J.E. Simon, eds. Timber Press, Portland, OR.

Kincaid, R., K. Johnson, J. Michal, S. Hulbert, W. Pan, J. Barbano, and A. Huisman, A. 2011. Biennial Canola for Forage and Ecosystem Improvement in Dryland Cropping Systems. *Advances in Animal Biosciences* 2: 457.

Kirkegaard, J., O. Christen, J. Krupinsky, and D. Layzell. 2008a. Break Crop Benefits in Temperate Wheat Production. *Field Crops Research* 107: 185–195.

Kirkegaard, J.A., S.J. Sprague, H. Dove, W.M. Kelman, S.J. Marcroft, A. Lieschke, G.N. Howe, and J.M. Graham. 2008b. Dual-Purpose Canola – A New Opportunity in Mixed Farming Systems. *Australian Journal of Agricultural Research* 59: 291–302.

Koenig, R.T. 2005. Dryland Winter Wheat. Eastern Washington Nutrient Management Guide. Washington State University Publication EB1987E. ***http://cru.cahe.wsu.edu/CEPublications/EB1987E/EB1987E.pdf***

Koenig, R., A. Hammac, and W. Pan. 2011. Canola Growth, Fertility and Development. Washington State University Extension Publication FS045E. ***http://cru.cahe.wsu.edu/CEPublications/FS045E/FS045E.pdf***

Long, D.S., F.L. Young, W.F. Schillinger, C.L. Reardon, J.D. Williams, B.L. Allen, W.L. Pan, and D.J. Wysocki. 2016. Development of Dryland Oilseed Production Systems in Northwestern Region of the USA. *BioEnergy Research* 9(1).

Lutcher, L.K, D.J Wysocki, M.K. Corp, and D.S. Horneck. 2013. Agronomic Guidelines for Flexible Cropping Systems in Dryland Areas of Oregon. Oregon State University Extension Publication EM 8999-E. ***https://catalog.extension.oregonstate.edu/em8999***

Lutcher, L.K., D.J. Wysocki, and M.D. Flowers. 2012. Performance of Hard Red Winter Wheat in Late-Planted No-Till Fallow. Pacific Northwest Extension Publication PNW 635. Oregon State University. ***https://catalog.extension.oregonstate.edu/pnw635***

Maaz, T.M., W.L. Pan and W.A. Hammac. 2016. Components of Improved Canola Nitrogen Use Efficiency with Increasing Water and Nitrogen. *Agronomy Journal* (in press).

Machado, S., B. Tuck, and C. Humphreys. 2004. Alternate Crops for Eastern Oregon: Research. ***http://cbarc.aes.oregonstate.edu/sites/default/files/Alternate_Crops_for_Eastern_Oregon2.pdf***

Machado, S., B. Tuck, and C. Humphreys. 2006a. Alternative Rotation Crops: Lentils. In 2006 Dryland Agriculture Research Annual Report. Oregon State University Extension CBARC Special Report 1068: 32–39.

Machado, S., C. Humphreys, B. Tuck, and M. Corp. 2006b. Seeding Date, Plant Density, and Cultivar Effects on Chickpea Yield and Seed Size in Eastern Oregon. *Crop Management Journal*.

Mahler, R.L. 2005a. Chickpeas. Northern Idaho Fertilizer Guide. University of Idaho Extension Publication CIS 823. ***http://www.extension.uidaho.edu/nutrient/pdf/Peas-Lentils/ChickpeasFG.pdf***

Mahler, R.L. 2005b. Lentils. Northern Idaho Fertilizer Guide. University of Idaho Extension Publication CIS 1083. ***https://www.cals.uidaho.edu/edcomm/pdf/CIS/CIS1083.pdf***

Mahler, R.L. 2015. Spring Peas. Northern Idaho Fertilizer Guide. University of Idaho Extension Publication CIS1084. ***https://www.cals.uidaho.edu/edcomm/pdf/CIS/CIS1084.pdf***

Mahler, R.L., and S.O. Guy. 2005. Spring Canola. Northern Idaho Fertilizer Guide. University of Idaho Extension Publication CIS 1012. ***https://www.cals.uidaho.edu/edcomm/pdf/CIS/CIS1012.pdf***

McGee, R.J. 2016. Breeding Fall Planted Pulse Crops. Pacific Northwest Direct Seed Association Annual Conference, January 2016. ***http://www.directseed.org/files/9514/5335/1350/Breeding_Fall_Planted_Pulse_Crops_Rebecca_McGee.pdf***

McGee, R.J., and K.E. McPhee. 2012. Release of Autumn-Sown Pea Germplasm PS03101269 with Food-Quality Seed Characteristics. *Journal of Plant Registrations* 6: 354–357.

McGee, R.J., J. Pfaff, S. Guy, and C. Chen. 2014. Developing Food Quality Autumn-Sown Legumes. ***http://css.wsu.edu/biofuels/files/2014/02/McGee2014OSDS.pdf***

McPhee, K.E., and F.J. Muelhbauer. 2005. Adaptation of Winter Grain Legumes to US Production Areas. ***http://pnwsteep.wsu.edu/directseed/conf2k5/pdf/mcphee.pdf***

McPhee, K.E., and F.J. Muelhbauer. 2007. Registration of 'Specter' Winter Feed Pea. *Journal of Plant Registrations* 1(2): 118–119.

Muehlbauer, F.J., and K.E. McPhee. 2007. Registration of 'Morton' Winter-Hardy Lentil. *Crop Science* 47(1): 438.

Muehlbauer, F.J., and D. Rhoades. 2016. A Brief History of Pulse Production. *Crops and Soils Magazine* March-April 2016: 16–19.

Muehlbauer, F.J., R.W. Short, R.J. Summerfield, K.J. Morrison, and D.G. Swan. 1981. Description and Culture of Lentils. Washington State University Extension Bulletin EB0957 (archived).

Murray, G.A., Kephart, K.D., O'Keeffe, L.E., Auld, D.L. and R.H. Callihan. 1987. Dry Pea, Lentil and Chickpea Production in Northern Idaho. University of Idaho Extension Bulletin 664.

NASS (National Agricultural Statistics Service). 2015. Barley Historic Data for Washington: 1882-2011.

Neely, C. 2010. The Effect of Forage Harvest on Forage Yield, Forage Quality, and Subsequent Seed Yield of Dual-Purpose Biennial Winter Canola (*Brassica napus* L.). University of Idaho Master's Thesis.

Nelson, H.R. 2016. Fall Planted Peas for Eastern Washington. Pacific Northwest Direct Seed Association Annual Conference, January 2016. ***http://www.directseed.org/files/3014/5335/1268/Fall_Planted_Peas_for_Eastern_Washington_Howard_Nelson.pdf***

NRCS (National Resource Conservation Service). 2014. CSP Energy Enhancement Activity. Using Nitrogen Provided by Legumes, Animal Manure and Compost to Supply 90 to 100% of Nitrogen Needs. ENR10.

Painter, K.M. 2016a. High Rainfall Enterprise Budget Worksheet. Based on Enterprise Budgets for the Dryland Grain Annual Cropping Region of the Pacific Northwest (in revision 2016). ***https://boundaryagblog.files.wordpress.com/2016/10/highrainfallenterprisebudget_5yearaverageprices_21oct2016.xlsx***

Painter, K.M. 2016b. Intermediate Rainfall Enterprise Budget Worksheet. Based on Connelly et al. 2016. ***https://boundaryagblog.files.wordpress.com/2016/10/intermediaterainfallenterprisebudget-5-year-average-crop-prices-2013-input-costs_21oct2016.xlsx***

Painter, K.M. 2016c. Low Rainfall Enterprise Budget Worksheet. Based on Connelly et al. 2015. ***https://boundaryagblog.files.wordpress.com/2016/12/lowrainfallenterprisebudget_tb09spreadsheet.xlsx***

Painter, K.M. 2016d. Enterprise Budgets for the Dryland Grain Annual Cropping Region of the Pacific Northwest (Submitted to University of Idaho Extension; currently in revision).

Pan, W.L., F.L. Young, T.M. Maaz, and D.R. Huggins. 2016. Canola Integration into Semi-Arid Wheat Cropping Systems of the Inland Pacific Northwestern USA. *Crop and Pasture Science* 67(4): 253–265.

Papendick, R.I., and D.E. Miller. 1977. Conservation Tillage in the Pacific Northwest. *Journal of Soil and Water Conservation* 32: 49–52.

Petrie, S. 2008. Identifying Spring Habit Specialty Barley Varieties for Direct-Seeding and Development of Winter Habit Forms: Final Report. In Solutions to Environmental and Economic Problems 2008 STEEP Annual Progress Report. University of Idaho, Oregon State University, Washington State University. ***http://pnwsteep.wsu.edu/annualreports/2008/index.htm***

Petrie, S., S. Machado, R. Johnson, L. Pritchett, K. Rhinhart, and B. Tuck. 2010. Adaptation and Yield of Spring and Fall Sown Safflower in Northern Oregon. Oregon State Extension CBARC Report. ***http://cbarc.aes.oregonstate.edu/sites/default/files/adaptation_and_yield_of_spring_and_fall_sown_safflower_in_northeastern_oregon.pdf***

Reckling, M., G. Bergkvist, C.A. Watson, F.L. Stoddard, P.M. Zander, R.L. Walker, A. Pristeri, I. Toncea, and J. Bachlinger. 2016. Trade-Offs between Economic and Environmental Impacts of Introducing Legumes into Cropping Systems. *Frontiers in Plant Science* (7): 669.

Rey, J.I., P.M. Hayes, S.E. Petrie, A. Corey, M. Flowers, J.B. Ohm, C. Ong, K. Rhinhart, and A.S. Ross. 2009. Production of Dryland Barley for Human Food: Quality and Agronomic Performance. *Crop Science* 49: 347–355.

Roberts, D., F.J. Fleming, C. Gross, T. Rush, E. Warner, C. Laney, B. Dobbins, D. Dobbins, R. Vold, A. Esser, D.P. Appel, and J. Clapperton. 2016. Cover Cropping for the Intermediate Precipitation Zone of Dryland Eastern Washington. In 2016 Dryland Field Day Abstracts. University of Idaho, Oregon State University, and Washington State University.

Rustgi, S., D. Von Wettstein, N. Wen, J. Mantanguihan, N.O. Ankrah, R. Brew-Appiah, R. Gemini, K.M. Murphy, and P. Reisenauer. 2015. Breeding Barley to Meet Demands of the Washington Growers. In 2015 Field Day Abstracts. Washington State University.

Schillinger, W.F. 2016. Seven Rainfed Wheat Rotation Systems in a Drought-Prone Mediterranean Climate. *Field Crops Research* 191: 123–130.

Schillinger, W.F., R. Jirava, J. Jacobsen, and S. Schofstoll. 2015. Late-Planted Winter Triticale in the Dry Region. In Regional Approaches to Climate Change for Pacific Northwest Agriculture: Climate Science Northwest Farmers Can Use. REACCH Annual Report Year 4: 30-31. University of Idaho, Washington State University, and Oregon State University.

Schillinger, W.F., R. Jirava, J. Jacobsen, and S. Schofstoll. 2016. Long-Term Safflower Cropping Systems Experiment near Ritzville, WA. In 2016 Field Day Abstracts. University of Idaho, Oregon State University, and Washington State University.

Schillinger, W., H. Johnson, J. Jacobsen, S. Schofstoll, A. Kennedy, and T. Paulitz. 2013. Winter Canola Rotation Benefit Experiment in the Intermediate Precipitation Zone. In 2013 Field Day Abstracts. Washington State University.

Schillinger, W.F., A.C. Kennedy, and D.L. Young. 2007. Eight Years of Annual No-Till Cropping in Washington's Winter Wheat-Summer Fallow Region. *Agriculture, Ecosystems & Environment* 120: 345–358.

Schillinger, W.F., and R.I. Papendick. 2009. Then and Now: 125 Years of Dryland Wheat Farming in the Inland Pacific Northwest. Washington State University Extension Publication EM004E. ***http://cru.cahe.wsu.edu/CEPublications/EM004e/em004e.pdf***

Schillinger, W.F., R.I. Papendick, S.O. Guy, P.E. Rasmussen, and C. van Kessel. 2003. Dryland Cropping in the Western United States. In Pacific Northwest Conservation Tillage Handbook Series No. 28, Chapter 2.

Schillinger, W.F., R.I. Papendick, and D.K. McCool. 2010. Soil and Water Challenges for Pacific Northwest Agriculture. In Soil and Water Advances in the United States. T.M. Zobeck and W.F. Schillinger, eds. SSA Special Publication 60.

Schillinger, W., T. Paulitz, B. Sharratt, A. Kennedy, W. Pan, S. Wuest, H. Johnson, J. Jacobsen, S. Schofstoll, and J. Hansen. 2014a. Dryland Irrigated Cropping Systems Research with Winter Canola, Camelina, and Safflower. In Washington Oilseed Cropping Systems Project 2014 Annual Progress Report. Washington State University. ***http://css.wsu.edu/biofuels/files/2012/09/Schillinger_Reg2_2014.pdf***

Schillinger, W.F., S.E. Schofstoll, and J.R. Allredge. 2012. Predicting Wheat Grain Yields Based on Available Water. Washington State University Extension Publication EM049E. ***http://lindstation.wsu.edu/files/2012/04/Available-water-and-wheat-yield-EM049E.pdf***

Schillinger, W.F., and D.L. Young. 2004. Cropping Systems Research in the World's Driest Rainfed Wheat Region. *Agronomy Journal* 96: 1182–1187.

Schillinger, W.F., D. Wysocki, T. Chastain, S.O. Guy, and R. Karow. 2014b. Camelina: Effects of Planting Date and Method on Stand Establishment and Seed Yield. Pacific Northwest Extension Publication PNW661. ***http://cru.cahe.wsu.edu/CEPublications/PNW661/PNW661.pdf***

Sharratt, B.S., and W.F. Schillinger. 2016. Soil Characteristics and Wind Erosion Potential of Wheat-Oilseed-Fallow Cropping Systems. *Soil Science Society of America Journal* 80: 704–710.

Sharratt, B.S., and W.F. Schillinger. 2014. Windblown Dust Potential from Oilseed Cropping Systems in the Pacific Northwest United States. *Agronomy Journal* 106(3): 1147–1152.

Smiley, R.W., and S. Machado. 2009. Pratylenchus Neglectus Reduces Yield of Winter Wheat in Dryland Cropping Systems. *Plant Disease* 93: 263–271.

Snapp, S.S., S.M. Swinton, R. Labarta, D. Mutch, J.R. Black, R. Leep, J. Nyiraneza, and K. O'Neil. 2005. Evaluating Cover Crops for Benefits, Costs and Performance within Cropping System Niches. *Agronomy Journal* 97: 322–332.

Sowers, K., D. Roe, and W. Pan. 2011. Oilseed Production Case Studies in the Eastern Washington High Rainfall Zone. Washington State University Extension Publication EM037E. ***http://cru.cahe.wsu.edu/CEPublications/EM037E/EM037E.pdf***

Sowers, K.E., R.D. Roe, and W.L. Pan. 2012. Oilseed Production Case Studies in the Eastern Washington Low to Intermediate Rainfall Zone. Washington State University Extension Publication EM048E. ***http://cru.cahe.wsu.edu/CEPublications/EM048E/EM048E.pdf***

Steury, D. 2014. Cover Crops, Soil Conservation, and Prevented Planting Acres. Regional Approaches to Climate Change for Pacific Northwest Agriculture: Climate Science Northwest Farmers Can Use. Annual Report Year 3: 24–25. University of Idaho, Washington State University, and Oregon State University. ***https://www.reacchpna.org/sites/default/files/AR3_4.1.pdf***

Sullivan, L.S., F.L. Young, R.W. Smiley, and J.R. Alldredge. 2013. Weed and Disease Incidence in No-Till Facultative Wheat in the Pacific Northwest, USA. *Crop Protection* 27: 932–942.

Thompson, W.H., and P.G. Carter. 2014. Cover Crop Water Consumption in Southeastern Washington Palouse. Poster in ASA, CSSA and SSSA International meeting. ***https://scisoc.confex.com/scisoc/2014am/webprogram/Paper84949.html***

Thorne, M.E., F.L. Young, W.L. Pan, R. Bafus, and J.R. Allredge. 2003. No-Till Spring Cereal Cropping Systems Reduce Wind Erosion Susceptibility in the Wheat/Fallow Region of the Pacific Northwest. *Journal of Soil and Water Conservation* 58(5): 250–257. ***http://www.jswconline.org/content/58/5/250.full.pdf+html***

Unger, P.W., D.W. Fryrear, and M.J. Lindstrom. 2006. Soil Conservation. In Dryland Agriculture, G.A. Peterson et al., eds. ASA-CSSA-SSSA, Madison, WI. *Agronomy Monograph* 23: 87–112.

Veseth, R. 1989. Small Red Lentil as a Fallow Substitute. In Crops and Varieties, Chapter 8, No. 10. PNW Conservation Tillage Handbook Series.

Wysocki, D.J., and M.K. Corp. 2002. Edible Mustard. Oregon State University Extension Publication EM8796. ***http://extension.oregonstate.edu/gilliam/sites/default/files/Mustard_em8796-e.pdf***

Wysocki, D.J., M.K. Corp, D.A. Horneck, and L.K. Lutcher. 2007. Irrigated and Dryland Canola. Nutrient Management Guide. Oregon State University Extension Publication EM8943-E. ***http://ir.library.oregonstate.edu/xmlui/bitstream/handle/1957/20480/em8943-e.pdf***

Wysocki, D.J., D.A. Horneck, L.K. Lutcher, J.M. Hart, S.E. Petrie, and M.K. Corp. 2006. Winter Wheat in Continuous Cropping Systems (Intermediate Precipitation Zone). Oregon State University Extension Publication FG 83. ***http://extension.oregonstate.edu/gilliam/sites/default/files/WW-CC_Med_Fert_Guide_FG83-E.pdf***

Yorgey, G., S. Kantor, K. Painter, L. Bernacchi, H. Davis, and D. Roe. 2016a. Enhancing Crop Diversity: Steve and Becky Camp. Farmer-to-Farmer Case Study Series. Pacific Northwest Extension Publication PNW690. Washington State University. ***http://cru.cahe.wsu.edu/CEPublications/PNW690/PNW690.pdf***

Yorgey, G., S. Kantor, K. Painter, D. Roe, H. Davis, and L. Bernacchi. 2016b. Flex Cropping and Precision Agriculture Technologies: Bill Jepson. Pacific Northwest Extension Publication PNW681. Washington State University. ***https://www.reacchpna.org/sites/default/files/flex_cropping_case.pdf***

Young, F.L., A.R. Alldredge, W.L. Pan, and C. Hennings. 2015. Comparisons of Annual No-Till Spring Cereal Cropping Systems in the Pacific Northwest Winter Wheat/Fallow Region. *Crop Forage and Turfgrass Management* 1:1.

Young F.L., D.S. Long, and J.R. Alldredge. 2012. Effect of Planting Methods on Spring Canola (*Brassica napus* L.) Establishment and Yield in the Low Rainfall Region of the Pacific Northwest. *Crop Management* 1:1.

Young, F.L, L. Port, and W.L. Pan. 2016a. Best Management Practices to Improve Low-Rainfall Oilseed Production. In 2016 Field Day Abstracts. University of Idaho, Oregon State University, and Washington State University.

Young, F.L., and M.E. Thorne. 2004. Weed Species Dynamics and Management in No-Till and Reduced-Till Fallow Cropping Systems for the Semi-Arid Agricultural Region of the Pacific Northwest, USA. *Crop Protection* 23: 1097–1110.

Young, F.L., D.K. Whaley, W.L. Pan, R.D. Roe, and J.R. Alldredge. 2014a. Introducing Winter Canola to the Winter Wheat-Fallow Region of the Pacific Northwest. *Crop Management* 13(1).

Young, F.L., D.K. Whaley, N.C. Lawrence, and I.C. Burke. 2016b. Feral Rye Control in Winter Canola in the Pacific Northwest. In 2016 Field Day Abstracts. University of Idaho, Oregon State University, and Washington State University.

Table 5-A1. Potential alternative crops for direct seeding in the inland Pacific Northwest by agroecological class.

	Annual Crop ●	Transition ▲	Grain-Fallow ■
Winter cereals			
Oats	5	3	2
Triticale	5	5	5
Winter broadleaf			
Fababean	3	2	2
Flax	4	3	3
Lentil	4	4	3
Lupine	3	3	3
Pea	4	3	3
Canola/rapeseed	5	5	5
Cool season spring cereals			
Oats	5	3	2
Triticale	5	5	5
Cool season spring broadleaf			
Chickpea	5	3	2
Crambe	5	4	3
Dry pea	5	3	2
Faba bean	3	2	1
Flax	4	3	3
Lentils	5	4	3
Lupine	4	4	3
Mustard	5	5	4
Canola/rapeseed	5	4	3
Warm season summer grasses			
Corn	4	3	3
Millet	4	4	3
Sorghum	4	4	3
Warm season summer broadleaf			
Buckwheat	4	3	1
Dry beans	5	3	1
Safflower	4	3	2
Soybean	3	2	1
Sunflower	4	2	2

1 = definitely not; 3 = possibly; 5 = definitely. Adapted from Guy and Karow (2009).

Table 5-A2. General soil water holding capacity (SWHC), soil organic matter (SOM), and crop productivity characteristics by agroecological class.

	Annual Crop ●			Transition ▲			Grain-Fallow ■		
SWHC (in/ft)[1]	2.0–2.4			1.8–2.2			1.6–2.0		
SOM (%)[1]	3–4			2–3			<1.5		
	Grain bu/ac	Grain lb/ac	Residue lb/ac	Grain bu/ac	Grain lb/ac	Residue lb/ac	Grain bu/ac	Grain lb/ac	Residue lb/ac
Small grains									
Winter wheat[1,2]	80–120 (90)	5400	7180	60–80; 30–40	4200; 2100	6090; 3549	40–60	3000	4440
Spring wheat[1,2]	75	4500	5580	55; 25	3300; 1500	4389; 2535	40	2400	3360
Winter barley[2]	125	6000	5700	93; 32	5200; 1800	6136; 4050	60	3500	5250
Spring barley[2]	110	5300	4929	83; 30	4700; 1700	4700; 2992	34	1900	3040
Winter triticale[3]	—	—	—	—	—	—	78	4890–6660	—
Hard red spring[4,5]	—	—	—	45	4700	—	—	—	—
Hard red winter[4,5]	—	—	—	73	4400	—	45	4700	—
Grain legumes									
Spring dry pea[2]	—-	2200	28601	—	1600	2560	—	—	—
Winter dry pea[6]	—	4000-5000	—	—	3500	—	—	1500	—

Table 5-A2 (continued). General soil water holding capacity (SWHC), soil organic matter (SOM), and crop productivity characteristics by agroecological class.

	Annual Crop ●			Transition ▲			Grain-Fallow ■		
SWHC (in/ft)[1]	2.0–2.4			1.8–2.2			1.6–2.0		
SOM (%)[1]	3–4			2–3			<1.5		
	Grain bu/ac	Grain lb/ac	Residue lb/ac	Grain bu/ac	Grain lb/ac	Residue lb/ac	Grain bu/ac	Grain lb/ac	Residue lb/ac
Spring Lentil[2]	—	16002	2720	—	1200x	2400	—	—	—
Winter lentil[6]	—	3000	—	—	—	—	—	—	—
Chickpea[6]	—	2000	—	—	—	—	—	—	—
Oilseeds									
Spring canola[7,8]	—	2000	5142	—	1500	3850	—	500	1285
Winter canola[7,8]	—	3000	7700	—			—	15005	
Camelina[9]	—	1500	3850	—	1050	2700	—	850	2185
Yellow mustard[10]	—	2000	5142	—	1500	—	—	<1200	—
Safflower[11]	—	3000	—	—	1500	—	—	500	—

Note: lb per bushel: wheat (60); triticale (56); barley (48); oats (32); sunflower (24-32); rapeseed (60); flax (60).
[1]Douglas et al. 1999; Table 5.2 p. 76. [2]Douglas et al. 1999; Table 5.3 p. 87. Residue calculated using residue-to-grain ratios for cereals and pea and lentil zones 2, 3, and 5. [3]Schillinger et al. 2015. [4]Connolly et al. 2016. [5]Connolly et al. 2015. [6]Ranges supported by various USDA variety trials. [7]Brown et al. 2009. [8]Oilseed residue values calculated using Grain*HI (0.20–0.35) = Biomass, and Residue = Biomass–grain. [9]Hulbert et al. 2012; Expected camelina yield = 70 lb/ac grain per inch annual precipitation. [10]J. Brown et al. 2005; Expected yellow mustard yield = 95 lb/ac grain per inch precipitation: 20” ppt = 1,900 lb/ac; 16” ppt = 1,520 lb/ac; 12” ppt = 1,140 lb/ac. [11]Petrie et al. 2010.

Table 5-A3. Typical nitrogen (N) requirements and relative water use.

	UNR lb N/bu	UNR lb N/100 lb	Relative Water Use[2,3]	
Soft white wheat[1]	2.7–3.6	4.5–6.0	Winter wheat, winter triticale, winter canola	Highest
Hard red wheat[2]	3.0–3.7	5.0–6.2		
Feed barley[2]	2	4.1		
Malt barley[2]	3	6.2	Sunflower, safflower, chickpea	↓
Triticale[3]	similar to SWWW	—		
Broadleaf			Spring cereal, spring mustard, canola, and camelina	↓
Canola[4]	—	7 (6.0–10.7)		
Camelina[5]	—	5–6	Lentil	↓
Safflower[6]	—	5		
Yellow mustard[7]	—	8–12	Pea	Lowest

[1]Koenig et al. 2005.
[2]Wysocki et al. 2006.
[3]Schillinger pers. comm.
[4]Koenig et al. 2011; Karow 2014.
[5]Hulbert et al. 2012.
[6]Armah-Agyeman et al. 2002.
[7]J. Brown et al. 2005; Davis and Wysocki 2010.

Chapter 6

Soil Fertility Management

Kristy Borrelli, Pennsylvania State University (formerly of University of Idaho)
Tai Maaz, Washington State University
William Pan, Washington State University
Paul Carter, Washington State University
Haiying Tao, Washington State University

Abstract

The inland Pacific Northwest's (PNW) warm, dry climate and deep soils make it ideal for producing high yields of high-quality wheat. Wheat can grow in some of the region's driest areas where other crops cannot. However, drastic topography and precipitation gradients result in variable growing conditions that impact crop yield and complicate nutrient management strategies. Interrelated climate, water, and nutrient dynamics drive wheat development, growth, and associated fertility recommendations; understanding these complex relationships will become increasingly important under changing climate conditions. Concerns related to nitrogen losses, soil erosion and nutrient runoff, and decreasing soil pH can result from improper use of fertilizers and further complicate management strategies. Farm-specific management approaches, regular soil testing, and detailed recordkeeping can help producers improve nutrient use, minimize nutrient losses and environmental degradation, and maintain high yields and quality.

Key Points

- Climate, weather, topography, and soil drive wheat productivity and soil fertility management strategies in the inland PNW.
- Variable landscapes and rainfall gradients affect crop fertilizer accessibility, nutrient use efficiencies, and crop growth. Inland PNW producers can achieve the best growing conditions for wheat by

tailoring management strategies to their specific field and within-field locations.

- Practices that maximize nitrogen use efficiency include fertilizer placement, fertilizer source, timing of application, and rates that match crop species' and varietal nitrogen needs.
- Over-use of fertilizers can result in harmful effects on air, water, and soil quality and negatively affect a producer's bottom line. Appropriate management strategies and regular soil testing can reduce nutrient loss and improve overall farm gains.
- Conserving soil water is imperative in dryland agricultural regions since available soil water directly drives wheat yields and nutrient availability. Increased water stress is likely under predicted climate scenarios. Management strategies that build up soil organic matter and improve soil health can buffer against crop nutrient and water loss.

Introduction

Maintaining soil fertility in wheat-based cropping systems is necessary to achieve satisfactory growth and yields, as well as protein and other quality factors in wheat. Complex interactions among multiple climate, weather, soil, and plant factors in the **inland PNW** can impact crop productivity and affect soil fertility management strategies across locations and within farms. Precipitation and temperature directly influence crop yields, the foundation for all fertilizer application rate recommendations, while the soil itself can affect nutrient availability, retention, and cycling. Plant nutrients are removed from the plant-soil system in the crop at harvest each year (Table 6-1), making it necessary to add many of them back at the beginning of each growing season to ensure marketable wheat yields and protein levels. Typically, wheat producers supply nutrients to their cropping systems by applying commercial fertilizers, and incorporating legumes into their rotations. On occasion, they might supply wheat with nutrients from organic fertilizer sources, although this is a relatively uncommon practice due to the distance of animal production in relation to wheat farms and the large quantities of amendments needed to supply adequate fertility to wheat and other cereal crops. An important concept to understand is that managing soil nutrients is more than simply considering how much fertilizer a crop needs to grow. It is a process that requires controlling a variety of fertility sources over time in complex and dynamic cycles. The challenge is meeting the overall need of the crop and applying nutrients at peak crop demand while continuing to sustain healthy soils and associated natural resources.

The goal of this chapter is to summarize the main factors that impact nutrient requirements and drive associated soil fertility management decisions for wheat-based cropping systems in the inland PNW. A more thorough understanding of how these factors interact and affect wheat yield and quality will help prepare producers and service providers to address regional soil fertility concerns, reduce negative effects to the environment, and achieve high yielding and quality wheat production.

Grower Considerations

Nitrogen Management

Nitrogen (N) is the nutrient that producers supply in the largest quantities each year in wheat-based cropping systems due to the high demand of N

Table 6-1. Nutrient supply and removal in harvested grain* across the main wheat-producing counties of the inland PNW.

Nutrient Source	N	P_2O_5	K_2O
		ton/yr	
commercial**	143,570	41,439	30,172
recovered manure***	1,377	2,013	6,242
biologically fixed by legumes***	25,322		
total nutrient supply	170,269	43,452	36,414
crop removed***	171,203	63,331	102,602
balance (supply - removed)	–934	–19,879	–66,188
removal ratio (removal/supply)	1.01	1.46	2.82

*Methods described in IPNI (2012). N = nitrogen; P_2O_5 = phosphorus; K_2O = potassium.
**1997, 2002, 2007, 2010-12 County level data interpolated and summarized by International Plant Nutrition Institute (IPNI) from fertilizer sales data collected by the Association of American Plant Food Control Officials (AAPFCO).
***Farm census data from USDA National Agricultural Statistical Service (USDA-NASS) Census of Agriculture, summarized by IPNI.

by cereal crops. Nitrogen is also the most deficient nutrient for all non-legume crops in the inland PNW, and wheat yields respond to additions of N fertilizer in most precipitation zones when N supplied by the soil is low (Rasmussen 1996). Therefore, N is a major focus of nutrient management plans and field recommendations (Pan et al. 1997). Although several forms of N exist in nature, nitrate-N (NO_3-N) and ammonium-N (NH_4-N) are the forms available to plants. The primary determinants of plant-available N are (1) release from soil organic matter by microbes (referred to as **mineralization**), (2) contributions from organic and inorganic N fertilizer sources, and (3) losses from the plant-available N pool (Figure 6-1) (Cassman et al. 2002). Nitrogen, a dynamic nutrient that is altered by biological soil processes as well as exposure to water and air, is particularly complicated to manage compared to nutrients that are less subject to transformations and losses. Nitrogen availability is influenced by precipitation, soil water, and field variability more than most nutrients as a result of its association with

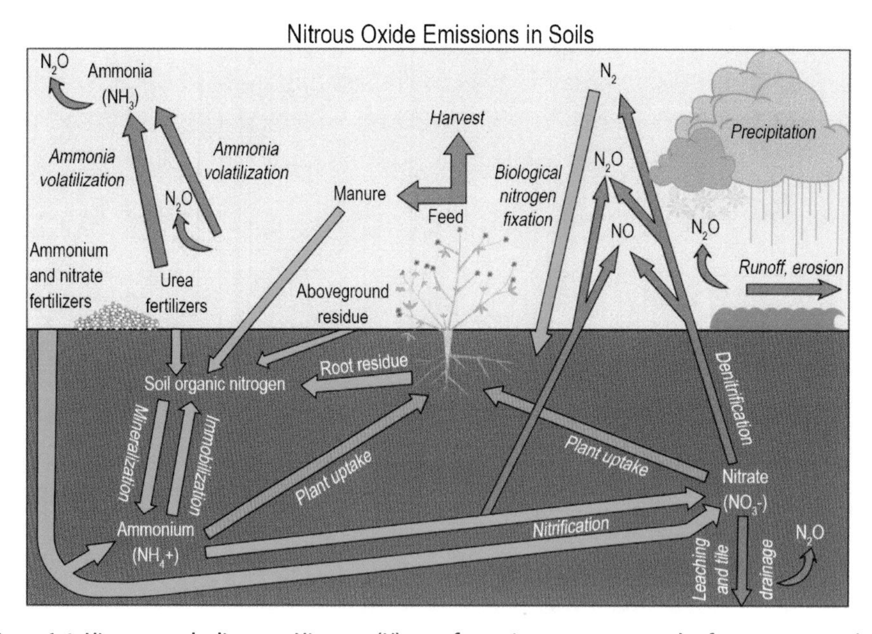

Figure 6-1. Nitrogen cycle diagram. Nitrogen (N) transformations occur as a result of associations with biological organisms, and N cycling processes are greatly impacted by temperature, heat, and other environmental conditions that affect biological metabolic activity. Plant-available N forms, nitrate-N (NO_3-N) and ammonium-N (NH_4-N), can be derived from inorganic fertilizers and organic sources. It is important to consider the complete N cycle to understand plant N uptake abilities as well as potentials for N loss. (Figure created by Nick Kennedy.)

biological organisms. Because climate and topographical factors impact N cycling and crop availability in the inland PNW (Pan et al. 1997), they are presented here with the best options for efficiently using N relative to the region's specific concerns.

Meeting nitrogen use efficiency – matching N availability with crop needs

In general, an efficient system is one that maximizes its outputs relative to its use of inputs, and efficient N use by a wheat crop means that the majority of N available throughout the growing season can be accounted for in the grain at harvest. If N is not taken up by the crop, it remains in the root zone of the soil. **Nitrogen use efficiency** (NUE) of an entire cropping system can be considered as the proportion of all N inputs that are (1) removed in harvested crop biomass, (2) contained in recycled crop residues, and (3) incorporated into residual soil N pools (organic and inorganic sources) (Cassman et al. 2002). All of these components are considered in nutrient management plans (Pan et al. 1997).

Nitrogen use efficiency can be partitioned into soil and plant processes governing **nitrogen uptake efficiency** (how efficiently the plant obtains N from the soil) and **nitrogen utilization efficiency** (how efficiently N is metabolized inside the plant) (Huggins and Pan 1993; Moll et al. 1982). Because NUE is the product of uptake and utilization efficiencies, inefficient use of N can occur because of low uptake of N by the plant, poor utilization by the plant to produce grain, or a combination of both.

Factors that reduce NUE include over-fertilization (too much N supplied for plant needs), sub-optimal yields (too little plant uptake of available N), and N losses (applied N not being used by the plant) (Sowers et al. 1994). Environmental factors that impact these variables can also reduce NUE. For instance, drought conditions can reduce yields, restrict the uptake of N by crops, and shorten the grain-filling period of crops (Maaz et al. 2016; Pan et al. 2007). Dry conditions impair root activity near the surface of the soil, thereby increasing the amount of soil N "stranded" and decreasing the uptake of available N. Nitrogen use efficiency can also be diminished with increasing annual precipitation, as N retention is reduced by denitrification and nitrate-N **leaching** from the root zone, particularly in shallow soils.

The best approach for maximizing NUE at economically optimal yields is to match fertilizer rates based on soil tests, as well as the timing of application with crop uptake needs, which can be difficult. Understanding current practices for applying fertilizers is necessary for knowing how to better meet NUE goals and addressing challenges.

Factors driving current nitrogen management strategies

Field and regional variability impacts on water and nitrogen relationships

Soil and climatic features across the region (e.g., precipitation gradient) as well as within fields (e.g., slope position), greatly impact N cycling dynamics and requirements that affect crop performance and yield potential (Pan et al. 1997). The topography of the inland PNW is quite diverse compared to many other regions (Mulla 1986; Rasumssen 1996), and most of the wheat production acreage is located on steep hillslopes (Busacca et al. 1985) (see Chapter 8: Precision Agriculture). Wheat productivity is determined largely by plant-available water and

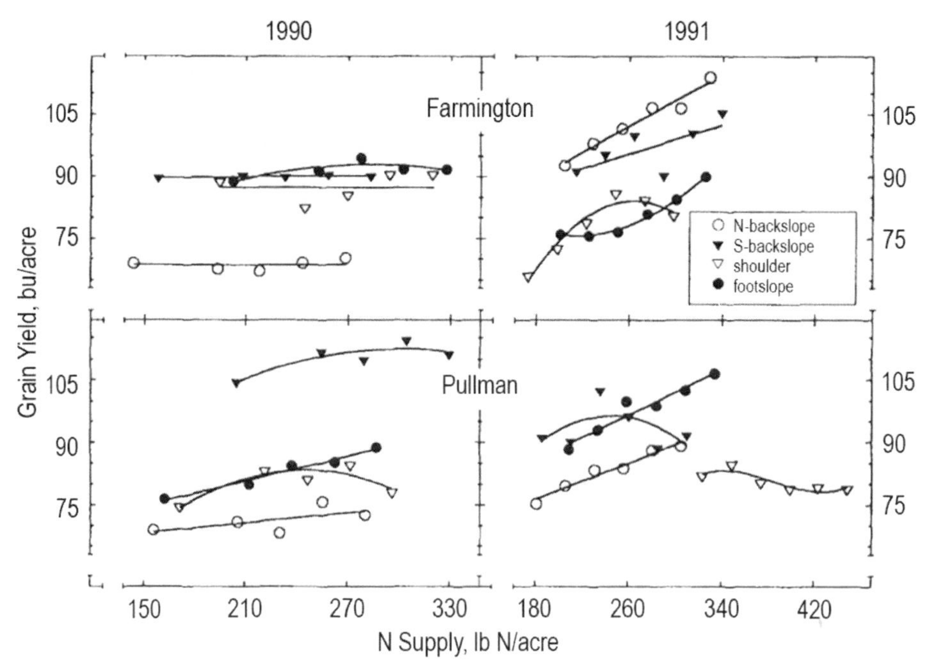

Figure 6-2. Grain yield versus nitrogen (N) supply (residual N plus mineralization plus fertilizer N) at different points along a hillslope at Farmington and Pullman, Washington, in 1990 and 1991. (Used with permission from Fiez et al. 1994b.)

the manner in which water, nutrients, and other crop growth factors are distributed across the landscape (Long et al. 2015). Specifically, water-holding capacity and rooting depth is often limited on eroded hilltops, and N losses and transformations can occur differently depending on field location (Fiez et al. 1994a; Long et al. 2015; Pan et al. 1997).

Especially notable in the hilly Palouse region of the inland PNW, differences in climate and soil productivity at different slope locations contributes to variations in wheat yield and protein levels across fields (Figure 6-2) (Fiez et al. 1994a; Fiez et al. 1995; Mulla 1992; Yang et al. 1998). As a result, there is high annual variability in the amount of N needed to attain optimal yields (Table 6-2) (Pan et al. 1997).

Relationships among precipitation, soils, fertilizers, and wheat yields have been integral to N recommendations since the adoption of commercial N fertilizers in the 1950s (Pan et al. 2007). In fact, quantitative relationships

Table 6-2. Yearly variation in the rate of nitrogen (N) producing the optimum grain yield of winter wheat, in a winter wheat-fallow rotation in Pendleton, Oregon, from 1963 to 1987.

		N Rate (lb/ac)			
Year	0	40	80	120	143
		Grain Yield (bu/ac)			
1963	36	**62**	55	42	37
1965	62	64	**71**	71	68
1967	42	**61**	54	45	39
1969	**54**	55	54	45	43
1971	51	79	**85**	77	67
1973	48	**55**	49	42	42
1975	51	**67**	59	57	57
1977	48	**54**	49	46	45
1979	42	**57**	55	45	49
1981	55	89	100	**104**	100
1983	42	62	86	**97**	92
1985	48	51	**57**	57	57
1987	57	79	**83**	74	71

[1]Grain yields in bold are at the N rate providing optimum yields.
Adapted from Rasmussen (1996).

among water, N, and wheat yield in the inland PNW were determined through extensive field experimentation in the 1950s (Jacquot 1953; Leggett 1959) and have been used ever since to estimate regional wheat yield potentials and soil fertility recommendations. Regional and annual variations increase the challenge for the inland PNW's producers to make appropriate N management decisions on their farms, which can be some of the most important and costly farm management decisions.

Precipitation and soil water

Wheat's responses to nutrients are especially sensitive to the amount of precipitation received during the growing season. Precipitation and soil water accumulated during the growing season influence soil nutrients and plant nutrition directly by (1) increasing crop yield potentials and (2) improving nutrient mobility and transformations in the soil. Wheat yields and total plant-available N are impacted by water availability, making weather-driven N recommendations necessary for dryland wheat production (Pan et al. 2007). Plant-available water also determines wheat protein content (Baker et al. 2004) in dryland environments. Nitrogen management affects the end-use quality of grain (Brown et al. 2005; Miller and Pan n.d.).

Nitrogen recommendations

The amount of N required to achieve wheat yield and protein goals is mainly dependent on the crop's yield potential and the class of wheat being produced. Nitrogen budget-based fertilizer guides are available to help producers estimate these values and develop a specific nutrient management plan for their own dryland wheat production systems in Washington (Koenig 2005), Oregon (Lutcher et al. 2005), and Idaho (Mahler 2007). (Various nutrient management guides exist for producing wheat in Oregon under different precipitation zones. Producers in Southern Idaho should also refer to the Southern Idaho Dryland Winter Wheat Production Guide. All guides are listed under Grower Resources later in this chapter.) Although they have been developed for different states, the basic recommendations and guidelines are similar for all three states. The standard components for calculating annual N fertilizer rates are the N supply needed based on (1) wheat yield potential of a specific site, (2) the amount of N required to achieve yield and protein goals for a desired wheat class, and (3) an inventory of soil N contributions (Figure 6-3).

Step 1: The amount of N fertilizer needed per acre of wheat can be calculated by multiplying the anticipated yield goal by the unit N required (UNR) for soft white winter or club wheat (2.7 lb N/bu) or hard red and white wheat varieties (3.0 lb N/bu). Yield goals and UNRs already account for plant available moisture and grain protein, respectively.

Yield Goal (bu/acre) x UNR (lb N/bu) = N fertilizer required (lb N/ acre)

Step 2: Account for residual soil N contributions in the soil profile to a depth of 4 to 5 feet. Residual N can be measured with soil tests and should be calculated prior to applying fertilizer.

Lbs. N fertilizer required per acre (from Step 1) – Residual Soil N (lb N/acre) = N fertilizer recommendation (lb N/acre)

Figure 6-3. Basic nitrogen (N) fertilizer rate calculation steps. Simplified calculations were developed from nutrient management guides (Koenig 2005; Lutcher et al. 2005; Mahler 2007).

1. *Wheat yield potential.* Wheat yield varies in response to weather, crop rotations, and other management variables. Wheat yields can vary from 25 to over 110 bushels per acre across the region. Correlations between available water and wheat grain yield have been measured extensively throughout the inland PNW and have been used to estimate yield potential in dryland wheat systems (Mahler 2007; Pan et al. 2007; Robertson et al. 2004). Yield goals can be estimated based on a producer's experience, from historical averages, or they can be calculated using pre-plant soil water and expected precipitation during the growing season (Koenig 2005). Estimated **water use efficiencies** (WUEs) for wheat are 7 bushels of soft white winter wheat per acre per inch of available water, and 6 bushels of wheat per acre per inch of available water (for all other wheat classes). Modern semi-dwarf wheat cultivars and improved agronomics have increased yields and therefore the WUE of wheat (Figure 6-4).

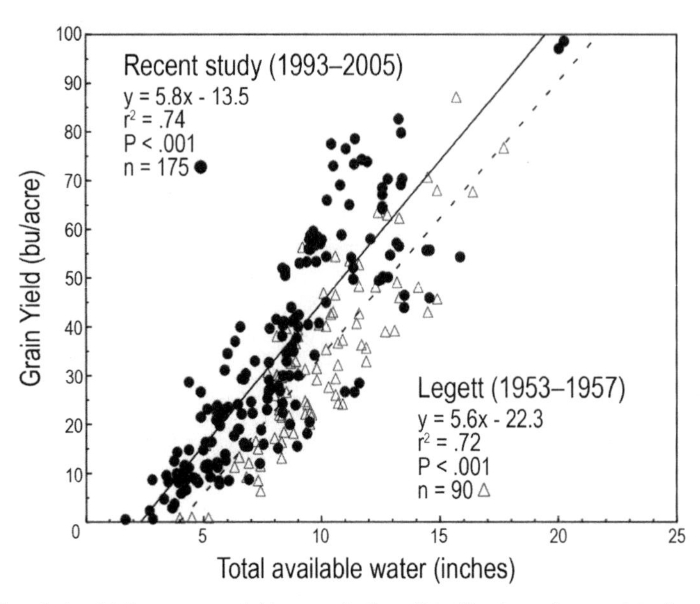

Figure 6-4. The relationship between available water in the soil profile plus spring precipitation and grain yield of dryland wheat in eastern Washington. Data were collected from 1953 to 1957 (dotted line, open triangles) and from 1993 to 2005 (solid line, filled circles). Grain yield data are from a combination of winter and spring wheat. The minimum threshold of available water required for wheat germination has decreased from 5.9 inches in 1953 to 1957 to 2.3 inches in the recent study (1993 to 2005) as a result of improved water use efficiency in newer cultivars of wheat. (From Schillinger et al. 2012.)

2. *Wheat class and protein*. Optimum grain protein levels range from 9% to 10% for soft white wheat and 11% to 14% for hard wheat classes (Brown et al. 2005; Koenig 2005; Lutcher et al. 2005). More specifically, market protein goals for hard wheat classes are 11.5% for hard red winter wheat, 12.5% for hard white wheat, and 14% for hard red spring wheat (Washington Wheat Commission Guides for Hard White Winter, Hard Red Spring, and Hard Red Winter Wheat). Meeting these protein values is critical to produce high-quality wheat and to avoid **dockage** and price reductions for grain at the elevator.
3. *Soil contributions of N*. Residual soil N in the rooting zone can come from both organic and inorganic sources and is one of the most complicated factors to predict and manage (see Chapter 4: Crop Residue Management and Chapter 7: Soil Amendments). Despite similar soil types, climatic conditions, and management

practices, residual soil N varies over time and location since plant-available forms of N are influenced by soil temperature, water content, and management practices. For example, flushes of N mineralization may occur after soil tillage, soil rewetting, or during freeze-thaw cycles.

Practices that promote **soil health** and increase soil organic matter (e.g., incorporating crop residues, manure, or compost) are going to have the greatest influence on supplying N from the soil. The amount of N released from organic sources like crop residues and manure depends on the amount of N they contain and rates of N mineralization. Nitrogen is released from organic compounds over a longer period of time than it is from inorganic sources since mineralization must occur. However, supplies of organic residues not only provide a slow-release source of N, but they also help build **soil structure** and retain water in the rooting zone (see Chapter 2: Soil Health). Improved soil **water holding capacity** can expedite mineralization and nutrient cycling in addition to remediating water stress that crops experience during dry and hot growing conditions.

It is important to consider N released from decomposing legume or other low **carbon-to-nitrogen** (C:N) ratio residues and other sources of soil organic matter. Koenig (2005) estimates that peas can contribute 10 to 20 pounds of N per acre (amounts vary based on pea biomass yield), lentils can contribute 10 pounds of N per acre when biomass is greater than 1,000 pounds per acre, and alfalfa can contribute as high as 50 pounds of N per acre (Table 6-3). These sources of N as well as soil organic matter are considered credits in the N fertilizer calculation (Koenig 2005). It is recommended to credit 20 pounds of N per acre for each 1% of organic matter under conventional tillage and 17 pounds of N per acre for each 1% up to 3% of organic matter under **direct seeding** systems.

Most inorganic N fertilizers enter the plant-available N pool rapidly since they are predominantly composed of simpler, plant-available nitrate-N, ammonium-N, or both. Using plant-available N sources, however, does not guarantee that N will not accumulate in the soil profile. It is possible to have elevated levels of residual N at harvest if inorganic fertilization additions exceed the annual yield requirements for that site. While residual fertilizer N can be recycled in subsequent seasons in inland PNW cropping systems (Maaz and Pan n.d.), over-fertilization can become particularly problematic when fertilizer N rates consistently exceed the amount of N exported in the grain, especially in landscape positions that are vulnerable to N losses.

Residual plant-available soil N, as the sum of nitrate-N and ammonium-N, should be accounted for in soil tests (0 to 5 feet deep or to a restricting layer). When crediting soil N sources, it is important to consider when the soil samples are collected. If soil samples are taken in the spring, up to half of the N released from soil organic matter annually has already occurred. Also, knowledge of crop rotation is important. **Immobilization** or N tie-up can occur when high amounts of cereal straw residues, with wide C:N ratios, are left on the field. Immobilization acts as a debit in fertilizer calculations. No debit is taken after summer **fallow**, but if wheat was the preceding crop, debits of 35 pounds of N per acre (winter wheat), 30 pounds of N per acre (spring wheat), and 25 pounds of N per acre (barley) should be taken.

Despite its complex nature and variability, residual soil N contributions are important to account for when making nutrient management decisions since inorganic N and N mineralization contributed to the majority of grain N at maturity (Sowers et al. 1994). Failing to account for plant-available N in the soil matrix can lead to over-fertilization and nutrient losses.

Factors that impact NUE

Timing application with growth

Crop management, such as residue retention, crop rotation, residual fertilizer from the previous crop, and fertilizer timing, placement, and

Table 6-3. Nitrogen (N) contributions from previous grain and forage legume crops.

Preceding Crop	Preceding Crop Yield (lb/ac)	N Credit (lb N/ac)
Peas	>2500	20
Peas	1500 to 2500	15
Peas	<1500	10
Lentils	>1000	10
Alfalfa	Any	50

Adapted from Koenig (2005).

source all affect NUE (Cassman et al. 2002; Dawson et al. 2008; Lea and Azevedo 2006; Maaz and Pan n.d.; Raun and Johnson 1999). To help producers manage N more effectively in their fields, researchers encourage them to make nutrient management decisions based on the **unit nitrogen requirement** (UNR) and NUE at economically optimal yield for each specific field and crop. Ideally, wheat producers want to identify the lowest UNRs to attain economically optimal yields.

Unit nitrogen requirements

The N requirement or the amount of N needed to produce 1 bushel of wheat is based on a grain protein goal in combination with plant-available soil water and N factors. This amount of N is also referred to as the UNR and is measured in pounds of N needed for each expected bushel of wheat. The UNR is the inverse of NUE (UNR = 1 ÷ NUE) at economically optimal yields (Fiez et al. 1994b). Once the UNR value is identified, it is multiplied by the expected yield (bushels per acre) to provide an N fertilizer application rate recommendation in pounds of N per acre (Step 1 in Figure 6-3). Recommended UNRs are typically 2.7 pounds of N per bushel for soft white winter and club wheat and 3 pounds of N per bushel for hard red or white varieties of wheat (Leggett 1959).

It is important to recognize that these UNR determinations were regionally averaged over different Washington wheat-growing areas (Hergert et al. 1997), and the data was typically produced from experiments conducted on gentle sloped or flat landscape positions for maximizing yield potential and for ease and accessibility for research plot equipment (F.E. Koehler, personal communication). In actuality, agricultural field landscapes exhibit differences in multiple factors that influence N supply, protein-yield relationships, and yield potential (Figure 6-2). Although current guidelines for nutrient management depend on these values, one should recognize that UNRs are variable among years and across landscape positions, ranging from as low as 2 pounds of N per bushel to as high as 3.5 pounds of N per bushel (Koenig 2005; Lutcher et al. 2005; Mahler 2007). Furthermore, these landscape differences do not always repeat their patterns with variable weather, which can result in variable UNRs over landscapes and years. For example, UNR values ranged from 1.8 to 3.9 pounds of N per bushel in a study conducted by Fiez et al. (1994b).

This strongly suggests that site-specific N recommendations cannot be extrapolated from regionally developed recommendation models (Pan et al. 1997).

In the inland PNW, the Extension nutrient management guides mostly take a "Liebig" approach to soft white winter wheat N recommendations (Figure 6-5), which assume that NUE is constant across a large range of yield potentials (Pan et al. 2016). Therefore, the main reason that inefficiencies in N use by wheat occur in the inland PNW is that fertilizer recommendations are calculated based on estimated UNR values as standards as opposed to actual field values. As a result, too little or too much fertilizer gets applied and the wheat plant either experiences N deficiency symptoms (e.g., poor growth and development or inadequate protein levels) or does not use all available N, making it susceptible to volatilization, runoff, or leaching loss. In contrast, a classic "Mitscherlich" approach (Figure 6-5) assumes that the NUE varies greatly across the range of yield potentials observed in the inland PNW, which may or may not be correct for a given crop, region, or farm. An alternative strategy is to integrate effects of yield potential and available water on the response of winter wheat to N, which combines the "Liebig" and "Mitscherlich" approaches. This integrated approach is currently being used to better

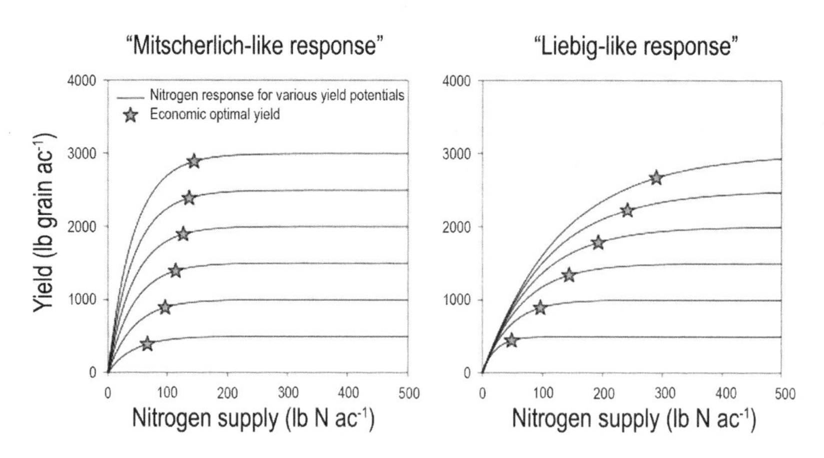

Figure 6-5. A classic Mitscherlich approach to nitrogen (N) recommendations assumes that nitrogen use efficiency (NUE) varies greatly across the range of yield potentials, whereas a Liebig approach assumes NUE is constant across a large range of yield potentials. An integrated approach between the two could improve accuracy in estimating N recommendations. (Adapted from Pan et al. 2016 and Pan 2015.)

understand changes in UNR across the region, and it has indicated that UNR decreased as precipitation increased along the 12 to 20 inch precipitation gradient (Maaz and Pan n.d.). However, additional studies are needed to characterize the potential decrease in NUE as precipitation exceeds 20 inches and the risk of soil N losses increase.

Accurately estimating UNR across fields is an ongoing concern and challenge (Fiez et al. 1994a; 1994b) due to landscape variability as well as an unrealistic requirement to sample a field as intensively as necessary for obtaining actual values. Fortunately, several strategies exist that can help improve N management and NUE, such as split application, **nitrification** inhibitors, and variable rate N applications. Variable N rate practices and technologies have been evolving as computer technologies have been incorporated with agricultural machinery, while remote sensing is a promising tool to delineate management zones based on crop performance indicators (Song et al. 2009). High-tech machines and software packages offer producers more opportunities to increase NUEs on their farms by using site-specific management (see Chapter 8: Precision Agriculture).

However, even without the use of high-tech farming equipment several strategies exist that can help improve N management strategies and NUE. Approaches to increasing NUE should integrate many known components of wheat production into one system (Raun and Johnson 1999). The following strategies can help improve NUE in most cropping systems.

Strategies that improve NUE

Fertilizer placement

Early season N availability is critical for yield and a moderate level of grain protein (Washington Wheat Commission Guide Hard Red Winter Wheat). Nitrogen must be available for plants to take up by the three-leaf stage (Figure 6-6) since the bud for the first tiller forms at this point and N is critical to support tiller formation from the very beginning (Waldren and Flowerday 1979). Most of the N required by wheat is taken up during vegetative growth and used to establish the yield potential (number of heads and kernels per head) (Washington Wheat Commission Guide Hard Red Winter Wheat). Vegetative N is later transported to the kernels to form protein during grain fill in mid to late summer.

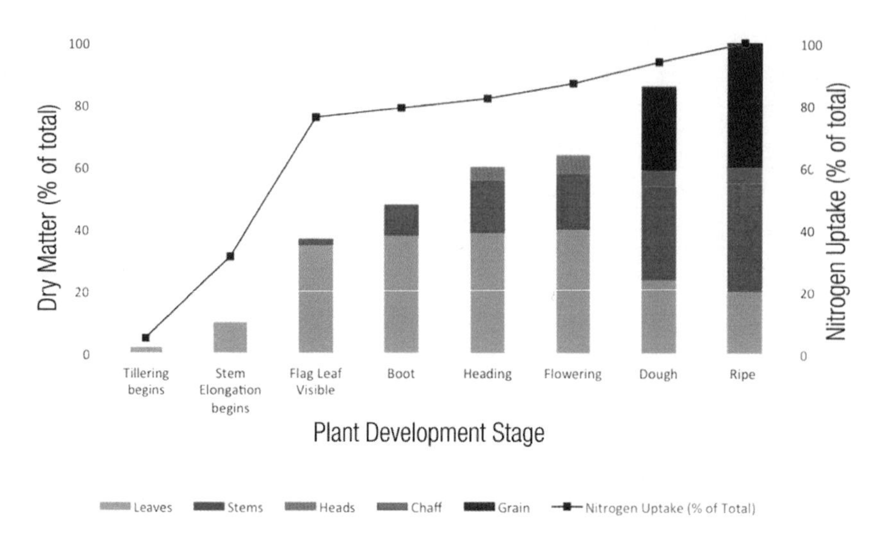

Figure 6-6. Nitrogen uptake by wheat. The total accumulated uptake of nitrogen by wheat estimated by growth stage. Timing of N application should occur earlier in the season when N uptake by the crop is high. As wheat matures, N uptake slows down as the crop metabolizes N to produce grain and protein. (Adapted with permission from Waldren and Flowerday 1979.)

Splitting N fertilizer applications between fall and spring allows producers to conservatively apply smaller amounts of N early in the growing season and then adjust rates and add additional fertilizer in the spring, only if the precipitation outlook is favorable for high wheat yields. Split (fall-spring) applications of N fertilizer are recommended in areas receiving approximately 21 inches of annual precipitation, whereas areas that experience heavy winter precipitation (greater than 24 inches) or have sandy soils might benefit from spring applications only (Mahler 2007). In higher precipitation areas, over 70% of the required fertilizer should be applied in spring (Mahler 2007). Lower NUE associated with all-fall N fertilizer application on winter wheat was associated with N loss, whereas spring-only and split (fall-spring) applications improved NUE as a result of N availability prior to the period of rapid plant uptake of N, which resulted in greater fertilizer recovery (Sowers et al. 1994). The efficient recovery of fertilizer applied in spring during this accelerated phase of N uptake suggests that there was intensive root activity in the surface soil between stem elongation and boot stage (Sowers et al. 1994). Mahler et al. (1994) found both winter wheat grain yield and NUE were greatest when applications were split between fall and spring compared to spring-only or fall-only applications. Under drier growing conditions, it is not necessary to split N applications, and full N rates can be applied in fall for both winter and spring wheat since leaching is less of a concern (Mahler 2007). In dry conditions, NUE is improved by earlier N application that allows fall and winter precipitation to move N lower into the crop root zone and reduces the likelihood of it getting stranded in dry soil.

Grain yield and NUE were not impacted by N source and placement (Mahler et al. 1994). Sowers et al. (1994) also did not see a difference in NUE when spring fertilizers were top-dressed or point-injected. Band application of N fertilizer increased wheat growth, N uptake, and yield in wheat in addition to reduced populations of wild oats (Koehler et al. 1987). Banding of N fertilizer is especially important for spring wheat since there is often not enough precipitation after seeding to move the fertilizer to the root zone (Koehler et al. 1987).

Wheat cultivar yield

Fertilizing with high rates of N to control grain protein is relatively inefficient as NUE decreases with increasing N levels, especially in dry soil conditions (Raun and Johnson 1999). Low crop vigor also decreases NUE. High crop yields help improve NUE because fast growing plants have root systems that more effectively utilize available soil N (Cassman et al. 2002; Pan et al. 2016). Modern wheat cultivars can have increased vigor and increased stress tolerance, as well as improved efficiencies for taking up water and nutrients. Schillinger et al. (2012) found that semi-dwarf varieties of wheat were able to begin grain production with less available water than taller varieties of wheat that were grown prior to 1960 (Figure 6-4). In relation, all aspects that impact crop health, and therefore growth, can also impact NUE, such as insect and weed management, water and temperature regimes, supplies of other nutrients, and use of the best-adapted cultivar (Cassman et al. 2002).

Crop rotations

The sequence and intensity of crops in rotation can influence N supply, NUE, and WUE. More intensive rotations, meaning growing more crops in a given period of time, can improve NUE by increasing WUE and maintaining soil health (Raun and Johnson 1999), especially in dryland systems like those found in the inland PNW. Nitrogen use efficiency can also increase when wheat follows legumes rather than when following fallow or continuous wheat (Baudaruddin and Meyer 1994). For instance, in the inland PNW, the yield potential of winter wheat was greater following spring peas rather than wheat, and winter wheat yields and protein levels were higher for a given amount of fertilizer due to a greater N supply and N uptake efficiency (Maaz and Pan n.d.).

Nitrogen loss and environmental impacts

Leaching

Nitrate-N leaching is a major pathway of N loss in wheat-based cropping systems in the inland PNW. Because it has a negative charge, nitrate-N does not adsorb to negatively charged particles in the same way that positively charged plant nutrients do (Havlin et al. 2005). Nitrate-N is also highly water soluble. As a result, nitrate-N can be lost to leaching when water filters through the soil profile. Wet soils with high residual soil N levels are especially susceptible to leaching losses. In addition, heavy precipitation immediately after urea fertilizer applications will cause substantial amounts of urea leaching loss.

Factors that affect leaching

Both annual and individual precipitation events greatly impact nitrate-N leaching. The risk of leaching from wheat fields in the inland PNW is greatest during winter and early spring when plant uptake of nitrate-N is low and precipitation is high. Leaching is greater when N fertilizer is applied in fall compared to spring (Cameron et al. 2013; Kyveryga et al. 2013). When nitrate-N accumulates in dry soil during the growing season, it can slowly leach down the soil profile following subsequent wet seasons (Cameron et al. 2013). It is therefore important to account for soil N that has been redistributed deep into the rooting zone when determining N fertilizer application rates in order to improve NUE. Leaching risks can also be high on coarse or sandy soils since nitrate-N can easily be lost through large soil pores. Improving soil health by increasing soil organic matter content (see Chapter 2: Soil Health) has the potential to reduce the risk of leaching by improving the soil's water holding capacity, **aggregate**

stability, and nutrient adsorption capacity. Nitrate-N leaching can also be reduced by matching crop N needs with soil type, N supply, and timing of application (e.g., avoiding periods of heavy precipitation).

Environmental impact of leaching

High levels of nitrate-N can be a source of water contamination in surface and groundwater (Schepers et al. 1991). Leaching or surface runoff of nitrate-N from soils to surface water can pollute water bodies through a process called eutrophication. Eutrophication results in high amounts of algal growth and depleted sources of oxygen in the water that can be deadly to fish and other aquatic organisms. Eutrophication is a devastating concern to many salt and fresh water sources across the globe, and these polluted areas or "dead zones" mainly occur downstream from areas that experience heavy agricultural production and over-use of fertilizers. Additionally, leaching or surface runoff of nitrate-N to drinking water sources can cause human and animal health issues. The US Environmental Protection Agency (EPA) has set the maximum contaminant level of nitrate-N at 10 ppm for the safety of drinking water (EPA 2016a). Drinking water exceeding this level has the potential to negatively impact health (Cameron et al. 2013).

Volatilization

Volatilization is the loss of ammonia-N (NH_3-N) gas to the atmosphere from soil. All ammonium and ammonia-based fertilizers, including manure, anhydrous ammonia (AA), aqua ammonia, urea, and urea ammonium nitrate (UAN) have the potential for volatilization (Jones et al. 2013).

Factors that affect volatilization

Water is necessary for transporting N fertilizer into the soil. When fertilizer has been surface- applied and precipitation is minimal, volatilization typically occurs during the first two to three weeks following application (Jones et al. 2007; Turner et al. 2012). High temperatures and high wind can exacerbate volatilization (Havlin et al. 2005; Jones et al. 2013) by drying out surface soils. Typically the risk of volatilization is greater on sandy soils, even with low fertilizer application rates and deep injection.

However, there is little loss from deep-injected fertilizers, particularly AA, to loamy soils (Mullen et al. 2000). Top-dressed urea fertilizers generally have the greatest potential for volatilization (Jones et al. 2007). One half-inch of precipitation or greater within the first 24 hours after surface application of urea or UAN can significantly reduce volatilization (Jones et al. 2013) by moving the fertilizer into the profile and reducing its contact with air.

Volatilization is also greater when N fertilizer is broadcast than when it is banded, and injecting or incorporating N fertilizer immediately after application can reduce volatilization. Shank banding AA fertilizer is the most common N fertilizer application before winter wheat seeding in the inland PNW to minimize volatilization losses (Pan et al. 1997). Because AA is often injected into dry soils, drought-prone field positions may exhibit greater ammonia loss than wetter zones, particularly when soil **compaction** prevents deep shanking (Pan et al. 1997). Other factors that increase the risk of volatilization include high concentrations of soil organic matter and crop residues as well as high soil **pH** and temperatures that increase soil ammonium-N content in dissolved soil water (Jones et al. 2007). Enhanced efficiency fertilizers, controlled-release products, and volatilization inhibitors can also be applied to fields to reduce the loss of ammonia-N (Jones et al. 2013).

Environmental impact of volatilization

Although the gaseous form of ammonia-N is not a greenhouse gas (GHG), it is an environmental concern because the deposition of ammonia-N plays a significant role in the formation of atmospheric particulate matter (PM2.5) that can reduce visibility and lead to atmospheric deposition of N in sensitive ecosystems. In the atmosphere, ammonia-N reacts rapidly with both sulfuric and nitric acids to form fine particles (NOAA 2000; Behera et al. 2013). According to the EPA National Air Pollution Trends Update in 1997, agriculture contributed 85% of US emissions, about 1/3 of which resulted from fertilizer applications.

Denitrification

Denitrification is a process mediated by soil microorganisms that converts nitrate-N to gaseous forms of N. Denitrification is controlled by oxygen concentration and available carbon and is highly variable in space and time (Pan et al. 1997). In particular, nitrous oxide (N_2O) can be lost to the atmosphere and act as a powerful GHG when denitrification converts N present in agricultural soils. Nitrous oxide has approximately 300 times the global warming potential of carbon dioxide (CO_2) for a 100-year timescale, and it accounted for approximately 6% of US GHG emissions (EPA 2016b). Seventy-nine percent of these emissions were from agricultural soil N fertilizer application and other management activities. Denitrification occurs under anaerobic conditions and is more likely to occur when soils high in residual N are saturated, since microbes actively convert nitrate-N to nitrous oxide under these conditions.

Factors that affect denitrification

Denitrification potential is high in anaerobic conditions when soil is saturated. In the inland PNW, the risk of denitrification is particularly high when surface soils remain saturated especially during snowmelt events, in spring or after heavy precipitation (especially on toeslopes), during frequent freeze and thaw events, or when tillage pans impede percolation. Denitrification is negligible when pH is lower than 5 but when pH levels near 6 to 6.5, nitrous oxide presents more than half of the N emissions. Denitrification increases rapidly when temperature is greater than 35°F and is highest at 77 to 95°F (Havlin et al. 2005). The presence of high amounts of residues with low C:N ratios (e.g., legume residues) and other sources of residual soil N can increase nitrous oxide emissions (Baggs et al. 2000) under certain conditions since this material is high in N and serves as a microbial food source.

Nitrate-N leaching, ammonia-N volatilization, and denitrification losses can influence crop N recovery, fertilizer N responses, and NUE. A more thorough overview of N cycling processes and the risk of losses in relation to climate change is available through the Regional Approaches to Climate Change (REACCH) PNW Agriculture and Climate Change Webinar Series available at ***https://www.reacchpna.org/seminars-nitrogen-series***.

Sulfur Management

Sulfur (S) is the second most deficient nutrient in crops after N. Similar to N reactions, biological transformations drive S cycles, which are dominated by organic and microbial soil processes. Due to its relationship with microorganisms, S has reactions that are regulated by soil, precipitation, and temperature. Plant roots absorb S in its oxidized form (SO_4-S) where it is synthesized into an essential component of protein. Although wheat only requires approximately a tenth as much S as it does N (Lutcher et al. 2005), S impacts crop yield and is necessary for producing high-quality baking flour and must be added when soil supplies are insufficient for meeting a crop's nutrient requirements.

Soil tests can help determine if S levels are deficient, but soil tests are less reliable for S than they are for other nutrients (Koenig 2005; Lutcher et al. 2005). In the top foot, Mahler (2007) recommends applying 20 pounds of S per acre for winter wheat when soil tests estimate SO_4-S levels are lower than 10 ppm. Others recommend more conservative application rates for winter wheat ranging from 10 to 15 pounds of SO_4-S per acre for the top foot (Lutcher et al. 2005) or 10 to 20 pounds of SO_4-S per acre for the top two feet (Koenig et al. 2005) when plant-available SO_4-S is between 0 and 8 ppm. For hard red wheat, 1 pound of SO_4-S is applied for each 5 pounds of N applied, up to 25 pounds SO_4-S per acre (Koenig 2005). Lutcher et al. (2005) suggest that producers should also consider adding S if winter wheat is seeded late in the fall, if the last application of S was more than 5 years ago, or if straw exists in greater-than-average quantities in the field.

Factors impacting sulfur requirements

Sulfur is most commonly applied as ammonium thiosulfate (ATS) liquid or ammonium polysulfide during N application. It can also be applied as dry gypsum, ammonium phosphate-sulfate, or ammonium sulfate (Rasmussen 1996). Some forms of S fertilizer (especially ATS) can damage seedlings when applied with the seed, but this problem can be avoided by placing the product adjacent to or below the seed. Wet winter conditions can cause leaching that moves SO_4-S below the root zone, making S deficiencies in winter wheat fairly common in early spring following a wet winter. Deficiency symptoms may disappear later in the season when root growth extends into the deeper layers and can reach S that has migrated. If elemental S fertilizer is used, it is important to know that it is not immediately plant-available. Most elemental S will not be available until 2 or 3 years after application (Lutcher et al. 2005). Higher rates of S may need to be added when a legume crop is following wheat (Koenig 2005). Sulfur fertilizer is also important to supply to canola when it is rotated with wheat in cropping systems (Koenig et al. 2011a).

Phosphorus Management

Phosphorus (P) is important for energy transfer in the plant and is therefore critical for seed formation and grain production in wheat. Plants absorb P as phosphate in the forms $H_2PO_4^-$ or HPO_4^{-2}. Sufficient supplies of P improve overall plant vigor, health, and development. Phosphorus is the third most frequently deficient nutrient in the inland PNW and its application can increase retention of tillers and accelerate crop maturity. Phosphorus deficiency is affected very little by tillage and cropping frequency, but becomes more prevalent when higher crop yields require more P for growth (Rasmussen and Douglas 1992).

Phosphorus recommendations are based on soil test values from the surface one-foot sample. Typically, the Bray or Mehlich method is used for testing acidic soils, and alkaline soils are tested using the Sodium Bicarbonate (Olsen) method. Testing method will impact fertilizer recommendations, and it is important to know which test was used by your lab. Also note that fertilizer P is expressed as P_2O_5. Therefore, P rates are generally expressed in pounds of P_2O_5 per acre. To convert P_2O_5 to elemental P, multiply by 0.44. Because P_2O_5 rates vary widely based on location, it is best to refer to nutrient management guides specific to your region for more detailed information. Specific nutrient management guides provide recommendations for increasing P rates when wheat is produced under reduced tillage (Mahler 2007) or when P fertilizer is broadcast (Koenig 2005).

The amount of P available in the soil solution for plant uptake is often very low due to its low mobility and high likelihood to become fixed to other soil particles. When it is in the form of organic compounds, P can be very stable and slow to mineralize, while in the inorganic form, P can form insoluble calcium, iron, or aluminum compounds. The greatest degree of P fixation occurs at very low and very high soil pH levels. Most P is plant-available when soil pH values are between 6 and 7. When pH values are below 5.5 (acidic soils), P forms compounds with iron and aluminum, and when pH levels are between 7.5 and 8 (alkaline soils), P binds with calcium (Havlin et al. 2005).

Because of its relative immobility in soils, placement of P is critical for reducing its sorption to soil colloids and optimizing its availability to crops. Banding P is recommended over broadcasting because P easily gets tied up by soil minerals and organic compounds, and P should be placed so that it can be easily accessed by the roots. It is best to place P fertilizer below and to the side of the seed since application of P in the seed zone has been associated with early seedling damage and reduced plant densities (Rusan and Pan 1998). Shallow placement of P can also strand it in dry surface soil, whereas deep placement (14 to 18 inches) of P increased its availability and allowed deep roots to access subsoil water later in the season, resulting in increased plant growth and grain yield (Rusan and Pan 1998). If P fertilizer is placed deeper, as recommended by this study, soil samples should be collected 2 feet deep at 1 foot increments to capture all residual soil P before making fertilizer recommendations.

Factors impacting phosphorus requirements

Phosphorus availability is greatly affected by soil pH and it becomes less accessible to crops as it precipitates with soil minerals at both low and high soil pH levels. Lack of soil water can also limit P mobility, especially late in the season in arid and semiarid environments. Crop root growth and activity are directly influenced by P supply and increased root proliferation helps the plant access more water and nutrients deeper in the soil profile. Because of these factors, limited P availability has negatively impacted wheat growth and yield when it was grown in dry soils (Rusan and Pan 1998) or on eroded hilltops (Guettinger and Koehler 1967; Pan and Hopkins 1991a), where water availability and rooting depth are restricted. Performance of wheat on eroded hilltops was improved by using **no-till** management (Pan and Hopkins 1991b) and by inoculating soil with symbiotic mycorrhizal fungi (Mohammad et al. 1995), which helped improve P uptake when supplies were low. In a late-planted, no-till fallow system, Lutcher et al. (2012) found that adding 11 and 33 pounds of P_2O_5 per acre increased overall wheat yields by 4.4% and 7.7%, respectively, compared to a control across sites in north-central Oregon and east-central Washington. Yield improvements were observed following P application when initial Sodium Bicarbonate soil tests were less than 12 ppm (Lutcher et al. 2012).

In many areas of the US, animal manure provides a source of organic P for use on crop fields. However, the geography of the inland PNW often separates animal production from the areas where wheat is grown, and animal manure is not a readily available fertilizer source for most of the wheat or other cereal crops that are produced. South-central Idaho is one exception. This region is dominated by dairy farms in combination with fields on high plateaus and very short growing seasons. Fertilizer costs cannot be recouped by low yield increases at these locations unless producers produce organically certified wheat (organic prices are 2 to 3 times of commodity wheat) and alfalfa hay, which also allows them to utilize locally available manure (Lorent et al. 2016). Research findings from these locations found that in-season application of 5 dry tons of composted dairy manure per acre was optimal to supply adequate P to crops (Hunter et al. 2012).

When applied in excess, P can lead to harmful losses to the environment, and similar to nitrate-N, it can lead to eutrophication. However, in contrast to nitrate-N, which is lost most commonly through leaching, P does not leach and its loss mainly occurs from erosion. Therefore, it is important to minimize erosion in order to reduce loss of P. Field vulnerability to P loss can be estimated based on site characteristics and management strategies using P indexes (field-scale qualitative assessment tools). Land managers concerned with P loss can access P indexes for agricultural phosphorus management using the Oregon/Washington Phosphorus Indexes (Sullivan et al. 2003) specific to sites east and west of the Cascade Mountain Range.

Potassium, Chloride, and Micronutrient Management

Although equally as important as N, S, and P for supporting healthy wheat growth and development, other nutrients are rarely added as fertilizers since their levels in the soil are generally sufficient for meeting the nutrient requirements of wheat in the inland PNW. Important aspects are mentioned below. However, due to low crop requirements for these nutrients, little information is available about them.

Potassium

Potassium (K) is an important macronutrient influencing plant-water relations and is supplied to them in its ionic form, K. Potassium is expressed as K_2O in fertilizers. To convert K_2O to K, multiply by 0.83. Silt loam textured soils that make up most of the inland PNW are rich in K-bearing minerals and generally test high or very high for available K. In general, soil K supplies are adequate across the region (Lutcher et al. 2005). Additions of K are typically not recommended for wheat; however, Mahler (2007) suggests incorporating K fertilizer into the soil during planting on eroded hilltops and knobs when soil tests indicate low levels. Applications of 50 to 100 pounds of K_2O per acre can be beneficial if soil test reports show less than 75 ppm K (sodium acetate test) or 90 ppm K (bicarbonate extract test) in the surface foot of soil (Koenig 2005). Responses to K fertilizers on eroded hilltops may only be obtainable when P deficiency is also addressed (Guettinger and Koehler 1967). Potassium is mostly contained in the straw and crop residue of wheat. In cropping systems where straw is baled and removed, K removal can be as much as 100 pounds of K_2O per ton of residue. Producers should carefully evaluate soil test levels of K on sandy textured soils or where straw or forage crops are being removed.

Chloride

Chloride (Cl) is important for biochemical processes in plants, and although wheat requires minimal amounts of this nutrient, dryland winter wheat responds when Cl is applied. On occasion, Cl may increase grain yield, test weight, and kernel size (Lutcher et al. 2005). Supplying wheat with Cl has also been noted to reduce the incidence of physiological leaf spot (PLS) and "take-all" root rot (see Chapter 10: Disease Management for Wheat and Barley) in winter wheat (Engle et al. 1997; Karow and Smiley 1997; Smiley et al. 1993). Some cultivars of winter wheat are more susceptible to PLS; however, if these diseases are a concern and if soil tests are low, 10 to 30 pounds of Cl per acre can be supplied to wheat, but should not be applied with the seed. Benefits from adding Cl at these rates can last for several years (Lutcher et al. 2005). Potassium chloride (KCl) is the most available form of Cl.

Micronutrients

Other micronutrients including Copper (Cu), Iron (Fe), Manganese (Mn), Molybdenum (Mo) and Boron (B) have not been found to cause responses to wheat in the inland PNW. Zinc (Zn) deficiencies can be found on severely eroded hillslopes, but Mahler (2007) and Lutcher et al. (2005) have not found additions of Zn to be economical for wheat. However, micronutrient status has not been widely investigated throughout the inland PNW, and it is likely that micronutrients may be limiting in some situations. If producers are applying any of these nutrients, leave untreated test strips to evaluate the application.

Soil pH and Liming

Decreasing soil pH (soil acidification) is a growing concern in eastern Washington and northern Idaho (Mahler 1985; McFarland and Huggins 2015; McFarland et al. 2015) and has serious implications for inland PNW cropping systems. Increasing soil acidity is becoming a crop production limitation in northern Idaho and other inland PNW cropland areas (Veseth 1987). Soil pH values have been measured below 5.2 throughout the Palouse region (Table 6-4). Aluminum becomes more soluble as soil pH declines below 5.2, causing the potential for aluminum toxicity in wheat. Acidic soils can also negatively impact wheat indirectly by interacting with wheat root diseases (Paulitz and Schroeder 2016) or herbicides (Raeder et al. 2016). However, wheat cultivars respond differently to acidic soils (Froese et al. 2015).

Table 6-4. Percentage of fields at different soil pH ranges in 1985 (Mahler 1985) collected in the eastern Washington and north Idaho region. Mahler's data is compared to a 2014 soil survey in Columbia County, Washington, Washington State University variety testing program sites, and other random sample sites from the eastern Washington region.

1985 Mahler survey of soil pH values surface 12 inches		2014 Columbia County survey of soil pH values surface 12 inches	
Soil pH	% of fields	Soil pH	% of fields
>6.0	9	>6.0	3
5.6–5.9	38	5.6–5.9	8
5.2–5.4	32	5.2–5.4	66
<5.2	21	<5.2	23

Soil acidity is a major limitation to soil productivity in much of the world and is considered the master variable by soil chemists (McBride 1994). Soil pH has a direct impact on many of the chemical and biological processes in the soil, causing yield reductions in many crops (Koenig et al. 2011b). Soil pH (concentration of hydrogen in solution) is the measure of hydrogen (H^{+}) ions (acid soils) or hydroxide (OH^{-}) ions (alkaline soils) in the soil solution (Horneck et al. 2007). Regionally, soil pH values are declining (becoming more acidic) because of applications of N fertilizers, plant nutrient uptake, and precipitation (Mahler 1985; McFarland and Huggins 2015; McFarland et al. 2015) with N being the largest contributor. The presence of ammonia-N or ammonium-N in the soil contributes to the presence of hydrogen ions that result from the conversion (nitrification) to nitrate by soil bacteria (Carter 2016). This process releases hydrogen ions into the soil. The hydrogen ions replace base cations (e.g., K^{+}, Mg^{+2}, Ca^{+2}) from the soil **cation exchange capacity** and lower the soil solution pH. An application of N placed on or in the soil normally is within the top 6 inches of the soil profile and often creates a stratified layer of acidity (Carter 2016).

Standard soil sampling methodologies (one foot samples) tend to overlook the problem of stratified acid soil layers that have developed in once neutral pH soils. Reduced tillage may intensify or narrow the thickness of layers of stratification compared to inversion tillage systems where more soil mixing occurs, diluting the acidity somewhat. Inversion tillage systems experience the same rate of acidification, but the soil is mixed in a larger volume and may not be as noticeable. Soil sampling procedures that collect the top foot of soil in increments of 3 inches or less have been adopted to try to identify stratification of acidity and nutrients that may have developed (Carter 2016). Soil pH meters have also been found to have good results for monitoring soil in the field (Carter 2016).

Liming of acid soils is a common practice on agriculture soils of the world that have low pH. Liming is the application of amendments containing calcium carbonate ($CaCO_3$) to react with the hydrogen ions to reduce the acidic conditions (Thompson et al. 2016a; 2016b; 2016c). Soil pH is the controlling factor in the availability of primary (Table 6-5) and secondary (not shown) soil nutrients. Other impacts of soil acidity include poor plant root development, herbicide persistence, reduced (or intensified)

Table 6-5. Fertilizer efficiency and availability at different soil pH levels.

Soil pH	% Fertilizer Efficiency			% Fertilizer Unavailable
	N	P	K	
7.0	100	100	100	0
6.0	89	52	100	20
5.5	77	48	77	33
5.0	53	34	52	54
4.5	30	23	33	71

Modified and used with permission from The Mosaic Crop Nutrition Fertilizer Use Guide (2016).

herbicide activity, reduced crop yield, increased drought stress, aluminum toxicity, reduced soil structure, and reduced water **infiltration**.

Liming applications should be considered whenever soil pH levels drop below critical levels, which varies depending on the crop type (Froese et al. 2015) and many soil-applied herbicides (Raeder et al. 2016). Some soil-applied herbicides require a minimum pH, which can be found listed on many herbicide labels.

As the soil pH continues to decline, larger application rates of lime should be applied. The soil pH scale is not a linear relationship; it is logarithmic, meaning a change from pH 7 to 6 is a 10× change in acidity, but a change from pH 7 to 5 is a 100× change in acidity. The application rate of liming products should be considered in the same manner because it will take a much larger application rate and more time to correct the lower pH (Thompson et al. 2016a; 2016b; 2016c).

Liming application rates and lime activity depend on the fineness of the ground limestone, purity of the product, and the availability of water. Finer particles react faster than coarse materials, which will require weathering to reach soil solution and neutralize hydrogen ions. Lime effectiveness is measured by purity, particle size, and chemical composition relative to calcium carbonate equivalency. The Oregon Lime Score is one such rating system, and the Calcium Carbonate Equivalent (CCE) scoring method is an expression of the acid-neutralizing capacity of a carbonate rock relative to that of pure calcium carbonate, expressed as a percentage. These scoring systems may not adequately rate some of the newer ultra-fine limestone products that are available. The finer particles are more mobile in the soil and react faster than typical agricultural liming products.

Soil testing labs will provide lime recommendations relative to the soil sample properties. These recommendations indicate the amount of CCE needed to raise the soil pH to a certain specified level (6.5) in the plow layer or top six inches of soil. Recommendation methods may not be standard between labs. If recommendations are large, split applications should be considered over multiple years. There are several lime recommendation procedures. Since declining pH is a relatively recent concern (e.g., last 30 years) (Mahler 1985) in the inland PNW, the most reliable procedures have yet to be tested or determined.

As a producer and/or landowner, it is important to monitor soil pH levels and make lime applications before pH levels get dangerously low in the top 6 inches of soil. Below 5.5 is certainly approaching a level of concern and should be considered for a corrective lime application. A recommendation might be as high as 4,000 pounds per acre of CCE lime to raise the pH to 6.5, but a split application would probably be in order.

It is more economical and more sustainable to maintain a soil pH of 6.5, which will allow more efficient use of fertilizer and soil-applied chemicals, healthier plants, better weed control, and optimized crop yields, leading to a more sustainable agriculture production system.

Soil pH can be highly variable across the landscape due to topographic and soil conditions. It is likely that most fields in wheat production in the inland PNW have variable liming rates to address variable soil pH conditions. Additional information about soil acidification in the inland PNW can be found at *www.smallgrains.wsu.edu*, including several factsheets and videos also listed in Grower Resources below.

Soil Sampling and Tests

Routine soil testing is an important practice for guiding producers to make appropriate soil fertility management decisions. Soil tests can accurately estimate factors such as nutrient deficiencies or excesses, nutrient holding capacity, soil pH, and organic matter content. Using actual values of available soil nutrients to determine fertilizer rates is one of the most important factors for improving nutrient use efficiencies, particularly in non-uniform environments like those found throughout much of the inland PNW.

Soil samples can be collected using composite sampling, grid sampling, or management zone sampling. Detailed explanations of these sampling procedures can be found in Collins (2012). Composite sampling is the least intensive sampling method and likely the most appropriate procedure for wheat producers sampling large-scale fields. Soil samples should be collected prior to planting and fertilizing the crop. Soil samples can be collected in spring or fall. An advantage of collecting soil in fall is that fall samples can determine how well the previous crop used the fertilizer it was supplied. If substantial nutrients are remaining in the soil after harvest, the plant did not use what was available and fertilizer rates should be adjusted. Errors of this type could occur if yield potential was estimated too high, residual soil contributions were not accurately accounted for, or other factors reduced yield and negatively impacted nutrient use (Koenig 2005). Post-harvest assessment of crop yield and protein performance is also encouraged in relation to N supply to help identify site-specific UNRs and improve NUE (Pan et al. 2007).

Soil test recommendations have been developed based on testing the top foot of soil with the exception of N and S. Because N and S are mobile nutrients, it is recommended to sample 0 to 5 feet, or to a restricting layer, for N and 0 to 2 feet for S, in one-foot increments. If practices that involve deeper placement of fertilizer are used for any nutrient, for example P, soil samples should be collected in one-foot increments up to that depth to improve accuracy of fertilizer recommendations. When sampling for pH, it is advisable to sample the top foot of soil in 3 or 4 inch increments to determine the degree of pH stratification, especially in long-term direct-seed fields.

Soil tests also help evaluate the long-term effectiveness of nutrient management strategies on a farm and detailed recordkeeping is encouraged. Carefully track annual fertilizer rates, additions of lime or other soil amendments, crop rotations and yields, and soil test data in a notebook, spreadsheet, or database. Multi-year records can provide a better understanding of nutrient dynamics within fields and can help indicate whether application rates and decision drivers, such as yield potentials, are accurate or need to be adapted. More in-depth details about appropriate soil sampling methods and test evaluations are provided in Soil Sampling (Mahler and Tindall 1997) and Soil Testing: A Diverse Guide for Farms with Diverse Vegetable Crops (Collins 2012). Fertilizer recommendations based on soil tests are specific to each nutrient, and producers should reference nutrient management guides for more specific information.

Conclusions

Interactions among climate, water, and nutrient dynamics impact wheat development, growth, and associated soil fertility management strategies in the inland PNW. Variability among these factors across the region is common, and diverse landscapes and local weather conditions complicate the choices that producers have to make regarding timing and application of nutrients. Environmental concerns associated with N losses, soil erosion, nutrient runoff, and soil acidification can result from improper use of fertilizers and make inland PNW wheat production systems less sustainable over the long-term. Because nutrient cycling is impacted by temperature, precipitation, and available soil water, understanding the importance of these variables in relation to current nutrient management decisions will help prepare producers to adapt to changes associated with future climate-related stresses. Practices that improve resource use efficiencies (e.g., fertilizers and water), build soil, and reduce pollution have the ability to decrease the negative environmental impacts associated with over-use of fertilizers while improving the resiliency of the soil and wheat production in the inland PNW.

Grower Resources

Nitrogen Management Webinar Series

https://www.reacchpna.org/seminars-nitrogen-series

A series of three, one-hour-long recorded video webinars addressing issues related to agricultural N and the environment.

- Nitrogen Cycling and Losses in Agricultural Systems
- Nitrous Oxide Emissions in Inland Pacific Northwest Cropping Systems

- Nitrogen Management and Climate Change Mitigation in Pacific Northwest Cropping Systems

Nitrogen and Post-Harvest Calculators

http://smallgrains.wsu.edu/soil-and-water-resources/

Dynamic tools based on the Dryland Winter Wheat Nutrient Management Guide that help calculate N supply.

Soil pH Video Series

http://smallgrains.wsu.edu/soil-and-water-resources/soil-acidification-in-the-inland-northwest/

A series of three, 6 to 10 minute videos discussing associated concerns and treatment of low soil pH.

- Soil pH – What it Looks Like
- Soil pH – How it Happens
- Soil pH – Managing it on the Farm

Extension Management Guides

Regionally specific Extension guides that provide recommendations and calculations for determining appropriate soil fertility management strategies in wheat-based cropping systems.

Nutrient Management

Dryland Winter Wheat Eastern Washington Nutrient Management Guide. 2005. Washington State University Extension Publication EB1987.

Nitrogen Management for Hard Wheat Protein Enhancement. 2005. Pacific Northwest Extension Publication PNW 578.

Northern Idaho Fertilizer Guide Winter Wheat. 2007. University of Idaho Extension Publication CIS 453.

Phosphorus Fertilization of Late-Planted Winter Wheat in No-till Fallow. 2012. Pacific Northwest Extension Publication PNW631.

Phosphorus Increases Wheat Yields. 1967. Washington State Cooperative Extension Service Publication EM 2940.

Southern Idaho Dryland Winter Wheat Production Guide. 2004. University of Idaho Extension Publication BUL0827.

Winter Wheat in Continuous Cropping Systems (high precipitation zone). 2007. Oregon State University Extension Service FG 84-E.

Winter Wheat in Continuous Cropping Systems (Intermediate precipitation zone). 2006. Oregon State University Extension Service FG 83-E.

Winter Wheat and Spring Grains in Continuous Cropping Systems (Low precipitation zone). 2006. Oregon State University Extension Service FG 81-E.

Winter Wheat in Summer-Fallow Systems (Intermediate precipitation zone). 2006. Oregon State University Extension Service FG 82-E.

Winter Wheat in Summer-Fallow Systems (Low precipitation zone). 2005. Oregon State University Extension Service FG 80-E.

Water Availability

Predicting Wheat Grain Yields Based on Available Water. 2012. Washington State University Extension Publication EM049E.

Nutrient Losses

Agricultural Phosphorus Management using the Oregon/Washington Phosphorus Indexes. 2003. Oregon State University Extension Service EM 8848-E.

Factors Affecting Nitrogen Fertilizer Volatilization. 2013. Montana State University Extension Publication EB0208.

Management of Urea Fertilizer to Minimize Volatilization. 2007. Montana State University and Washington State University Extension Publication EB173.

Soil pH and Liming

Acid Soils: How Do They Interact with Root Diseases? 2016. Washington State University Extension Publication FS195E.

Acidifying Soil for Crop Production: Inland Pacific Northwest. 2007. Pacific Northwest Extension Publication PNW599.

Agricultural Lime and Liming, Part 1: Introduction Agricultural Lime and Liming. 2016. Soil Acidification Series. Washington State University Extension Publication FS212E.

Agricultural Lime and Liming, Part 2: Laboratory Testing to Determine Lime Requirements. 2016. Soil Acidification Series. Washington State University Extension Publication FS217E.

Agricultural Lime and Liming, Part 3: Aglime Product Selection and Comparison Calculator User Guide. Soil Acidification Series. Washington State University Extension Publication FS213E.

How Soil pH Affects the Activity and Persistence of Herbicides. 2016. Soil Acidification Series. Washington State University Extension Publication FS189E.

Recommended Crop Species and Wheat Varieties for Acidic Soil. 2015. Soil Acidification Series. Washington State University Extension Publication FS169E.

Soil Acidity and Aluminum Toxicity in the Palouse Region of the Pacific Northwest. 2011. Washington State University Extension Publication FS050E.

Soil pH and Implications for Management: An Introduction. 2015. Soil Acidification Series. Washington State University Extension Publication FS170E.

Using a pH Meter for In-Field Soil pH Sampling. 2016. Soil Acidification Series. Washington State University Extension Publications. FS205E.

Soil Sampling and Testing

Soil Testing: A Diverse Guide for Farms with Diverse Vegetable Crops. 2012. Washington State University Extension Publication EM050E.

Soil Sampling. 1997. University of Idaho Extension Bulletin 704.

Other

Organic Small Grain Production in the Inland Pacific Northwest: A Collection of Case Studies. 2016. Pacific Northwest Publications PNW683. Washington State University.

Canola, Growth, Development, and Fertility. 2011. Washington State University Factsheet FS045E.

Physiological Leaf Spot and Chloride. 1997. Oregon State University Extension Service Crop Science Report EXT/CRS 108.

References

Baggs, E.M., R.M. Rees, K.A. Smith, and A.J.A. Vinten. 2000. Nitrous Oxide Emission from Soils after Incorporating Crop Residues. *Soil Use Management* 16: 82–87.

Baker, D.A., D.L. Young, D.R. Huggins, and W.L. Pan. 2004. Economically Optimal Nitrogen Fertilization for Yield and Protein in Hard Red Spring Wheat. *Agronomy Journal* 96: 116–123.

Baudaruddin, M., and D.W. Meyer. 1994. Grain Legume Effects on Soil Nitrogen, Grain Yield, and Nitrogen Nutrition of Wheat. *Crop Science* 34: 1304–1309.

Behera, S.N., M. Sharma, V.P. Aneja, and R. Balasubramanian. 2013. Ammonia in the Atmosphere: A Review on Emission Sources, Atmospheric Chemistry and Deposition on Terrestrial Bodies. *Environmental Science Pollution Research International* 20: 8092–8131.

Brown, B., M. Westcott, N. Christensen, B. Pan, and J. Stark. 2005. Nitrogen Management for Hard Wheat Protein Enhancement. Pacific Northwest Extension Publication PNW578. University of Idaho.

Busacca, A.J., D.K. McCool, R.I. Papendick, and D.L. Young. 1985. Dynamic Impacts of Erosion Processes on Productivity of Soils in the Palouse. In Proceedings of the National Symposium on Erosion and Soil Productivity. American Society of Agricultural Engineering St. Joseph, MI.

Cameron, K.C., H.J. Di, and J.L. Moir. 2013. Nitrogen Losses from the Soil/Plant System: A Review. *Annals of Applied Biology* 162: 145–173.

Carter, P.G. 2016. Using a pH Meter for In-Field Soil pH Sampling. Washington State University Extension Publication FS205E.

Cassman, K.G., A.R. Dobermann, and D.T. Walters. 2002. Agroecosystems, Nitrogen-Use Efficiency, and Nitrogen Management. *Ambio* 31: 132–140.

Collins, D. 2012. Soil Testing: A Guide for Farms with Diverse Vegetable Crops. Washington State University Extension Publication EM050E.

Dawson, J.C., D.R. Huggins, and S.S. Jones. 2008. Characterizing Nitrogen Use Efficiency in Natural and Agricultural Ecosystems to Improve the Performance of Cereal Crops in Low-Input and Organic Agricultural Systems. *Field Crops Research* 107: 89–101.

Engle, R.E., P.L. Bruckner, D.E. Mathre, and S.K.Z. Brumfield. 1997. A Chloride-Deficient Leaf Spot Syndrome of Wheat. *Soil Science Society of America Journal* 61: 176–184.

EPA (Environmental Protection Agency). 2016a. Table of Regulated Drinking Water Contaminants. ***https://www.epa.gov/ground-water-and-drinking-water/table-regulated-drinking-water-contaminants#Inorganic***

EPA (Environmental Protection Agency). 2016b. Overview of Greenhouse Gasses. ***https://www.epa.gov/ghgemissions/overview-greenhouse-gases***

Fiez, T.E., B.C. Miller, and W.L. Pan. 1994a. Winter Wheat Yield and Grain Protein Across Varied Landscape Positions. *Agronomy Journal* 86: 1026–1032.

Fiez, T.E., B.C. Miller, and W.L. Pan. 1994b. Assessment of Spatially Variable Nitrogen Fertilizer Management in Winter Wheat. *Journal of Production Agriculture* 7: 17–18; 86–93.

Fiez, T.E., W.L. Pan, and B.C. Miller. 1995. Nitrogen Use Efficiency of Winter Wheat among Landscape Positions. *Soil Science Society of America Journal* 59: 1666–1671.

Froese, P.S., A.H. Carter, and M.O. Pumphrey. 2015. Recommended Crop Species and Wheat Varieties for Acidic Soils. Washington State University Extension Publication FS169E.

Guettinger, D.L., and F.E. Koehler. 1967. Phosphorus Increases Wheat Yields. Washington State Cooperative Extension Service Publication EM2940.

Havlin, J.L., J.D. Beaton, S.L. Tisdale, and W.L. Nelson. 2005. Soil Fertility and Fertilizers: An Introduction to Nutrient Management. New Jersey. Pearson Education, Inc.

Hergert, G.W., W.L. Pan, D.R Huggins, J.H. Grove, T.R. Peck, and G.L. Malzer. 1997. The Adequacy of Current Fertilizer Recommendations for Soil Specific Management. In The State of Site-Specific Management for Agriculture, F.J. Pierce and E.J. Sadler, eds. ASA, CSSA, SSSA, Madison, WI.

Horneck, D., D. Wysocki, B. Hopkins, J. Hart, and R. Stevens. 2007. Acidifying Soil for Crop Production: Inland Pacific Northwest. Pacific Northwest Extension Publication PNW599. Oregon State University.

Huggins, D.R., and W.L. Pan. 1993. Nitrogen Efficiency Component Analysis: An Evaluation of Cropping System Differences in Productivity. *Agronomy Journal* 85: 898–905.

Hunter, L.A., C.L. Falen, A. Moore, A. Falen, and C. Kinder. 2012. Phosphorus and Potassium Availability from Dairy Compost in a Dryland, High-Elevation Organic Alfalfa System. *Journal of the NACAA* 5(2).

IPNI (International Plant Nutrition Institute). 2012. A Nutrient Use Information System (NuGIS) for the U.S. Norcross, GA. ***www.ipni.net/nugis***

Jacquot, H.D. 1953. Annual Cropping with Nitrogen in the Intermediate Rainfall Areas of Eastern Washington. Washington Agricultural Experiment Station Stations Circular No. 214.

Jones, C.A., R.T. Koenig, J.W. Ellsworth, B.D. Brown, and G.D. Jackson. 2007. Management of Urea Fertilizer to Minimize Volatilization. Montana State University Extension Publication EB173.

Jones C., B.D. Brown, R. Engel, D. Horneck, and K. Olson-Rutz. 2013. Factors Affecting Nitrogen Fertilizer Volatilization. Montana State University Extension Publication EB0208.

Karow, R., and D. Smiley. 1997. Physiological Leaf Spot and Chloride. Oregon State University Extension Service Crop Science Report EXT/CRS 108.

Koehler, F.E., V.L. Cochran, and P.E. Rasmussen. 1987. Fertilizer Placement, Nutrient Flow and Crop Response in Conservation Tillage. In STEEP Conservation Concepts and Accomplishments, L.F. Elliot, ed. Washington State University, Pullman, WA.

Koenig, R.T. 2005. Dryland Winter Wheat Eastern Washington Nutrient Management Guide. Washington State University Extension Publication EB1987.

Koenig, R.T., W.A. Hammac, and W.L. Pan. 2011a. Canola, Growth, Development, and Fertility. Washington State University Extension Publication FS045E.

Koenig, R.T., K. Schroeder, A. Carter, M. Pumphrey, T. Paulitz, K. Campbell, and D. Huggins. 2011b. Soil Acidity and Aluminum Toxicity in the Palouse Region of the Pacific Northwest. Washington State University Extension Publication FS050E.

Kyveryga, P.M., P.C. Caragea, M.S. Kaiser, and T.M. Blackmer. 2013. Predicting Risk from Reducing Nitrogen Fertilization using Hierarchical Models and On-Farm Data. *Agronomy Journal* 105: 85–94.

Lea, P.J., and R.A. Azevedo. 2006. Nitrogen Use Efficiency. 1. Uptake of Nitrogen from the Soil. *Annals of Applied Biology* 149: 243–247.

Leggett, G.E. 1959. Relationships between Wheat Yield, Available Moisture and Available Nitrogen in Eastern Washington Dryland Areas. Washington Agricultural Experiment Bulletin 609.

Long, D.S., J.D. Whitmus, R. Engel, and G.W. Brester. 2015. Partial Budget Analysis of Variable-Rate Nitrogen Application on Dryland Spring Wheat. *Agronomy Journal* 107: 1055–1067.

Lorent, L., D. Roberts, and I. Burke. 2016. Organic Small Grain Production in the Inland Pacific Northwest: A Collection of Case Studies. Pacific Northwest Extension Publication PNW683. Washington State University.

Lutcher, L.K., D.A. Horneck, D.J. Wysocki, J.M. Hart, S.E. Petrie, and N.W. Christensen. 2005. Winter Wheat in Summer-Fallow Systems (Low Precipitation Zone). Oregon State University Extension Service Fertilizer Guide FG 80-E.

Lutcher, L.K., N.W. Christensen, W.F. Schillinger, D.J. Wysocki, and S.B. Wuest. 2012. Phosphorus Fertilization of Late-Planted Winter Wheat in No-Till Fallow. Pacific Northwest Extension Publication PNW631. Oregon State University.

Maaz, T., W. Pan, and W. Hammac. 2016. Influence of Soil Nitrogen and Water Supply on Canola Nitrogen Use Efficiency. *Agronomy Journal* 108: 2099–2109.

Maaz, T.M., and W.L. Pan. n.d. Residual Fertilizer Nitrogen and Cropping Sequence Impact Rotational Nitrogen Use. *Agronomy Journal* (in review).

Mahler, R.L. 1985. Long-Term Acidification of Farmland in N ID and E WA. *Communications in Soil Science and Plant Analysis* 16: 83–95.

Mahler, R.L., F.E. Koehler, and L.K. Lutcher. 1994. Nitrogen Source, Timing of Application, and Placement: Effects on Winter Wheat Production. *Agronomy Journal* 86: 637–642.

Mahler, R.L., and T.A. Tindall. 1997. Soil Sampling. University of Idaho Cooperative Extension System Bulletin 704.

Mahler, R.L. 2007. Northern Idaho Fertilizer Guide Winter Wheat. University of Idaho Extension Publication CIS 453.

McBride, M.B. 1994. Soil Acidity. In Environmental Chemistry of Soils, M.B. McBride, ed. Oxford University Press, New York, NY.

McFarland, C., and D.R. Huggins. 2015. Acidification in the Inland Pacific Northwest. In Crops and Soils Magazine. American Society of Agronomy, Madison, WI.

McFarland, C., D.R. Huggins, and R.T. Koenig. 2015. Soil pH and Implications for Management: An Introduction. Washington State University Extension Publication FS170E.

Miller, B., and B. Pan. n.d. Practical Ways to Manage Grain Protein. Washington Wheat Commission Publication, Spokane, Washington.

Mohammad, M.J., W.L. Pan, and A.C. Kennedy. 1995. Wheat Responses to Vesicular-Arbuscular Mycorrhizal Fungal Inoculation of Soils from Eroded Toposequence. *Soil Science Society of America Journal* 59: 1086–1090.

Moll, R.H., E.J. Kamprath, and W.A. Jackson. 1982. Analysis and Interpretation of Factors Which Contribute to Efficiency of Nitrogen Utilization. *Agronomy Journal* 74: 562.

Mosaic Crop Nutrition. 2016. Fertilizer Use Guide. ***http://www.cropnutrition.com/efficient-fertilizer-use-guide***

Mulla, D.J. 1986. Distribution of Slope Steepness in the Palouse Region of Washington. *Soil Science Society of America Journal* 50: 1401–1406.

Mulla, D.J. 1992. A Comparison of Winter Wheat Yield and Quality Under Uniform versus Spatially Variable Fertilizer Management. *Agriculture, Ecosystems & Environment* 38: 301–311.

Mullen, R.W., G.V. Johnson, W.R. Raun, and B.M. Howell. 2000. Simulating Volatilization Losses from Anhydrous Ammonia Applications: A Simple Laboratory Exercise. *Journal of Natural Resources Life Science Education* 29: 107–110.

NOAA (National Oceanic and Atmospheric Administration). 2000. Atmospheric Ammonia: Sources and Fate. A Review of Ongoing Federal Research and Future Needs. Air Quality Research Subcommittee Meeting Report. ***http://www.esrl.noaa.gov/csd/AQRS/reports/ammonia.pdf***

Pan, W.L., and A.G. Hopkins. 1991a. Plant Development, and N and P Use of Winter Barley I. Evidence of Water Stress-Induced P Deficiency in an Eroded Toposequence. *Plant and Soil* 135: 9–19.

Pan, W.L., and A.G. Hopkins. 1991b. Plant Development, and N and P use of Winter Barley II. Responses to Tillage and N Management Across Eroded Toposequences. *Plant and Soil* 135: 21–29.

Pan, W.L., D.R Huggins, G.L. Malzer, and J.L. Smith. 1997. Variable Soil-Plant Relationships: Implications for Site-Specific Nitrogen Management. In The State of Site-Specific Management for Agriculture, F.J. Pierce and E.J. Sadler, eds. ASA, CSSA, SSSA, Madison, WI.

Pan, W.L., W. Schillinger, D. Huggins, R. Koenig, and J. Burns. 2007. Fifty Years of Predicting Wheat Nitrogen Requirements in the Pacific Northwest U.S.A. In Managing Crop Nitrogen for Weather. International Plant Nutrition Institute, Norcross, GA 30092-2837 USA.

Pan, W.L. 2015. The Roots of Soil Fertility. Invited Presentation, Leo M. Walsh Soil Fertility Distinguished Lectureship ASA-SSSA-CSSSA annual meetings, Minneapolis, MN. ***https://scisoc.confex.com/scisoc/2015am/videogateway.cgi/id/23020?recordingid=23020***

Pan, W.L., T. McClellan Maaz, W.A. Hammac, V.A. McCracken, and R.T. Koenig. 2016. Mitscherlich-Modeled, Semi-Arid Canola Nitrogen Requirements Influenced by Soil Nitrogen and Water. *Agronomy Journal* 108: 884–894.

Paulitz, T., and K. Schroeder. 2016. Acid Soils: How Do They Interact with Root Diseases? Washington State University Extension Publication FS195E.

Raeder, A.J., D. Lyon, J.B. Harsh, and I. Burke. 2016. How Soil pH Affects the Activity and Persistence of Herbicides. Washington State University Extension Publication FS189E.

Rasmussen, P.E., and C.L. Douglas, Jr. 1992. The Influence of Tillage and Cropping-Intensity on Cereal Response to Nitrogen, Sulfur, and Phosphorus. *Fertilizer Research* 31: 15–19.

Rasmussen, P.E. 1996. Fertility Management in Dryland Conservation Cropping Systems of the Pacific Northwest. *American Journal of Alternative Agriculture* 11: 108–114.

Raun, W.R., and G.V. Johnson. 1999. Improving Nitrogen Use Efficiency for Cereal Production. *Agronomy Journal* 91: 357–363.

Robertson, L.D., S.O. Guy, and B.D. Brown. 2004. Southern Idaho Dryland Winter Wheat Production Guide. University of Idaho Extension Publication BUL0827.

Rusan, M.J., and W. Pan. 1998. Wheat Yield and P Placement Under Dryland Conditions. In Triticeae III, A.A. Jaradat, ed. Science Publishers, Inc., U.S.

Schepers, J.S., K.D. Frank, E.E. Alberts, and M.G. Moravek. 1991. Maize Production Impacts on Groundwater Quality. *Journal of Environmental Quality* 20: 12–16.

Schillinger, W.F., S.E. Schofstoll, and J.R. Aldredge. 2012. Predicting Wheat Grain Yields Based on Available Water. Washington State University Extension Publication EM049E.

Smiley, R.W., W. Uddin, P.K. Zwer, D.J. Wysocki, D.A. Ball, and T.G. Chastain. 1993. Influence of Crop Management Practices on Physiologic Leaf Spot of Winter Wheat. *Plant Disease* 77: 803–810.

Song, X., J. Wang, W. Huang, L. Liu, G. Yan, and R. Pu. 2009. The Delineation of Agricultural Management Zones with High Resolution Remotely Sensed Data. *Precision Agriculture* 10: 471–487.

Sowers, K.E., B.C. Miller, and W.L. Pan. 1994. Optimizing Yield and Grain Protein in Soft White Winter Wheat with Split N Applications. *Agronomy Journal* 86: 1020–1025.

Sullivan, D.M., and R.G. Stevens. 2003. Agricultural Phosphorus Management Using the Oregon/Washington Phosphorus Indexes. Oregon State University Extension Service EM 8848-E.

Thompson, W.H., C. McFarland, T. Brown, and D.R. Huggins. 2016a. Agricultural Lime and Liming, Part 1: Introduction to Agricultural Lime and Liming. Washington State University Extension Publication FS212E.

Thompson, W.H., C. McFarland, T. Brown, and D.R. Huggins. 2016b. Agricultural Lime and Liming, Part 2: Laboratory Testing to Determine Lime Requirements. Washington State University Extension Publication FS217E.

Thompson, W.H., C. McFarland, T. Brown, and D.R. Huggins. 2016c. Agricultural Lime and Liming, Part 3: AgLime Product Selection and Comparison Calculator User Guide. Washington State University Extension Publication FS213E.

Turner, D.A., R.E. Edis, D. Chen, J.R. Freney, and O.T. Denmead. 2012. Ammonia Volatilization from Nitrogen Fertilizers Applied to Cereals in Two Cropping Areas of Southern Australia. *Nutrient Cycling in Agroecosystems* 93: 113–126.

Veseth, R. 1987. Yield Losses Resulting from Soil Acidity. STEEP Pacific Northwest Conservation Tillage Handbook 6(8). Washington State University, Pullman, WA.

Waldren, R.P., and A.D. Flowerday. 1979. Growth Stages and Distribution of Dry Matter, N, P, and K in Winter Wheat. *Agronomy Journal* 71: 391–397.

Washington Wheat Commission. n.d. Hard White Wheat Protein: Nitrogen and Protein Management Guide. Washington Wheat Commission, Spokane, Washington.

Washington Wheat Commission. n.d. Hard Red Spring Wheat: Nitrogen and Protein Management Guide. Washington Wheat Commission, Spokane, Washington.

Washington Wheat Commission. n.d. Hard Red Winter Wheat: Nitrogen and Protein Management Guide. Washington Wheat Commission, Spokane, Washington.

Yang, C., C.L. Peterson, G.J. Shropshire, and T. Otawa. 1998. Spatial Variability of Field Topography and Wheat Yield in the Palouse Region of the Pacific Northwest. *Transactions of the American Society of Agricultural Engineering* 41: 17–27.

Chapter 7

Soil Amendments

Georgine Yorgey, Washington State University
William Pan, Washington State University
Rakesh Awale, Oregon State University
Stephen Machado, Oregon State University
Andy Bary, Washington State University

Abstract

Soil amendments can improve the sustainability of agricultural systems by building soil carbon and improving numerous other soil health indicators, including soil structure, water infiltration and retention, bulk density, nutrient availability, and microbial activity. This chapter covers some general considerations relevant to the use of soil amendments followed by a discussion of biosolids, animal manures, biochar, and black liquor. Topics covered include amendment composition, application rates, yield impacts, grain quality impacts, soil health benefits, nutrient loss concerns, and potential contaminants.

Key Points

- Although they are only rarely used in the dryland systems of the Inland Pacific Northwest, soil amendments can provide a range of soil health benefits by building soil carbon, reducing bulk density, and improving soil structure, water infiltration and retention, and nutrient availability.

Research results are coded by agroecological class, defined in the glossary, as follows:

● Annual Crop ▲ Annual Crop-Fallow Transition ■ Grain-Fallow

- Biosolids can be used by conventional producers, but not certified organic producers, and may be available at relatively low cost, though supply is limited. Biosolids applied to agricultural soils at agronomic rates and in accordance with current guidelines can result in equivalent or greater yields than equivalent use of chemical fertilizers.
- Manures can be an important resource for building or maintaining soil health for producers in proximity to concentrations of livestock. Manures with higher nutrient concentrations may provide some nutrients, particularly for certified organic dryland producers, though cost can be an issue.
- Biochar has the potential to positively impact pH and other soil health indicators, and can also improve productivity. However, economics currently limit its use in dryland agriculture. Likewise, black liquor has intriguing potential to improve soil health, and may become more feasible if paper production facilities are established in the inland Pacific Northwest.

Introduction

As discussed in Chapter 2: Soil Health, historical soil carbon (C) losses under agricultural cultivation in the **inland Pacific Northwest** (PNW) have been severe. At the Columbia Basin Agriculture Research Center (CBARC) near Pendleton, Oregon, conventional winter wheat-summer fallow depleted soil organic C by approximately 35% in the first 50 years after cultivation began, and up to 63% over 80 years of cultivation (Rasmussen et al. 1998; Ghimire et al. 2015). ■ Similar patterns of C loss contribute to reduced **soil health** across the dryland region (Brown and Huggins 2012).

Soil amendments, when economically and practically viable, could play a powerful role in slowing C losses and rebuilding soil C. In Douglas County, Washington, long-term **biosolids** applications every four years at any of three rates increased total soil C (Figure 7-1). These results are particularly striking in light of the fact that in the grain-fallow region, where a crop is only being grown every other year (and therefore residues are only being added to the soil every other year), soil C levels are unlikely to be maintained, even when tillage is reduced or eliminated (Machado 2011; Gollany et al. 2013). ■

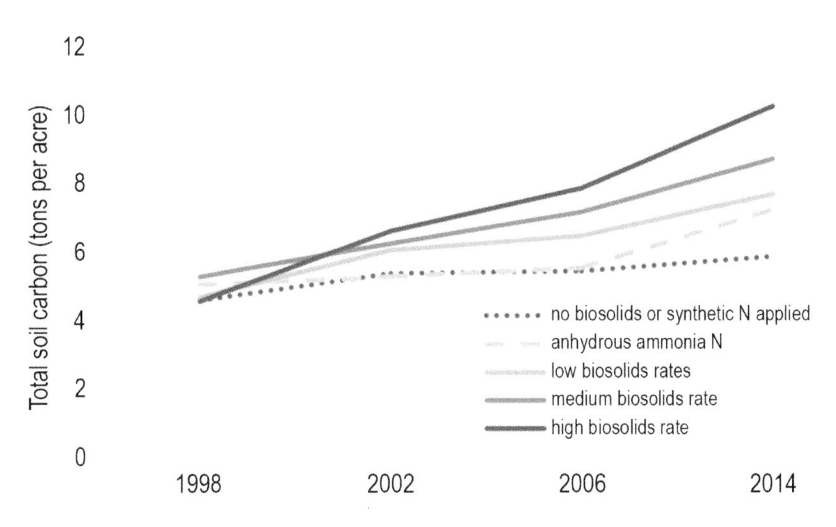

Figure 7-1. Total soil carbon measured in the top 3.9 inches of soil in a winter wheat-fallow system in Douglas County, Washington, after additions of biosolids every four years at a rate of 2.2, 3.1, or 4.0 dry tons per acre (green, blue, and purple solid lines). These rates were compared to no nutrient additions (red dotted) or addition of anhydrous ammonia at a rate of 50 lb per acre (yellow dashed); N = nitrogen. (Data from Pan et al. n.d.) ■

Beyond C, soil amendments can improve numerous physical, chemical, and biological properties of soil, and thus can be an important strategy for improving soil health and sustaining agricultural production over the long term (Figure 7-2). Amendments can improve **soil structure**, water **infiltration** and retention, nutrient availability, and microbial activity, while reducing **bulk density** (Brown et al. 2011; Cogger et al. 2013; Reeve et al. 2012; Wuest et al. 2005). Additional information on these factors can be found in Chapter 2: Soil Health.

Recycling organic C and plant nutrients contained in organic materials can also contribute to mitigating climate change (Brown et al. 2010). First, the buildup and storage of soil organic matter (SOM) draws carbon dioxide out of the atmosphere. Second, amendment-based plant nutrients can substitute for synthetic fertilizers, which generate greenhouse gases when they are produced. Although amendments often require transport, this has a relatively small effect on overall greenhouse gas balances (Brown et al. 2010). Meanwhile, impacts of soil amendments on emissions of nitrous oxide, a powerful greenhouse gas, from soils are likely to be complex when amendments substitute for conventional fertilizers. To date, these

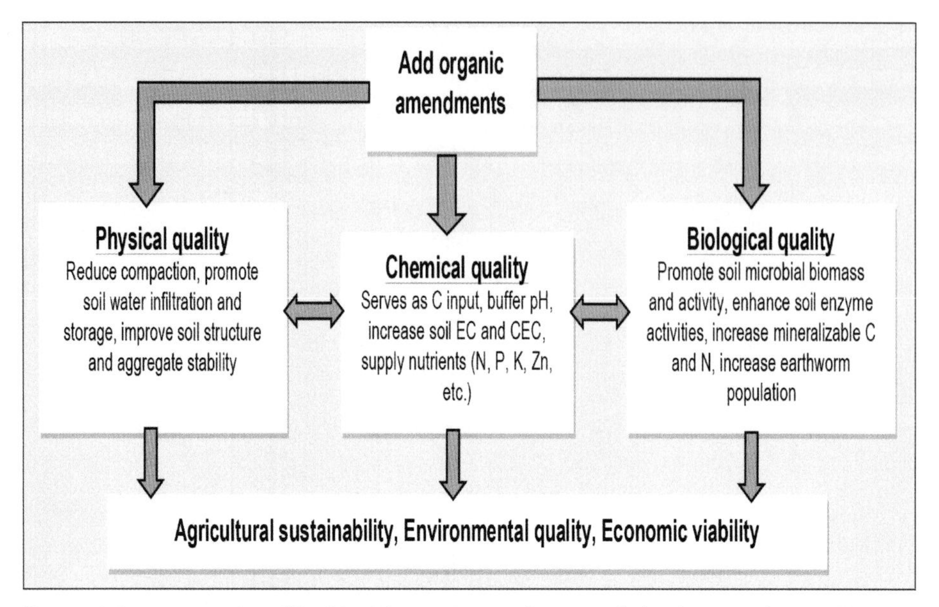

Figure 7-2. Improvement in soil health with organic amendments in dryland crop production. EC = electrical conductivity; CEC = cation exchange capacity; N = nitrogen; P = phosphorus; K = potassium; Zn = zinc; C = carbon. (Adapted from Cogger et al. 2013; Reeve et al. 2012; and Wuest et al. 2005.)

impacts are not well understood.

This chapter covers some general considerations relevant to use of soil amendments and discusses individual amendment products with a goal of providing an understanding of the situations in which they are most likely to be beneficial, along with some key factors related to their use. The chapter also points readers to Extension resources that provide more detailed information on biosolids and manures—the amendments growers are currently most likely to choose to apply. In addition to biosolids and manures, **biochar** and paper-manufacturing wastes ("black liquor") are briefly described, as these amendments may become relevant to some dryland growers in the future given technology advances and changing production economics.

Considerations in Using Amendments

Unlike traditional (synthetic) fertilizers, which have a consistent formulation and contain only the nutrients spelled out in the product description, amendments can be variable in nature and contain a suite of elements. A single amendment type, such as cattle manure, can vary substantially in composition, with differences resulting from different inputs (e.g., animal diet), differences in processing, and from seasonal or other types of variability. This makes testing of amendments using an appropriate lab prior to application critical.

In most cases, amendments provide C as well as a suite of plant macro and micronutrients. When amendments provide primarily C, with relatively low concentrations of nutrients, they should be used as soil conditioners to build SOM and improve soil health. In contrast, amendments that have adequate amounts of nutrients, in forms that are available to plants and with timing that matches crop needs, can also be used as the primary nutrient source for a crop.

The **carbon-to-nitrogen** (C:N) ratio is an important related factor that contributes to determining whether an amendment should be used primarily as a nutrient source or as a soil conditioner. Generally, amendments with a low C:N ratio (<15:1) decompose quickly and release nitrogen (N) rapidly in soil, while the decomposition of high C:N materials (>15:1) is slow, with the available N in the amendment and soil immobilized by microorganisms for their own needs as they carry out decomposition (Gale et al. 2006). Figure 7-3 illustrates that the addition of high C:N organic materials can temporarily diminish available N for crops due to higher microbial activity and greater assimilation of N by microbes for their growth. Later, ongoing scarcity of available N causes some soil microbes to die, releasing N. To address N **immobilization**, extra N fertilizer may need to be applied. A list of some common organic materials with their rough C:N composition is shown in Table 7-1.

When some amendments are applied to meet crop N needs, they provide other nutrients in excess. Therefore, growers applying amendments should evaluate the potential for nutrient losses, especially when applying amendments repeatedly over time. More information on the pathways of phosphorus (P) and N losses to the environment, and the factors contributing to risk of such losses, is provided in Chapter 6: Soil Fertility Management.

Amendment quality, including the amount of other nutrients, may also impact how much C from the amendment is stored in soils. In Pendleton,

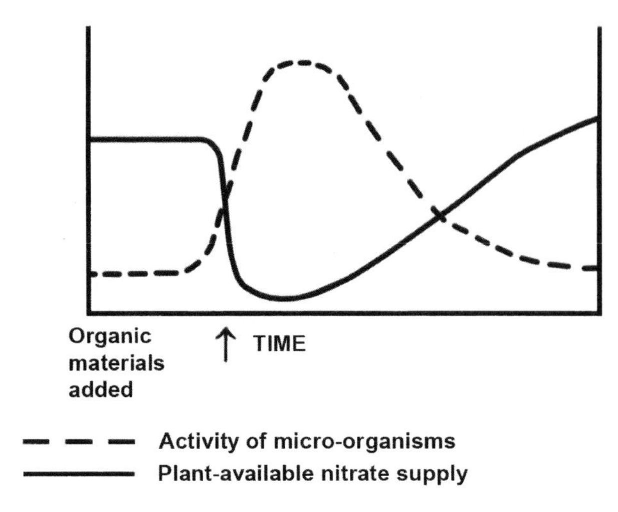

Figure 7-3: Generalized diagram showing temporary loss of plant-available soil nitrogen upon addition of organic materials with a high carbon-to-nitrogen (C:N) ratio. (Used with permission from Glewen 2016.)

Table 7-1. Selected common organic amendments and their approximate carbon-to-nitrogen (C:N) ratios.

Organic amendments	C:N[a]
High in C (relative to N)	
Sawdust	400
Wheat, oat, or rye straw	80
Green rye	36
Dairy separated solids	32
Alfalfa hay	20
Dairy solids compost	18
High in N (relative to C)	
Stockpiled dairy manure	15
Clover and alfalfa (early)	13
Beef manure compost	11
Poultry litter	10
Soil	10–12

[a]Nitrogen is always 1 in the C:N ratio.
Adapted from Bary et al. 2016; and Magdoff and van Es 2010.

Oregon, a field experiment comparing different amendments suggested that biosolids and un-aged cattle manure were substantially more efficient at sequestering C than other amendments, including alfalfa, wood sawdust, composted and uncomposted wheat residues, Brassica residues, sucrose, and cotton linters (Table 7-2). ■ Stable soil C gain appeared very closely related to the content of P as well as sulfur (S) in the amendments (Figure 7-4). Other important mechanisms that have been proposed for such differences in carbon storage efficiency include the amount of enhancement in primary productivity and the microbial processing that amendments have undergone (e.g., for biosolids or composted manures) (Brown et al. 2011; Cogger et al. 2013; Pan et al. n.d.).

A variety of other important factors also influence decisions about amendment use, including cost, availability, transportation, public acceptance, application methods, and the potential for contaminants. These vary from amendment to amendment, so are discussed in the subsequent sections.

Table 7-2. Effect of amendment type on soil organic carbon (C) accumulation in the surface (0–9.8 inches) of a silt loam soil near Pendleton, Oregon. Amendments were applied at similar C rates for five years and sampled 7 years after final amendment application. ■

Amendment (2230 lb C/ac)	Sequestration Efficiency[a]
	%
Municipal biosolids	49
Cattle manure (no bedding)	21
Alfalfa feed pellets	14
Wood sawdust	11
Composted wheat residue	11
Brassica residue	10
Wheat residue	9
Sucrose	5
Cotton linters	3

[a]Sequestration efficiency is calculated by increase in soil organic C compared with the treatment receiving no amendment, divided by the amount of C applied.
Adapted from Wuest and Reardon 2016; see also Wuest and Gollany 2012.

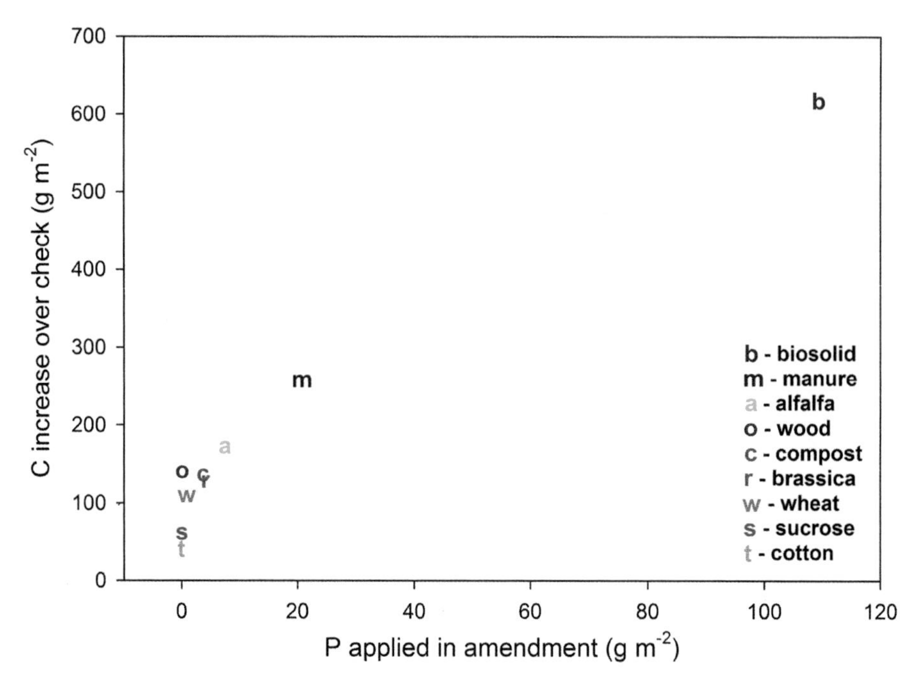

Figure 7-4. Phosphorus (P) applied in the amendment compared to amendment carbon (C) remaining in the soil seven years later. Soil C increase was calculated by subtracting soil organic C measured in the treatments receiving no amendments and is the average of the main plots treatments ($n = 8$). Amendments were applied at a rate of 250 g C per m^2 each year for five years (total 1250 g C per m^2; equivalent to 5.6 tons per acre). (Adapted from Wuest and Reardon 2016.)

Biosolids

Biosolids are materials produced by municipal wastewater treatment of organic solids, transformed through the treatment process into a product that is made up of living and dead wastewater treatment microorganisms, small inorganic particles, and insoluble compounds. Although there are other possible ways to manage biosolids, the majority are land-applied to recycle nutrients.

The biosolids most often used as soil amendments for dryland wheat are Class B biosolids (CFR Title 40). These biosolids have been treated to substantially reduce the level of biological pathogens and meet Environmental Protection Agency (EPA) standards for regulated contaminants including metals, some of which are required plant nutrients. Class B biosolids can be applied to crops whose edible parts do not make contact with the soil, as long as the applications are more than 30 days prior to harvest. In contrast, Class A biosolids meet more stringent requirements for pathogen reductions and can be used more widely, including in garden or landscaping applications in residential or commercial areas.

Grains are among the most common receiving crops for Class B biosolids in the inland PNW. Acreage in a grain-fallow rotation is particularly flexible for receiving biosolids because of the wide window for application during the **fallow** year (Sullivan et al. 2015). ■ In 2015, eastern Washington croplands received almost 44,000 dry tons of biosolids applied to roughly 15,500 acres (Peter Severtson, personal communication). Of this, 93% was applied to wheat, with the balance applied to grass hay, corn, hops, and other crops.

Because biosolids are a byproduct that must be managed by wastewater treatment facilities, they may be available at no cost or reduced cost to producers. In some cases, municipalities charge a transportation fee, application fee, or a fee equal to the N value of the biosolids (Sullivan et al. 2015). While permitting is required for all Class B biosolids, this is normally taken care of by the wastewater treatment plant or the private company that applies the biosolids (Weaver 2013).

The Extension publication Fertilizing with Biosolids (Sullivan et al. 2015) covers many practical aspects of biosolids applications, including additional information on nutrients, **pH**, and soil health considerations; how to use university fertilizer guides with biosolids application; and obtaining needed site approvals.

Composition, Nutrients, and Application Rates

Biosolids supply organic matter, plant macronutrients, and micronutrients, including those listed in Table 7-3 as well as copper (Cu), boron (B), molybdenum (Mo), zinc (Zn), and iron (Fe). Potassium (K) is notably absent. As indicated in Table 7-3, levels of organic matter and nutrients can vary considerably. When applied to meet crop N needs, biosolids generally provide P in excess of crop needs, though only about 20% to 60% of this P is plant-available (Ippolito et al. 2007; Cogger et al. 2013; Sullivan et al. 2015). Particularly for one-time applications, P may provide benefits to dryland cropping in the inland PNW if soils are deficient. Sulfur, Zn,

Table 7-3. Biosolids organic matter and macronutrients (dry weight basis, total elemental). Not all elemental content is plant-available.

	Usual Range (%)[a]	
Nutrient	**Low**	**High**
Organic Matter	45	70
Nitrogen (N)	3.0	8.0
Phosphorus (P)[b]	1.5	3.5
Sulfur (S)	0.6	1.3
Calcium (Ca)	1.0	4.0
Magnesium (Mg)	0.4	0.8
Potassium (K)[b]	0.1	0.6

[a]Usual range for freshly digested biosolids. Lagooned biosolids, composted biosolids, and alkaline-stabilized biosolids typically have lower nutrient concentrations.
[b]P and K are expressed on an elemental basis. Use the following conversion factors to convert to units used for fertilizer marketing: To get P_2O_5 (phosphate), multiply P by 2.29. To get K_2O (potash), multiply K by 1.2.
Reproduced with permission from Sullivan et al. 2015.

Fe, and other micronutrients present in biosolids may also be beneficial when amounts in soils are below desired levels (Koenig et al. 2011).

Biosolids application rates in the inland PNW are generally based on crop N requirements, and typically range from 2.2–3.6 tons per acre of dry weight biosolids every 2 to 4 years in wheat-fallow rotations (Cogger et al. 2013). ■ Biosolids contain ammonium-N, available to crops immediately after application. They also contain organic N, which must be converted (mineralized) over time before it is available to plants. Most biosolids do not contain nitrate-N (Sullivan et al. 2015). The Worksheet for Calculating Biosolids Application Rates in Agriculture (Cogger and Sullivan 2007) describes the calculation of agronomic application rates based on biosolids analysis, estimates of ammonium-N retained after application, organic N mineralized from the current and previous biosolids applications, crop N requirements, site information, and regulatory limits for trace element application. The values in this worksheet are based on short-term studies of biosolids additions in tilled grain-fallow rotations across a range of environmental conditions in the PNW (Sullivan et al. 2009; Cogger et al. 1998). For **no-till** systems, rates may need to be adjusted to account for higher rates of ammonia volatilization from unincorporated biosolids (Barbarick et al. 2012).

Yield Impacts

Biosolids applications can produce equivalent or better grain yields than typical applications of inorganic N in tilled and no-till wheat systems (Sullivan et al. 2009; Koenig et al. 2011; Barbarick et al. 2012). Preliminary results (4 years) suggest this is also true for biosolids that are applied to wheat grown with **conservation tillage** with an undercutter (Schillinger et al. 2015). ■

When yields are increased for dryland wheat compared to inorganic N, this is often attributed to the P or S provided by the biosolids (Koenig et al. 2011; Ippolito et al. 2007; Cogger et al. 2013). Other possible factors include improved soil physical properties, or the fact that N supplied by biosolids is made available gradually and thus may limit vegetative growth, reducing the potential for moisture stress and associated reductions in grain yield (Koenig et al. 2011).

Meanwhile, higher biosolids rates can lead to yield loss through lodging or excessively vigorous vegetative growth that leads to moisture stress (Cogger et al. 2013; Mantovi et al. 2005; Cogger et al. 1998). Application rates for biosolids take into account the amount of N available from all sources, minimizing this risk, though Sullivan et al. (2009) also recommend that growers who are applying biosolids choose varieties that are resistant to lodging.

Grain Quality Considerations

When making decisions about biosolids use, growers should be aware that biosolids applications generally raise grain protein, with a 0–13% increase in protein in the first and second crops following biosolids application (Cogger et al. 2013; low and medium rates). Practical experience across the region suggests that this increase is not generally so great that it leads to a negative impact on prices received (Andy Bary, personal communication).

It has also been suggested that biosolids applications might be more beneficial for hard red and hard white wheat since greater protein content is valued for these wheat types (Sullivan et al. 2009). While this may be true, biosolids should not be assumed to increase grain protein concentration for hard wheats when biosolids are applied immediately before planting (Koenig et al. 2011). From a grain protein perspective, application to hard wheats during the fallow year may be preferred if fallow is part of the rotation.

Soil Health Benefits

Biosolids applications can meaningfully increase soil C when used over time, and particularly when combined with other strategies such as reducing tillage and maximizing residue production and retention. Across regional dryland systems, the increase in total soil C after repeated biosolids applications has been up to 49–77% of the C added in biosolids, larger than most other types of amendments (Wuest and Gollany 2012; Wuest and Reardon 2016; Cogger et al. 2013, Pan et al. n.d.). ■ Factors that contribute to the impact of biosolids on long-term C levels likely include the balanced nutrients provided, increased primary productivity in comparison to conventionally fertilized soils, and the microbial processing that biosolids have undergone, as microbial C can be a major part of stabilized SOC. Generally speaking, the C benefit may be smaller if SOC levels are already high, as a steady state is approached or achieved (Lal 2001; Cogger et al. 2013; Brown et al. 2011).

Over the long term, biosolids may decrease bulk density, increase soil **water holding capacity**, and benefit other measures of soil health (Brown et al. 2011; Cogger et al. 2013) (Table 7-4). ■ Although Table 7-4 indicates a decrease in pH after biosolids application, soil pH can be increased or decreased depending on whether or not alkaline materials are used in the biosolids process (Sullivan et al. 2015).

Biosolids also generally increase soil **aggregation**, and it would generally be expected that this would reduce wind erosion (Neilsen et al. 2003; Wallace et al. 2009). This question is being explored in existing research on biosolids application in Lind, Washington, under both **conventional tillage** and conservation tillage with an undercutter (Sharratt et al. 2016). ■

Table 7-4. Improvement in soil physical and chemical properties at 0–3.9 inches after applying biosolids compared to chemical nitrogen (N) fertilizer or no N application in Douglas County, Washington. Little change occurred below the tillage zone (3.9 inches). ■

Treatment	Applied Rate[a]	Cumulative Rate	Total C	Total N	Bulk Density	pH	EC[b]	Olsen-P[c]
	lb/ac		%		$g\ cm^{-3}$		$dS\ m^{-1}$	ppm
Zero N			0.84a*	0.08a	1.26a	6.1a	0.08a	15a
Anhydrous Ammonia	50		0.94a	0.09a	1.22a	5.7b	0.08ab	16a
Biosolids	ton/ac							
Low	2.2	10.7	1.39b	0.13b	1.09b	5.5bc	0.13bc	74b
Medium	3.1	15.2	1.69c	0.15c	1.05bc	5.4c	0.18c	114c
High	4.0	20.1	1.64c	0.14c	1.02c	5.4c	0.16c	128c

[a]Anhydrous ammonia was applied every other year. Biosolids were applied every fourth year from 1994–2010.
[b]EC = electrical conductivity
[c]Olsen-P: soil test based on extraction with bicarbonate; extractable phosphorus (P) pool correlates to plant available P.
*Means within a column followed by different letters are significantly different ($P < 0.05$) by protected least significant difference.
Adapted from Cogger et al. 2013.

Nutrient Loss Considerations

Though there are clear potential benefits from biosolids applications, one potential concern is whether biosolids applications can increase losses of N and P from agricultural systems. There has generally been only a small effect on soil nitrate-N after one-time biosolids applications to dryland wheat systems (Sullivan et al. 2009; Koenig et al. 2011). A recent analysis of N rates associated with repeated applications of biosolids to dryland wheat found that roughly 35% of the added N was stored in the top 3.9 inches of soil, while 24% to 37% was removed in grain (Pan et al. n.d.). ■ The remaining 28–41% was assumed to reside in the subsoil of the rooting zone as soil N, as root N in biomass, or lost to nitrate **leaching** or N volatilization. Increases in soil N may benefit crops by providing a pool of N that can be drawn on in addition to fertilizer N supplies, but can also indicate a higher risk of N loss.

Biosolids applications have also led to increases in bicarbonate-extractable P, with evidence of limited downward movement of P in both tilled and direct seed systems (Cogger et al. 2013; Ippolito et al. 2007; Barbarick et al. 2012). In areas where P is deficient, P loss is likely a relatively low concern for one-time applications at agronomic rates. Those carrying out repeated applications should test soil levels regularly and evaluate the potential for P loss under their local conditions (Sullivan et al. 2015). Phosphorus loss tends to be associated with soil erosion, and is most problematic when fields are in proximity to a water body.

Contaminants

Municipal wastewater facilities that produce biosolids treat wastewater from household and industrial facilities, which may contain various contaminants including metals, pathogens, antibiotics, some industrial and household chemicals, odorants, and aerosols. Contaminants that are not degraded during the biosolids treatment process are present in the resulting biosolids.

Historically, heavy metals were a concern in biosolids. However, concentrations of metals in biosolids have fallen sharply over the last 40 years since the passage of the Clean Water Act, and are no longer present in biosolids at concentrations that could cause human, animal, or environmental health issues (Mitchell et al. 2016; Sullivan et al. 2015). Concentrations of these metals in land-applied biosolids are regulated and monitored, and concentrations must be below federal limits that are set based on risk assessments.

Meanwhile, study of other known and emerging contaminants is ongoing, but the current available evidence suggests that biosolids applied to agricultural soils at agronomic rates and in accordance with current guidelines present low to minimal levels of risk from pathogens, antibiotics, industrial and household chemicals, and pharmaceuticals (Mitchell et al. 2016). The Extension publication Guide to Biosolids Quality (Mitchell et al. 2016) reviews the literature on potential contaminants in biosolids in more detail.

Manures

An increasing interest in improving soil health, the rise of fertilizer costs, and the unique nutrient needs of organic producers have all contributed to renewed interest in manure amendments in dryland systems. Meanwhile, applying manure to agricultural lands that are in need of nutrients or organic matter could help reduce nutrient overloading concerns for dairies, feedlots, and poultry operations. Across the main wheat-producing counties of the inland PNW, an estimated 1,377 tons of N, 2,013 tons of phosphate (P_2O_5), and 6,242 tons of potash (K_2O) is available in recovered manures (IPNI 2012).

However, economics currently prevents widespread use of manure in dryland agriculture. Manure is heavy relative to its nutrient content, and therefore relatively expensive to transport. Across the inland PNW, animal production is often practiced in concentrated areas, far from where the bulk of the dryland wheat is grown. South central Idaho is one exception. This region is dominated by dairy farms in combination with dryland fields on high plateaus. Production of organically certified wheat (with prices that are 2 to 3 times those of commodity wheat) and alfalfa hay in these areas allows for utilization of locally available manure. Further information about certified organic wheat practices in the inland PNW can be found in Organic Small Grain Production in the Inland Pacific Northwest: A Collection of Case Studies (Lorent et al. 2016).

Two other related practices are also receiving some interest, driven in part by these barriers to manure use. Particularly in wetter areas of the region, there is re-emerging interest in grazing **cover crops** or residues, a process that contributes to nutrient cycling through manure deposition by grazing animals. Meanwhile, some dryland organic producers use crop rotation with N-fixing crops such as alfalfa or pulses as the primary means to add N to their soils rather than manures (Lorent et al. 2016). For more information, see Chapter 5: Rotational Diversification and Intensification.

Manure can be applied in raw or aged form, or can be processed before land application. Potential treatments include primary and secondary solids separation, anaerobic digestion, composting, and nutrient recovery. Treatments result in products that may be quite different than raw manure. Sometimes more than one treatment process is used in sequence; for example, anaerobic digestion followed by separation of fiber and separation of fine solids. Some of these processes are active areas of research and commercial development.

Composting is the most common manure treatment. When manure is composted, it is managed in a way that allows microorganisms to decompose manure and bedding in the presence of air. Composts used in organic production need to be produced following the rules of the National Organic Program that specify, among other things, initial C:N ratios, time, and temperature requirements that must be achieved during composting. Composts can be applied to food crops without restriction.

Separated dairy solids (primary separation) are generated on many regional dairies by utilizing screens and rotary and screw presses to separate out easily settled fibrous solids (Ma et al. n.d.). Primary separated solids may be composted after separation. Meanwhile, secondary solids separation, in the early stages of commercialization in 2016, can be used following primary solids separation, with or without anaerobic digestion. This secondary process focuses on very fine, suspended solids that are clay-like in nature.

The Extension publication Fertilizing with Manure and Other Organic Amendments (Bary et al. 2016) covers a range of practical topics of interest to dryland producers considering manure use, including composition of different types of manures, manure testing, calculating application rates, applying manure, manure storage, and long-term effects of manure application. For those growing dryland forage crops, see also Manure Application Rates for Forage Production (Downing et al. 2007).

Composition, Nutrients, and Application Rates

Manures contain N, P, K, and other plant nutrients. Nutrient content of manures and manure-based products can vary widely, depending on the type of animal, diet, manure handling and storage, treatments applied, and other factors. Typical values for some of the more common uncomposted manure types are provided in Table 7-5, but these values should be used with caution. Testing following established methods is critical. While some manures have enough N (generally more than 2% N on a dry weight basis) to be used primarily as fertilizers to meet crop nutrient needs, others have less concentrated nutrients and should be primarily seen as soil builders (Bary et al. 2016). For example, separated dairy solids with total N of 1.4% and a C:N ratio of 32 are mainly soil builders.

When nutrients are the main goal, understanding the forms of N in manure is important to choosing appropriate application rates and methods. When excreted, somewhere in the range of 50% of dairy or beef manure N may be in the form of ammonium (Ketterings et al. 2005). Ammonium can be lost through volatilization when manure is exposed to air, including during storage, and after land application, if not immediately incorporated into the soil. Ammonium can also be taken up by crops, either directly or after conversion to nitrate in the soil. Because of these dynamics, N contribution is generally greater for manures that are tilled-in compared to those left on the surface. Manure also contains organic N. Some forms of organic N break down quickly, while stable organic N compounds in manure can take as long as 5 years or longer to mineralize into the ammonium and nitrate forms that are available to plants (Russelle et al. 2016; Moore and Ippolito 2009).

The Manure Management Planner (***http://www.purdue.edu/agsoftware/mmp/***), developed by Purdue University, may be helpful for calculating uncomposted manure application rates to meet crop needs while protecting

Table 7-5. Typical nutrient content, dry matter, bulk density, and nitrogen (N) availability for uncomposted animal manures. Nutrients, dry matter, and density can all vary widely depending how manure is handled. Values for additional manure types can be found in Bary et al. 2016.

Type	N	P[a]	K[a]	Dry Matter	Density	Total N	C:N Ratio[b]	Available N[c]	
	lb/ton as-is[d]			%	lb/cu yd	% dry weight		% of total N	lb/ton as-is
Broiler with litter	56	27	39	68	850	4.1	11	40 to 60	22–34
Beef	14	4	12	28	1400[e]	2.4	15	15 to 30	2–4
Stockpiled dairy manure	18	6	35	50	1000	1.9	15	10 to 20	2–4
Dairy cow separated solids (primary separation)	6	1	3	21	1100	1.4	32	–5 to 10[f]	< 1

[a]Phosphorus (P) and potassium (K) are expressed on an elemental basis. Use the following conversion factors to convert to units used for fertilizer marketing: To get P_2O_5 (phosphate), multiply P by 2.29. To get K_2O (potash), multiply K by 1.2.
[b]Carbon-to-nitrogen ratio
[c]Available N is an estimate of the amount of N that becomes available to plants during the first growing season after application.
[d]"As-is" is typical dry mater content for solid manure stored under cover.
[e]Estimate based on value for manures and composts with high moisture (low dry matter).
[f]Negative value indicates N immobilization (conversion from available to unavailable form).
Used with permission from Bary et al. 2016. Data sources include Gale et al. 2006; Brown 2013; Moore et al. 2015; and unpublished PNW data from A. Bary, C. Cogger, D. Sullivan, and A. Moore.

surface and ground water quality. The tool uses existing state-specific fertilizer recommendations and information for estimating manure N availability from Extension and National Resources Conservation Service. While the tool can use standard manure production and nutrient values, supplying the tested nutrient content and volume will greatly improve accuracy. The tool provides suggested application rates, as well as a P index and N leaching index to give insight about the risk of nutrient losses to the environment.

In comparison to most raw manures, composted manures have less plant-available N (Table 7-6), as most of the easily mineralizable forms of N are converted to more stable organic forms or lost as ammonia gas during the composting process. Generally speaking, this means that composts are better used to build SOM and improve tilth, rather than as organic fertilizers (Gale et al. 2006; Bary et al. 2016). Composting also increases the amount of stabilized C, and makes the amendment more uniform and easier to apply.

In contrast to primary separated solids, secondary solids have a clay-like nature with smaller particle size and higher nutrient content, particularly P (Table 7-7). Research to date has focused on applications that provide P at agronomic rates to cropping systems such as potatoes that have high P requirements (e.g., Collins et al. 2016). Note that fine solids generated with systems utilizing polymers such as polyacrylamide may be incompatible with organic certification, though there is ongoing investigation regarding the use of natural polymers (Mehta et al. 2015).

Grain Quality Considerations

In irrigated systems in southern Idaho, excessive applications of manure elevated protein levels beyond desirable amounts for soft white wheat and barley, and also caused lodging (Moore 2016).

Soil Health Benefits

Manure applications are better able to maintain and increase SOM than crop residues, both regionally (Machado et al. 2011; Wuest and Gollany 2012; Wuest and Reardon 2016) and globally (Edmeades 2003). Elsewhere in the western US, one-time applications of composted manure have also

Table 7-6. Typical nutrient content, dry matter, bulk density, and nitrogen (N) availability for composted animal manures.

Type	N	P[a]	K[a]	Dry Matter	Density	Total N	C:N Ratio[b]	Available N[c]	
	lb/ton as-is[d]			%	lb/cu yd	% dry weight		% of total N	lb/ton as-is
Broiler litter "compost"[e]	44	26	33	57	900	3.8	10	30 to 40	12–18
Beef manure compost	18	7	22	64	900	1.4	11	0 to 10	2–4
Separated dairy solids compost (primary separation)	10	1	3	25	1400	2.1	18	0 to 10	0–1

[a]Phosphorus (P) and potassium (K) are expressed on an elemental basis. Use the following conversion factors to convert to units used for fertilizer marketing: To get P_2O_5 (phosphate), multiply P by 2.29. To get K_2O (potash), multiply K by 1.2. [b]Carbon-to-nitrogen ratio. [c]Available N is an estimate of the amount of N that becomes available to plants during the first growing season after application. [d]"As-is" is typical dry mater content for compost at the point of sale or use. [e]Broiler litter sold as "compost" is often not fully composted because no water was added to facilitate microbial decomposition during the process.

Used with permission from Bary et al. 2016. Values for additional composted manure types can be found in Bary et al. 2016. Data sources include Gale et al. 2006; Larney et al. 2006; and unpublished PNW data from A. Bary, C. Cogger, and D. Sullivan.

Table 7-7. Elemental composition of two anaerobic digestion (AD) recovered fine solids products and one commercial manure-based fertilizer.

	C	N	P	K	S	Ca	Mg	Fe	pH
Fertilizer Amendment	---------- % ----------								
AD Dairy fine solids 2-3-1	19.7	15	14	13	14	62	14	8	8.0
AD Poultry fine solids 4-6-2	20.7	39	26	18	13	82	25	2	7.7
Commercial poultry fertilizer (2-8-1, Perfect Blend)	25.6	18	34	10	30	70	7	1	6.3

C = carbon; N = nitrogen; P = phosphorus; K = potassium; S = sulfur; Ca = calcium; Mg = magnesium; Fe = iron. Adapted from Collins et al. 2016.

raised SOC, with effects lasting at least 16 years after application (Reeve et al. 2012). Beyond improved SOM, manure applications have been shown to increase soil water infiltration and stability of soil compared to synthetic N fertilizer, with an associated increase in earthworm and mycorrhizal fungi activity (as measured by glomalin, a fungal glycoprotein) (Wuest et al. 2005) (Table 7-8). ■

Table 7-8. Effect of 70 years of organic amendments compared to synthetic nitrogen (N) fertilizers on soil properties of the surface soil (0–5.9 inches) in a winter wheat-summer fallow rotation near Pendleton, Oregon. ■ Differences between chemical fertilizer and manure and pea vine treatments were highly significant for all soil parameters shown here except earthworm counts.

Soil Parameters	Organic Amendments		Synthetic N Fertilizer	No Fertilizer
	Manure	Pea Vines		
	100 lb N/ac[a]	30 lb N/ac	80 lb N/ac	0 lb N/ac
Total carbon (C) (%)	1.590	1.260	1.170	1.090
Total N (%)	0.135	0.103	0.092	0.088
Wet Soil Stability				
Whole soil (proportion)	0.48	0.41	0.35	0.30
1–2 mm aggregates (proportion)	0.83	0.69	0.65	0.56
Water Infiltration				
Percolation (cubic inches per hour)	1.06	0.95	0.84	0.73
Ponded infiltration (inches per hour)	5.53	4.09	1.49	1.46
Total glomalin[b] (%)	0.259	0.235	0.214	0.213
Earthworm count (per square meter, sampled 9.8 inches deep)	107	120	60	67

[a]Pounds of N shown is on a per-crop cycle basis for all treatments.
[b]Glomalin is a glycoprotein produced by arbuscular mycorrhizal fungi. Its concentration has been associated with the abundance of water-stable aggregates, and it incorporates potentially large pools of soil C and N.
Adapted from Wuest et al. 2005; this table is also Table 2-3.

In the inland PNW, Cox et al. (2001) used compost (85% by volume animal manure and bedding, along with 10% coal ash and 5% food and landscaping waste) to restore eroded Palouse hilltops. Composts benefitted bulk density in some years, reduced soil impedance (low soil impedance is associated with improved root growth), increased water-stable aggregates, and increased total soil C. Compost also increased extractable P and available K, two nutrients that are usually less available on eroded hilltops (Pan and Hopkins 1991). After N immobilization was overcome, yields improved compared to untreated controls, likely through improvements in soil fertility, soil structure, and perhaps water infiltration.

Nutrient Loss Considerations

When manures are applied to crops to meet N needs, P is typically applied at rates 3 to 6 times greater than the crop can use (Moore and Ippolito 2009), while K is also usually in excess of plant needs. Loss of P and K will likely not be an issue if these nutrients are deficient in soils or for single applications, but soil levels should be monitored every three to five years, especially if applications are repeated over time (Bary et al. 2016).

Excess P and K can be managed in some cases through the inclusion of dryland forage crops, such as alfalfa, in the rotation. Forage plants can take up additional K as soil concentrations increase, benefitting soil nutrient balances—and alfalfa also has high P requirements. However, growers should note that excess K concentrations in forage can cause health problems for cattle, with suggested limits of 2% on a dry weight basis (Moore and Ippolito 2009).

Contaminants

Pathogens can be a concern with manures, though composting or anaerobic digestion can reduce (but do not eliminate) these concerns. As of late 2016, The Food and Drug Administration (FDA), as part of the Food Safety Modernization Act, which applies to produce normally eaten raw, was conducting a risk assessment and extensive research on the number of days needed between applications of raw manure as a soil amendment and harvesting to minimize the risk of contamination (FDA

2015). While the Food Safety Modernization Act specifically applies to fruits and vegetables normally eaten raw, it is possible that this guidance, when published, may also have spillover impacts on other crops. In the meantime, the FDA stated that it did not object to farmers complying with the USDA National Organic Program standards. Meanwhile, the National Organic Program standards, which would apply to organic dryland producers, requires a 90-day interval between the application of raw manure for crops that are produced for human consumption and whose edible parts do not come in contact with the soil (CFR Title 7). (A 120-day interval is required for crops whose edible parts come in contact with the soil.)

The Food Safety Modernization Act also addresses compost production, establishing limits on detectable amounts of bacteria (including *Listeria monocytogenes*, *Salmonella* spp., fecal coliforms, and *Escherichia coli* "E. coli" 0157:H7) for processes used to treat biological soil amendments, including manure (FDA 2015). Stabilized compost prepared using approved methods that conform to these standards can be applied without a specified waiting period before harvest, but must be applied in a manner that minimizes the potential for contact with produce during and after application.

Weed seed is another contaminant of concern, as weed seeds can remain viable after passing through an animal's digestive tract. Composting manure at sufficiently high temperatures will kill weed seeds, but quality control must be sufficient to ensure that all the manure is exposed to these conditions.

Manures can also contain antibiotics that may persist during manure storage and even after land application (Kuchta and Cessna 2009; Aga et al. 2005; Schlusener et al. 2003). However, concentrations are generally low, and the potential for environmental impacts is not well understood. Composting and anaerobic digestion effectively lower concentrations of some but not all antibiotics (Ramaswamy et al. 2010; Kim et al. 2012; Mitchell et al. 2013; Mohring et al. 2009).

Growers who are carrying out repeated applications of dairy manure over time should be aware of a low risk of copper toxicity. Copper sulfate ($CuSO_4$) from cattle foot baths is washed out of dairy barns along with manure and into wastewater lagoons. Moore and Ippolito (2009) suggest testing soils every 2 to 3 years if dairy manure applications are ongoing, and ceasing copper additions if soils tests indicate greater than 50 ppm DTPA-extractable copper.

Biochar

Biochar is charcoal-like material that is generated when organic materials are heated in oxygen-limited environments (Ronsse et al. 2013). Biochar has been receiving growing attention as a C-rich soil amendment that can improve measures of soil health. Recent meta-analyses of worldwide data from field experiments indicate that benefits exist under at least some agricultural conditions, with suggestions that important mechanisms that may include a liming effect, improved ability to retain nutrients, improved soil water holding capacity, and perhaps improved soil structure (Jeffery et al. 2011; Biederman and Harpole 2013; Liu et al. 2013). As of late 2016, biochar was available in limited commercial quantities from suppliers across the PNW (Tom Miles, personal communication).

Biochar can be made from a wide range of biomass feedstocks, with lignocellulosic materials such as forestry or agricultural residues among the most common choices (Suliman 2015). The physical and chemical characteristics of biochar depend on the feedstocks used, the temperature and other conditions under which the biochar is produced, and the pre-treatments and post-treatments applied (Zhao et al. 2013; Ronsse et al. 2013; Suliman et al. 2017) (Table 7-9). Variations in performance depending on feedstock and soil type have been seen in greenhouse studies using biochar from the Palouse region of eastern Washington (Naff silt loam, Palouse silt loam, and Thatuna silt loam) (Streubel et al. 2011).

Application of biochar derived from forest wastes has benefitted pH and wheat yields under dryland wheat cropping in the inland PNW (Machado and Pritchett 2014) (Figure 7-5). At this site, applications of 10 tons per acre or more of biochar increased grain yield by 26% to 33%. Application above 10 tons per acre did not result in any additional significant yield increases, and biochar application did not influence test weight. Applying this alkaline biochar (pH 10.6) increased soil pH by a factor of 0.21 at the highest rate. While these results are encouraging, separate analysis

Table 7-9. Variation in elemental composition of four biochars made from different feedstocks, with pyrolysis at 500°C (932°F). In addition to the elements listed here, biochars also contained calcium, magnesium, iron, boron, copper, manganese, and zinc.

Biochar Source	Ash	C	N	P	K	S	pH
	% by weight						
Switchgrass	20.81 (0.8)[a]	59.2	1.99	0.47 (0.014)	3.28 (0.2)	0.11 (0.002)	9.4
Anaerobically digested fiber	15.44 (0.1)	65.8	2.23	0.76 (0.032)	1.17 (0.1)	0.30 (0.002)	9.3
Softwood bark	5.40 (0.2)	72.7	0.35	0.047 (0.003)	0.10 (0.01)	0.023 (0.001)	7.6
Wood pellets	1.16 (0.1)	78.2	0.13	0.022 (0.003)	0.10 (0.01)	0.017 (0.002)	7.2

C = carbon; N – nitrogen; P = phosphorus; K = potassium; S = sulfur.

[a]Values in parentheses are the standard deviations.

Adapted from Streubel et al. 2011.

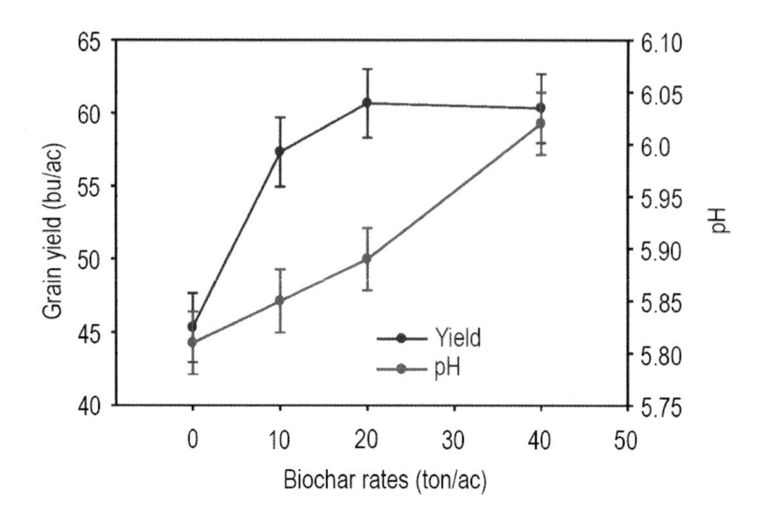

Figure 7-5. Biochar effects on soil pH and winter wheat yield in Athena, Oregon. (Adapted from Machado et al. unpublished; this figure is also Figure 2-10).

suggests that biochar is not economical if only the liming impacts are considered (Granatstein et al. 2009; Galinato et al. 2011).

Ongoing research efforts are investigating whether biochars can be engineered to improve agronomic performance and economics. For example, biochar can be "charged" with nutrients (e.g., by absorbing manure effluent and associated nutrients) or oxidized (made to lose electrons). Recent laboratory studies have indicated that Quincy sandy soils from the irrigated region of Washington amended with biochar that had been oxidized by exposure to air held significantly more water than soils amended with non-oxidized biochar, with both out-performing non-amended soils (Suliman et al. 2017).

Paper Manufacturing Wastes

Among other uses, residues from cereal and grass seed systems can be used for papermaking. (See Chapter 4: Crop Residue Management.) Straw fibers are typically pulped with sodium hydroxide (NaOH) under pressure, a process that produces a large quantity of "black liquor," an organic byproduct that has traditionally been discharged into waterways where it can contribute to pollution. Possible alternative uses for the waste include as a soil amendment and K source. Greenhouse and field studies in irrigated corn in Washington state suggested that black liquor could be an effective fluid liming material and could increase soil biological activity as well as wet stable aggregates (Xiao et al. 2006; Xiao et al. 2007).

Conclusion

Soil amendments, when economically and practically viable, could play an important role in improving soil health in dryland systems by increasing soil C, reducing bulk density, and improving nutrient availability, soil structure, water infiltration and retention, and microbial activity.

Among amendments, biosolids applications can produce equivalent or better grain yields than typical applications of inorganic N in tilled and no-till wheat systems. When applied to meet the N needs of crops, biosolids also provide organic matter, P, S, Zn, Fe, and other micronutrients that may be beneficial to wheat when amounts in soils are below desired levels. Biosolids can be a relatively low-cost amendment because some costs are generally borne by the biosolids producers—though the supply of biosolids is limited because they are the byproduct of waste treatment. Those applying biosolids to wheat should be aware that application during the fallow year may increase protein levels compared to conventional fertilizers.

Manures are also effective soil amendments, though producers should carefully distinguish between manures with high enough N content to be used to provide nutrients to crops, and those that act mainly as a soil conditioner, building SOM. In comparison to most raw manures, composted manures have less plant-available N, and thus should be used as a soil conditioner. Nutrient content of uncomposted and composted manures can vary widely, depending on the type of animal, diet, manure handling and storage, treatments applied, and other factors.

When biosolids or manures are applied at agronomic rates for N, excess P is normally applied (for both biosolids and manures) and excess K is applied (for manures). This is usually a relatively low concern for one-time applications to dryland soils, but producers carrying out repeated applications should regularly test soil levels. Likewise, although concerns about contaminants such as pathogens and weed seeds exist, these can usually be managed.

Two other amendments, while not currently practical, may be of increasing interest in the future. Biochar, a charcoal-like material, has been receiving some attention as an amendment that could improve soil health through a liming effect, nutrient retention, water holding capacity, and enhanced soil structure. Application of biochar derived from forest wastes has benefitted pH and wheat yields under dryland wheat cropping in the PNW, though cost remains a barrier to use. Meanwhile, black liquor is another organic waste that may be used as a soil amendment to improve physical soil structure, raise pH, and provide K.

Resources and Further Reading

Bary, A., C. Cogger, and D. Sullivan. 2016. Fertilizing with Manure and Other Organic Amendments. Pacific Northwest Extension Publication PNW0533. Washington State University. Cogger, C.G., and D.M. Sullivan. 2007. Worksheet for Calculating Biosolids Application Rates in Agriculture. Pacific Northwest Extension Publication PNW0511e. Oregon State University.

Lorent, L., D. Roberts, and I. Burke. 2016. Organic Small Grain Production in the Inland Pacific Northwest: A Collection of Case Studies. Pacific Northwest Extension Publication PNW683. Washington State University.

Sullivan, D.M., C.G. Cogger, and A.I. Bary. 2015. Fertilizing with Biosolids. Pacific Northwest Extension Publication PNW508. Oregon State University.

References

Aga, D.S., S. O'Connor, S. Ensley, J.O. Payero, D. Snow, and D. Tarkalson. 2005. Determination of the Persistence of Tetracycline Antibiotics and Their Degradates in Manure-Amended Soil Using Enzyme-Linked Immunosorbent Assay and Liquid Chromatography-Mass Spectrometry. *Journal of Agricultural and Food Chemistry* 53: 7165–7171.

Barbarick, K.A., J.A. Ippolito, J. McDaniel, N.C. Hansen, and G.A. Peterson. 2012. Biosolids Application to No-Till Dryland Agroecosystems. *Agriculture, Ecosystems & Environment* 150: 72–81.

Bary, A., C. Cogger, and D. Sullivan. 2016. Fertilizing with Manure and Other Organic Amendments. Pacific Northwest Extension Publication PNW0533. Washington State University. ***http://cru.cahe.wsu.edu/CEPublications/pnw533/pnw533.pdf***

Biederman, L.A., and W.S. Harpole. 2013. Biochar and Its Effects on Plant Productivity and Nutrient Cycling: A Meta-Analysis. *GCB Bioenergy* 5(2): 202–214.

Brown, C. 2013. Available Nutrients and Value for Manure from Various Livestock Types. Ontario Ministry of Agriculture and Food Publication 13-043.

Brown, S., A. Carpenter, and N. Beecher. 2010. A Calculator Tool for Determining Greenhouse Gas Emissions for Processing and End Use. *Environmental Science & Technology* 44(24): 9509–15.

Brown, S., K. Kurtz, A. Bary, and C. Cogger. 2011. Quantifying Benefits Associated with Land Application of Organic Residuals in Washington State. *Environmental Science & Technology* 45(17): 7451–58.

Brown, T.T., and D.R. Huggins. 2012. Soil Carbon Sequestration in the Dryland Cropping Region of the Pacific Northwest. *Journal of Soil and Water Conservation* 67(5): 406–15.

CFR (Code of Federal Regulations). Title 7, Part 205, Subpart C. Organic Production and Handling Requirements. ***http://www.ecfr.gov/cgi-bin/text-idx?SID=77eab826b09884b72f4966184e75f4bf&mc=true&node=se7.3.205_1203&rgn=div8***

CFR (Code of Federal Regulations). Title 40, Chapter 1, Subchapter O, Part 503. Standards for the Use and Disposal of Sewage Sludge. ***http://www.ecfr.gov/cgi-bin/retrieveECFR?gp=2&SID=3ba5c96eb4bfc5bfdfa86764a30e9901&ty=HTML&h=L&n=pt40.30.503&r=PART***

Cogger, C.G., A.I. Bary, A.C. Kennedy, and A.M. Fortuna. 2013. Long-Term Crop and Soil Response to Biosolids Applications in Dryland Wheat. *Journal of Environmental Quality* 42: 1872–1880.

Cogger, C.G., and D.M. Sullivan. 2007. Worksheet for Calculating Biosolids Application Rates in Agriculture. Pacific Northwest Extension Publication PNW0511e. Oregon State University.

Cogger, C.G., D.M. Sullivan, A.I. Bary, and J.A. Kropf. 1998. Matching Plant-Available Nitrogen from Biosolids with Dryland Wheat Needs. *Journal of Production Agriculture* 11: 41–47.

Collins, H.P., E. Kimura, C.S. Frear, and C. Kruger. 2016. Phosphorus Uptake by Potato from Fertilizers Recovered from Anaerobic Digestion. *Agronomy Journal* 108(5): 2036–2049.

Cox, D., D. Bezdicek, and M. Fauci. 2001. Effects of Compost, Coal Ash, and Straw Amendments on Restoring the Quality of Eroded Palouse Soil. *Biology and Fertility of Soils* 33(5): 365–72.

Downing, T.W., D. Sullivan, J. Hart, and M. Gamroth. 2007. Manure Application Rates for Forage Production. Oregon State University Extension Service. ***http://whatcom.wsu.edu/ag/documents/dairy/ManureAppRates_OSUem8585-e.pdf***

Edmeades, D.C. 2003. The Long-Term Effects of Manures and Fertilizers on Soil Productivity and Quality: A Review. *Nutrient Cycling in Agroecosystems* 66(2): 165–180.

FDA (Food and Drug Administration). 2015. Food Safety Modernization Act: Final Rule on Produce Safety. ***http://www.fda.gov/downloads/Food/GuidanceRegulation/FSMA/UCM472887.pdf***

Gale, E.S., D.M. Sullivan, C.G. Cogger, A.I. Bary, D.D. Hemphill, and E.A. Myhre. 2006. Estimating Plant-Available Nitrogen Release from Manures, Composts, and Specialty Products. *Journal of Environment Quality* 35(6): 2321.

Galinato, S.P., J.K. Yoder, and D. Granatstein. 2011. The Economic Value of Biochar in Crop Production and Carbon Sequestration. *Energy Policy* 39(10): 6344–6350.

Ghimire, R., S. Machado, and K. Rhinhart. 2015. Long-Term Crop Residue and Nitrogen Management Effects on Soil Profile Carbon and Nitrogen in Wheat-Fallow Systems. *Agronomy Journal* 107(6): 2230–2240.

Glewen, K. 2016. What Kind of Harm Can Come from Soil Organic Matter? From Soils: Soil Organic Matter. Plant and Soil Sciences eLibrary. ***http://passel.unl.edu/pages/informationmodule.php?idinformationmodule=1130447040&topicorder=1&maxto=8&minto=1***

Gollany, H.T., A.M. Fortuna, M.K. Samuel, F.L. Young, W.L. Pan, and M. Pecharko. 2013. Soil Organic Carbon Accretion vs. Sequestration Using Physicochemical Fractionation and CQESTR Simulation. *Soil Science Society of America Journal* 77(2): 618.

Granatstein, D., C.E. Kruger, H. Collins, S. Galinato, M. Garcia-Perez, and J. Yoder. 2009. Use of Biochar from the Pyrolysis of Waste Organic Material as a Soil Amendment. Final Project Report, Center for Sustaining Agriculture and Natural Resources, Washington State University, Wenatchee, WA.

IPNI (International Plant Nutrition Institute). 2012. A Nutrient Use Information System (NuGIS) for the U.S. Norcross, GA. ***www.ipni.net/nugis***

Ippolito, J.A., K.A. Barbarick, and K.L. Norvell. 2007. Biosolids Impact Soil Phosphorus Accountability, Fractionation, and Potential Environmental Risk. *Journal of Environmental Quality* 36(3): 764–772.

Jeffery, S., F.G.A. Verheijen, M. van der Velde, and A.C. Bastos. 2011. A Quantitative Review of the Effects of Biochar Application to Soils on Crop Productivity Using Meta-Analysis. *Agriculture, Ecosystems & Environment* 144(1): 175–87.

Ketterings, Q.M., G. Albrecht, K. Czymmek, and S. Bossard. 2005. Nitrogen Credits from Manure. Cornell University Cooperative Extension. Agronomy Fact Sheet Series, Fact Sheet 4. ***http://cceonondaga.org/resources/nitrogen-credits-from-manure***

Kim, K.R., G. Owens, Y.S. Ok, W.K. Park, D.B. Lee, and S.I. Kwon. 2012. Decline in Extractable Antibiotics in Manure-Based Composts During Composting. *Waste Management* 32: 110–116.

Koenig, R.T., C.G. Cogger, and A.I. Bary. 2011. Dryland Winter Wheat Yield, Grain Protein, and Soil Nitrogen Responses to Fertilizer and Biosolids Applications. *Applied and Environmental Soil Science* 2011: 1–9.

Kuchta, S.L., and A.J. Cessna. 2009. Lincomycin and Spectonomycin Concentrations in Liquid Swine Manure and their Persistence During Simulated Manure Storage. *Archives of Environmental Contamination and Toxicology* 57: 1–10.

Lal, R. 2001. World Cropland Soils as a Source or Sink for Atmospheric Carbon. *Advances in Agronomy* 71: 145–191.

Larney, F.J., D.M. Sullivan, E. Buckley, and B. Eghball. 2006. The Role of Composting in Recycling Manure Nutrients. *Canadian Journal of Soil Science* 86: 597–611.

Liu, X., A. Zhang, C. Ji, S. Joseph, R. Bian, L. Li, G. Pan, and J. Paz-Ferreiro. 2013. Biochar's Effect on Crop Productivity and the Dependence on Experimental Conditions—A Meta-Analysis of Literature Data. *Plant and Soil* 373(1-2): 583–94.

Lorent, L., D. Roberts, and I. Burke. 2016. Organic Small Grain Production in the Inland Pacific Northwest: A Collection of Case Studies. Pacific Northwest Extension Publication PNW683. Washington State University.

Ma, J., C. Frear, and G. Yorgey. n.d. Approaches to Nutrient Recovery from Dairy Manure. Washington State University Extension Publication.

Machado, S. 2011. Soil Organic Carbon Dynamics in the Pendleton Long-Term Experiments: Implications for Biofuel Production in Pacific Northwest. *Agronomy Journal* 103(1): 253–60.

Machado, S., and L. Pritchett. 2014. Biochar Effects on Wheat Productivity in Chemical Fallow. Poster Presentation. Presented at the Regional Approaches to Climate Change for Pacific Northwest Agriculture Annual Meeting, Richland, WA, March 5.

Magdoff, F., and H. van Es. 2010. Building Soils for Better Crops, 3rd ed. Sustainable Soil Management Handbook Series Book 10. Sustainable Agriculture Research and Education (SARE) Program. United Book Press. ***http://www.sare.org/Learning-Center/Books/Building-Soils-for-Better-Crops-3rd-Edition***

Mantovi, P., G. Baldoni, and G. Toderi. 2005. Reuse of Liquid, Dewatered, and Composted Sewage Sludge on Agricultural Land: Effects of Long-Term Application on Soil and Crop. *Water Research* 39(2-3): 289–96.

Mehta, C.M., W.O. Khunjar, V. Nguyen, S. Tait, and D.J. Batstone. 2015. Technologies to Recover Nutrients from Waste Streams: A Critical Review. *Critical Reviews in Environmental Science and Technology* 45(4): 385–427

Mitchell, S.M., J.L. Ullman, A.L. Teel, R.J. Watts, and C. Frear. 2013. The Effects of the Antibiotics Ampicillin, Florfenicol, Sulfamethazine, and Tylosin on Biogas Production and Their Degradation Efficiency during Anaerobic Digestion. *Bioresource Technology* 149: 244–252.

Mitchell, S.M., G.G. Yorgey, and C.E. Kruger. 2016. Producer Guide to Biosolids Quality. WSU Extension Publication FS192E. ***http://cru.cahe.wsu.edu/CEPublications/FS192E/FS192E.pdf***

Mohring, S.A.I., I. Strzysch, M.R. Fernandes, T.K. Kiffmeyer, J. Tuerk, and G. Hamscher. 2009. Degradation and Elimination of Various Sulfonamides during Anaerobic Fermentation: A Promising Step on the Way to Sustainable Pharmacy? *Environmental Science & Technology* 43(7): 2569–2574.

Moore, A. 2016. Manure and Environmental Impacts. 2016 Idaho Potato Conference. University of Idaho. ***https://www.uidaho.edu/cals/potatoes/conferences/idaho-potato-conference/proceedings***

Moore, A., M. de Haro-Marti, and L. Chen. 2015. Sampling Dairy Manure and Compost for Nutrient Analysis. Pacific Northwest Extension Publication PNW673. University of Idaho.

Moore, A., and J. Ippolito. 2009. Dairy Manure Field Applications: How Much is Too Much? University of Idaho Extension Publication CIS 1156.

Neilsen, G.H., E.J. Hogues, T. Forge, and D. Neilsen. 2003. Surface Application of Mulches and Biosolids Affect Orchard Soil Properties after 7 Years. *Canadian Journal of Soil Science* 83(1): 131–137.

Pan, W.L., and A.G. Hopkins. 1991. Plant Development, and N and P Use of Winter Barley: II. Responses to Tillage and N Management Across Eroded Toposequences. *Plant Soil* 135: 21–29.

Pan, W.L., L.E. Port, Y. Xiao, A. Bary, and C.G. Cogger. n.d. Soil Carbon and Nitrogen Fractionation and Balances During Long-Term Biosolids Applications to Semi-Arid Wheat-Fallow.

Ramaswamy, J., S.O. Prasher, R.M. Patel, S.A. Hussain, and S.F. Barrington. 2010. The Effect of Composting on the Degradation of a Veterinary Pharmaceutical. *Bioresource Technology* 101: 2294–2299.

Rasmussen, P.E., S.L. Albrecht, and R.W. Smiley. 1998. Soil C and N Changes under Tillage and Cropping Systems in Semi-Arid Pacific Northwest Agriculture. *Soil and Tillage Research* 47(3-4): 197–205.

Reeve, J.R., J.B. Endelman, B.E. Miller, and D.J. Hole. 2012. Residual Effects of Compost on Soil Quality and Dryland Wheat Yield Sixteen Years after Compost Application. *Soil Science Society of America Journal* 76(1): 278–285.

Ronsse, F., S. van Hecke, D. Dickinson, and W. Prins. 2013. Production and Characterization of Slow Pyrolysis Biochar: Influence of Feedstock Type and Pyrolysis Conditions. *GCB Bioenergy* 5: 104–115.

Russelle, M., K. Blanchet, G. Randall, and L. Everett. 2016. Nitrogen Availability from Liquid Swine and Dairy Manure: Results of On-Farm Trials in Minnesota. University of Minnesota Extension. ***http://www.extension.umn.edu/agriculture/nutrient-management/nitrogen/nitrogen-availability-from-liquid-swine-and-dairy-manure/***

Schillinger, W., C. Cogger, A. Bary, and B. Sharratt. 2015. Biosolids in Conservation Tillage: Trials for Washington's Winter Wheat-Summer Fallow Region. Conference Presentation. Biofest 2015, Chelan, WA. September 20–22.

Schlusener, M.P., K. Bester, and M. Spiteller. 2003. Determination of Antibiotics Such as Macrolide, Ionophores, and Tiamulin in Liquid Manure by HPLC-MS/MS. Analytical and Bioanalytical Chemistry 375: 942–947.

Sharratt, B., W. Schillinger, A. Bary, and C.G. Cogger. 2016. Fine Particulate Emissions after Applying Biosolids to Agricultural Land. Abstract. ASA-CSSA-SSSA meeting, Nov. 6–9: Phoenix, AZ.

Streubel, J.D., H.P. Collins, M. Garcia-Perez, J. Tarara, D. Granatstein, and C.E. Kruger. 2011. Influence of Contrasting Biochar Types on Five Soils at Increasing Rates of Application. *Soil Science Society of America Journal* 75(4): 1402.

Suliman, W. 2015. Toward an Understanding of the Role of Biochar as an Agro-Environmental Tool: Potential for Controlled Water Release, Bacterial Retention, and Greenhouse Gas Emissions. PhD Dissertation. Washington State University, Pullman, WA.

Suliman, W., J.B. Harsh, N.I. Abu-Lail, A.M. Fortuna, I. Dallmeyer, and M. Garcia-Perez. 2017. The Role of Biochar Porosity and Surface Functionality in Augmenting Hydrologic Properties of a Sandy Soil. *Science of the Total Environment* 574: 139–147.

Sullivan, D.M., A.I. Bary, C.G. Cogger, and T.E. Shearin. 2009. Predicting Biosolids Application Rates for Dryland Wheat across a Range of Northwest Climate Zones. *Communications in Soil Science and Plant Analysis* 40(11-12): 1770–1789.

Sullivan, D.M., C.G. Cogger, and A.I. Bary. 2015. Fertilizing with Biosolids. Pacific Northwest Extension Publication PNW508. Oregon State University.

Wallace, B.M., M. Krzic, T.A. Forge, K. Broersma, and R.F. Newman. 2009. Biosolids Increase Soil Aggregation and Protection of Soil Carbon Five Years after Application on a Crested Wheatgrass Pasture. *Journal of Environment Quality* 38(1): 291.

Weaver, M. 2013. Researchers Study Biosolids' Benefits for Growers, "Yuck Factor." Capital Press. ***http://www.capitalpress.com/content/mw-Biosolids-061413-art***

Wuest, S.B., T.C. Caesar-TonThat, S.F. Wright, and J.D. Williams. 2005. Organic Matter Addition, N, and Residue Burning Effects on Infiltration, Biological, and Physical Properties of an Intensively Tilled Silt-Loam Soil. *Soil and Tillage Research* 84: 154–167.

Wuest, S.B., and H.T. Gollany. 2012. Soil Organic Carbon and Nitrogen after Application of Nine Organic Amendments. *Soil Science Society of America Journal* 77(1): 237.

Wuest, S.B., and C.L. Reardon. 2016. Surface and Root Inputs Produce Different Carbon/Phosphorus Ratios in Soil. *Soil Science Society of America Journal* 80(2): 463.

Xiao, C., M. Fauci, D.F. Bezdicek, W.T. McKean, and W.L. Pan. 2006. Soil Microbial Responses to Potassium-Based Black Liquor from Straw Pulping. *Soil Science Society of America Journal* 70(1): 72.

Xiao, C., R. Stevens, M. Fauci, R. Bolton, M. Lewis, M.W. McKean, D.F. Bezdicek, and W.L. Pan. 2007. Soil Microbial Activity, Aggregation and Nutrient Responses to Straw Pulping Liquor in Corn Cropping. *Biological Fertility of Soils* 43: 709–719.

Zhao, L., X. Cao, O. Masek, and A. Zimmerman. 2013. Heterogeneity of Biochar Properties as a Function of Feedstock Sources and Production Temperatures. *Journal of Hazardous Materials* 256-257C: 1–9.

Chapter 8

Precision Agriculture

Bertie Weddell, Washington State University
Tabitha Brown, Washington State University
Kristy Borrelli, Penn State University (formerly of University of Idaho)

Abstract

Precision agriculture, or site-specific farming, is a management approach that addresses farm- and field-scale variability in order to improve crop production by efficiently matching resource inputs with crop needs. Advances in satellite and computer technologies provide producers with many opportunities to observe, measure, and respond to the needs of their crops by addressing site-specific problems in their fields. In the inland Pacific Northwest (PNW), dryland cereal producers use precision management in many ways, such as section control to reduce herbicide application and Global Positioning Systems (GPS) for reducing overlap. The use of these technologies is relatively straightforward and the payoffs are clear. In this chapter, we primarily focus on a more complex precision management strategy: variable application of nitrogen (N) fertilizer, because N is a main limiting factor in cereal production systems in the inland PNW. The region's climate and topography cause variation in wheat yields within and across fields and therefore in N fertilizer requirements. Synchronizing N supply with crop N demand is a major challenge for improving fertilizer use efficiencies and reducing N losses, which makes it a critical priority for the global research agenda. Precision agriculture technologies provide opportunities to manage complex N, water, and crop

Research results are coded by agroecological class, defined in the glossary, as follows:

● Annual Crop ▲ Annual Crop-Fallow Transition ■ Grain-Fallow

interactions. This chapter presents information on the principles and assumptions behind precision agriculture, types of precision agriculture tools and equipment available, steps involved in implementing variable rate N application, and sources of support for making decisions about managing variability.

Key Points

- Precision agriculture is a site-specific management approach that uses technology to manage field variability and achieve specific goals such as crop yield, percentage of protein, and nitrogen use efficiency.
- Precision agriculture assumes that variability in the major factors that affect crop yield and quality can be accurately measured at scales relevant to farm management and that the resulting information can be used to improve the efficiency of crop input use.
- If the above assumptions are met, precision agriculture strategies might result in a win-win-win scenario with improved crop yields and quality, higher economic returns, and decreased environmental impacts from excessive inputs.
- Precision agriculture technologies provide the ability to monitor crop and field variability and help diagnose agronomic problems that occur across fields and years.
- Decisions about adoption of precision agriculture involve the consideration of economic, agronomic, technical, environmental, and social factors.
- Additional research is needed to evaluate the combined effects of precision agriculture on productivity, profit, and environmental quality.

Introduction

What is Precision Agriculture?

Precision agriculture is the management of farm- and field-scale variability to achieve explicit goals such as improved crop yield, grain quality, and **nitrogen use efficiency** (NUE) by more accurately matching crop needs with specific input requirements. Dryland cereal producers use precision management in many ways, such as section control to reduce herbicide application and Global Positioning Systems (GPS) for reducing overlap. This strategy utilizes data from information technologies with high spatial and temporal resolution, combined with grower knowledge, to inform management decisions that take into account the high degree of variability associated with agricultural production (National Research Council 1997). Precision agriculture has been described as the use of the right input in the right amount at the right time and in the right location (Mulla 2013). These concepts (also known as the 4 Rs) can be used alone or in combination. Improved precision agriculture technologies allow producers to measure multiple interacting variables and to potentially use this information to maximize their profits, use resources efficiently, and minimize environmental damage.

Precision agriculture is a multi-step process that involves: (1) gathering information on spatial and temporal cropland variability, (2) applying that information to develop site-specific management strategies that use precise techniques to match fertilizer, pesticide, or water inputs to crop needs, (3) assessing the resulting benefits and costs, and (4) repeating and/or adjusting management on the basis of lessons learned. Initial mapping provides information on spatial variation, whereas the comparison of multi-year sequences of data provides information on changes over time, which is becoming increasingly important in responding to climate change. (For more information on climate change, see Chapter 1: Climate Considerations.)

History of Precision Agriculture

The concept of tailoring management to spatial variation in soil and other environmental conditions was introduced in the 1920s (Linsley and Bauer 1929), but at that time technology for implementing this approach was not available. By the 1980s, the typical farm size had increased in the **inland PNW** (Duffin 2007; Jennings et al. 1990) and farms encompassed more within-farm variability. Increased awareness of high spatial and temporal variability in crop performance on large-scale farms continues to heighten the importance of precision agriculture. Ecological, economic, and social factors all contribute to high levels of interest in precision agriculture, especially in the developed world. (For information on federal policies that promote precision agriculture, see Chapter 12: Farm Policies and the Role for Decision Support Tools.)

Variation in geology, soil, terrain, water availability, microclimate, and biota is high across landscapes in the inland PNW. In addition, variability within fields is high. This variation, combined with diverse management legacies and inconsistent weather patterns, creates heterogeneous growing conditions. In the inland PNW, conversion of **conventional tillage** systems to **direct seeding** has further increased variability due to tillage management impacts on soil properties, pests, rotations, and crop performance (Huggins 2004).

The global need to increase food production without increasing negative environmental impacts from agriculture is another reason why interest in precision agriculture continues to increase (Cassman et al. 2002; Tilman et al. 2011). More precise and efficient farming of land that is already in production is critical. By matching inputs such as fertilizers and pesticides to site-specific conditions that regulate crop demand, precision agriculture has the potential to help producers improve crop productivity and economic returns. Applying resources when and where they are needed can also reduce negative environmental effects (National Research Council 1997) by improving efficient use of these resources.

Relationship of N Fertilizer to Climate Change

Excessive use of N fertilizer can increase the risk of losses of N as nitrous oxide and result in accelerated rates of reactive N entering and cycling through ecosystems. The agricultural sector is the largest contributor to rising nitrous oxide emissions in the US, and nitrous oxide emissions from agricultural soil management are the largest source of agricultural greenhouse gas emissions (EPA 2013).

The Need to Address Field-Scale Variability

In order to precisely match agricultural inputs to crop needs, it is essential to understand the impact of variability in factors that influence crop growth and development. Soil and crop properties that vary over time as well as across space are especially difficult to diagnose (Huggins 2004).

Present-day soils result from natural processes of soil formation as well as the effects of land use and management practices (Busacca and Montgomery 1992). Geographic diversity in terrain, soil, and climate across the inland PNW results in different dryland cereal production systems that are mapped as **agroecological classes** (AECs). (See Chapter 1: Climate Considerations for details on AECs.) Effective soil depth can be limited by soil profile layers with high clay content and low permeability (Pan and Hopkins 1991). In the driest, westernmost parts of the region, calcic horizons and duripans rich in lime and silica limit productivity. In areas with intermediate precipitation, impermeable clay-rich horizons are limiting; and in the easternmost part of the inland PNW, seasonally perched water tables overlying clay-rich horizons limit rooting depth and cause lateral water flow (Figure 8-1) (Busacca and Montgomery 1992). In the Annual Crop AEC, steep topography creates varied microclimates that affect both the need for and response to nutrients (Fiez et al. 1994a; 1994b). (See Chapter 6: Soil Fertility Management.) For example, north-facing slopes in the Annual Crop AEC are wetter and have seasonally perched water with subsurface lateral flow (Brooks et al. 2012).

Landscape-specific processes affect local environmental conditions such as slope, soil depth, and the presence of an impermeable layer of soil (Figure 8-2). For example, across a single hill, variations in slope, aspect, and soil

Figure 8-1. Lateral water movement on a hillside near Troy, Idaho. (Photo: Erin Brooks, with permission.)

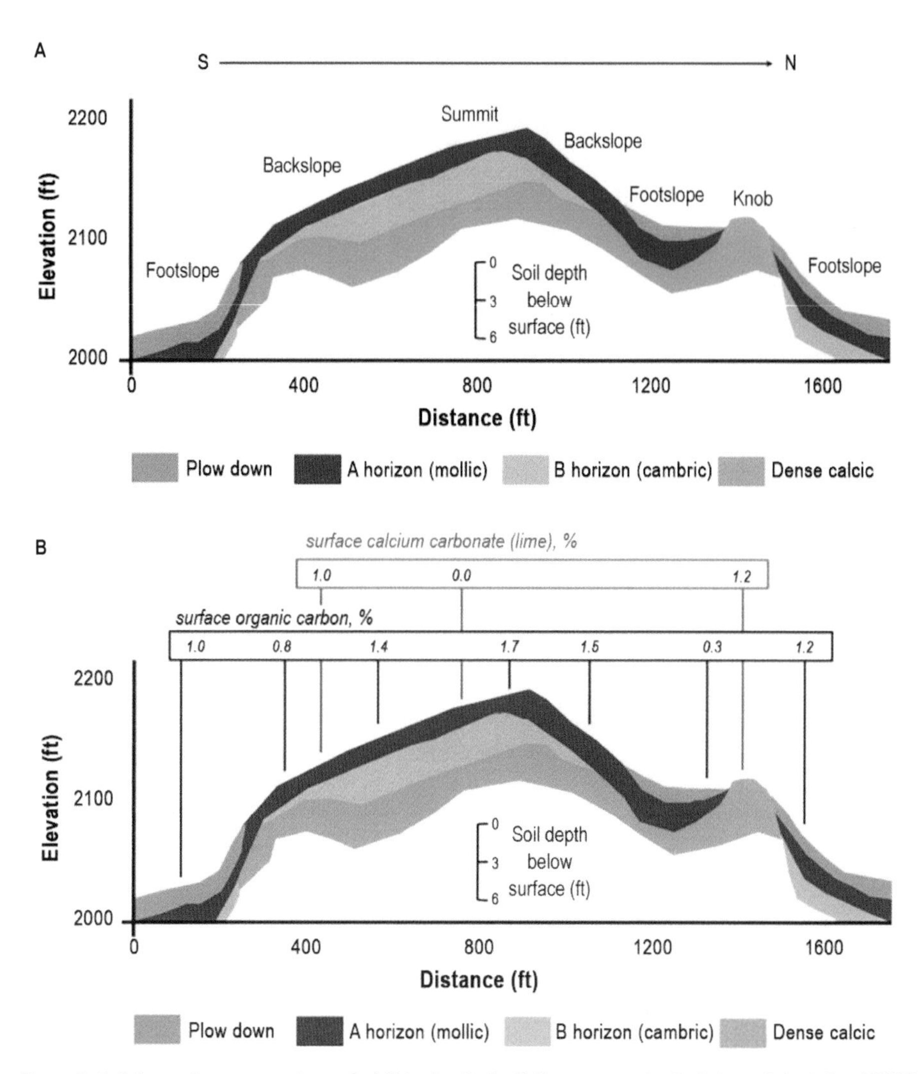

Figure 8-2. Schematic cross-sections of a hill in the Grain-Fallow agroecological class of the inland PNW. Note that different scales are used for horizontal distance, elevation, and subsurface depth. (A) Soil profiles at different terrain positions; (B) Amount of calcium carbonate and organic carbon in the top 4 inches of soil at selected points, indicated by vertical lines, across the same hill. (Adapted from Busacca and Montgomery 1992.) ■

texture affect a suite of interrelated variables including microclimate, snow accumulation, runoff, erosion, ponding, soil **water holding capacity**, and evaporation. In addition, management decisions about tillage, fertilization, rotations, and residue treatment affect soil properties such as nutrient availability and uptake efficiency, soil organic matter accumulation and decomposition, soil **bulk density**, **pH**, and crop rooting depth. (See Chapter 6: Soil Fertility Management for a more detailed discussion of factors that affect productivity.) The result is a complex within-farm mosaic of areas differing in yield potential (Busacca et al. 1985; Ibrahim and Huggins 2011; Mulla 1986; Rodman 1988). (See Chapter 2: Soil Health; Chapter 4: Crop Residue Management; Chapter 5: Rotational Diversification and Intensification; and Chapter 6: Soil Fertility Management.)

Considerable variability within fields and across years has been reported by researchers and farmers in the inland PNW. Some of this variation can be inferred from county soil surveys, but the coarse scale (1:20,000) of these maps obscures the fine scale heterogeneity of multiple characteristics. Figure 8-3 compares a Whitman County soil survey map of a field at the Washington State University Cook Agronomy Farm to a soil survey and a soil organic carbon map of the same field using finer resolution. Detailed field-scale maps reveal more variability and are therefore more useful for identifying spatial patterns and prescribing site-specific management. Year-to-year and within-field variability continue to generate interest in site-specific rather than uniform rates of agricultural inputs (Huggins 2010; Huggins and Pan 1993). Precision agriculture in the inland PNW can potentially address constraints on agricultural productivity due to the impacts of landscape variation and management, such as soil texture, compaction, effective rooting depth, drainage, acidification, and erosion. (More information can be found in Chapter 6: Soil Fertility Management; Chapter 2: Soil Health; and Chapter 3: Conservation Tillage Systems.)

Steps in the Process of Site-Specific Management

Precision management involves the following steps carried out by growers in consultation with industry, research, and Extension advisors as necessary (Figure 8-4):

- Specify goals (e.g., grain yield or protein concentration) of the operation. It is important to match the scales of measurement and management as closely as possible for crops, soils, and terrain. For instance, it is inefficient to collect data on a scale that is finer than what is treatable by available equipment. If an input can be applied only on a scale of feet, then it is wasteful to measure it on a scale of inches (Pierce and Nowak 1999).

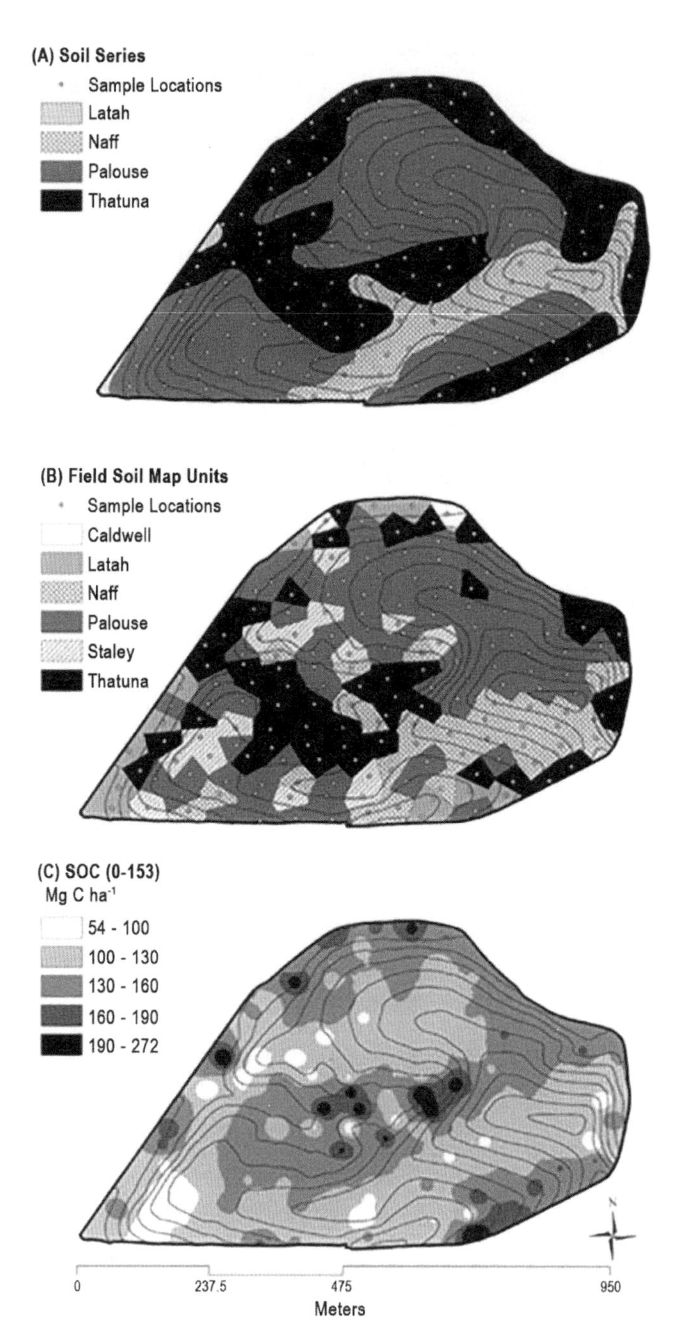

Figure 8-3. Maps of (A) soil series, (B) field soil map units, and (C) soil organic carbon (SOC; at 0–60 inches) for the 92-acre Washington State University Cook Agronomy Farm. Data for B and C were obtained by sampling alternating points from a systematic, non-aligned grid of 369 geo-referenced sample locations at a resolution of ±10 feet. (Source: Huggins and Uberaga 2010.)

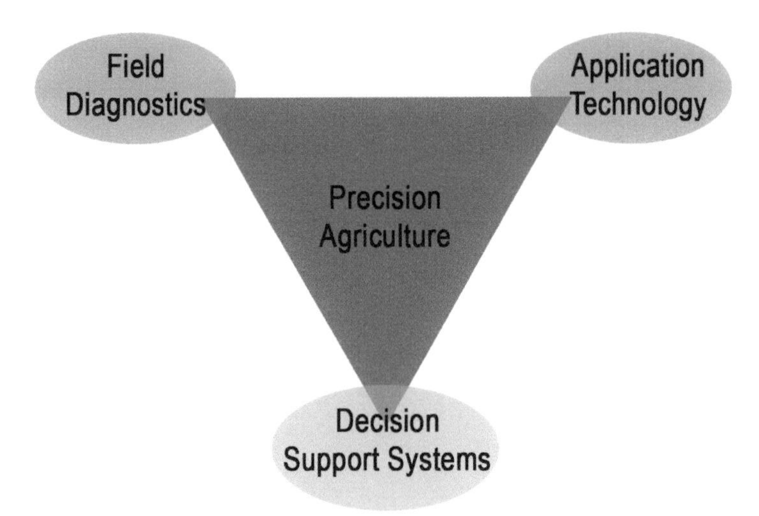

Figure 8-4. Key elements of precision agriculture. (Adapted from Huggins 2015.)

- Obtain accurate, fine-scale data on within-field variability that influences desired outcomes.
- Use the resulting data to generate multi-layer maps that illustrate factors influencing crop yield.
- Use the information from the preceding step to diagnose crop needs and develop prescription maps using computer software. These maps prescribe site-specific management zones (SSMZs) or areas that are relatively homogeneous with regard to yield-controlling attributes (Table 8-1). Management zones should not be confused with the much larger agronomic zones defined by Douglas et al. (1992). Collectively, SSMZs form the basis of a precise management plan (Corwin 2013). Decisions about how many SSMZs should be recognized are crucial. If too few zones are prescribed, areas that are substantially different will be grouped together and variable management will not address site-specific variability effectively. If too many zones are designated, areas that differ only slightly will receive different treatments, and management will be more complicated and expensive than necessary, or controllers will not be able to react to the small distances or time units that are specified. The challenge associated with making this decision is one reason why producers who are

Table 8-1. Examples of crop inputs that are commonly applied to site-specific management zones using variable rate technology.

Crop inputs applied to site-specific management zones	Sources of information for zone delineation
Gypsum	Grower knowledge, yield patterns, EC_a maps, soil tests for pH and Na
K	Topography, grid or directed soil sampling, soil survey maps, EC_a maps
Lime	pH, soil texture
Manure	Soil texture, organic matter, yield patterns, bare soil photos, nitrate nitrogen, crop canopy reflectance
N	Soil texture, organic matter, yield patterns, bare soil photos, nitrate nitrogen, crop canopy reflectance
P	Topography, grid or directed soil sampling, soil survey maps, EC_a maps
Herbicide	Weed maps, soil organic matter, soil texture
Irrigation	Soil texture, topography, yield zones, EC_a-directed soil sampling
Seeds	Historical yield maps and topsoil depth

Note: K = potassium; N = nitrogen; P = phosphorus; EC_a = apparent soil electrical conductivity. Adapted from Corwin 2013.

developing prescription maps for variable application often seek decision support from a precision agriculture consultant.

- Implement the plan during the growing season by applying inputs at the variable rates and locations specified in the precision management plan. Application rates can be fine-tuned using information from field scouting and ongoing data analysis.
- During harvest, use yield monitors to collect information about crop quantity and quality (e.g., grain protein concentration).
- After one year of using precision agriculture, evaluate how well management goals were met. Modify management plan if necessary using the results from the first year's trial to inform this decision. Evaluating the response of areas managed with precision agriculture may require a check strip (not managed with precision agriculture) for comparison.

- Repeat the above steps as many times as necessary to achieve desired site-specific results. Remember that throughout this process equipment suppliers and researchers are available to provide decision support.

Research on Site-Specific N Management

Importance of Site-Specific N Management

Nitrogen is typically the most limiting nutrient in crop production. **Nitrogen supply** and demand are affected by soil chemical, physical, and biological processes. Nitrogen requirements are affected by crop N need and the amount of N supplied from soil and crop residues. Worldwide recovery of N in harvested cereal crops is estimated to be only 33% of total N applied (Raun and Johnson 1999). Poor N recovery is due to multiple factors that can include biological N **immobilization**, losses occurring from **leaching** and volatilization, and inefficient crop uptake and utilization. Improving the efficiency of N utilization is of concern because movement of N beyond agroecosystem boundaries contributes to degradation of air and water and because growers seek to reduce costs from wasteful inputs. (See Chapter 6: Soil Fertility Management.)

Management strategies that increase NUE can mitigate greenhouse gas emissions. (See Chapter 4: Crop Residue Management and Chapter 6: Soil Fertility Management.) The potential for site-specific N application to reduce nitrous oxide emissions has been experimentally demonstrated in the field (Sehy et al. 2003). The significance of reductions in applied N is controversial because several interacting factors influence N transformations in soil (Venterea et al. 2012). However, recent analyses and field work have suggested that emissions of nitrous oxide increase exponentially with N input (Hoben et al. 2011; Shcherbak et al. 2014) rather than linearly. For dryland farmers of the inland PNW, precision agriculture is an important strategy for climate change **mitigation**. Agricultural N management practices that reduce N fertilizer application rates without reducing crop yields could potentially reduce agricultural nitrous oxide emissions, generate greenhouse gas offsets, and enhance overall environmental quality.

Under current greenhouse gas offset programs, emission offset credits generated by agricultural N management actions are based on reducing annual N fertilizer application rates for a given crop without reducing yield. Because reductions in N fertilizer entail economic risk for producers due to possible yield depression from under-applied N, economic incentives are important. Programs that offer incentives for N emission reductions are a promising strategy for reducing regional emissions (Brown 2015; Ward 2015); however, in interviews of 33 growers from across inland PNW drylands, insufficient financial incentives and excessive paperwork were commonly cited as barriers to involvement in nutrient management plans (Ward 2015).

Principles of Precision N Management

Precision management of N fertilizer means synchronizing N application with variability in crop N demand. To do this, it is necessary to understand crop requirements for N (Fiez et al. 1995) and crop response to applied N (Huggins and Pan 1993). Fiez et al. (1995) concluded that recommended N application rates on north-facing backslopes at a study site in the Annual Crop AEC were excessive because of low fertilizer uptake and high losses of N. However, landscape position alone was not the best predictor of site-specific N fertilizer needs in this study. ●

Regional fertility guides (Koenig 2005; Mahler 2007) are not appropriate for developing site-specific N management prescriptions because they do not account for variability in crop **unit nitrogen requirement** and N supply (Pan et al. 1997). Some landscape positions have high soil N supply but low yield potential and low response to N fertilizers, while other areas have high yield potential, lower N supply, and good response to applied N. In addition to landscape position, yield and crop performance are influenced by interacting effects of soil, water, weather, nutrient availability, crop variety, and management (Mulla et al. 1992; Pan and Hopkins 1991). (See Chapter 6: Soil Fertility Management for a discussion of processes that affect N availability.)

Patterns in Site-Specific Yield Responses to N Management

In a 2007 on-farm study near Colfax, Washington, in the Annual Crop AEC, Huggins (2010) compared yield benefits from variable rate nitrogen

(VRN) applications to uniform N rates at different landscape positions. The farmer identified low- and high-yielding areas of the field, and uniform and variable urea fertilizer N inputs were compared at selected geo-referenced locations across the field. Positive yield response to VRN application tended to be most pronounced in the low-yielding areas, which was likely due to relatively shallow soils and low potential to store soil water (Figure 8-5). Reducing the N fertilizer rate by 40% in the lower yielding areas increased hard red spring wheat yield by 25% compared to uniform N application. Previous application of excessive N in such areas may have increased soil water consumption during vegetative growth and left less water available during grain filling. Conversely, increasing N rates in higher yielding areas by 63% increased hard red spring wheat yields by 12% compared to uniform N application. Areas with relatively high yields tended to occur on relatively flat uplands, which likely had deep soil and high water holding capacity (Huggins 2010). ●

In this example, two N application rates, one for low-yielding areas and one for high-yielding areas, were adequate to improve efficiency compared to a single application rate. On the basis of this study, reducing N rates leading to increasing yields on low-yielding locations would be expected to result in greater economic returns. The economic effects of increasing N inputs on higher yielding areas is less clear and would probably depend on additional factors such as wheat and N prices. The results of this on-farm research trial suggest that site-specific characteristics such as slope and soil type combined with grower knowledge have potential as decision aids for precision N applications, especially because the required data can be generated by growers with just a yield and protein monitor, GPS, and VRN technologies.

Refining Predictions of Site-Specific Responses to N Management

Although insight into these landscape patterns in response to N application are useful, producers need tools with a higher degree of site-specificity for predicting responses to VRN. Developing grower-oriented, field-scale decision support tools to evaluate spatial variability in crop performance and assess site-specific management strategies is an emerging focus of research in the inland PNW. Brown (2015), Huggins et al. (2010), and

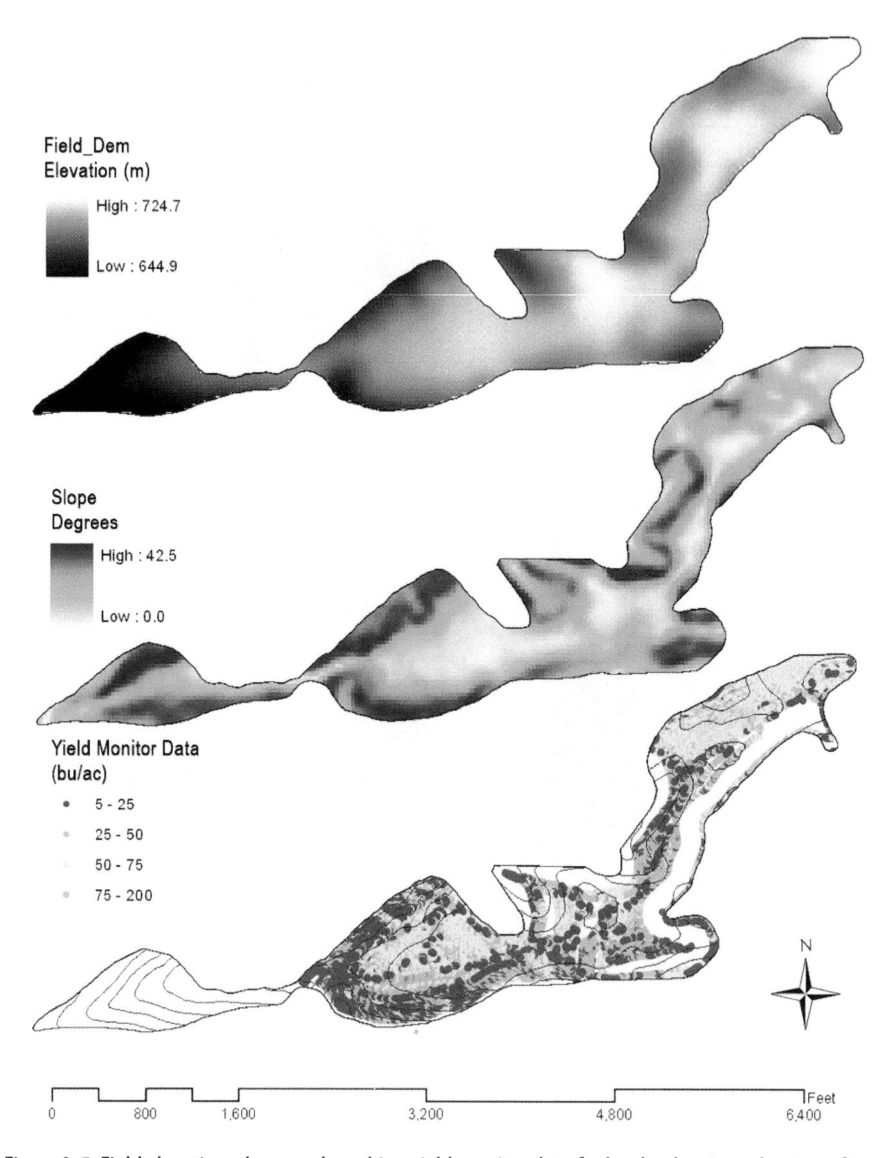

Figure 8-5. Field elevation, slope, and combine yield monitor data for hard red spring wheat at a farm near Colfax, Washington. (Dem = digital elevation model; Source: Huggins 2010.)

Taylor (2016) have developed tools for identifying crop performance with regard to increasing productivity as well as NUE (Table 8-2).

For example, using the key developed by Brown (2015) winter wheat can be separated into five performance classes (Figure 8-6). This classification

Table 8-2. Performance criteria for assessing wheat responses to variable application of nitrogen in the Annual Crop agroecological class. ●

Criteria	Crop	Location
GPC, N uptake efficiency, N retention efficiency	Hard red spring wheat	CAF, near Pullman
Yield, GPC, NUE	Soft white winter wheat	CAF, near Pullman
Protein, N balance index (N_g/N_f)	Soft white winter wheat	Near Walla Walla; CAF, near Pullman

GPC = grain protein concentration; NUE = nitrogen use efficiency; N_g = grain N; N_f = N fertilizer application rate; CAF = Cook Agronomy Farm. All locations are in Washington. Adapted from Brown 2015, Huggins et al. 2010, and Taylor 2016.

allows for post-harvest interpretation of crop performance that diagnoses limitations to N supply or crop uptake. In addition, the information can be used to guide future site-specific N management decisions for optimizing yield, yield quality, and efficient use of N supply.

Available Technology

Remote Sensing

Remote sensors of soil or crop properties may be proximal (hand-held or tractor-mounted sensors), aerial (mounted on airplanes or unmanned aerial vehicles), or satellite-mounted. Remote sensing uses interactions between electromagnetic energy and soil or plant material. Applications of electromagnetic radiation in remote sensing involve non-contact measurement of reflected electromagnetic radiation. Precision agriculture systems use primarily visible, near infrared (NIR), infrared, and thermal sensor data (Figure 8-7A). Plant pigments absorb visible light of specific wavelengths and reflect radiation that is not absorbed. Measuring light absorption at different wavelengths enables detection of active crop growth (i.e., chlorophyll a and chlorophyll b activity).

Spectral indices use ratios of plant reflectance in the visible and NIR regions to assess characteristics of plant canopies and soils. For example, in a study near Adams, Oregon, of the effects of the number of yellow flowers per unit area and leaf area index (LAI) on canopy spectral reflectance of

Figure 8-6. Dichotomous key to classification of soft white winter wheat performance on the basis of components of N utilization. (Adapted from Brown 2016.)

Step	NUE Criteria	Performance Class
1	N Utilization Efficiency is equal to or greater than 45	
	If yes, go to Step 2	
	If no, go to Step 5	
2	N Uptake Efficiency is equal to or greater than 0.5	
	If yes, Step 3	
	If no, Step 4	
3	N Use Efficiency is less than 30	
	If yes, ...	Class 1
	If no, ...	Class 2
4	....	Class 3
5	N Uptake Efficiency is equal to or greater than 0.5	
	If yes, ...	Class 4
	If no, ...	Class 5

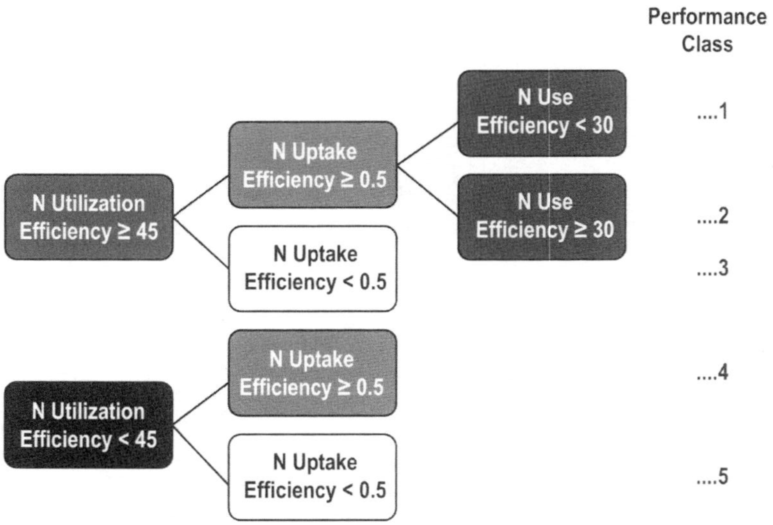

N Utilization Efficiency ≥ 45
N Uptake Efficiency ≥ 0.5
N Use Efficiency < 30
N Use Efficiency ≥ 30
N Uptake Efficiency < 0.5
N Utilization Efficiency < 45
N Uptake Efficiency ≥ 0.5
N Uptake Efficiency < 0.5

Performance Class
....1
....2
....3
....4
....5

spring canola, the ratio of NIR and blue light was suitable for estimating LAI during flowering (Sulik and Long 2015). (For information on the use of spectral indices to predict crop residue cover and density, see Chapter 4: Crop Residue Management.)

Green vegetation absorbs most colors of visible light (except green) and reflects a large portion of incoming NIR wavelengths. Consequently, senesced, diseased, or sparse vegetation reflects more visible light and less NIR radiation (Figure 8-7B). The normalized difference vegetation index

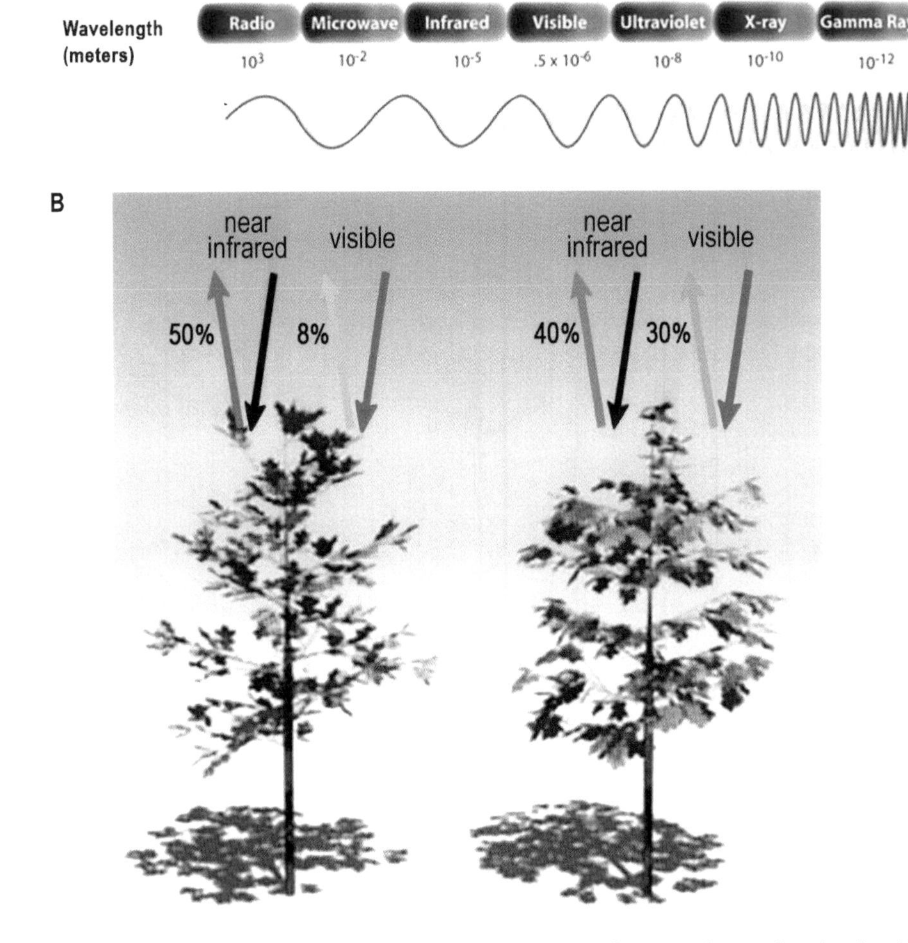

Figure 8-7. (A) The electromagnetic spectrum. (B) Differences in reflectance of near infrared and visible light by green and unhealthy or senesced vegetation. Note that the green foliage reflects less visible light than the yellowish-brown leaves. (Sources: NASA 2016a; 2016b. Illustration by Robert Simmon.)

(NDVI) uses a ratio of red and NIR spectral absorption to assess spatial and temporal variability patterns in photosynthetic activity. The NDVI and other spectral indices are used to study crop density, development, and reproductive capacity during the growing season.

Hyperspectral remote sensing collects reflectance data over a wide spectral range at small increments. Because hyperspectral imaging can be collected across a large range of wavelengths and at fine spatial resolution, it is useful for understanding spatial and spectral variability in reflectance for bare or vegetated ground (Mulla 2013).

Electrical conductivity (EC) sensors measure the electromagnetic energy of soil (the ability of soil to conduct an electrical current), which depends on total solute concentration (salinity) (Rhoades et al. 1989). EC is one of the measurements used most frequently in precision agriculture research to characterize spatial and temporal variation in properties that often affect crop yield. Sensors for continuous, real-time proximal sensing of soil EC have been used for mapping spatial patterns in soil clay content, salinity, soil moisture, and cation exchange capacity (Corwin and Lesch 2005). The EC of bulk soil, referred to as apparent soil electrical conductivity (EC_a), has three components (Figure 8-8).

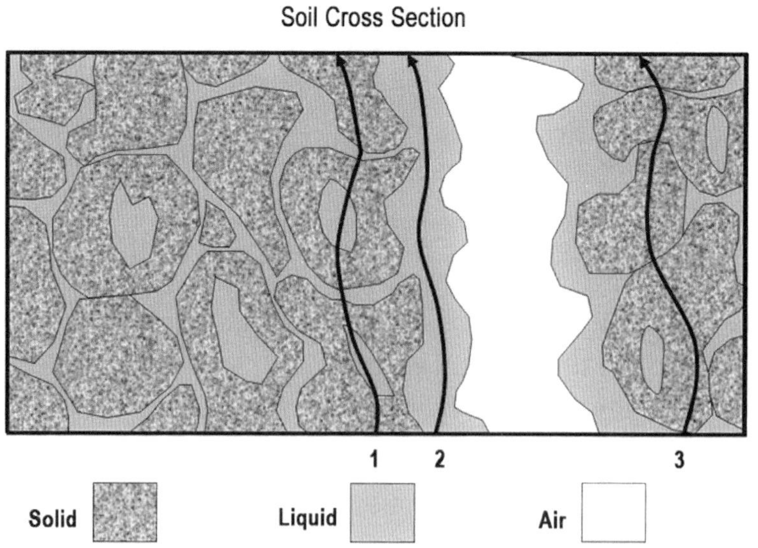

Figure 8-8. Three pathways for measuring apparent electrical conductivity of soil. (1) Solid-liquid phase, (2) liquid phase, and (3) solid phase. (Source: Corwin and Lesch 2005.)

Remote sensing of thermal radiation emitted by leaves and canopies has been used to estimate water stress (Cohen et al. 2005). This is done using canopy temperature as an indicator of water status.

Geospatial Referencing

A geographic information system (GIS) uses computer software and hardware to capture, store, manipulate, manage, and display geographically referenced information. When a GIS database is referenced to a base map or base data layer, geographic data can be projected onto a flat paper or screen. Data from many sources can be used, including soil or crop sampling, topographic surveys, digitized maps or photographs, and information from sensors.

Most GIS uses either vector or raster spatial data. Vector data uses coordinates to represent point, line, or polygon features on maps. Raster data are displayed as discrete picture elements termed pixels and can include any information displayed and stored as pixels, including aerial photography and scanned images.

The USDA Farm Service Agency provides growers with hard or soft copies of GIS imagery. Farmers can also obtain GIS data from some agricultural supply companies or build their own GIS project using software, such as Farm Works or SMS, and tractor or combine data collected during field operations. Some basic GIS software is free, but the cost of advanced GIS software may be substantial. Because of the technical complexity of some GIS applications, decision support personnel often play an important role in interpreting, displaying, and archiving of GIS data.

The Global Positioning System, or GPS, is a network (termed Navstar) of satellites put in orbit by the US Department of Defense. As of September, 2016, there were 31 active satellites (***http://www.gps.gov/systems/gps/space/***) transmitting one-way radio signals giving satellite position (latitude, longitude, and elevation) and time to users. GPS receiver equipment receives signals from the satellites and uses this information to calculate the user's three-dimensional position. The use of GPS in conjunction with GIS allows real-time data to be combined with accurate position information. This enables manipulation and analysis of large amounts of geospatial data.

Models

Models make it possible to analyze large data sets and make projections about outcomes if specific assumptions are met. It is important to keep in mind, however, that the projections of a model are not the same as experimentally demonstrated results. The more a model's assumptions are tested and supported by test results, the more robust we can consider the model.

Use of Precision Agriculture Technology in the Inland PNW

Precision Farm Equipment

A variety of products that use precision agriculture technology are available to producers. Some of the most common types are listed below. For more information, see Yorgey et al. (2016).

- Combine monitors consist of sensors connected to GPS receivers. This equipment typically monitors yield at different locations within a field. With several years' data, the accumulated information can be used to generate prescription maps that divide fields into different zones for variable management. In addition to yield, combine sensors can monitor other variables, such as grain moisture content, protein content, and straw yield (Reyns et al. 2002).
- Aerial infrared crop images are another tool that can be used to develop prescription maps for fertilizer application. On infrared images, dense and vigorous vegetation is bright red, whereas less vigorous plants are lighter red or grey. Photos taken during crop growth can thus be used to delineate zones of higher and lower potential yield or plant biomass.
- Spatial soil mapping using measurements of soil EC_a can also be used to develop prescription maps. For example, ground conductivity meters developed by Geonics, Ltd. contain a transmitting coil and a receiving coil. The meter is placed directly on the ground where the transmitting coil generates small currents that are sensed by the receiver coil (McNeill 1980). A meter can

be coupled with GPS and with appropriate software to generate spatially specific data about EC_a.

- Variable rate fertilizer applicator systems use information corresponding to a sensed position to adjust fertilizer application rates as the applicator system moves across a field (Stombaugh and Shearer 2000). A variable rate applicator adjusts the rate of fertilizer that is delivered, using specifications from a zone map and a management plan. This technology is suited to fields with a wide range of yield potentials or fields with variation in residual levels of nutrients. Nitrogen is the nutrient most commonly applied at variable rates, but phosphorus, sulfur, and other nutrients can also be applied at variable rates.
- Auto-steer systems use the Differential Global Positioning System (DGPS), a highly accurate satellite system (***http://www.gps.gov/systems/augmentations/***) to detect the location of equipment and steer it across the field in a way that reduces overlap in passes across the field. Auto-steer systems can generally be retrofitted onto existing equipment.
- Section controllers improve the efficiency of input application by automatically shutting down operations (often a section of a piece of equipment) at locations (such as field edges and areas that overlap with previously covered ground) where additional application would be wasteful. Applications such as seed, fertilizer, and herbicide can be fine-tuned in this way. Section controllers can be retrofitted onto existing equipment.

Adoption of Precision Agriculture Technology

The benefits and risks associated with precision agriculture adoption depend on a variety of technical, geographic, economic, social, and cultural factors. Many growers use some precision agriculture technology but have not delineated SSMZs. For example, auto-steer was one of the first precision agriculture technologies to be adopted in the inland PNW and remains popular (Table 8-3). This technology has been shown to pay for itself rather quickly in dryland regions because it reduces overlap in input applications and requires little or no decision support or system component integration (McBratney et al. 2005).

Table 8-3. Responses to producer surveys in 2011 and 2012 about specific precision agriculture technologies.

Precision Agriculture Tools	Use		Have and do not use	Do not have	No response
	2011	*2012*	*2012*	*2012*	*2012*
GPS Guidance (vision-based)	46.8	65.6%	1.7%	29.3%	3.4%
Variable Fertilizer Applicator	N/A	39.4%	4.0%	53.1%	3.4%
Yield Monitor	N/A	34.3%	4.7%	57.6%	3.4%
Precision Agriculture Software	N/A	24.7%	3.8%	68.1%	3.4%
Aerial Crop Imagery	N/A	20.2%	2.6%	73.6%	2.4%
Variable Seeding Equipment	N/A	20.1%	3.2%	73.2%	3.4%
Spatial Soil Mapping	N/A	13.2%	3.4%	79.6%	3.8%
Auto-steer System (DGPS-based)	36.6%	N/A	N/A	N/A	N/A
Variable Rate N Application	20.4%	N/A	N/A	N/A	N/A
Section Controllers	25.5%	N/A	N/A	N/A	N/A

N/A = question not asked. See Bell (2000) for discussion of difference between vision-based guidance and DGPS-based auto-steer systems. Adapted from Mahler et al. 2014 and Gantla et al. 2015.

On the other hand, adoption of VRN technology requires a greater commitment. Since N fertilizer is the most expensive variable input cost for growers, a reduction in N application could increase gross returns. Ward (2015) estimated a 1.72 lb per acre reduction in N loss could be achieved under a precision N management plan based on a crop model simulation that compared variable and uniform N fertilizer application. Yet, other economic analyses of variable rate application of plant nutrients to small grains in inland PNW drylands have produced mixed results. Several studies reported that VRN increased profitability in some situations but not in others. In a hypothetical case analysis in the Annual Crop AEC near Farmington and Pullman, Washington (Fiez et al. 1994a), use of experimentally determined unit N requirements increased net returns from winter wheat by as much as $14.80 per acre, but Taylor (2016) reported that at three sites in the Annual Crop AEC of eastern Washington the economic benefits of VRN management depended on yield potential. ● Reducing N rates in low-yielding field locations resulted in N fertilizer savings with VRN management, whereas increasing N rates in high-yielding zones resulted in no yield benefit but decreased NUE and economic returns. In addition to the cost of precision agriculture technology, the price of wheat and fertilizer as well as the likelihood of realizing a yield increase should be regularly evaluated to gain insight into how VRN might provide the least economic risk in high-yielding zones.

Mahler et al. (2014) summarized University of Idaho surveys of grower adoption of precision agriculture technology over three decades. The surveys, conducted in 1981, 1996, and 2011, showed that adoption increased markedly between 1996, when use of these technologies was estimated at 10%, and 2011. Mahler et al. (2014) concluded that variable rate systems were more popular in drier areas, and that younger farmers and farmers on relatively large farms were most likely to adopt new technologies. In 2012, the University of Idaho's Social Science Research Unit surveyed a representative sample of producers in dryland farming counties of northern Idaho, eastern Washington, and northeastern Oregon about their use of technology (Gantla et al. 2015). Responses from the 2011 and 2012 surveys are summarized in Table 8-3. The proportion of farmers using GPS guidance increased markedly between 2011 (47%) and 2012 (66%). In 2012, over one-third of respondents reported having and using variable fertilizer applicator and yield monitoring technology. Few producers (less than

5% for each of seven technologies) reported having precision agriculture technology and not using it, but more than two-thirds of respondents did not have technology for spatial soil mapping, aerial crop imagery, variable seeding equipment, and precision agriculture software. Among growers who said they did not have precision agriculture technology, the most commonly selected reasons cited were that equipment was "too expensive" (62%) and/or "not cost-effective for my operation" (59.9%). In addition, about one-quarter of respondents indicated the technologies were "not worth the investment of new capital" and that they were "difficult to learn to operate and maintain" (Gantla et al. 2015).

Things to Consider when Making Decisions about Precision Agriculture

When making decisions about whether to add precision agriculture equipment and technology to a farming operation, it is important to consider several questions. This is true whether the producer is a novice or has already experimented with precision agriculture and is considering further changes. Most growers will want to give some thought to the following interacting issues:

Impacts on your finances. How much will the initial investment in precision agriculture technology cost? Will financial support be available? What are the projected maintenance costs? How long will it take for projected reductions in input costs to offset initial costs? Will there be costs associated with getting help from experts? How will fuel costs and other farm costs such as fertilizer inputs change when the new technology is implemented? Will improvements in grain yield and quality boost your farm's income?

Impacts on your agroecosystem. What specific factors limit productivity on your farm? Can the limiting factor(s) be measured accurately and treated effectively? Which factors can you control and which ones do you have to live with?

Impacts on labor and time. How much time will be required to install new equipment? How long will it take to gather the data needed for making management decisions? How long will it take to gain the expertise needed to use precision technology? Once the system is up and running, will implementation save time? If you participate in a financial assistance program, will the associated paperwork be time-consuming?

Impacts on grower knowledge. How many people on your farm will need to gain the knowledge necessary to use precision agriculture? How difficult or time-consuming will it be to acquire this knowledge?

Impacts on the environment. How will site-specific management affect off-site flows of fertilizer, insecticides, herbicides, and/or water? How will precision agriculture adoption affect on-farm vehicular passes and related soil compaction and water movement? Will long-term, site-specific management improve the quality of your soil?

Impacts on society. What effects would widespread adoption of precision agriculture be likely to have on your community? What about the long-term stability of your farm? Will adoption of precision agriculture affect whether younger generations decide to stay on the family farm or leave home to look for work elsewhere?

Because precision agriculture is based on site-specific practices, field results at one locale may not apply to other sites, even within the same AEC. The same is true for growers: what works for one person won't necessarily work for another. Grower intuition, skills, and priorities are important for evaluating whether to adopt precision agriculture.

Conclusions: Challenges and Future Directions

Precision agriculture technology and practices are rapidly developing. Spatial resolution, return frequency, and spectral resolution have improved dramatically in the past 25 years (Mulla 2013), as have data storage and analysis capabilities. But complex, highly precise technology does not always translate to appropriate management.

In the inland PNW, research on N dynamics has led to improved understanding of the effects of site-specific spatial and temporal variability on crop growth and development. This information has been used to develop criteria to rank wheat into performance classes that can be used to predict response to applied N (Figure 8-6). Research on regional hydrology has clarified vertical and horizontal movement of water, nutrients, and soil. Such information forms the basis for site-specific diagnosis and treatment

Farmer-to-Farmer Case Study Series

Ron Jirava, farms near Ritzville, Washington, in the Grain-Fallow AEC where he receives about 11.5 inches of annual precipitation. Jirava prefers not to invest heavily in variable rate technology because he hasn't "been convinced that, in an area where moisture's the limiting factor," precise mapping would be worth the economic investment (Mallory et al. 2000; ***https://www.youtube.com/watch?v=z2fdxvANWsA&list=UUO_J3MbC2_x772upBPM-CvQ&index=29***). ■

Eric Odberg, who farms near Genesee, Idaho, in the Annual Crop AEC where annual precipitation is about 22 inches, has been using VRN for several years and is pleased with the results (Yorgey et al. 2016; ***http://cru.cahe.wsu.edu/CEPublications/PNW691/PNW691.pdf***). With VRN, Odberg says, "You're applying less nitrogen out there, which is better for the environment. You're not putting on excess, which would just go into the groundwater and into our rivers and streams."

Odberg had two things working in his favor as he transitioned to VRN technology. First, he was able to reduce some initial equipment costs by participating in cost-share programs such as USDA-NRCS's Environmental Quality Incentives Program and the Conservation Security Program. Using auto-steer with a 90 lb per acre application saved Odberg $3.97 per acre, while varying his application rate saved an additional $4.54 per acre. Second, Odberg sought help from experts. He cautions other growers against being intimidated by precision agriculture technology. On the other hand, because his system is complex and requires a skilled operator, Odberg feels he needs to be the one to operate the machinery. Thus, his benefits in precision and cost savings come with a cost in lost flexibility. ●

The Farmer-to-Farmer Case Study Series can be accessed at ***https://www.reacchpna.org/case_studies***.

of constraints on crop yield and quality. This foundation sets the stage for further research on how to match inputs to crop needs.

In particular, additional research on the following agronomic issues has the potential to increase scientific understanding of precision agriculture in the inland PNW:

- Additional field testing of performance classes for wheat response to applied N.

- Site-specific approaches to addressing limiting factors other than N (such as P, S, pH, pests, and pathogens).
- Research on the challenges of using precision agriculture in the Annual Crop-Fallow Transition and Grain-Fallow AECs. ▲ ■

Currently agronomic research on precision agriculture in the inland PNW is ahead of evaluation. Additional research assessing the impacts of precision management on society and the environment is needed. For example, attention to these areas would increase our understanding of the wider context of precision agriculture:

- Additional economic analyses of specific conditions under which precision agriculture improves net returns to producers.
- Studies of the environmental effects of precision agriculture in the inland PNW (e.g., does nitrate loss to the environment decrease with precision agriculture management?).
- Studies of the interactions between agronomic, environmental, economic, and social effects of precision agriculture in the inland PNW (i.e., can precision agriculture result in win-win-win scenarios, and, if so, under what circumstances?).

Finally, programs that continue to promote communication among growers as well as between growers and support personnel are essential for promoting appropriate and effective precision agriculture in the inland PNW.

Resources

AgBiz Logic: Farm Decision Tools for Changing Climates

https://www.reacchpna.org/sites/default/files/tagged_docs/6b.2.pdf

AgWeatherNet

http://www.weather.wsu.edu

GeoCommunity Sources for GIS and Mapping Software

http://software.geocomm.com/viewers/

http://spatialnews.geocomm.com/features/viewers2002/

Geospatial Data Gateway

https://gdg.sc.egov.usda.gov/

National Coordination Office for Space-Based Positioning, Navigation, and Timing, 2016. The Global Positioning System.

http://www.gps.gov/systems/gps/

REACCH Precision Agriculture Resources for Farmers

https://www.reacchpna.org/Precision_Agriculture_Resources_for_Farmers

REACCH Nitrogen Management Webinar Series

https://www.reacchpna.org/seminars-nitrogen-series

REACCH Farmer-to-Farmer Case Studies

https://www.reacchpna.org/case_studies

US Global Positioning System Agriculture Applications

http://www.gps.gov/applications/agriculture/

USDA Aerial Photography Field Office

http://www.apfo.usda.gov

USDA Natural Resources and Conservation Service: Incentive Programs and Assistance for Producers

http://www.nrcs.usda.gov/wps/portal/nrcs/detail/national/climatechange/resources/?cid=stelprdb1043608

WSU Extension Learning Library

http://extension.wsu.edu/learn/?keyword=precision+ag&posts_per_page=6

References

Bell, T. 2000. Automatic Tractor Guidance Using Carrier-Phase Differential GPS. *Computers and Electronics in Agriculture* 25(2): 53–66.

Brooks, E.S., J. Boll, P.A. McDaniel, and H. Lin. 2012. Hydropedology in Seasonally Dry Landscapes: The Palouse Region of the Pacific Northwest USA. In Hydropedology: Synergistic Integration of Soil Science and Hydrology 329–350, H. Lin, ed. Elsevier, Waltham, MA.

Brown, T. 2015. Variable Rate Nitrogen and Seeding to Improve Nitrogen Use Efficiency. Ph.D. Dissertation, Washington State University, Pullman.

Busacca, A.J., D.K. McCool, R.I. Papendick, and D.L. Young. 1985. Dynamic Impacts of Erosion Processes on Productivity of Soils in the Palouse. *Proceedings of the National Symposium on Erosion and Soil Productivity* 1984: 152–169. American Society of Agricultural Engineers.

Busacca, A.J., and J.A. Montgomery. 1992. Field-Landscape Variation in Soil Physical Properties of the Northwest Dryland Crop Production Region. *Proceedings of the 10th Inland Northwest Conservation Farming Conference* 1992: 8–19.

Cassman, K.G., A. Dobermann, and D.T. Walters. 2002. Agroecosystems, Nitrogen-Use Efficiency, and Nitrogen Management. *Ambio: A Journal of the Human Environment* 31(2): 132–140.

Cohen, Y., V. Alchanatis, M. Meron, Y. Saranga, and J. Tsipris. 2005. Estimation of Leaf Water Potential by Thermal Imagery and Spatial Analysis. *Journal of Experimental Botany* 56(417): 1843–1852.

Corwin, D.L. 2013. Site-Specific Management and Delineating Management Zones. In Precision Agriculture for Sustainability and Environmental Protection 135-157, M.A. Oliver, T.F.A. Bishop, and B.P. Marchant, eds. Routledge, Abingdon, Oxon, UK.

Corwin, D.L., and S.M. Lesch. 2005. Apparent Soil Electrical Conductivity Measurements in Agriculture. *Computers and Electronics in Agriculture* 46(1): 11–43.

Douglas, Jr., C.L., R.W. Rickman, B.L. Klepper, and J.F. Zuzel. 1992. Agroclimatic Zones for Dryland Winter Wheat Producing Areas of Idaho, Washington, and Oregon. *Northwest Science* 66(1): 26–34.

Duffin, A.P. 2007. Plowed Under: Agriculture and Environment in the Palouse. University of Washington Press, Seattle.

EPA (US Environmental Protection Agency). 2013. Inventory of US Greenhouse Gas Emissions and Sinks: 1990–2011. Reports and Assessment EPA 430-R-13-001. ***https://www.epa.gov/ghgemissions/sources-greenhouse-gas-emissions.***

Fiez, T.E., B.C. Miller, and W.L. Pan. 1994a. Assessment of Spatially Variable Nitrogen Fertilizer Management in Winter Wheat. *Journal of Production Agriculture* 7(1): 86–94.

Fiez, T.E., B.C. Miller, and W.L. Pan. 1994b. Winter Wheat Yield and Grain Protein Across Varied Landscape Positions. *Agronomy Journal* 86(6): 1026–1032.

Fiez, T.E., W.L. Pan, and B.C. Miller. 1995. Nitrogen Use Efficiency of Winter Wheat among Landscape Positions. *Soil Science Society of America Journal* 59(6): 1666–1671.

Gantla, S., J. Gray, L. McNamee, L. Bernacchi, K. Borrelli, B. Mahler, M. Reyna, B. Foltz, S. Kane, and J.D. Wulfhorst. 2015. Precision Agriculture Technology and REACCH. Regional Approaches to Climate Change in Pacific Northwest Agriculture, Supported by a National Institute for Food & Agriculture Competitive Grant, Award # 2011-68002-30191.

Hoben, J.P., R.J. Gehl, N. Millar, P.R. Grace, and G.P. Robertson. 2011. Nonlinear Nitrous Oxide (N_2O) Response to Nitrogen Fertilizer in On-Farm Corn Crops of the US Midwest. *Global Change Biology* 17(2): 1140–1152.

Huggins, D. 2004. Application of Precision Farming Strategies to Direct Seed Systems: The Cunningham Agronomy Farm. Seventh Annual Northwest Direct Seed Cropping Systems Conference and Trade Show, Pendleton, OR. Steep III. ***http://pnwsteep.wsu.edu/directseed/index.htm***.

Huggins, D. 2010. Site-Specific N Management for Direct-Seed Cropping Systems. In Climate Friendly Farming: Improving the Carbon Footprint of Agriculture in the Pacific Northwest, Kruger et al., eds. CSANR Research Report 2010-001. Pullman: Washington State University. ***http://csanr.wsu.edu/pages/Climate-Friendly-Farming-Final-Report/***.

Huggins, D. 2015. PowerPoint presentation for Washington Grain Commission. Spokane, WA.

Huggins, D.R., and W.L. Pan. 1993. Nitrogen Efficiency Component Analysis: An Evaluation of Cropping System Differences in Productivity. *Agronomy Journal* 85(4): 898–905.

Huggins, D., W. Pan, and J. Smith. 2010. Yield, Protein and Nitrogen Use Efficiency of Spring Wheat: Evaluating Field-Scale Performance. CSANR Research Report, 2010-001.

Huggins, D.R., and D.P. Uberuaga. 2010. Field Heterogeneity of Soil Organic Carbon and Relationships to Soil Properties and Terrain Attributes. In Climate Friendly Farming: Improving the Carbon Footprint of Agriculture in the Pacific Northwest, Ch. 14, Kruger et al., eds. CSANR Research Report 2010-001. Pullman: Washington State University. ***http://csanr.wsu.edu/pages/Climate-Friendly-Farming-Final-Report/***.

Ibrahim, H.M., and D.R. Huggins. 2011. Spatio-Temporal Patterns of Soil Water Storage under Dryland Agriculture at the Watershed Scale. *Journal of Hydrology* 404(3): 186–197.

Jennings, M.D., B.C. Miller, D.F. Bezdicek, and D. Granatstein. 1990. Sustainability of Dryland Cropping in the Palouse: An Historical View. *Journal of Soil and Water Conservation* 45(1): 75–80.

Koenig, R.T. 2005. Dryland Winter Wheat: Eastern Washington Nutrient Management Guide. Washington State University Extension Publication EB1987.

Linsley, C.M., and F.C. Bauer. 1929. Test Your Soil for Acidity. Agricultural Experiment Station Circular 346. Urbana: University of Illinois. ***http://hdl.handle.net/2142/33105***.

Mahler, R.L. 2007. Northern Idaho Fertilizer Guide: Winter Wheat. University of Idaho Extension Publication CIS 453.

Mahler, R.L., S. Wilson, and B. Shafii. 2014. The Adoption of New Technologies in Dryland Farming Regions of Idaho, Washington, and Oregon. *Natural Sciences Education* 43(1): 102–108.

Mallory, E.B., R.J. Veseth, T. Fiez, R.D. Roe, and D.J. Wysocki. 2000. Ron Jirava Farm Case Study, Direct Seeding in the Inland Northwest Series. Pacific Northwest Extension Publication PNW528. Washington State University.

McBratney, A., B. Whelen, T. Ancev, and J. Bouma. 2005. Future Directions of Precision Agriculture. *Precision Agriculture* 6(1): 7–23.

McNeill, J.D. 1980. Electromagnetic Terrain Conductivity Measurement at Low Induction Numbers. Technical Note TN-6. Ontario: Geonics Limited.

Mulla, D.J. 1986. Distribution of Slope Steepness in the Palouse Region of Washington. *Soil Science Society of America Journal* 50(6): 1401–1406.

Mulla, D.J. 2013. Twenty-Five Years of Remote Sensing in Precision Agriculture: Key Advances and Remaining Knowledge Gaps. *Biosystems Engineering* 114(4): 358–371.

Mulla, D.J., A.U. Bhattia, M.W. Hammond, and J.A. Benson. 1992. A Comparison of Winter Wheat Yield and Quality under Uniform versus Spatially Variable Fertilizer Management. *Agriculture, Ecosystems & Environment* 38(4): 301–311.

NASA (National Aeronautics and Space Administration). 2016a. Earth Observatory. Measuring Vegetation (NDVI & EVI), Normalized Difference Vegetation Index (NDVI). ***http://earthobservatory.nasa.gov/Features/MeasuringVegetation/measuring_vegetation_2.php***.

NASA (National Aeronautics and Space Administration). 2016b. Electromagnetic Spectrum. ***https://mynasadata.larc.nasa.gov/images/EM_Spectrum3-new.jpg***.

National Research Council. 1997. Precision Agriculture in the 21st Century: Geospatial and Information Technologies in Crop Management. National Academies Press, Washington, D.C.

Pan, W.L., and A.G. Hopkins. 1991. Plant Development, and N and P Use of Winter Barley I. Evidence of Water Stress-Induced P Deficiency in an Eroded Toposequence. *Plant and Soil* 135(1-2): 9–19.

Pan, W.L., D.R. Huggins, G.L. Malzer, C.L. Douglas, Jr., and J.L. Smith. 1997. Field Heterogeneity in Soil-Plant Nitrogen Relationships: Implications for Site-Specific Management. In The State of Site-specific Management for Agriculture 81–99, F.J. Pierce and E.J. Sadlers, eds. ASA, CSSA, and SSSA, Madison, WI.

Pierce, F.J., and P. Nowak. 1999. Aspects of Precision Agriculture. *Advances in Agronomy* 67(1): 1–85.

Raun, W.R., and G.V. Johnson. 1999. Improving Nitrogen Use Efficiency for Cereal Production. *Agronomy Journal* 91(3): 357–363.

Reyns, P., B. Missotten, H. Ramon, and J. De Baerdemaeker. 2002. A Review of Combine Sensors for Precision Farming. *Precision Agriculture* 3(2): 169–182.

Rhoades, J.D., N.A. Manteghi, P.J. Shouse, and W.J. Alves. 1989. Soil Electrical Conductivity and Soil Salinity: New Formulations and Calibrations. *Soil Science Society of America Journal* 53(2): 433–439.

Rodman, A.W. 1988. The Effect of Slope Position, Aspect, and Cultivation on Organic Carbon Distribution in the Palouse. M.S. Thesis, Washington State University, Pullman.

Sehy, U., R. Ruser, and J.C. Munch. 2013. Nitrous Oxide Fluxes from Maize Fields: Relationship to Yield, Site-Specific Fertilization, and Soil Conditions. *Agriculture, Ecosystems & Environment* 99(1-3): 97–111.

Shcherbak, I., N. Millar, and G.P. Robertson. 2014. Global Metaanalysis of the Nonlinear Response of Soil Nitrous Oxide (N_2O) Emissions to Fertilizer Nitrogen. *Proceedings of the National Academy of Sciences* 111(25): 9199–9201.

Stombaugh, T., and S. Shearer. 2000. Equipment Technologies for Precision Agriculture. *Journal of Soil and Water Conservation* 55(1): 6–11.

Sulik, J.J., and D.S. Long. 2015. Spectral Indices for Yellow Canola Flowers. *International Journal of Remote Sensing* 36(10): 2751–2765.

Taylor, S.E. 2016. Precision Nitrogen Management: Evaluating and Creating Management Zones Using Winter Wheat Performance. M.S. Thesis, Washington State University, Pullman.

Tilman, D., C. Balzer, J. Hill, and B.L. Befort. 2011. Global Food Demand and the Sustainable Intensification of Agriculture. *Proceedings of the National Academy of Sciences* 108(50): 20260–20264.

Venterea, R.T., A.D. Halvorson, N. Kitchen, M.A. Liebig, M.A. Cavigelli, S.J. Del Grosso, P.P. Motavalli, K.A. Nelson, K.A. Spokas, B.P. Singh, C.E. Stewart, A. Ranaivoson, J. Strock, and H. Collins. 2012. Challenges and Opportunities for Mitigating Nitrous Oxide Emissions from Fertilized Cropping Systems. *Frontiers in Ecology and the Environment* 10(10): 562–570.

Ward, N.K. 2015. Improving Agricultural Nitrogen Use through Policy Incentivized Management Strategies: Precision Agriculture on the Palouse. M.S. Thesis, University of Idaho, Moscow.

Yorgey, G.G., S.I. Kantor, K.M. Painter, H. Davis, and L.A. Bernacchi. 2016. Precision Nitrogen Application: Eric Odberg (Farmer-to-Farmer Case Study Series). Pacific Northwest Extension Publication PNW691. Washington State University.

Chapter 9

Integrated Weed Management

Ian Burke, Washington State University
Kendall Kahl, University of Idaho
Nicole Tautges, Washington State University
Frank Young, USDA-ARS and Washington State University

Abstract

Integrated weed management (IWM) strategies are critical for effective long-term management of weeds in the agroecosystem. Knowledge of weed biology is critical for successful long-term IWM, as is integration of multiple methods of weed management. Methods of weed management include preventative, mechanical, cultural, and chemical inputs. Weed managers should develop a management plan that incorporates knowledge of weed biology, consideration of inputs, and effective method evaluation. A good competitive crop will always be the best weed management practice, and a sequence of successful crop rotations are critical for managing weeds in the inland Pacific Northwest (PNW).

Key Points

- An IWM approach depends on knowledge and application of ecological principles, an understanding of plant interference and weed-crop competition, and the appropriate use

Research results are coded by agroecological class, defined in the glossary, as follows:

● Annual Crop ▲ Annual Crop-Fallow Transition ■ Grain-Fallow

of preventative, cultural, mechanical, and chemical weed management strategies.

- Herbicides are an effective tool for managing weeds in inland PNW grain production, but they should be used judiciously and in combination with other strategies in order to implement a weed management program that is effective, economical, and prevents the development or spread of herbicide-resistant weed biotypes.
- Examples of IWM strategies for problematic weeds of inland PNW grain production are presented for downy brome, Russian thistle, jointed goatgrass, and Italian ryegrass.
- Anticipated climate change may impact weed management through earlier maturity of weed species, variation in environmental conditions affecting weed-crop competition, shifting ranges of weed species, and indirectly through changes in cultural practices and cropping systems.

Introduction

Effective weed management is achieved by manipulating the crop-weed relationship so that crop growth is maximized while weed growth is minimized or prevented. Weed control tactics are often applied singly without purposeful consideration of the elements of crop production contributing to weed control. IWM is a decision support system for assisting a grower in identification, selection, and use of weed control tactics singly or integrated into a management strategy. An IWM system typically consists of four components: (1) knowledge and application of ecological principles, (2) knowledge of plant interference and crop-weed competition, (3) use of **thresholds**, (4) integration of several weed control techniques, including selective herbicides (Zimdahl 2013).

Weeds are fundamentally different **pests** than insects or diseases. **Integrated pest management** approaches for insects and diseases often include development of host plant resistance or fundamental understanding of beneficial predator-prey relationships. No such efforts have yielded success for weed management in crop production—crop breeders do not specifically focus on improving crop competitiveness with weeds nor have we discovered ways to efficiently capitalize on plant defense mechanisms with weeds, such as allelopathy. Instead, growers have primarily relied on synthetic chemical weed management inputs and tillage to manage weeds on a broad basis, resulting in widespread **herbicide resistance** and soil erosion problems.

IWM systems include not only chemical inputs but also cultural and mechanical inputs (Figure 9-1). An IWM system considers the biology of the weed and crop, informing both short- and long-term management plans based on weed life cycle or seed longevity, for instance. IWM systems also consider the broader impacts of weeds on crop production—an integrated management strategy should include cost-benefit analyses that take into account not only the interests of and impacts on growers, but also on society and the environment (Norris et al. 2003). Such a system is not possible based on the available information and inputs, but rather represents an objective to aspire to. The objective of this chapter is to review the major components of an IWM system for the **inland PNW** small grains production region.

Knowledge and Application of Ecological Principles

Importance of Weed Ecology for Weed Management

Ecology is the study of the interactions between organisms and their environment. Weed ecology gives special emphasis to the adaptive mechanism that enables weeds to survive and prosper under conditions of extreme disturbance, ideally with the goal of identifying specific characteristics or traits targetable for management decisions. The most successful weed management programs are developed on a foundation of understanding weed ecology. Fundamental aspects of weed ecology that lend themselves to management include, but are not limited to, weed response to climate, weed life cycles, weed seed biology, and seed dispersal (Table 9-1).

Knowledge of weed ecology and life cycles enables managers to exploit vulnerable stages of a weed's life cycle and to use targeted control methods during those life stages. For instance, common lambsquarters cannot recover from mowing when they are small, but later in their life cycle they can resprout from the base after a mowing. Additionally, root reserves of

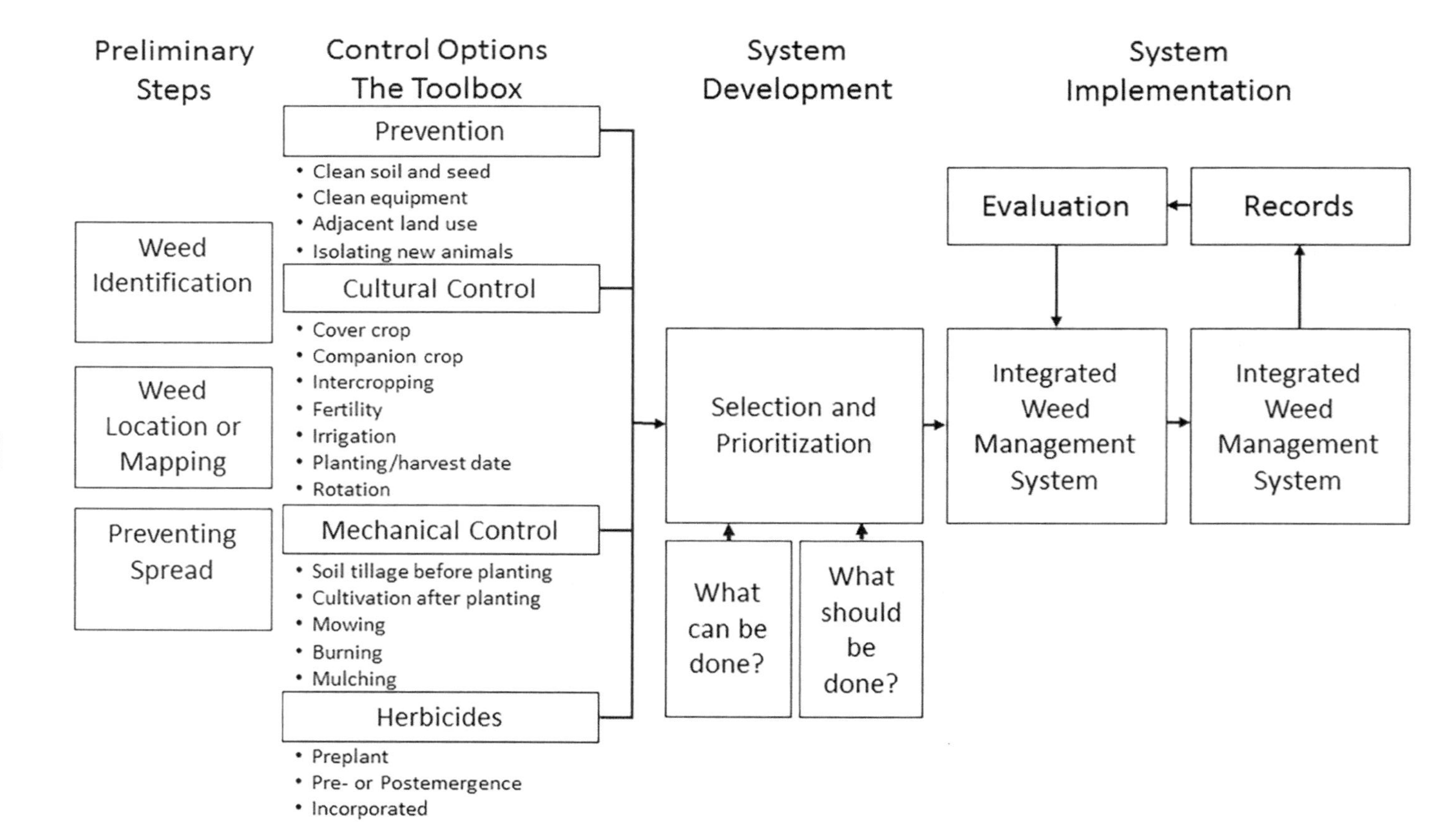

Figure 9-1. A conceptual model for an Integrated Weed Management System.

Table 9-1. Dispersal mechanisms, dormancy type and length, management to induce germination, seed production and seed persistence for common weeds in inland PNW dryland small grain production systems.

Weed	Dispersal mechanism	Dormancy type/length	Management–induced germination	Seed production	Seed persistence	
Grasses						
Downy brome	Humans/ livestock/simple dehiscence/ wind	Physiological/3 months to >1 year	Tillage/fire/ overgrazing	Up to 10,000 seeds per plant	<4 years	
Italian ryegrass	Humans/ livestock/simple dehiscence	Physiological/				
0 to 3 years	No-till	100,000 to 300,000 seeds per plant	Up to 3 years			
Rattail fescue	Humans/simple dehiscence	Physiological/0 to 2 years	No-till	0.6 to 11.2 g per plant	Up to 3 years	Ball et al. (2008)
Jointed goatgrass	Humans/ livestock/simple dehiscence	Physiological/0 to 1 year	Tillage	Up to 3,000 seeds per plant	Up to 5 years	Donald and Ogg (1991)
Cereal rye	Humans/simple dehiscence	Physiological/0 to 1 year	None	Up to 50 per plant	<2 years	

Table 9-1 (continued). Dispersal mechanisms, dormancy type and length, management to induce germination, seed production and seed persistence for common weeds in inland PNW dryland small grain production systems.

Weed	Dispersal mechanism	Dormancy type/length	Management–induced germination	Seed production	Seed persistence	
Broadleaves						
Common lambsquarters	Humans/ livestock/simple dehiscence	Physiological/0 to 3 years	Tillage, soil nitrate	Up to 176,000 seeds per plant	Up to 20 years	Harrison (1990); Bassett and Crompton (1978)
Prickly lettuce	Humans/ livestock/simple dehiscence/ wind	Physiological/0 to 3 years	Tillage/fire/ overgrazing	100 to 200,000 seeds per plant	Up to 2 years	Weaver and Downs (2003)
Russian thistle	Humans/simple dehiscence/ wind	Physiological/0 to 2 years	None	Up to 100,000 seeds per plant	Up to 3 years	Beckie and Francis (2009)
Kochia	Humans/simple dehiscence/ wind	Physiological/0 to 10 years	Tillage, overgrazing	2,000 to 25,000 seeds per plant	Up to 3 years	Friesen et al. (2009)
Redroot pigweed	Humans/ livestock/simple dehiscence/ wind	Physiological/0 to 3 years	Tillage, soil nitrate	200 to 100,000 seeds per plant	Up to 30 years	Costea et al. (2004)
Mayweed chamomile	Human/ livestock/water	Physiological/0 to 2 years	Tillage	5,000 to 17,000 seeds per plant	Up to 11 years	Gealy et al. (1985); Lyon et al. (2017)

perennial weeds can be depleted via multiple, repeated tillage or mowing operations during a season.

Weeds and Climate

Weed seed germination and plant development are strongly dictated by climate. The presence of soil moisture and sufficient diurnal fluctuation of soil temperature stimulate seed germination of many weed species. During vegetative growth, plants use temperature and day length as cues to begin the reproduction cycle. Seasonal variation in climate drives variation in weed management input timing in the same way it drives crop production inputs. Climate variability can be a contributing factor to seasonal successes and failures in weed management. Weed species express a large amount of phenotypic plasticity and are usually adaptable to a wide range of climates. **Adaptation** to climate likely occurs on a local level, including at the field scale.

Weed Life Cycles

Plant life cycles are classified into three types: annuals that reproduce by seed in a single calendar year, biennials that reproduce by seed in more than one calendar year, and perennials that live three or more years and reproduce by seed or perennial survival structures. Knowledge of weed life cycles facilitates identification of the approach and assessment of the commitment required to manage them. Annuals and biennials require fundamentally different approaches than perennial weeds because of their respective life strategies. Perennial weeds typically produce a lower quantity of seed, or sometimes none at all, relying on the perennial survival structures. Annuals, by contrast, reproduce strictly by seed. Seed management approaches can be quite different from perennial propagule management approaches.

Weed Reproduction and Dispersal

Weed species reproduce and disperse by seeds, with the exception of perennial weeds like field bindweed and Canada thistle. Weeds are capable of producing a vast quantity of seed. One of the most prolific seed producer in inland PNW agronomic systems is common lambsquarters, which is capable of producing up to 176,000 seeds per plant (common lambsquarters plants average far less when growing in competition with a crop, though). Grass weeds generally produce less seeds per plant, but often occur in high enough densities to produce very high seed loads per unit area. In agronomic settings where weeds are growing in competition with crops, weed seed production tends to be much lower, but still significant.

In addition to producing large amounts of seed, newly produced weed seed are usually initially dormant. Seed dormancy is essentially seed dispersal through time. Dormant seed, when presented with conditions for germination, do not germinate. The consequence of weed seed dormancy is that a grower is often managing weeds arising from seed set in multiple previous years. There are three primary types of dormancy: physiological, physical, and developmental. Weed seed with physiological dormancy require light to germinate. Physical dormancy is conferred through a hard seed coat that is largely impervious to moisture. Developmental dormancy, most common in grasses, is dormancy conferred when weed seeds are not fully developed when they are shed. Combinations of two or even all three can be found in plants. Managing weed seed with consideration of seed dormancy and seed numbers often involves thoughtful application of tillage to place seed where it cannot germinate, or to stimulate germination. For example, tilling at night in the absence of light nearly stops the germination of weed seed with physiological dormancy. When there is a failure to control weeds and a substantial amount of weed seed is deposited in the soil, inversion tillage could be used to bury the seed to a depth where it cannot successfully emerge. Repeated inversion tillage will bring the seed back to the surface, though. Alternately, tillage or irrigation can be used to stimulate germination prior to planting.

Finally, many weed seeds typically have specialized dispersal mechanisms that enable long-distance movement. Some of our most common weed species are tumbleweeds, where the entire plant is adapted for long distance dispersal via wind. Weed seed can also be transported by humans, particularly as a contaminant in crop seed, hay, and movement of implements, livestock, and other agricultural commodities. Some weed species, like jointed goatgrass and cereal rye, have seed of similar shape and size as wheat, making it very hard to separate the seed of the two species, which facilitates dispersal by growers. Weed seed production, dormancy, and dispersal mechanisms are key factors in the development of IWM strategies for a given weed species.

The Weed Seedbank

The species composition and density of weed seed in soil, called the **weed seedbank**, varies considerably and are tightly linked to past cropping history, seed durability and dormancy. The species found in the weed seedbank varies from field to field and even in areas within fields. Most weed seedbanks are dominated by a few species and these species represent the primary weed pests of the cropping system. A lower fraction of the seedbank—as much as 20%—is composed of species that were formerly dominant, potentially dominant, or simply adapted to the area but not current production practices (Zimdahl 2013). A small fraction of the seedbank consists of very old germinable seeds or newly introduced seeds, with the balance of the seedbank composed of seed deposited in the previous two years. New seeds can enter the seedbank following multiple pathways, but most seed is deposited in the weed seedbank from weeds producing seed within a field. Weeds growing with crops produce far less seed than weeds in open spaces due to crop competition, management inputs, and other factors. Although seed production in most weeds can be reduced by management inputs, seed production will likely remain substantial enough to maintain or increase the seedbank. Seed also enter the seedbank from external sources that can include farm equipment, contaminated crop seed, or long-distance dispersal by wind, animals, and water. Although typically much smaller numbers of seed are introduced from external sources, those sources are the way new species are introduced. Movement of weed seed by combines and other harvest equipment is of particular concern (Currie and Peeper 1988).

The abundance of weed seed in the soil can range widely and is dependent on past success or failure to manage weeds (Zimdahl 2013). Weed seed abundance can range from 300 to 350,000 seeds per square yard (Koch 1969), equating to 1.2 million to 1.4 billion seeds per acre. The abundance of weeds in the weed seedbank in the PNW are similar in size—densities have been reported ranging from ~300 to ~5,000 seeds per square yard in low precipitation zone field sites (Thorne et al. 2007). ■ As aboveground weed productivity increases, the composition of the weed flora and the seedbank changes as well. Mayweed chamomile and Italian ryegrass densities were as high as 28,274 and 77,547 seeds per square yard in fields near Pullman, Washington (Unger et al. 2012). ●

Once in the soil, a host of processes act to reduce the germinable seed bank. Not all of the seeds germinate at once. There is a growing list of seed predators, including earthworms, some carabid beetles and other insects, birds, and small mammals that are known to utilize weed seed as a food source. Seeds are also exposed to the same disease pressure as crop seed, and soil microbial community activity can play a large role in the persistence of certain weeds in the seedbank. Managing for weed seed disease and predator pressure can be difficult, is not well understood, and is an area of active research. Steep declines are observed in seedbank populations when effective integrated management strategies are employed. Unfortunately, even a 95% reduction in the seedbank results in a germination rate that is still a problem to be managed. Furthermore, if there is a single failure to manage a weed cohort in a season, the seedbank is effectively replenished from the resulting seed production.

Interactions between Weeds and Other Crop Pests

Weeds often act as alternate or reservoir hosts for diseases and insects. Management of weeds to facilitate management of diseases and insects is critical for a successful integrated pest management program (Cook et al. 2000). Consideration for the **green bridge** effect and weed-pest relationships are covered in other chapters of this book. In most cases, a good IWM plan, where weeds are effectively managed, will reduce or mitigate weeds as alternate hosts for diseases and insects.

Plant Interference and Crop-Weed Competition

Interference is an alteration of crop growth due to the presence of another plant. Interference is most commonly a negative effect, but it can be a neutral or positive effect in certain situations. Of particular concern to growers is the negative interference called competition.

Competition is where a weed utilizes water, light, or nutrients to the detriment of crop plants. Competition is strongly influenced by crop and weed density, but also by relative proportion and spatial arrangement of the species in the interaction. Density is the number of individuals per unit area, and proportion is the relative ratio or abundance of each species in the interaction. Spatial arrangement is usually quite random for weeds, but somewhat less so for crops. Competition is strongly related to vigor of the species involved and the timing of germination and establishment. Manipulating all of these factors purposefully with conscious knowledge of the biological relationships forms the basis of cultural weed management.

In order to relate the complex interactions associated with competition, weed scientists have devised a series of thresholds that relate the density of weeds or timing of competition to the crop yield loss that results. As weeds increase in density, the proportion of the total biomass production is shifted toward weed biomass. The resulting crop yield loss is quantifiable and can be assigned a monetary value. The basis for the array of thresholds is the law of constant final yield, which states that the biomass production of a given area is critically linked to the resources available, and that a wide range of plant densities (except for very low densities) will result in the same yield of biomass. By understanding and manipulating crop density and reducing or eliminating weed growth, crop biomass is the only biomass produced in a given area.

Thresholds in Weed Management

There are three primary types of thresholds: damage, period, and economic thresholds. Thresholds are expressed in density or weed biomass per unit area—in the inland PNW, biomass is often a stronger predictor of yield loss than density. Damage thresholds, the simplest type of threshold, quantify the weed population when there is a detectable negative crop response. Economic thresholds take damage thresholds and determine the economic damage that equals the cost of a management input. Economic thresholds assist growers in deciding on the return on their management input investment. Period thresholds define a time span during crop production where crops are **susceptible** to yield loss (as indicated in Figure 9-2, a conceptual period threshold for winter wheat).

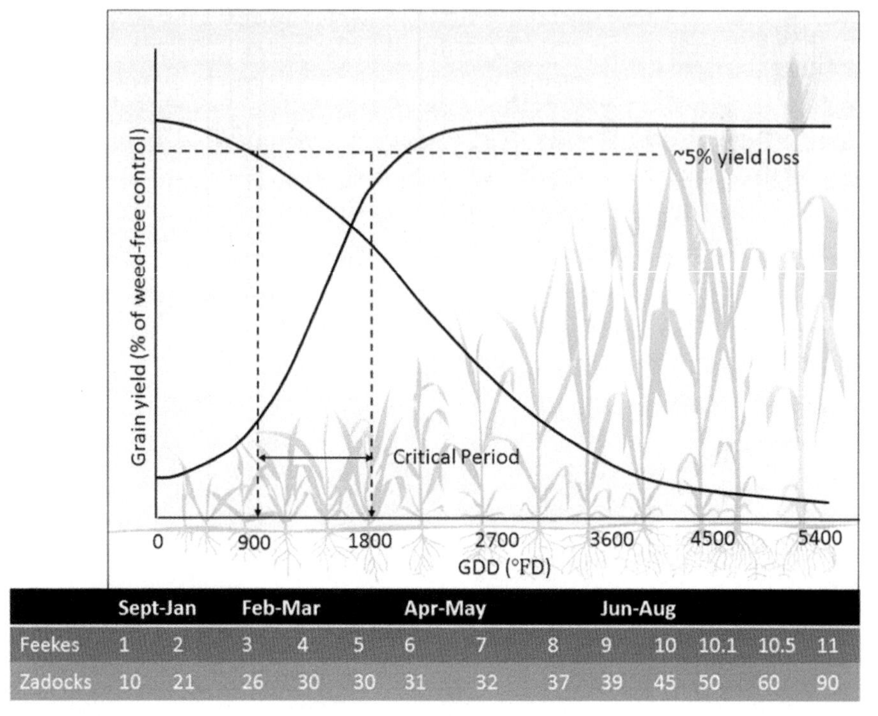

Figure 9-2. The critical period of weed control in winter wheat based on growing degree days, wheat growth stages, or calendar months. Year-to-year variation in climate, and climate change in general, can alter the relationship significantly. The critical period of weed control often occurs early in crop growth and development, and although weeds present in crops after the critical period can cause harvest losses and dockage, they won't reduce yield. The critical period of weed control is based on two components, indicated by the pair of solid black lines: the weed-free period and the weed-infested period. (Adapted from Welsh et al. 1999.)

Thresholds can change considerably from year to year even in the same crop as resources and growing conditions vary. Often, damage thresholds can be much less than 1 weed per square yard, and in any case, growers usually always apply a herbicide at some point during the growing season (99% of inland PNW wheat crops are treated with a herbicide during the growing season). Consequently, thresholds are seldom employed to make herbicide decisions in the inland PNW in the same way that thresholds are used to manage insect pests (see Chapter 11: Insect Management Strategies). Unlike insects and diseases, weeds always exceed thresholds stimulating treatment.

Weed Management Inputs

Preventative Weed Management Components of IWM

Preventative weed management refers to all activities that will reduce or eliminate the opportunity for new species to enter an area or for weeds present in an area to persist. Examples of preventative weed management include using clean weed seed-free crop seed for planting, cleaning harvesters and tillage equipment frequently and particularly between fields, and preventing the reproduction of weeds. The most recent development in preventative weed management is harvest weed seed control. A grower invention, harvest weed seed control involves managing seed that has not shattered out of the seed head. A wide range of approaches have been developed, largely in Australia, where early efforts focused on windrowing and burning chaff. Chaff carts of several different types were also developed, and the chaff was then burned. The most recent development is a chaff management system paired with a hammer mill that pulverizes the weed seed (Harrington and Powles 2012). The management of weed seed is an important yet unrealized opportunity for IWM systems in the inland PNW.

Cultural Weed Management Components of IWM

Crop rotation

Crop rotation is a critical component of IWM systems. Weeds typically associate with certain crops—wild oat or jointed goatgrass in winter or spring wheat, for example. Weeds can also adapt to management philosophies, like rattail fescue in **no-till** systems. When the same crop is produced repeatedly in the same area, weeds that associate with that crop typically increase, as the same environmental and cultural conditions that facilitate crop growth facilitate weed success. Over several rotations, the population of adapted weeds can become large.

Intensive crop rotation maximizes opportunities for varying competitive attributes among crop species (e.g., growth form and rate, life cycle length, water and nutrient use efficiency, and nitrogen fixation) (Buhler 2002), reducing the ability for weed species to become abundant. A diverse cropping system inherently includes varying seeding dates, crop life cycle, herbicide modes of action, herbicide application timing (pre-plant, postemergence, pre-harvest, or post-harvest), crop residues, and soil disturbance, and it provides a means of managing weeds by reducing weed densities and reliance on herbicides (Derksen et al. 2002). Extending rotations to 4 years or more improves suppression of most common weeds (Ogg and Seefeldt 1999; Tautges et al. 2016), particularly if rotation includes winter and spring varieties of both cereal and non-cereal crops (Blackshaw 1994; Derksen et al. 2002; Moyer et al. 1994). Different crops compete differently with weeds: barley, oats, and triticale are the most competitive, wheat and canola are marginally less competitive (a spring hard red wheat is the least competitive wheat), and pulses are among the least competitive crops. The more competitive the crop, the less dependent weed management is on other inputs, including herbicides.

Rotational flexibility is strongly influenced by climate and soil, and the climate in the PNW limits rotational opportunities to just a few crops (see Chapter 5: Rotational Diversification and Intensification). In the Grain-Fallow **agroecological class** (AEC) (a large area of central Washington and adjacent north central Oregon), climate (principally annual precipitation) limits rotation to primarily a winter wheat-summer fallow sequence where the summer **fallow** functions to facilitate weed management inputs and store a portion of the winter precipitation (Schillinger and Papendick 2008). ■ IWM approaches in grain-fallow rotations are in many ways similar to a more conventional two year rotation. In areas of the inland PNW with more precipitation, more intensive and complex rotations are practiced. Above ~15 inches of precipitation, a winter wheat-spring cereal or spring oilseed-summer fallow rotation is typical (Annual Crop-Fallow Transition AEC) ▲, and above ~18 inches of precipitation, crops are grown every year (Annual Crop AEC) ● and summer fallow is replaced by a spring pulse or spring oilseed rotation. As rotations are intensified and diversified, IWM systems become more complex and more effective.

Increasing in-season crop competitiveness

Plant characteristics that generally increase crop competitiveness with weeds during early stages of crop growth and development include rapid germination, root and early vegetative development and vigor, rapid canopy closure, increased leaf area index (unit leaf area per unit ground area), leaf duration, crop canopy height, and allelopathic properties (Buhler 2002; Callaway 1992; Pester et al. 1999). Rapid shading of the ground and crop resource capture increases early season crop competitiveness. Facilitating rapid canopy coverage of the ground reduces weed seed germination. Yield loss declines the longer weeds are kept from establishing and competing with the crop for resources. Increasing the competiveness of the crop is among the least expensive IWM inputs. Inputs that increase competitiveness are usually associated with "good farming" and are typically activities that would occur in the course of crop production. Thinking about how those very simple decisions, like choice of cultivar, seeding rate, and placement of fertility inputs, affect crop competitiveness and weed management are integral to an IWM system.

Competitive cultivars

The choice of crop cultivar can affect the competiveness of the crop. Cultivars often vary in early season vigor, leaf growth and size, and crop height-even how the leaves of wheat nod can affect competiveness. Plant breeders note such information but it is not widely reported, so growers are encouraged to explore variety trials not only for highest yield but for growth and development throughout the season when selecting crop cultivars for competitiveness. Often the highest yielding varieties, where the **harvest index** has been significantly shifted toward seed production, are not the most competitive. A competitive cultivar is typically combined with other cultural practices to encourage rapid crop emergence and canopy cover, ultimately resulting in increased resource capture by crop plants over weeds (Andrew et al. 2015; Harker et al. 2011). Sacrificing some yield potential by choosing a cultivar with a lower harvest index may result in greater yields because of less weed pressure over time.

Seeding rate

Most growers are familiar with the economically optimal population for crop production. Essentially, for a given piece of ground there is an ideal crop plant population that results in the greatest yield for the minimum investment in seed. Plant more seed above that, and a grower incurs greater expense in seed for no increased yield return. Plant below the economically optimal population, and yield declines (Schillinger 2005). The economically optimal crop population is usually lower than the most competitive population, often by 30 to 100%. Although seeding rate is usually optimized for wheat, pulse seeding rates are often too low for generating competition with weeds (Manuchehri 2012). Planting more seed and growing more plants per unit area reduces the space for weeds to establish, and intensifies the competition for resources not only between crops and weeds, but also among crop plants. Seed rate and crop population are among the most inexpensive ways to modify the competitiveness of crops with weeds.

Row spacing

Like plant population, optimizing row spacing functions to increase crop competitiveness by facilitating rapid canopy development. Row spacing is harder to manipulate as growers seldom are interested in purchasing multiple seed drills or planters to utilize different row spacing. Row spacing is set when the seed drill or planter is set up, usually based on experience or local practice, and all crops are planted to that row spacing. In the Grain-Fallow AEC, row spacing is wider than 15 inches as a moisture conservation tool. ■ In general, row spacing tends to decline as yearly precipitation increases, and the Annual Crop AEC in the inland PNW are planted to 7-inch rows or less. ●

For wheat, 7-inch row spacing in the high precipitation zone is ideal. There is a growing body of evidence that 7-inch rows are too wide for pulse production. In work in Australia, lentil yields were optimized and competition maximized using 3.5 inch row spacing. Dry pea would likely also benefit from planting in row spacing less than 7 inches (Borger et al. 2016).

Fertilization

Timing and location of fertilizer applications maximize nutrient availability to crop plants and reduce nutrient availability to weeds, improving the competitiveness of the crop (Buhler 2002). Banding fertilizer is the typical method used at the time of planting. Winter wheat yields and nitrogen uptake increase when fertilizer is banded 2 inches below seed; weed biomass and nitrogen uptake decrease under banded compared to broadcast applications of nitrogen (Cochran et al. 1990; Reinertsen et al. 1984; Veseth 1985). An in-depth discussion of fertilizer placement for crop production is included in Chapter 6: Soil Fertility Management.

Varying seeding dates

Timing of inputs primarily refers to delaying planting dates to manage weeds that germinate and emerge in the fall or spring. Delaying the planting date allows an application of tillage or herbicides to control weeds that emerge before planting the crop. Varying when many of the typical inputs occur can have a profound impact on weeds that germinate and emerge in a single flush of seedlings, but is less effective for managing weeds that germinate and emerge in multiple flushes. The Mediterranean climate in the inland PNW also limits the utility of delaying the timing of planting for weed management inputs, as a delay in planting usually results in yield loss.

Mechanical Weed Management Components of IWM

Mechanical weed management includes the familiar tillage, but also includes inputs like flooding and applications of heat. Stubble burning is a practice that was once widely applied in both wheat and grass seed production in the PNW. Burning stubble was usually applied as a stubble management input, but had the positive effect of significantly reducing the number of viable seeds. Effectiveness was related to the duration and intensity of the burn, but reductions in seed deposition and viability could be significant. Jointed goatgrass seed numbers were reduced 43% to 64% and reduced seed viability by 95% to 100% (Young et al. 1990; Young et al. 2010). Fire is widely employed in Australia, where growers windrow chaff and then burn it to destroy weed seed.

Tillage or cultivation of the soil is among the oldest forms of weed management. In addition to weed management, cultivation is an important input for seedbed preparation, improving soil physical conditions, precipitation **infiltration**, and incorporation of amendments like fertilizers or herbicides (Radosevich et al. 2007). Cultivation manages weeds by ripping, tearing, or burying them. Small weeds are more susceptible than large weeds, and annuals are usually more easily managed by cultivation than perennial weeds. Weather conditions can influence the outcome of a tillage input—warm, dry weather is preferable to dry the weeds out after cultivation. Tillage is classified into two types: primary and secondary.

Primary tillage is used to prepare the soil for planting. The equipment used in primary tillage can vary widely from very shallow tillage or deep tillage to break a plow pan. Primary tillage implements include the moldboard, disk, chisel, or subsoiler. When used properly, moldboard and disk plows can reposition weed seed on the surface of the soil to the depth of the tilled soil. Secondary tillage is used primarily for weed control, and as a consequence, timing of secondary tillage can vary widely. Secondary tillage equipment seldom disturbs more than the surface of the soil, as deeper tillage would damage crop roots. Tools for secondary tillage include different types of harrows, shovels, sweeps, chisels, rotary hoes, and rodweeders.

Weed control by cultivation is achieved by burying seedlings and small annual weeds with the soil thrown over them through the action of the tillage tools and uprooting them, or severing their roots, resulting in death by desiccation. Care must be taken in the secondary tillage operation so the roots or aboveground parts of the crop plants are not injured. Cultivation too late in the season may injure the root system and make the crop more susceptible to drought. Finally, cultivation may bring up weed seed in the soil profile and place them in a zone conducive to germination.

Secondary tillage has some advantages. In particular, there is a wide selection of implements, and large areas can be economically weeded. The disadvantage of machine tillage is that it does not control weeds growing in the seed row—a problem that may be solved by robotics. Weeds in crops planted in wide rows can be controlled reasonably well, but weeds within crop drill rows usually require hand hoeing or the use of specialized equipment. Depending on how aggressive the secondary tillage, it can control many small annual weeds, but does not work well on perennial weed species. In practice, the variable topography and recommended soil conservation practices (see Chapter 3: Conservation Tillage Systems) preclude the use of in-crop cultivation for weeding in the inland PNW dryland grain production areas.

Chemical Weed Management Components of IWM

Herbicides are sometimes erroneously perceived as the only solution to managing weeds, leading to the idea that weed management is a simple process. Although herbicides are a critical component for a successful IWM system, there are many factors that contribute to successful weed management, and some successful IWM systems do not use herbicides. Often, the element offering the most visible and striking contribution to weed control is the herbicide component of the system. However, when discussing chemical weed management, it is important to remember that many other factors contribute to weed management, including the integration of preventative, cultural, and mechanical inputs that have been previously discussed.

Herbicide selection should not be a decision made in haste. Herbicide selection should be part of planning for the entire crop production process. A major difficulty in selecting herbicides is the focus on trade names over active ingredient names. The proliferation of trade names—the commercial identifier or trademark designation—causes confusion because a single active ingredient can be sold under a variety of different trade names. Roundup, for instance, is a ubiquitous trade name, but the common name (or active ingredient) is glyphosate. The active ingredient is what is responsible for phytotoxicity in a herbicide formulation. While trade names can change, the active ingredient in a herbicide does not change. Many products are available that contain one or more of the same active ingredients for use in wheat. As a result, knowing the common name for a herbicide provides a basis for assessing products and finding the most cost-effective source of a herbicidal active ingredient.

Herbicides are important inventions with significant positive and negative attributes. First and foremost, herbicides facilitate the management of weeds in the drill row where mechanical inputs fail to reach. Some are selective and, when used appropriately, do not harm the crop or other desirable organisms, as crops are **herbicide tolerant**. Herbicides have facilitated the adoption of minimum or no-till systems on a regional and national basis. They reduce time and labor and facilitate early planting by allowing growers to manage weeds when tillage cannot be performed because of soil moisture. Herbicides have dramatically increased the area a single person is able to farm—they are highly efficient labor-saving tools. Herbicides have negative attributes as well, including the potential to injure the crops in which they are used and the potential for off-target injury due to movement away from the site of application. Herbicide use can restrict rotational crops, often significantly. Herbicides have contaminated ground and surface water in areas of the US. Residues of herbicides can remain in and on the crops in which they are used. The large majority of pesticides used in the US are herbicides, and it is testimony to their safety and efficacy that the vast majority of herbicide applications are applied safely and effectively.

Herbicides are used primarily as broadcast applications in the inland PNW. They are usually applied preplant (before the crop is planted), preemergence (after the crop is planted, but before emergence), or postemergence (after crop and weed emergence). Spot treatments are also common. Less commonly used are band applications (applied over the crop row only) or directed applications (applied around the base of the crop, where the crop has a woody stem). Postemergence herbicides are applied with surfactants, spray additives that facilitate transport across the cuticle of weeds.

Herbicides are classified by their mode of action. Mode of action refers to the way a herbicide affects a plant. Inland PNW crop protection systems rely on a very narrow subset of the available herbicide modes of action: the growth regulators, the amino acid synthesis inhibitors, the fatty acid synthesis inhibitors, the photosystem inhibitors, the PROTOX inhibitors, and the seedling growth inhibitors. Knowledge of herbicide modes of action is useful for managing the risk of developing herbicide resistance (see Table 9-2 to assess likelihood of herbicide resistance development on a per species basis). Managing herbicide resistance is critically important, as no new herbicide modes of action have been discovered since the early 1990s.

Herbicide resistance testing, and understanding mechanisms of resistance to herbicides, will be critical to future weed managers and IWM systems. Herbicides are currently applied to entire fields with little knowledge of the status of resistance within the fields. Instead, field managers and growers

Table 9-2. An assessment matrix for likelihood of herbicide resistance development. To self-assess, it is critical to know how many herbicide active ingredients in a mixture have activity on the weed being managed, and ultimately how many modes of action are employed for the control of a single species.

Management Option	Low	Moderate	High
Herbicide mix or rotation in cropping system	>2 modes of action	2 modes of action	1 mode of action
Weed control in cropping system	Cultural, mechanical, and chemical	Cultural and chemical	Chemical alone
Use of same mode of action per season	Once	More than once	Many times
Cropping system	Full rotation	Limited rotation	No rotation
Resistance status to mode of action	Unknown	Limited	Common
Weed infestation	Low	Moderate	High
Control in last 3 years	Good	Declining	Poor

Adapted from Moss 1998.

select herbicides based on cost and on broadly successful treatments. In the future, managers will likely be testing for resistance broadly and using a range of old and new herbicides based on knowledge of the response of different populations within fields. Employing generic inexpensive herbicides where they are effective, and doing so with knowledge of the resistance status of the weed being managed, has the potential for reducing the economic cost of managing weeds.

Societal and Environmental Considerations

Public perceptions and expectations often influence the management tactics that are available to growers of a given commodity. For example, in the inland PNW, transgenic glyphosate-resistant wheat was discarded as a management tactic because of the inability to manage volunteer transgenic wheat in rotation. There were also concerns on the part of the primary overseas markets, who indicated that they would not accept transgenic wheat. Public policy at multiple levels can also be influenced by public perceptions and expectations—sometimes such influence is driven by emotion rather than science. Such influence can limit tactics available for use in IWM systems. Educating the public on the scientific basis of IWM is critical for enabling the greatest possible suite of inputs and limiting misdirected public policy (Norris et al. 2003).

Pesticide use comes with risk. Herbicides are strictly regulated, and the herbicide label sets forth directions for use that ensure that residues that enter the food system are only present at, or lower than, EPA-mandated levels. Although chronic long-term exposure to traces of correctly applied pesticides can be of concern, the risk associated with most exposures is viewed as very low (Norris et al. 2003). Food-borne diseases, malnutrition, non-pesticide related environmental contaminants, and naturally occurring toxicants are all considered more important than pesticide residues when making a risk versus benefit analysis. Nevertheless, public perception of pesticides continues to be negative. It is critical that those that use pesticides do so according to label directions.

Environmental issues in the inland PNW related to IWM are considerable. The practice of tillage in the inland PNW directly contributes to the highest rates of erosion in the US. As a result of concerns over airborne dust, PM10 standards (particulate matter standards that apply to particles 10 micrometers or less) were developed that regulate the reduction of tillage, increasing reliance on herbicides. Herbicide drift is a major concern in the inland PNW. As orchard and vineyard acreage increase, limits on the application of volatile formulations of herbicides are increasing. Certain volatile formulations of 2,4-D are currently prohibited in the state, and there are very specific limitations on when less volatile formulations can be applied. The burning of agricultural fields, a weed and crop residue management practice, has been very tightly regulated, limiting the length of grass seed field productivity. When a weed management tactic is lost, it often places pressure on alternative tactics, or there is a reduction in productivity until effective alternatives are found. Ultimately, social concerns have had, and will continue to have, a profound influence on IWM systems in the inland PNW.

Developing an IWM System

A systematic approach to developing a weed management plan should be followed to implement an IWM system (Figure 9-1). Steps to developing and executing a weed management plan could include (1) problem diagnosis, (2) program selection, (3) program execution, and (4) method evaluation (Ross and Lembi 1999).

Problem Diagnosis

The first step in designing an IWM system is a systematic review of both the problem weeds to manage and the environmental conditions they are to be managed in. Although not exhaustive, the list should capture most of the important factors for consideration. A practitioner is encouraged to identify additional components based on local knowledge.

Weed identification

Identification of the species present in the area to be managed is critical for success. As noted previously, knowledge of the biology and ecology of the weed species facilitates biologically based management strategies.

Weed abundance

The abundance and economic importance of each species in the field, ideally based on economic thresholds, should be determined. Identification of both important species with high abundance as well as those that are in small areas or in low abundance should be performed. Scouting in-season and in the fall is a critical part of the process as weeds present at harvest will likely produce seed and thus be problems for at least the following season (usually for at least 2 years, and sometimes more than 10 years). Scouting also facilitates an early detection rapid response approach. Depending on the weed complex and the production system, management options may be very limited. Fields usually contain a mixture of broadleaf and grass weeds. Interestingly, an effective IWM system is often indicated by low populations of a diversity of weed species.

New weeds

Weeds not present in an area or field before must be recognized and managed. Small patches of a new species is indicative of an invasion process just beginning. A preventive program is critical to managing weed species that could become serious weed problems, such as Italian ryegrass, downy brome, jointed goatgrass, prickly lettuce, mayweed chamomile, field bindweed, Russian thistle, or common lambsquarters, and will be an excellent return on the investment in time and energy if these weeds are prevented from invading a non-infested area. Growers cannot assume that these weeds will be easily and economically controlled with modern herbicides. For example, numerous biotypes of Italian ryegrass **resistant** to both ALS (Group 2/B) and ACCase (Group 1/A) herbicides occur in the high precipitation zone. Widespread resistance to the Group 2/B herbicides also occurs in downy brome. Mechanical and cultural methods have to be employed to manage weeds resistant to available herbicides.

Soil

Understanding soil chemistry is essential when working with soil-active herbicides. Additionally, certain weeds associate with specific soil environments or soil types—some weeds can tolerate acidic soil conditions, for example. Soil texture (% sand, silt, and clay) affects soil **water holding capacity** and, as a consequence, the rate of herbicide and water movement. Soil conditions while planting, tilling, and applying herbicides can be critical for success of the input. Surface moisture, plant residue, and soil surface roughness all affect the outcome of a soil-applied herbicide. Different soil types within the same field can result in reduced selectivity in areas of the field where the binding of the herbicide to soil is weak. Plants under water stress do not respond to herbicides, and functional soil drainage in wet areas would be considered an input for weed management.

Texture and organic matter

Consider the soil texture and organic matter. Both can impact the safe use of soil-applied herbicides, particularly in large fields with variable soil found in the PNW. Eliminate those choices that do not fit the soil type or types prevalent in a given field. Guidance on use of inputs based on texture and organic matter are found on herbicide labels or in Extension publications.

Soil pH

The soil **pH** determines the chemical charge of some herbicides, and each of those herbicides responds to pH in a different way. Soil pH strongly affects the amount of dissolved herbicide in the soil water fraction. In general, herbicides sensitive to soil pH vary in their solubility, and pH can affect the persistence, or lack of persistence, of the herbicide in the environment. Although too complex to fully address here, a PNW publication is available (Raeder et al. 2015).

Erosion potential

The inland PNW has some of the highest soil erosion rates in the world, so systems for maintaining crop residues in place are common. However, maintaining crop residues often precludes the use of machine tillage inputs, with a few exceptions. Herbicide incorporation by tillage cannot be used in no-till production. No-till or minimum tillage systems increase dependence on herbicides. Eliminating soil disturbance often changes the species of weeds to be managed.

Crop rotation

The previous and planned future crop rotations are critical to consider. Integral to any crop rotation is the herbicide use in each rotation, as certain herbicides can limit rotational flexibility. Canola is an excellent example of a crop that is highly sensitive to certain herbicides. Crop rotation is essential for IWM, and growers are encouraged to use methods that allow them to be flexible.

Program Selection

Selection of inputs should be based on effectiveness of weed control and cost. Rather than relying on a single input like a herbicide, the goal is to purposely choose and apply a variety of components (Ross and Lembi 1999). Consider the following when selecting elements of an IWM program.

Economic return

The cost of the various inputs should be considered in the context of economic return. Using economic thresholds can help assess the return on investment of various IWM system components.

Management system

The input needs to be compatible with the current system, and the manager needs to have the capability to execute the input.

Equipment

Equipment needs to be available and of a size that makes the operation feasible in a limited time period.

Custom services

Both field managers and consultants need to be reliable.

Time

Can the input be effectively applied over the required area in the window of opportunity available?

Operational capability

Can staff accomplish the operation?

Crop and management system

Is the input compatible with the long term rotational plan?

Identification of the problem weeds to be managed the following year facilitates a narrowing of management options. For example, the list of prospective herbicides for a crop could be narrowed on the basis of available herbicides that actually have activity on the weeds present in the field. Charts of herbicide efficacy are usually available (see the PNW Weed Handbook or the WSU Small Grains website in Additional Resources for such charts). Create a list of active ingredients that best fit the weeds to be managed, and then identify the herbicide products that contain those active ingredients. The herbicides should be selected on a field-by-field basis—not the entire farm (in the future it will be on a subfield basis). The use of specific herbicide programs for individual fields may not be practical, and growers usually group fields with similar weed problems. However, given the size of farms in the PNW, the blanket application of a few products to an entire farm is likely wasteful. Modern farm management software should allow growers to quickly and easily target problem fields with more expensive herbicides and manage less weedy areas with less expensive but equally effective herbicides.

Value of early season programs

Early season programs are critical for spring-sown crops, as weeds germinate and emerge at nearly the same time as the crops. Management inputs should be planned to keep spring-sown crops weed free for the first weeks of growth. Pulses often require a weed-free period of 6 weeks or longer. Ideally, the plan should include alternatives for if management inputs fail in the first few weeks of crop growth.

Rotation of herbicide modes and sites of action

Use herbicides with different modes and sites of action for weed management. There is a growing body of evidence that using herbicides with different modes of action that have activity on the same weed is a better resistance management strategy than rotating between modes of action. Although there are not enough options to effect this strategy for grass weed control, it is easily accomplished for broadleaf weed control.

Herbicide resistance is a very complex phenomenon, and rotation of herbicide families or modes of action is an oversimplified solution. Many herbicides and modes and sites of action are used on more than one crop. For example, growing conventional canola and a pulse in rotation is not an effective herbicide rotation because the same herbicide mode of action is employed to manage grass weeds. Using two herbicide modes of action to control each weed species is the ideal approach to manage resistance, but such an approach is not possible in some cases. Routine herbicide resistance testing is part of the IWM approach for wheat producers in Europe and Australia and should be considered essential for PNW growers, too. Testing for resistance allows growers to choose herbicides with the knowledge that they will be effective.

Herbicide resistance

Managing for herbicide resistance is often only addressed after resistance develops. Resistance to herbicides is a symptom of a flawed IWM system.

The solution is often a careful evaluation of the system and deployment of additional control tactics (e.g., doing what should have been done prior to the development of resistance). The list of management inputs to choose from does not change once herbicide resistance occurs. Therefore, growers are encouraged to be proactive. Develop and execute an IWM plan before resistance occurs. A herbicide-resistant weed is a much higher priority for management.

Program Execution

Execution is a critical step in an IWM system. Three factors are essential: (1) operations must be completed in a timely fashion, (2) the right equipment must be used, and (3) equipment must be correctly operated (Ross and Lembi 1999).

Appropriate follow-up

Monitoring the outcome of weed management inputs should be planned as part of the system. By carefully recording and observing the outcome of each management input, growers can assess whether to continue the management practice or to alter or forgo it altogether.

Weed managers are encouraged to be realistic and to recognize that not all the weeds need to or can be controlled. By integrating multiple management inputs into a complex program, weed management can be achieved successfully and for multiple seasons. Managers are often overly optimistic and rely on a few inputs over a long period of time. Products and technologies are often over-promoted, and managers always seem to be interested in the one input that will take care of everything. A sensible, rational, realistic long-term IWM plan is the most effective strategy for managing weeds.

Method Evaluation

A careful assessment of currently available management inputs is the final step in devising an IWM plan. There are a wide range of resources available, including Extension material based on field trials. Weed scientists, agricultural Extension specialists, agricultural consultants, and industry representatives are excellent sources of information on new or novel management inputs. The inputs should be evaluated for specific criteria (Ross and Lembi 1999).

Past weed management systems and results

A careful evaluation of past experience must be conducted every year. Management inputs that were effective should be retained, while those that were not effective should be reconsidered. Careful assessment of weed populations at the field level should be mapped, and the mapped populations monitored. Shifts in weed species composition are often observed when a change in management tactics is applied. Often it takes 10 years or more for the shift to occur.

Effectiveness

Consider the effectiveness on each weed species present.

Consistency

How consistent is the outcome? Variation in the outcome of a system is an indication that the system is vulnerable to failure. Sometimes, evidence of consistency is as subtle as choosing to use a higher rate of a herbicide because of a sense of failure from the previous year or application.

Fit within the individual system

Will the management input fit into the current system? Incompatibility comes in many forms. Tillage is, of course, incompatible with no-till systems. Pesticide compatibility can also be a concern, particularly when systems use a single pass to apply multiple pesticides. Although the pesticides may be compatible in mixture, the complexity of such mixtures can complicate troubleshooting when there is problem, and there is increased risk of injury and reduced weed control.

Flexibility

If the timing of an input is critical for success, consider that weather in the inland PNW is quite variable, and the window of opportunity for application of a management tactic is often only a day. Can the input be applied in such a tight window based on climate and crop growth and development?

IWM of Selected Problematic Weeds of the inland PNW

Downy Brome ▲ ■

Downy brome (*Bromus tectorum*) is a winter annual and **facultative** spring annual grass that develops early in the season, usually flowering in April and May. Downy brome reaches seed maturity at an average of 1,000 **growing degree days** (a measure of thermal time) in the inland PNW (Ball et al. 2004; Lawrence and Burke 2015). No-till spring cropping helps control downy brome because downy brome plants that establish during the fall and winter can be controlled with a herbicide or tillage before spring seeding (Thorne et al. 2007). Average downy brome seed persistence is 2 to 3 years (Thorne et al. 2007), though some seeds can persist longer. Dormancy is thought to be more complex than previously realized, and there is a wide range of dormancy periods (Hauvermale et al. 2016). Downy brome seed deposited in the soil seedbank can be reduced when light tillage is used in combination with fall or spring herbicide applications (Yenish et al. 1998; Young et al. 2014), but such an approach seldom completely solves the problem.

Cultural management of downy brome usually requires manipulation of crop rotation, tillage, and nutrient management. Growers should avoid excessively early seeding because it can promote yield-reducing disease and insect pests in grain-fallow systems (including barley yellow dwarf, wheat streak mosaic, dryland foot rot, Cephalosporium stripe strawbreaker foot rot, stripe rust, Russian wheat aphid, greenbug aphid, and others, see Chapter 10: Disease Management for Wheat and Barley and Chapter 11: Insect Management Strategies). If rains occur just before the planned planting date, delay seeding, wait for downy brome to emerge, and apply a non-selective herbicide or tillage before seeding. However, avoid seeding later than the optimal planting date to avoid yield loss due to other climatic and growing season conditions (Yenish et al. 1998).

General IWM principles should be applied: plant competitive, vigorous varieties; avoid using excessive nitrogen; top dress only when needed; and deep band nitrogen to limit availability to downy brome. Crop rotation to spring crops or fall-seeded broadleaf crops, such as winter canola or winter pea, facilitates use of in-crop grass herbicides in order to minimize downy brome seed production (Yenish et al. 1998). Alternative crop/fallow rotations for downy brome management include winter wheat-fallow-spring wheat, winter canola-fallow, or other permutations of cropping sequence that facilitates the use of effective spring inputs. Rotating out of winter cereals for a minimum of 3 years is key to managing infestations (Yenish et al. 1998).

Use a combine chaff spreader to distribute seed, and then harrow or perform other light tillage (tine harrows or skew treaders) soon after harvest to increase seed-soil contact and subsequent germination when fall precipitation occurs. Wheat yield loss is most severe when downy brome germinates within 21 days of wheat emergence; beyond this point, only extremely high-density downy brome reduces wheat yield (Rydrych 1974; Blackshaw 1993). When downy brome emerges within 14 days after wheat emergence, downy brome densities of 24 plants per yard and 65 plants per yard can reduce wheat yield by 10% and 20%, respectively (Stahlman and Miller 1990).

Integrated management of downy brome using herbicides is very challenging due to the lack of options. Herbicides for downy brome management are limited to the Group 2/B herbicides: pyroxsulam, sulfosulfuron, propoxycarbazone, mesosulfuron, chlorsulfuron, and imazamox, which form the basis for herbicidal control of downy brome. Interestingly, all Group 2/B herbicide products are labeled for control in fall applications, but only spring suppression. Most growers apply these herbicides postemergence in the spring to control both fall- and spring-emerging downy brome, and to ensure that replanting can be accomplished if needed. Suppression is the most common outcome of such applications.

Applying a downy brome growing degree days model to climate change model projections for the inland PNW indicate that downy brome may reach seed maturity between 15 to 25 days earlier by mid-century (2031–2060) (Lawrence and Burke 2015). Late flowering biotypes of downy brome are predominantly located in the Palouse region (eastern Washington). Early flowering biotypes dominate the central Washington wheat region. Eastern Washington (late-flowering biotype region) is projected to undergo the greatest amount of change in growing degree day accumulation. Early flowering biotypes may spread to the east due to the changes in growing degree day accumulation, as they are better adapted to warmer springs and milder winters. Several early flowering biotypes have herbicide resistance to ALS-inhibiting herbicides (Lawrence and Burke 2015). Earlier downy brome development and greater spring precipitation may limit the opportunity for in-field spring treatment of downy brome. Herbicide effectiveness may also decline with the spread of resistant biotypes (Lawrence and Burke 2015).

Russian Thistle ▲ ■

Russian thistle (*Salsola tragus*) is a summer annual broadleaf weed. Russian thistle is most troublesome in spring crops but can be a problem in winter wheat and during fallow periods (Young 1998; Schillinger and Young 2000; Thorne et al. 2007). Preventive management of Russian thistle is important, and includes controlling populations along borders and non-cropped areas because of its "tumbleweed capability" of very long distance dispersal (Young et al. 1995). Russian thistle germinates in the early spring through late summer, and flowers all summer and into fall until a killing frost (Young et al. 1995). Matured seed requires a fall after-ripening period. After an after-ripening period, Russian thistle seed can germinate under a wide range of conditions in the spring, even at relatively low temperatures (37°F to 42°F) (Young and Evans 1972; Thorne et al. 2007; Young and Thorne 2004). Russian thistle can be controlled within a few years by preventing seed production (Thorne et al. 2007), assuming no new introductions occur. The majority of seed viability declines within 2 years—management should focus on preventing seed production or new introductions (Young et al. 1995). A significant amount of flowering occurs after harvest, and Russian thistle can regrow after being cut by a combine. Russian thistle exhibits an indeterminate growth habit, and will continue to grow and set seed until temperatures drop below 25°F.

Russian thistle has high **water use efficiency**. If left to grow post-harvest, Russian thistle can reduce soil water storage for the next crop in rotation (Young et al. 1995) and usually causes the greatest damage under drought conditions, in thin crop stands, or if the crop is planted late (Young et al. 1995). One Russian thistle plant can use up to 18.5 gallons of soil water while growing in a spring wheat crop. Post-harvest growth uses an additional 26.5 gallons and accumulates significant biomass until killing frost occurs late October (Schillinger and Young, 2000).

In-crop management is usually accomplished with non-selective and selective herbicides, depending on timing. Both pre-harvest and post-harvest control is achieved with a non-selective herbicide. During fallow, control is achieved with herbicides or with tillage, usually with a rodweeder or an undercutter sweep (usually operated to a depth of 4 inches, which is more effective than 2 inches). Timely application of tillage is essential, as larger Russian thistle can survive rodweeding.

Management in the fallow year usually includes spring non-inversion tillage with wide-blade sweeps followed by 2 to 3 secondary treatments with a rodweeder to help retain surface residue (Schillinger and Young 2000). For spring crops, spring wheat should be planted early and in narrow row spacing (6 to 9 inches) to increase crop competitiveness. Spring wheat cultivars with rapid and prostrate early growth increase spring cropping for controlling Russian thistle (Schillinger and Young 2000). Post-harvest treatment using an undercutter V-sweep can consistently kill Russian thistle, eliminate seed production, retain more soil water in the fall post-harvest, and produce greater spring wheat yields in the following year compared to using only a post-harvest treatment of paraquat + diuron, alone. Paraquat treatment allowed greater soil water extraction post-harvest and allowed production of an average of 370 seeds per square yard from lower branches where herbicide did not penetrate (Schillinger 2007).

Russian thistle may not be as great a problem in no-till spring crop rotations as it is in tillage-based spring crops. Summer annual broadleaf weeds are usually most successful in systems that include intensive tillage (Derksen 1993; Thorne et al. 2007).

Sensor-based (e.g., a Weed Seeker by Patchen or other systems) herbicide applications are feasible for post-harvest control of Russian thistle in arid and semi-arid regions of the PNW (Riar et al. 2011). Greater than 90% control was achieved with a light-activated sensor-controlled sprayer using paraquat + diuron, and chemical use was reduced by 42% compared to broadcast applications. Control with glyphosate + 2,4-D was unacceptable regardless of applicator type. Sensor-based technology can be easily calibrated for Russian thistle because its bright-colored leaves contrast greatly with the brown-gold of the soil-wheat stubble post-harvest background. Small plants (<3 inches tall and <1.5 inches diameter) were missed by the sensor applicator, but these plants were not mature or flowering, and additional growth or seed production after the herbicide treatment was thought to be insignificant (Young et al. 2008).

Jointed Goatgrass ● ▲ ■

Jointed goatgrass (*Aegilops cylindrica*) is a winter annual grass weed of importance in small grain production systems because it is closely related to wheat. Jointed goatgrass, as a winter annual weed, is very competitive with winter wheat. Climate can affect how jointed goatgrass competes with wheat. In a dry year, winter wheat yield loss ranged from 55% to 84%, while in a wet year, grain yield was reduced by 30% to 40% (Ogg and Seefeldt 1999).

Rotations that avoid winter cereals for more than 3 years in wet climates (>18 inches precipitation) ●, or more than 6 years in drier climates (<18 inches precipitation) ■ are required for reducing goatgrass seed germination to less than 0.1% (Cook and Veseth 1991). Integrated planting strategies for winter wheat that can help reduce jointed goatgrass in grain-fallow regions include planting competitive varieties, increasing seeding rates and seed sizes, and fertilizing nitrogen, sulfur, and starter phosphorus with the seed (Young et al. 2010). ■ Combining a tall wheat cultivar and increased seeding rate with nitrogen fertilizer banded by the seed reduced jointed goatgrass densities 45% to 60% compared to conventional practices (Young et al. 2010). In all regions, the spring wheat-fallow-winter wheat rotation reduced jointed goatgrass **dockage** most consistently (Young et al. 2010). In the PNW, winter wheat varieties with rapid height gain and greater height are more competitive against jointed goatgrass and reduce jointed goatgrass seed production, particularly in drier years (<12 inches precipitation) (Ogg and Seefeldt 1999). ■

Spring crop rotations are important for jointed goatgrass management and are effective at preventing viable weed seed production. Spring-germinated jointed goatgrass does not compete well with spring wheat. Delaying spring wheat planting by 13 to 15 days generally did not decrease yields significantly and has been shown to prevent the production of viable seed from spring-germinated jointed goatgrass in plot studies in various precipitation zones across the PNW. Use of fall and spring tillage and herbicide applications before planting spring crops controlled jointed goatgrass plants and spikelets over time (Young et al. 2003)

Jointed goatgrass can be selectively managed by growing imazamox-resistant wheat without causing crop injury (Ball et al. 1999). There are considerable plant-back restrictions in the inland PNW following an application of imazamox, effectively forcing the grower to use imazamox-resistant rotational crops to minimize crop injury in rotation. The system is also expensive. As a consequence, it is often far more economical to hand-weed small populations when discovered and only use imazamox-resistant wheat for large, dense populations as part of an integrated management strategy.

Italian Ryegrass

Italian ryegrass (*Lolium perenne* var. *multiflorum* Lam.) is a cool season bunchgrass and a major annual weed in inland PNW cropping systems (Hulting et al. 2012). Italian ryegrass causes economic losses because it competes with winter wheat, can contribute to cereal lodging, and can cause lower harvest grain quality and higher dockage (Hulting et al. 2012).

Long-term planning is essential to managing Italian ryegrass. It is found primarily in the high precipitation zone in eastern Washington. The seed are viable for 3 to 5 years, depending on placement in the soil. Deep burial will likely reduce seed persistence, but shallow incorporation may increase it. The seed are moderately persistent on the seed head, making it feasible to destroy the seed after it is set but before dehiscence. An infestation can result from germination of only 2% to 4% of seed present in the soil (Unger et al. 2012). Italian ryegrass is competitive with winter wheat for nutrients, water, space, and light (Carson et al. 1999; Hashem et al. 1998; Hulting 2014). Cultural inputs for Italian ryegrass management are necessary for satisfactory management. Crop rotation is the most effective input for managing Italian ryegrass, and a 4-year or longer rotation is likely required. However, no long-term rotational studies have been conducted to address Italian ryegrass management in the inland PNW. The winter wheat-spring wheat-spring pulse crop rotation is likely inadequate for managing ryegrass. A second consecutive year of a broadleaf crop, like canola, would increase the efficacy of crop rotation for management of ryegrass, but only if the ryegrass is effectively managed in the pulse and spring wheat crops. Glyphosate resistant spring canola is an excellent tool for Italian ryegrass management in a rotation. Conventional canola is no different than a pulse since the same herbicide mode of action would be used in each crop.

Planting date, seeding rate, row spacing, a competitive cultivar, and fertilizer placement all play a role in managing Italian ryegrass. Growers tend to think in terms of seed weight per acre but are encouraged to begin to think about plants per square foot, and the seed required to achieve a competitive crop plant density. The crop plant population required for optimal economic return is usually considerably less than the seeding rate required for optimal crop competition. Pulse seeding rates are likely half of what they need to be for achieving a competitive stand. Typical row spacing for lentil and pea are also likely too narrow.

The most compelling new tool for seed management is the Harrington Seed Destructor, a tool that destroys the seed in the harvest process. Developed to manage a similar weed called rigid ryegrass, the Harrington Seed Destructor grinds weed seed during the harvest process (Harrington and Powles 2012).

No-till and minimum tillage have increased reliance on postemergence herbicide applications in winter wheat production in order to manage Italian ryegrass and broadleaf weeds. As a result, Italian ryegrass populations in the PNW (and other regions) have developed cross- and multiple-herbicide resistance to a number of herbicide groups, including ACCase inhibitors (Group 1/A), acetolactate synthase–inhibiting herbicides (Group 2/B), photosystem II inhibitors (Group 5), glyphosate (Group 8/G), and very-long-chain fatty acid synthesis inhibitors (Group 15/K) (Hulting et al. 2012; Perez-Jones et al. 2007; Rauch et al. 2010).

Chemical inputs must include effective herbicides—often an effective herbicide program includes both preeemergence and spring-applied postemergence herbicides. Growers should make sure that the ryegrass under management is not resistant to the selected herbicide treatments. Products containing pyroxasulfone or flufenecet are required for successful management of Italian ryegrass in winter wheat as Italian ryegrass has not yet developed widespread resistance to these herbicide materials. They are applied in the fall according to the label. The product that contains flufenecet (Axiom) also contains metribuzin, and therefore cannot be applied until the wheat has two true leaves. Metribuzin can improve control achieved with pyroxasulfone. Pyroxasulfone can be applied immediately after the winter wheat is planted as a preemergence treatment. Continued experimentation with pyroxasulfone use in winter wheat will provide better understanding of its utility for control of other important grass and broadleaf weeds in cereal cropping systems. Integrated Italian ryegrass management strategies would be necessary in order to maintain efficacy of pyroxasulfone in the long term (Hulting et al. 2012).

Postemergence herbicides should be rotated. Group 1/A and Group 2/B herbicides cannot be mixed to improve Italian ryegrass control because of antagonism, so rotating products containing pyroxsulam and mesosulfuron with pinoxaden or clodinafop is critical. Herbicides used for Italian ryegrass management should be applied alone—applying grass weed management herbicides with broadleaf herbicides reduces efficacy on grass weeds. Plant size should be monitored carefully, and an assessment of the likelihood of continued germination should be made prior to the decision to apply herbicides. Herbicides should be applied when the target species is of appropriate size. Don't wait for the broadleaf weeds to germinate, or vice versa. Utilizing full rates of preemergence herbicides for grass control in broadleaf crops is also essential. Employing effective chemical weed management tools in rotation by utilizing metolachlor or dimethenamid is critical for long-term management of Italian ryegrass.

Summary

IWM strategies are critical for effective long-term management of weeds in the agroecosystem. Growers are encouraged to learn more about the biology and ecology of the weeds they are managing, and to use that knowledge to exploit vulnerable life stages of the weed for control. Growers should think critically about inputs employed for crop production in the context of weed management. Herbicides are an important part of an IWM strategy, but no herbicide can keep a crop weed-free for the entire season. A good competitive crop will always be the best weed management practice, and a sequence of successful crop rotations are critical for managing weeds in the inland PNW.

Additional Resources

Weed Management Handbook

https://pnwhandbooks.org/weed

The Weed Management Handbook is part of the Pacific Northwest Pest Management Handbooks publication series. Updated yearly, the site contains weed management information on most crops produced in the PNW.

Oregon State University Weed Science webpage

http://horticulture.oregonstate.edu/group/weed-science

http://cropandsoil.oregonstate.edu/group/weed-science

The Oregon State University Weed Science webpage is a clearinghouse for weed science-related information. The site contains comprehensive information for weed managers in Oregon as well as links to resources for the PNW.

Washington State University Weed Resources webpage

http://smallgrains.wsu.edu/weed-resources

The Washington State University Weed Resources webpage is a clearinghouse for weed science-related information. Part of the Wheat and Small Grains website, the Weed Resources webpage includes timely and frequently updated information on common weeds, regional PNW publications from both Oregon State University and the University of Idaho, weed identification services, herbicide resistance testing services, and decision support tools.

University of Idaho Weed Resources websites

http://www.cals.uidaho.edu/weeds2/IWR/iwr-v6_website/

https://www.uidaho.edu/cals/kimberly-research-and-extension-center/weed-science

University of Idaho hosts two websites: one focused on invasive weeds and one focused on southern Idaho weed management.

Herbicide Labels

CDMS: ***http://www.cdms.net***

Green Book: ***http://www.greenbook.net/***

Herbicide labels can be found at two sites in searchable databases. The CDMS site has a very powerful advanced search that requires registration.

National and Regional Organizations that Host Weed Science Information

American Society of Agronomy (ASA)
http://www.agronomy.org/

Weed Science Society of America (WSSA)
http://www.wssa.net/

Western Society of Weed Science (WSWS)
http://www.wsweedscience.org/

Western Region Sustainable Agriculture Research and Education (WSARE)
http://www.westernsare.org/

Education and Online Lessons

Crop Adviser Institute
http://www.cai.iastate.edu/

Online Crop Technology Lessons
http://croptechnology.unl.edu/pages/

Herbicide Resistance

International Survey of Herbicide Resistant Weeds
http://www.weedscience.org

Herbicide Resistance Action Committee
http://www.hracglobal.com/

Weed ID Resources and Weed Photos

PNW Weed Management Weed ID collection
http://pnwpest.org/pnw/weeds?weeds/id/index.html

UC IPM Weed Photo Gallery
http://www.ipm.ucdavis.edu/PMG/weeds_common.html

WSSA Weed Photo Album
http://wssa.net/wssa/weed/weed-identification/

References

Andrew, I.K.S., J. Storkey, and D.L. Sparkes. 2015. A Review of the Potential for Competitive Cereal Cultivars as a Tool in Integrated Weed Management. *Weed Research* 55(3): 239–248.

Ball, D.A., S.M. Frost, L. Fandrich, C. Tarasoff, and C. Mallory-Smith. 2008. Biological Attributes of Rattail Fescue (*Vulpia myuros*). *Weed Science* 56(1): 26–31.

Ball, D.A., S.M. Frost, and A.I. Gitelman. 2004. Predicting Timing of Downy Brome (*Bromus tectorum*) Seed Production Using Growing Degree Days. *Weed Science* 52(4): 518–524.

Ball, D.A., F.L. Young, and A.G. Ogg, Jr. 1999. Selective Control of Jointed Goatgrass (*Aegilops cylindrica*) with Imazamox in Herbicide-Resistant Wheat. *Weed Technology* 13(1): 77–82.

Bassett, I.J., and C.W. Crompton. 1978. The Biology of Canadian Weeds: 32 *Chenopodium album* L. *Canadian Journal of Plant Science* 58(4): 1061–1072.

Beckie, H.J., and A. Francis. 2009. The Biology of Canadian Weeds: 65 *Salsola tragus* L. (updated). *Canadian Journal of Plant Science* 89(4): 775–789.

Blackshaw, R.E. 1994. Rotation Affects Downy Brome (*Bromus tectorum*) in Winter Wheat (*Triticum aestivum*). *Weed Technology* 8(4): 728–732.

Blackshaw, R.E. 1993. Downy Brome (*Bromus tectorum*) Density and Relative Time of Emergence Affects Interference in Winter Wheat (*Triticum aestivum*). *Weed Science* 41(4): 551–556.

Borger, C.P.D., G. Riethmuller, and M. D'Antuono. 2016. Eleven Years of Integrated Weed Management: Long-Term Impacts of Row Spacing and Harvest Weed Seed Destruction on *Lolium rigidum* Control. *Weed Research* 56(5): 359–366.

Buhler, D.D. 2002. 50th Anniversary-Invited Article: Challenges and Opportunities for Integrated Weed Management. *Weed Science* 50(3): 273–280.

Carson, K.H., H.T. Cralle, J.M. Chandler, T.D. Miller, R.W. Bovey, S.A. Senseman, and M.J. Stone. 1999. *Triticum aestivum* and *Lolium multiflorum* Interaction During Drought. *Weed Science* 47(4): 440–445.

Callaway, M.B. 1992. A Compendium of Crop Varietal Tolerance to Weeds. *American Journal of Alternative Agriculture* 7(4): 169–180.

Cochran, V.L., L.A. Morrow, and R.D. Schirman. 1990. The Effect of N Placement on Grass Weeds and Winter Wheat Responses in Three Tillage Systems. *Soil and Tillage Research* 18(4): 347–355.

Cook, R.J., B.H. Ownley, H. Zhang, and D. Vakoch. 2000. Influence of Paired-Row Spacing and Fertilizer Placement on Yield and Root Diseases of Direct-Seeded Wheat. *Crop Science* 40(4): 1079–1087.

Cook, R.J., and R.J. Veseth. 1991. Wheat Health Management. APS Press, St. Paul.

Costea, M., S.E. Weaver, and F.J. Tardif. 2004. The Biology of Canadian Weeds: 130 *Amaranthus retroflexus* L., *A powellii* S. Watson and *A. hybridus* L. *Canadian Journal of Plant Science* 84(2): 631–668.

Currie, R.S., and T.F. Peeper. 1988. Combine Harvesting Affects Weed Seed Germination. *Weed Technology* 2(4): 499–504.

Derksen, D.A., R.L. Anderson, R.E. Blackshaw, and B. Maxwell. 2002. Weed Dynamics and Management Strategies for Cropping Systems in the Northern Great Plains. *Agronomy Journal* 94(2): 174–185.

Donald, W.W., and A.G. Ogg, Jr. 1991. Biology and Control of Jointed Goatgrass (*Aegilops cylindrical*): A Review. *Weed Technology* 5(1): 3–17.

Friesen, L.F., H.J. Beckie, S.I. Warwick, and R.C. Van Acker. 2009. The Biology of Canadian Weeds: 138 *Kochia scoparia* (L.) Schrad. *Canadian Journal of Plant Science* 89(1): 141–167.

Gealy, D.R., F.L. Young, and L.A. Morrow. 1985. Germination of Mayweed (*Anthemis cotula*) Achenes and Seed. *Weed Science* 33(1): 69–73.

Harrison, S.K. 1990. Interference and Seed Production by Common Lambsquarters (*Chenopodium album*) in Soybeans (*Glycine max*). *Weed Science* 38(2): 113–118.

Harker, K.N., J.T. O'Donovan, R.E. Blackshaw, E.N. Johnson, F.A. Holm, and G.W. Clayton. 2011. Environmental Effects on the Relative Competitive Ability of Canola and Small-Grain Cereals in a Direct-Seeded System. *Weed Science* 59(3): 404–415.

Harrington, R.B., and S.B. Powles. 2012. Harrington Seed Destructor: A New Nonchemical Weed Control Tool for Global Grain Crops. *Crop Science* 52(3): 1343–1347.

Hashem, A., S.R. Radosevich, and M.L. Roush. 1998. Effect of Proximity Factors on Competition between Winter Wheat (*Triticum aestivum*) and Italian Ryegrass (*Lolium multiflorum*). *Weed Science* 46(2): 181–190.

Hauvermale, A., N.C. Lawrence, and I.C. Burke. 2016. Variation in Phenology and Vernalization Requirements of *Bromus tectorum* Collected from the Small Grain Production Region of the PNW. *Weed Science Society of America* 56: 211.

Hulting, A.G., J.T. Dauer, B. Hinds-Cook, D. Curtis, R.M. Koepke-Hill, and C. Mallory-Smith. 2012. Management of Italian Ryegrass (*Lolium perenne* ssp. *multiflorum*) in Western Oregon with Preemergence Applications of Pyroxasulfone in Winter Wheat. *Weed Technology* 26(2): 230–235.

Koch, W. 1969. Influence of Environmental Factors on the Seed Phase of Annual Weeds, Particularly from the Point of View of Weed Control. Habilitations-schrift Landw. Hochech. Universität Hohenheim, Arbeiten der Universität Hohenheim 50: 20.

Lawrence, N., and I.C. Burke. 2015. Variation in Phenology of Downy Brome. *Proceedings of the Western Society of Weed Science* 68: 40.

Lyon, D., I.C. Burke, A.G. Hulting, and J. M. Campbell. 2017. Integrated Management of Mayweed Chamomile in Wheat and Pulse Crop Production Systems. Washington State University Extension.

Manuchehri, M.R. 2012. The Relative Competitiveness of Spring Crops in a Dryland Organic System in Eastern Washington. Master's Thesis, Washington State University.

Moss, S.R. 1997. Strategies for the Prevention and Control of Herbicide Resistance in Annual Grass Weeds. In Weed and Crop Resistance to Herbicides 283–290, R. De Prado, J. Jorrin, and L. Garcia-Torres, eds. London: Kluwer Academic Publishers.

Moyer, J.R., E.S. Romann, C. W. Lindwall, and R.E. Blackshaw. 1994. Weed Management in Conservation Tillage Systems for Wheat Production in North and South America. *Crop Protection* 13(4): 243–259.

Norris, R.F., E.P. Caswell-Chen, and M. Kogan. 2003. Concepts in Integrated Pest Management. New Jersey, Prentice Hall: 2003.

Ogg, Jr., A.G., and S.S. Seefeldt. 1999. Characterizing Traits which Enhance the Competitiveness of Winter Wheat (*Triticum aestivum*) Against Jointed Goatgrass (*Aegilops cylindrica*). *Weed Science* 47(1): 74–80.

Perez-Jones, A., K.W. Park, N. Polge, J. Colquhoun, and C.A. Mallory-Smith. 2007. Investigating the Mechanisms of Glyphosate Resistance in *Lolium multiflorum*. *Planta* 226(2): 395–404.

Pester, T.A., O.C. Burnside, and J.H. Orf. 1999. Increasing Crop Competitiveness to Weeds through Crop Breeding. *Journal of Crop Production* 2(1): 31–58.

Reinertsen, M.R., V.L. Cochran, and L.A. Morrow. 1984. Response of Spring Wheat to N Fertilizer Placement, Row Spacing, and Wild Oat Herbicides in a No-Till System. *Agronomy Journal* 76(5): 753–756.

Radosevich, S.R., J.S. Holt, and C.M. Ghersa. 2007. Ecology of Weeds and Invasive Plants: Relationship to Agriculture and Natural Resource Management. John Wiley and Sons.

Raeder, A.J., D. Lyon, J. Harsh, and I.C. Burke. 2015. How Soil pH Affects the Activity and Persistence of Herbicides. Washington State University Extension Publication FS189E.

Rauch, T.A., D.C. Thill, S.A. Gersdorf, and W.J. Price. 2010. Widespread Occurrence of Herbicide-Resistant Italian Ryegrass (*Lolium multiflorum*) in Northern Idaho and Eastern Washington. *Weed Technology* 24(3): 281–288.

Riar, D.S., D.A. Ball, J.P. Yenish, and I.C. Burke. 2011. Light-Activated, Sensor-Controlled Sprayer Provides Effective Postemergence Control of Broadleaf Weeds in Fallow. *Weed Technology* 25(3): 447–453.

Ross, M.A., and C.A. Lembi. 1999. Applied Weed Science. Prentice Hall, Upper Saddle River, N.J.

Rydrych, D.J. 1974. Competition between Winter Wheat and Downy Brome. *Weed Science* 22(3): 211–214.

Schillinger, W.F. 2007. Ecology and Control of Russian Thistle (*Salsola iberica*) after Spring Wheat Harvest. *Weed Science* 55(4): 381–385.

Schillinger, W.F., and F.L. Young. 2000. Soil Water Use and Growth of Russian Thistle after Wheat Harvest. *Agronomy Journal* 92(1): 167–172.

Schillinger, W.F., and R.I. Papendick. 2008. Then and Now: 125 Years of Dryland Wheat Farming in the Inland Pacific Northwest. *Agronomy Journal* (100)3: S-166.

Schillinger, W.F. 2005. Tillage Method and Sowing Rate Relations for Dryland Spring Wheat, Barley, and Oat. *Crop Science* 45(6): 2636–2643.

Stahlman, P.W., and S.D. Miller. 1990. Downy Brome (*Bromus tectorum*) Interference and Economic Thresholds in Winter Wheat. *Weed Science* 38(3): 224–228.

Tautges, N.E., I.C. Burke, K. Borrelli, and E.P. Fuerst. 2016. Competitive Ability of Rotational Crops with Weeds in Dryland Organic Wheat Production Systems. *Renewable Agriculture and Food Systems* 32(1): 57–68.

Thorne, M.E., F.L. Young, and J.P. Yenish. 2007. Cropping Systems Alter Weed Seed Banks in Pacific Northwest Semi-Arid Wheat Region. *Crop Protection* 26(8): 1121–1134.

Unger, R., M.E. Swanson, I.C. Burke, D.R. Huggins, E.R. Gallandt, and S. Higgins. 2012. The Effects of Crop Rotation and Terrain Attributes on the Weed Seed Bank. *Proceedings of the Western Society of Weed Science* 65: 173.

Veseth, R. 1985. Deep Banding Fertilizer: A Weed Management Tool! In Pacific Northwest Conservation Tillage Handbook Series No. 4.

Weaver, S.E., and M.P. Downs. 2003. The Biology of Canadian Weeds: 122 *Lactuca serriola* L. *Canadian Journal of Plant Science* 83(3): 619–328.

Welsh, J.P., H.A.J. Bulson, C.E. Stopes, R.J. Froud-Williams, and A.J. Murdoch. 1999. The Critical Weed-Free Period in Organically-Grown Winter Wheat. *Annals of Applied Biology* 134(3): 315–320.

Yenish, J., R. Veseth, A.C. Ogg, D.C. Thill, D. Ball, F. Young, E. Gallandt, D. Morishita, C. Mallory-Smith, D. Wysocki, and T. Gohlke. 1998. Managing Downy Brome under Conservation Tillage Systems in the Northwest Cropping Region. In Pacific Northwest Conservation Tillage Handbook Series No. 15.

Young, J.A., and R.A. Evans. 1972. Germination and Establishment of Salsola in Relation to Seedbed Environment. I. Temperature, Afterripening, and Moisture Relations of *Salsola* Seeds as Determined by Laboratory Studies. *Agronomy Journal* 64(2): 214–218.

Young, F.L., A.G. Ogg, Jr., and P.A. Dotray. 1990. Effect of Postharvest Field Burning on Jointed Goatgrass (*Aegilops cylindrica*) Germination. *Weed Technology* 4(1): 123–127.

Young, F.L., J.P. Yenish, G.K. Launchbaugh, L.L. Mcgrew, and J.R. Alldredge. 2008. Postharvest Control of Russian Thistle (*Salsola tragus*) with a Reduced Herbicide Applicator in the Pacific Northwest. *Weed Technology* 22(1): 156–159.

Young, F.L., J.P. Yenish, D.L. Walenta, D.A. Ball, and J.R. Alldrege. 2003. Spring-Germinating Jointed Goatgrass (*Aegilops cylindrica*) Produces Viable Spikelets in Spring-Seeded Wheat. *Weed Science* 51(3): 379–385.

Young, F., R. Veseth, D. Thill, W. Schillinger, and D. Ball. 1995. Russian Thistle Management under Conservation Systems in Pacific Northwest Crop-Fallow Regions. In Pacific Northwest Conservation Tillage Handbook Series No. 16.

Young, F.L., D.A. Ball, D.C. Thill, J.R. Alldredge, A.G. Ogg, and S.S. Seefeldt. 2010. Integrated Weed Management Systems Identified for Jointed Goatgrass (*Aegilops cylindrica*) in the Pacific Northwest. *Weed Technology* 24(4): 430–439.

Young, F.L., A.G. Ogg, and J.R. Alldredge. 2014. Postharvest Tillage Reduces Downy Brome (*Bromus tectorum* L.) Infestations in Winter Wheat. *Weed Technology* 28(2): 418–425.

Young, F.L., and M.E. Thorne. 2004. Weed-Species Dynamics and Management in No-Till and Reduced-Till Fallow Cropping Systems for the Semi-Arid Agricultural Region of the Pacific Northwest, USA. *Crop Protection* 23(11): 1097–1110

Zimdahl, R.L. 2013. Fundamentals of Weed Science, 4th ed. London: Academic Press.

Chapter 10

Disease Management for Wheat and Barley

Elizabeth Kirby, Washington State University
Timothy Paulitz, USDA-ARS and Washington State University
Timothy Murray, Washington State University
Kurtis Schroeder, University of Idaho
Xianming Chen, USDA-ARS and Washington State University

Abstract

Many pathogens can limit yield potential in the dryland cereal-based production region of the inland Pacific Northwest (PNW). The region's diverse biogeographical factors, including soil type, temperature, and precipitation, and production system variables, including crop genetics, tillage, residue management, rotation and other cropping practices, affect the incidence, risk, and severity of crop disease. This chapter provides an overview of several key wheat and barley pathogens, conditions or practices that favor disease, and integrated management practices. Climate change, with predicted shifts in temperature and precipitation patterns, will also influence crop disease dynamics in the region, but currently, only limited information is available.

Key Points

- Successful disease management relies on understanding pathogen distribution, environmental conditions, and cropping practices that favor disease incidence or severity, relative potential for economic crop damage to occur, and the appropriate use of integrated management strategies.

- The PAMS integrated management approach utilizes prevention, avoidance, monitoring, and suppression strategies. Genetic resistance or chemical controls are not available for many soilborne pathogens, and growers rely on cultural practices to favor plant health.
- Adoption of new technologies and cropping practices may have a greater impact on wheat and barley diseases than climate change in the near future. System-wide monitoring of crop response is an important tool to determine if changes in cropping practices or climate effects reduce the effectiveness of current management strategies.
- There is much uncertainty regarding the impact of climate change on disease incidence and severity in PNW cereal production. Climate change effects could accelerate, extend, or slow typical disease cycles, or favor the introduction of new diseases.

Introduction

Overview of Pathogens Affecting Inland PNW Cereal Production Regions

Historically, more than 30 wheat and barley diseases have decreased profitability in the dryland, wheat-based cropping region of the **inland PNW**. Small grain pathogens and plant parasitic nematodes reduce grain yield and quality by damaging roots, stems, leaves, or grain heads. Fungal pathogens cause the greatest economic damage to small grains globally and, in the inland PNW, are the second most challenging biotic factor after weeds. Foliar diseases, such as stripe rust, and several soilborne pathogens have the potential to cause severe crop losses.

For a disease to develop, three factors must be present: a virulent pathogen, a **susceptible** crop host, and environmental conditions favorable to development of the disease. Complex interactions among these factors determine the frequency and severity of a disease. Pathogen inoculum may be airborne, present in or on seed, in soil, in infected living host tissue or residue, or vectored from plant to plant, usually by insects. Pathogens are highly sensitive to changes in moisture and temperature. Foliar diseases are typically favored by high canopy humidity and free water, whereas root and stem diseases caused by soilborne fungal pathogens are often favored by cool, moist soils.

Variation in weather and climate, soil properties, and agronomic practices modify the host-pathogen environment, affecting pathogen distribution and population, and the potential risk of specific crop diseases. Cool season small grains are well-adapted to the region's Mediterranean-type climate and diverse biogeographical conditions. Typical warm, dry summers limit some foliar diseases that are more of a concern in other US wheat-growing regions, whereas cool, wet springs favor soilborne diseases, especially in high residue **conservation tillage** systems. Cool, wet conditions also favor stripe rust. For further information on the region's diversity and climate, see Chapter 1: Climate Considerations.

Foliar and head diseases

Diseases caused by fungal pathogens of small grains can lead to economic losses when unmanaged and conditions are favorable for development. Producers have successfully reduced the impact of many foliar wheat and barley diseases (e.g., smuts, stripe rust) using integrated genetic, cultural, and chemical management strategies. For example, over the past 50 years, the inoculum of smut pathogens, once widely distributed across the region, has been effectively reduced to very low levels due to the use of pathogen-free seed, **resistant** varieties, and fungicide seed treatments. These practices have reduced the incidence of common bunt, flag smut, loose smut, and dwarf bunt diseases. However, growers have become increasingly reliant on seed treatment for control as many of the current commercial wheat cultivars are susceptible to smut pathogens. Common bunt disease has emerged as a concern in organic production systems, highlighting a need for continued screening for resistant cultivars and for research on alternative seed treatments suitable for organic production (Matanguihan et al. 2011). Stripe rust damage, caused by the fungus *Puccinia striiformis*, can be managed by the integrated use of resistant varieties and fungicide when predicted to be severe. Stripe rust continues to be one of the most important foliar diseases of wheat and barley in the inland PNW and is discussed in detail later in this chapter.

Viral diseases

Historically, *Barley yellow dwarf virus* (BYDV) has been the most important viral pathogen of small grains in the PNW. Transmission of the virus is dependent on infected aphid vectors; therefore, the primary discussion of this disease is found in Chapter 11: Insect Management Strategies. The virus is widespread and has many hosts including barley, wheat, oats, corn, and grasses that can serve as inoculum reservoirs. Barley yellow dwarf (BYD) disease causes the greatest damage to winter wheat, barley, and oats; less damage occurs on spring-planted grains. BYDV-infected plants may also be more susceptible to root rot diseases. Although total field loss can occur, estimated average losses are less than 10%. Eliminating volunteer crop and grassy weed hosts and the **green bridge** effect reduces primary inoculum density. Control of aphid vectors, delayed fall planting, and use of cultivars with some resistance to BYDV can also limit economic impact.

Wheat streak mosaic virus, vectored by the wheat curl mite, is also discussed in Chapter 11: Insect Management Strategies. *Soilborne wheat mosaic virus* is a relatively recent discovery in the PNW. This pathogen, vectored by a soilborne fungus, is discussed later in this chapter.

Root-infecting fungal pathogens and nematodes

Root, crown, stem, and vascular diseases caused by root-infecting fungal pathogens and plant-parasitic nematodes significantly impact small grain production across the region and can be a barrier to producer adoption of **direct seeding**. The effects of soilborne diseases are most evident under dryland conditions because plants with damaged roots are less efficient at water and nutrient uptake than healthy plants, and predisposed to drought stress and nutrient deficiencies. Reduced tillage, increased residue levels, and cool, moist soils favor some soilborne pathogens and may increase the risk of disease.

This chapter focuses on many economically important soil and residue-borne pathogens and nematodes. In contrast to seed-transmitted, airborne, or insect-vectored pathogens, soil and residue-borne pathogens and nematodes have limited management options. Genetic resistance and chemical control options are often lacking and growers rely on cultural techniques to manipulate the crop environment to favor plant health and growth. Breeding efforts have produced commercial varieties with genetic resistance to some root-infecting diseases such as eyespot, snow molds, and Cephalosporium stripe; however, no locally adapted varieties are available with resistance to Pythium and Rhizoctonia root rots or take-all (Paulitz et al. 2009). Registered fungicide seed treatments are effective against many seed-transmitted pathogens but may provide only short-term suppression, or no control, of several soilborne pathogens, and most root-infecting pathogens cannot be controlled with foliar chemical applications. Nematicides are not registered for use. Crop rotation, or **fallow**, with at least one year out of wheat, barley, or other host crops can adequately reduce inoculum levels for some diseases, such as take-all, depending on the environment and how efficiently cereal residues break down; however, other diseases such as eyespot and Cephalosporium stripe require longer rotations. Long rotations may not be an economical or effective management tool for situations where multiple years away from wheat or barley are required, or for small grain pathogens with multiple hosts including legumes or oilseeds.

Effects of Climate Change on Cereal Pathogens and Disease

There are many unanswered questions about the potential effects of climate change on wheat and barley diseases in the PNW. Better understanding the conditions most favorable to pathogens and the development of disease will help growers adapt and minimize risk. Evolving cropping practices, technologies, and economic factors are likely to have a greater impact on our regional crop production systems than climate change, at least in the near future.

Complex interactions between crop host, pathogen, and environment make it challenging to predict the impact of climate change on the distribution of crop pathogens, the risk and severity of disease, and management guidelines. The main climate factors are variations in precipitation, temperature, and increased atmospheric carbon dioxide concentration. Any disease may become more important if climate tips the balance in favor of the pathogen. For example, *Pythium* would be favored by the predicted cooler, wetter early spring conditions.

Predicted decreases in late-spring and early-summer precipitation, in tandem with elevated temperatures, will make it more difficult for root-damaged crops to obtain water and nutrients during the warmer summer months, increasing economic risk. Milder winter temperatures may favor inoculum survival (e.g., stripe rust), whereas precipitation variability or earlier drought stress may predispose crops to disease or slow disease progress. Climate variability may also result in new pathogens or races of endemic pathogens emerging in the region.

Inadequate information has so far limited the opportunity to model regional climate impacts on wheat and barley diseases and potential crop losses in the inland PNW. In the future, data from recent baseline population surveys of fungi and nematodes across the region can be used to improve modeling, our understanding of pathogen response to changing environments, and management decisions. Effects are expected to be site-specific; yield response will depend on direct effects on pathogens, indirect effects caused by the host crop, and grower adaptation of management strategies.

PAMS Integrated Pest Management Strategies for Small Grain Pathogens

Producers seek to balance economic, crop health, and environmental constraints. Understanding production limits, setting affordable yield goals, and minimizing environmental and nutritional stresses on the crop support success. Targeted use of integrated genetic, cultural, chemical, and biological management tools to Prevent, Avoid, Monitor, and Suppress (**PAMS**) crop disease can eliminate or reduce the impact of many wheat and barley pathogens. Many useful management strategies have been identified and implemented in the inland PNW; using multiple strategies improves the odds for profitable management. The USDA-Natural Resources Conservation Service (NRCS 2010) adopted the PAMS approach to site-specific **integrated pest management** planning, an integral part of the Environmental Quality Incentives Program. This section presents a general overview of several PAMS integrated disease management practices; specific management options are presented later in the chapter for each of the pathogens discussed.

Prevention

Excluding a pathogen from a non-infested field is the first and most economical line of defense against crop disease. Many small grain pathogens are already widespread in dryland PNW fields, thus prevention is not applicable.

Field sanitation

Conscientious field and equipment hygiene reduces movement of soil or residue-borne pathogens from infested fields into clean fields.

Clean seed

Use of pathogen-free seed prevents the introduction of new seedborne diseases.

Avoidance

Avoidance is the use of cultural practices to avoid **pest** populations that already exist in a field to reduce the risk of disease.

Seed quality

Use of fresh, high-quality, pathogen-free seed promotes seedling establishment, vigor, and health.

Planting a non-host crop

Planting a non-host crop avoids infection and disease development, reducing inoculum of a specific pathogen.

Host crop resistance

Host crop resistance is the ability of a host crop (e.g., a cultivar) to inhibit growth and reproduction of a pathogen. Cultivars that suppress or prevent reproduction of a pathogen are classed as resistant; those that allow moderate to high rates of reproduction are susceptible. Tolerance is the ability to endure infection by a pathogen without serious damage or yield loss. Resistance may be race-specific (resistant to some but not all races of a pathogen) or race non-specific (resistant to all races). Planting a susceptible, **tolerant** variety can reduce yield loss of the current crop but does not limit reproduction or inoculum that can affect subsequent crops.

Use of resistant crops can be considered either an avoidance or suppression strategy, depending on the degree of resistance. Planting varieties with a high degree of resistance (i.e., no infection occurs) is an example of an avoidance strategy, whereas planting varieties that slow disease progress, if an infection occurs, is an example of a suppression strategy.

Monitoring

Identification and quantification

Effective disease management relies on accurate diagnosis and quantification of pathogens and timely application of control measures. It is helpful to identify pathogens to the level (i.e., genus, species, or race) that may affect management decisions. Root diseases are often difficult to diagnose based on aboveground symptoms. Sampling methods vary by pathogen.

Monitoring and recordkeeping

Understanding pathogen populations prior to planting can help determine risk and support crop choice and management decisions. Once visible symptoms caused by root-infecting pathogens appear in a crop, typically no actions are available to suppress disease development. However, tracking symptoms and severity by management unit during the current growing season informs subsequent management decisions.

Thresholds

Economic damage or action **thresholds** are based on population studies, forecast models, visual symptoms, field history, and yield correlation. Action guidelines are available for only a few cereal pathogens in the PNW (e.g., stripe rust, eyespot) where foliar fungicides can suppress damage. Correlation of pathogen population densities and predicted yield loss (e.g., the take-all pathogen) are needed to support pre-plant management decisions such as crop selection. Crop damage by parasitic nematodes is expected when pre-plant populations reach defined levels. However, precise yield loss is difficult to correlate with populations for most root-infecting pathogens because of complex environmental interactions across the region.

Forecasting models

Forecasting models are available for stripe rust.

Suppression

Growers can reduce or eliminate pathogen populations, disease severity, or crop damage using cultural, mechanical, chemical, or biological suppression practices.

Cultural

Green bridge control

Eliminating volunteer host crops and weeds is a first-defense management tool for reducing inoculum density. Timing cultivation or herbicide application with a sufficient period between application and planting to prevent the pathogen from bridging to the crop is important.

Host crop resistance

Planting crop varieties with varying degrees of resistance can prevent infection, slow disease progress if infection occurs, and reduce inoculum, limiting in-crop damage and risk to subsequent host crops. (See the Avoidance section.)

Rotation

Inoculum density of most root-infecting pathogens increases or decreases with the frequency of host crops in rotation. Using rotation to suppress disease is most effective when alternate, non-host crops are available, precipitation is not limiting, and conditions promote rapid residue decomposition. Clean fallow can adequately reduce some wheat disease inoculum (e.g., take-all) but may not meet conservation goals.

Planting dates

Planting dates influence crop growth and development, and severity of many cereal diseases. It is difficult to create precise planting date guidelines due to the diverse conditions across the inland PNW. Optimal planting dates should be site-specific to account for variation in landscape position, moisture, and temperature. In general, early fall planting favors pathogens causing stripe rust and BYD as well as several soilborne diseases including take-all, eyespot, Fusarium crown rot, and Cephalosporium stripe. Later fall planting favors Pythium root rot, Rhizoctonia root rot, and snow molds. Late planting tradeoffs include decreased grain yields and increased potential for soil erosion following fallow.

Nutrient management

Adequate nutrition optimizes crop health and profitability. Fertilizer rates should be based on site-specific yield potential (see Chapter 6: Soil Fertility Management). Placing fertilizer with the seed or deep-banding nitrogen (N), phosphorus (P), and sulfur (S) below the seed at the time of planting can offset the yield-limiting effects of Pythium and Rhizoctonia root rots and take-all; nutrients placed adjacent to roots help seedlings overcome early nutrient deficiencies caused by root pruning.

Mechanical

Conservation tillage and residue management

Crop choice, available moisture, temperature, and tillage affect biomass and production, residue decomposition, and pathogen survival. Greater surface residue creates cooler, moister soils at planting that can favor seed and root-infecting pathogens including Pythium root rot, Rhizoctonia root rot, and take-all. Cephalosporium stripe and Fusarium crown rot have had variable responses to tillage systems, whereas eyespot can be reduced under conservation tillage. Stripe rust and BYD are typically unaffected by tillage. Annual cropping regions have larger biomass and grain yield potentials compared to grain-fallow systems. However, greater precipitation supports faster residue decomposition. Adapting equipment to spread chaff evenly and using high-disturbance openers can reduce risk of infection by residue-borne pathogens. For more information on residue management, see Chapter 4: Crop Residue Management.

Chemical suppression

Foliar fungicides

Foliar fungicides, in combination with other management strategies such as planting resistant cultivars, are effective for a few cereal diseases such as stripe rust and eyespot.

Seed treatment

Planting pathogen-free seed helps eliminate the need for chemical treatment of seedborne diseases. However, where soil or residue-borne fungal pathogens are present, seed treatment is a relatively low-cost suppression strategy and is particularly effective for pathogens with short disease infection periods (e.g., smuts). Fungicide treatments (metalaxyl, mefenoxam) in tandem with careful planting practices, such as monitoring root zone moisture at planting, can protect seedlings from damping off and rot caused by *Pythium* species. Root pathogens with the ability to infect plants over a longer time period have shown less response to seed treatments. Some fungicide seed treatments including difenoconazole, tebuconazole, or triadimenol (not registered in Washington) can temporarily suppress root diseases caused by *Fusarium* spp. and take-all disease. Newer chemistries (tebuconazole, sedaxane, pyraclostrobin, penflufen) have improved short-term suppression of some root diseases such as Rhizoctonia root rot. Systemic fungicides are effective for a longer time period than contact fungicides. Although helpful in controlling some diseases, systemics are less helpful in controlling root rots because the active materials move upward through the seedling rather than downward into the roots. There are no seed treatments available for bacterial or viral diseases. Systemic insecticide seed treatments can help control aphid vectors of BYDV.

Biological

Suppressive soils

Suppressive soils are defined as "soils in which the pathogen does not establish or persist, establishes but causes little damage, or establishes and causes disease for a while but thereafter the disease is less important even though the pathogen may persist in the soil" (Weller et al. 2007). Take-all decline and Rhizoctonia bare patch suppression are examples of natural, microbial-based mechanisms of defense against root-infecting pathogens. In many areas of the US, growers have been able to maintain long-term suppression of take-all in continuous wheat production, especially under irrigation. Periods of fallow or rotation away from continuous cereals reduces suppressiveness of take-all decline. It takes several years for soils to develop suppressiveness to *Rhizoctonia* (Schillinger et al. 2014), thus it is not a practical management strategy.

Microbiological control

Currently no effective commercial biological controls are available for field use.

Selected Pathogens of Inland PNW Dryland Cereal Production Systems: Research and Management Implications

Recent studies have improved our understanding of the regional distribution of wheat and barley pathogens and agro-climatic and crop production factors that influence the risk of disease and affect management decisions. This chapter summarizes several key dryland diseases and management strategies, particularly those that may be impacted by **conservation cropping** practices or have limited control options. Table 10-1 illustrates

Table 10-1. Cropping system practices that can impact (+) disease management in the PNW or that have no effect or are not available (–).

	Cultural practices	Variety selection	Chemical control	
			Foliar	Seed
Stripe rust	+	+	+	–
Eyespot	+	+	+	–
Cephalosporium stripe	+	+	–	–
Rhizoctonia root rot	+	–	–	–/+
Fusarium crown rot	+	–/+	–	–/+
Pythium root rot	+	–	–	+
Snow molds	+	+	–	–
Barley yellow dwarf	+	–	–	+
Take-all	+	–	–	–
Cereal cyst nematode	+	+	–	–
Root-lesion nematode	+	+	–	–

Note: Gray boxes indicate greatest impact. Adapted from Murray 2016.

Table 10-2. Cultural management practices that impact disease incidence.

	Seeding date		Tillage		Green bridge	Fertility	Soil pH	Crop rotation
	Winter	Spring	MinTill	NoTill				
Stripe rust	↓L	↓E	–	–	+	+	–	–
Eyespot	↓L	–	↓	↓	–	–	–	–
Cephalosporium stripe	↓L	–	–/+	↑	–	+	+	+
Rhizoctonia root rot	+/–E	+/–L	–/+	↑	+	–/+	–	–
Fusarium crown rot	↓L	–	–	↑↓	–	+	–	–/+
Pythium root rot	↓E	↓L	–	↑	+	–/+	–	–
Snow molds	↓E	↓L	–	–	–	–	–	–
Barley yellow dwarf	↓L	↓E	–	–	+	+	–	–
Take-all	↓L	–	–	–	+	+	+	+
Cereal cyst nematode	–	–	–	–	+	–	–	+
Root-lesion nematode	–	–	–	–	+	–	–	–/+

Note: (↑) Practice can favor pathogen; (↓) Practice can reduce risk; E = early and L = late; (+) Practices can impact management; (-) Practices do not impact management or are unavailable; Tillage impact is relative to conventional tillage. Gray boxes indicate greatest impact.
Adapted from Murray 2016.

general management components (cultural practices, variety selection, or chemical control) that impact management, have no effect on management, or are not available. Table 10-2 summarizes specific cultural practices that may favor or reduce disease.

It is beyond the scope of the chapter to address all small grain cereal and broadleaf rotation crop diseases that occur in the dryland PNW. Excellent resources are available for additional detail on multiple wheat and barley diseases and their management, including the Compendium of Wheat Diseases and Pests (Bockus et al. 2010), the Compendium of Barley Diseases (Mathre 1997), and Diseases of Small Grain Cereal Crops: A Colour Handbook (Murray et al. 2009b). Readers should refer to the PNW Plant Disease Management Handbook especially for chemical suppression information. Additional local resources are listed in the Resources and Further Reading section.

Stripe Rust

Background, causal agents, and distribution

Rusts are the most serious foliar diseases of small grains in the PNW. Stripe rust occurs on wheat, barley, rye, and various cultivated and wild grasses. Stripe rust is caused by the *Puccinia striiformis* species that is divided into different special forms (*formae speciales*) based on specialization to different cereal crops. For example, wheat stripe rust is mostly caused by *P. striiformis* f. sp. *tritici* and barley stripe rust is mostly caused by *P. striiformis* f. sp. *hordei*. The forms of wheat stripe rust and barley stripe rust can infect some barley and some wheat varieties, respectively, but do not cause severe diseases on the other crop. Therefore, stripe rust that develops in wheat fields generally does not impact barley crops and vice versa. Many grass species are highly susceptible to both the wheat and barley stripe rust forms and, when infected, can provide an inoculum reservoir (Table 10-3).

Barley stripe rust is a relatively new disease in the US. The disease reached the inland PNW by 1995 causing localized severe damage in the late 1990s. From 2001 to 2009, researchers observed up to 40% yield losses in experimental and commercial fields on susceptible varieties in eastern Washington. Yield losses on commercial barley varieties ranged from 0%

Table 10-3. Stripe rust characteristics and management options for dryland cereal producers.

Stripe rust	
Background • Causal agents: *Puccinia striiformis* • Source: wind dispersal of inoculum • Wide distribution across region • Hosts: wheat, barley, grasses • High risk: wet fall, warm winter; warm, wet early spring; cool, wet late-spring and summer. Barley has lower risk than winter and spring wheat.	**Economic impact** • Potential losses of 0–45% on commercial wheat varieties with varying degrees of resistance; potential losses up to 90% on highly susceptible-check varieties **Management options** • Host resistance • Fungicides • Eliminate green bridge • Avoid early-fall planting
Key diagnostics • Irregular patches with yellow-orange rust pustules on seedling leaves; stripes with pustules on leaves, leaf sheath, and glumes on adult plants • Presence of yellow-orange spore powder distinguishes stripe rust from Cephalosporium stripe, physiological leaf spot, and BYD • Resistant varieties may have white necrotic stripes with or without rust pustules on mature leaves	**Ongoing research** • Forecasting • Monitoring occurrence, severity, and distribution • Identifying pathogen races and distribution, virulence, and frequency • Developing new resistant varieties • Fungicide testing and variety response

to 26% during that period, with average annual losses of 12% and 11% measured in 2002 and 2005, respectively. In general, stripe rust epidemics on barley are not as widespread and damaging as stripe rust epidemics on wheat because mainly spring barley is grown while both winter and spring wheat are grown. Also, barley-growing regions are scattered while wheat-growing regions are more contiguous; plus, barley matures faster. Thus, the barley stripe rust pathogen has a much shorter season to infect and develop on barley, and the much longer period between barley crops reduces pathogen survival.

Key diagnostic features

Stripe rust is more easily recognized when the disease is fully developed, but it is important to identify the disease as early as possible to implement appropriate controls. Infection can occur throughout the growing season. Once the pathogen infects plants, it takes one week to several months to show symptoms, depending upon temperatures. Symptoms first appear as chlorotic patches on leaves. These symptoms are hard to recognize as stripe rust as they appear similar to infections by other pathogens, abiotic stress, or chemical injuries. Recognizable signs are tiny yellow-orange uredinia (pustules), which are clustered in patches (not stripes) on seedling leaves, and form stripes between leaf veins in adult plant stages, usually starting at stem elongation (Figure 10-1). Stripe rust can be confused with Cephalosporium stripe, physiological leaf spot, BYD, or even cereal leaf beetles, especially from a distance, but it is easily distinguished from these by the rust spore powders. Pustules contain powdery urediniospores which may be rubbed off on fingers. Pustules

Figure 10-1. Stripe rust pustules in a commercial winter wheat field near Lamont, Washington, on November 8, 2016. (Photo by Xianming Chen.)

also occur on leaf sheaths, glumes, developing grains, and awns when the disease is severe. In the late plant growth stage, black telia form; at this stage no more infectious urediniospores are produced. Compared to a susceptible check variety, resistant varieties have different responses to stripe rust infection, ranging from no symptoms or rust signs, to various sizes or lengths of necrotic patches (on seedlings) or stripes (on adult plant leaves) without or with rust pustules, to a relatively low number of rust pustules.

Disease cycle and conditions that favor the pathogen

The three most important environmental factors governing the risk of stripe rust include temperature, moisture, and wind. The stripe rust pathogen prefers cool environments and causes more damage when the fall is wet; winter is warm; early spring is warm and wet; and later spring and summer are cool and wet. Urediniospores need a minimum of 3 hours of a dew period to germinate and penetrate plants; extended dew periods favor greater infection rates. During wet autumn conditions, urediniospores in the air, either produced in the region or blown by wind from other regions, can infect emerged wheat plants. The wetter the weather and the larger the plants, the more infection will occur. Urediniospore germination and infection can occur when temperatures are from just above 33°F to about 73°F, and best when temperatures are between 45°F and 54°F. Once the fungus grows into plant tissue, moisture is not an affective factor until producing new spores (sporulation). The period from infection to sporulation is called the latent period, and temperature is the most important factor during this time. In the PNW, stripe rust fungus lives as mycelium in the plant tissue during the winter time. The fungus can survive when temperatures are above 23°F; colder temperatures reduce survival. In general, if temperatures are below 14°F for about three days, the fungus will die. Snow cover, especially during cold spells, helps the rust fungus survive. Wind-kill can eliminate rust fungus as it kills wheat leaves. Winter hardy varieties can help rust fungus survive in plant tissue when temperatures are above 14°F (Ma et al. 2015). The latent period can take from two weeks when temperatures are optimal (59–75°F) for growth and sporulation to more than five months when temperatures are below 40°F. The warmer the winter is, the more rust survival. In the PNW, stripe rust fungus can occasionally produce spores during winter time. Because winter weather conditions, mainly temperatures and snow cover, determine the level of stripe rust survival, epidemics can be predicted for the PNW using forecast models based on historical weather and yield loss data (Sharma-Poudyal and Chen 2011).

With daily average temperatures in early spring constantly above 40°F, stripe rust will start to sporulate, and new spores will infect wheat plants. Warm, wet conditions speed up sporulation and increase new infections. Late-spring and summer high temperatures and dry conditions are not favorable for infection. Cool, wet conditions increase infection rates, prolong the crop growth season and rust season, resulting in more yield losses. Also, this will shorten the period from crop maturity to the emergence of the fall-planted wheat crops, increasing the chance for infection of the fall-planted crops. During late summer, the stripe rust fungus survives on spring wheat crops and volunteer plants. The major limiting factor is temperature. In general, when temperatures are above 74°F, no infection will occur, although the fungus can survive as mycelium in plant tissue for two to three weeks or as airborne urediniospores for up to a month. Cool, wet conditions help the rust survive summer temperatures by continually infecting plants. Generally, when daily high temperatures are above 95°F for a few days, stripe rust fungus will die even in plant tissue. The big difference between daily maximum and minimum temperatures during summer in the PNW allows stripe rust fungus to survive relatively well compared to many other regions in the US and the world. This is another reason why stripe rust is a more frequent disease in the PNW, in addition to the relatively mild winter and growth of both winter and spring wheat crops. Urediniospores survive well under cool and dry conditions, and they can infect fall-planted wheat crops when dew occurs on leaves, potentially leading to another cycle of rust fungus survival and development.

Potential effects of climate change

Changes in temperature and precipitation patterns, particularly during the growing season, and, to a lesser extent, mean annual changes, can affect stripe rust incidence. To some extent, the fungus is capable of adapting to climate change and continuing to survive and reproduce. Hotter, drier summers would limit epidemics that might occur later in the growing season; however, milder winters would enhance rust survival and lead to earlier infections. Also, less snow cover could result in increased wheat susceptibility to winter-kill, reducing inoculum. The need to continue to develop and grow stripe rust-resistant varieties will remain a priority regardless of direction of climate change.

Stripe rust management strategies

Prevention

The stripe rust pathogen is in the region and cannot be kept away. However, it is always a good idea to prevent transport of exotic races or strains of the pathogen from other regions. Clothing and footwear should be changed after visiting a field with stripe rust.

Avoidance

The best approach to prevent stripe rust is to grow varieties with a high level of resistance (varieties with a stripe rust rating of 1 or 2 in the most recent Buyer's Guide). However, appropriate late planting of winter wheat may avoid stripe rust infection in the fall. Similarly, later planting of spring wheat may shorten rust season as the normally hot and dry weather conditions are not favorable for stripe rust. However, spring wheat planted too late can suffer serious yield reduction as hot and dry weather conditions are not good for grain filling.

Monitoring

Good monitoring allows for timely fungicide application and avoids unnecessary use of fungicide. Fungicide application is more effective when disease is in the very early stage. The general recommendation is to apply a fungicide before stripe rust reaches 5%, at least no later than 10% incidence (percentage of leaves or plants with rust). Application is generally not recommended if (1) no rust can be found in the field, unless the field is planted with moderately susceptible or susceptible varieties (stripe rust ratings of 5 to 9), (2) stripe rust has been occurring in nearby areas, and (3) weather has been and will be favorable for disease. As fungicides vary in efficacy and duration of effectiveness (20 to 40 days depending upon chemicals), growers should begin checking fields about 2 to 5 weeks after application to determine if another application is needed. Varieties with race-specific resistance may become susceptible if new virulent races occur in the region. Monitoring fields planted with varieties having this type of resistance can prevent unexpected damage. Annual forecast data alert growers to the potential severity of stripe rust.

Suppression

Resistance

Planting resistant cultivars is the most effective control. Most wheat varieties grown in the PNW have some level of resistance, but not all have adequate levels of resistance when stripe rust is severe. Growing these varieties reduces rust damage. For example, under the extremely severe stripe rust epidemic in 2011, the susceptible check variety (not commercially grown) had more than 90% yield loss, while commercial variety losses ranged from 0% to 43% at an average of 21% yield loss without fungicide application. The commercial winter wheat varieties with various levels of resistance suppressed potential yield losses of over 90% to 21% (Chen 2014). Similarly, commercially grown spring wheat varieties were able to reduce potential yield loss from 45% to 15% on average. Several barley varieties also have some degree of resistance. An example of a stripe rust-resistant hard red winter wheat compared to a susceptible club wheat is shown in Figure 10-2.

Figure 10-2. Stripe rust damage on a susceptible club wheat (right) compared to a resistant hard red winter wheat, cv. 'Farnum' (left). (Photo by Xianming Chen.)

Chemical

When a variety does not have adequate resistance, fungicide should be used to suppress the disease and reduce yield loss. Several triazole (Group 3), succinate dehydrogenase inhibitor (SDHI) (Group 7), and strobilurin (Group 11) fungicides are labeled for control of stripe rust and information can be found in the PNW Disease Management Handbook. Check the labels for their rates, total amount that can be used in a growing season, and the latest stage by which they can be used. In the PNW, usually either one or two applications at the time of herbicide application (early application) and/or at the flag leaf to flowering stage (late application) are needed depending upon how early stripe rust starts and how fast the disease develops. If stripe rust starts early, early application is needed in fields grown with moderately susceptible or susceptible varieties (ratings 5–9) to reduce over-wintered rust and prevent new infection in the early growing season. Early application is easy to do as it is usually through ground application and adds no additional application cost because of mixing with herbicide. Often, this early application is not necessary if the disease starts developing after the herbicide application time. It is more important to use fungicide to protect crops for the grain-filling period. It is critical to make a decision for applying or not applying fungicides before flowering stage based on variety susceptibility, yield potential, disease pressure, and weather conditions, as most labeled chemicals cannot be used after flowering.

For more information on stripe rust, go to the USDA-ARS and Washington State University stripe rust website at ***http://striperust.wsu.edu/***.

Rhizoctonia Root Rot and Bare Patch

Background, causal agents, and distribution

Rhizoctonia is a soilborne parasitic fungus that can attack root systems of wheat and barley, pruning and rotting the roots and inhibiting their ability to take up water and nutrients. As a result, plants are stunted and show decreased yields. This disease was first documented in Australia in the 1930s and discovered in the PNW in the mid-1980s.

Most *Rhizoctonia* species have a wide host range and will attack cereal crops and volunteers, broadleaf rotation crops, and grassy weeds (Table 10-4).

Table 10-4. Rhizoctonia root rot characteristics and management options for dryland cereal producers.

Rhizoctonia root rot and bare patch	
Background • Causal agents: *Rhizoctonia solani* AG-8; *R. oryzae* • Source: infested soil, residue • Wide distribution • Wide host range: cereals, grasses, rotation crops • High risk: cool, wet, spring conditions	**Economic impact** • 10–20% potential grain yield loss **Management options** • Eliminate green bridge • Starter fertilizer placed below or with seed • Residue management (higher disturbance seed openers in no-till systems; fallow)
Key diagnostics • Chronic: field areas of uneven plant height • Acute: field patches with extreme stunting • Sunken brown lesions on girdled or severed roots (spear points)	**Ongoing research** • Distribution and impact surveys • Resistance screening • Suppressive soils, biocontrol • Fungicide efficacy (sedaxane)

Adapted from Schroeder 2014.

The most virulent *Rhizoctonia* causing root rot and bare patch of wheat and barley is *R. solani* AG-8; others can cause more mild symptoms. The species *R. solani* contains numerous groups, called anastomosis groups (AGs). Although very virulent on wheat, AG-8 can also attack roots of other rotation crops such as pea, lentil, and canola. Other groups of *R. solani* have been isolated from roots, including AGs 2-1, 4, 5, 9, and 10. Many of these have been tested in Washington State, but do not appear to cause major diseases on wheat, although they are pathogenic on other broadleaf crops such as pea and canola and may colonize wheat roots. According to recent surveys, AG-8 seems to be found in the PNW but not in the other wheat-growing areas of the US, including the upper and lower Midwest.

Another pathogen is *R. oryzae*. This pathogen is more severe as a seed and seedling rotter and can reduce plant stand under high inoculum conditions, as well as cause root rot. It forms very distinct microsclerotia on the roots that are pink-orange in color. One other group that has been isolated from wheat and barley are the binucleate *Rhizoctonia* species, also known as *Ceratobasidium*. There are many subgroups, but most have not been shown to be pathogenic on wheat, except for *R. cerealis*, which causes sharp eyespot on the lower stems of wheat.

Key diagnostic features

Soilborne pathogens are often difficult to diagnosis based on aboveground symptoms. However, *Rhizoctonia* does cause some distinct symptoms. Bare patch, the most acute form of the disease, is easily recognized by large patches in the field of severely stunted wheat or barley. These patches are irregular to circular, extending up to 10–20 feet in diameter (Figure 10-3). The plants in the patch are stunted, have delayed maturity, and may be yellow or purple in color from nutrient deficiency. The patches appear about one month after emergence when the plants in the healthy area continue to grow but the plants in the patches stop growing. Symptoms are more distinct on spring-planted wheat since the pathogen is more active under the cooler, wet conditions of spring.

Figure 10-3. Irregular patches in wheat field caused by *Rhizoctonia solani* AG-8. (Photo by Timothy Paulitz.)

In the more chronic disease phase, instead of patches, the stand will be of uneven height, with tall plants next to smaller stunted plants, giving the field a wavy appearance. In areas of stunted plants, the plant cover is reduced, more ground is visible, and weeds may be more of a problem.

The pathogen attacks the seminal and crown roots of seedlings. This causes a characteristic spear-tipping or tapered tips of roots, where the growing tip of the root is killed. The pathogen causes a brown rot/lesion on the root. In other parts of the root, the outer layer (cortex) is killed, leaving the stele or central vascular system intact. This area is usually brown in color and gives the roots a pinched appearance.

Under severe conditions, the entire seedling can be killed, usually when the plants are young, since young plants are more susceptible than older plants. These seedlings rot fairly quickly and are often difficult to see. Overall, barley is more susceptible than wheat, and shows more symptoms. In terms of economic loss, soilborne pathogens have been documented to cause 3–12% yield loss in wheat; with bare patch, the yield is essentially zero in the patches, and up to 10–20% of the field can be covered by patches in severe situations (Cook et al. 2002).

Disease cycle and conditions that favor the pathogen

Unlike many fungi, *Rhizoctonia* does not produce spores. Thus, it survives primarily as thick-walled hyphae in decaying roots, or as a multicellular structure composed of thick cells called a sclerotium. This inoculum can survive for one to two years in dry or frozen soils. When the root grows adjacent to an infected root or sclerotium in the spring, the fungus is stimulated to germinate, and forms a network of mycelium. These strands attach to the root, penetrate the root, produce enzymes that kill the root, and proceed to grow up and down the root system. Once the root is killed, the fungus can continue to grow on the dead roots as a saprophyte. *Rhizoctonia* distribution is affected by several environmental factors. It is favored by moist soil in the spring, and cool temperatures (50–60°F). The fungus can grow a considerable distance from the inoculum source and makes a network that spreads through the soil, causing patches. This disease is favored by reduced tillage, or direct seeding, but also occurs with **conventional tillage** (Schroeder and Paulitz 2008). Studies have documented that about 2 years after conversion from conventional to

no-till, the disease can become more severe and cause bare patches. However, after no-till has been continued for 7–10 years, the disease then declines which may be a result of natural suppression mediated by natural microflora in the soil. There is some evidence that the pathogen is favored by more sandy soils, possibly due to larger pore sizes and ease of hyphal spread. In eastern Washington, studies have shown the highest incidence of AG-8, and also the appearance of bare patches, is found in the **lower precipitation** zones of the wheat-summer fallow areas of Ritzville, Lind, and Connell, as well as the Dayton-Walla Walla area. *R. oryzae* is more evenly distributed across eastern Washington (Okubara et al. 2014). Finally, sulfonylurea (SU) and imidazolinone (IMI) herbicides may predispose cereals to infection by *Rhizoctonia*.

Potential effects of climate change

The potential effect of climate change on Rhizoctonia root rot is unknown at this time.

Rhizoctonia management strategies

Prevention

The pathogen is already widely distributed, and is not seed transmitted. However, use of fresh, certified seed contributes to overall seedling vigor and health.

Avoidance

There are no resistant wheat or barley varieties at the present time and crop rotation is not effective.

Monitoring

Tracking symptoms by management units during the current growing season is critical to making informed management decisions for the following growing season. There are no effective control actions to suppress the disease once symptoms appear in a crop. Sampling for positive identification of *R. solani* or *R. oryzae* aids decision-making. Historically, quantification of most of the soilborne pathogens has been very difficult. However, recently developed molecular methods of quantification (real-time PCR) and identification are now available and the technology has been transferred to commercial labs (e.g., Western Labs in Parma, Idaho). Bare patch can be monitored visually; remote sensing may be a useful tool for monitoring.

Suppression

Resistance

There is no resistance in any commercially available varieties. However, recent research has identified promising germplasm derived from synthetic wheats, selected in the field under high inoculum conditions (Mahoney et al. 2016).

Cultural practices

Cultural practices are the primary pathway for managing this disease. The most important strategy is green bridge management and appropriate herbicide timing. *Rhizoctonia* can also attack the roots of volunteer crop and grassy weeds in the fall and spring. When these plants are killed by herbicides such as glyphosate (Roundup), the herbicide shuts down the defense pathways in the plant. Thus, *Rhizoctonia* can act as a saprophyte and quickly colonize the dying weed in high levels. If the new crop is planted soon after spraying out, *Rhizoctonia* can bridge or spread from these dying roots to the new crop, causing extensive damage. However, if there is a suitable interval between spraying of weeds and planting, disease is reduced because natural microbial activity will reduce the *Rhizoctonia* inoculum. Research has shown the ideal interval to be about two to three weeks. Tillage may reduce the disease, possibly because of breaking up hyphal networks, but this has other disadvantages such as increased soil erosion, decreased organic matter and **soil health**, and increased fuel inputs. In no-till systems, high-disturbance seed openers such as chisel openers, as opposed to low-disturbance disk openers, may reduce disease. Fallow has been shown to reduce inoculum, but after one year, disease can still occur, although reduced. Rotation has not been shown to be effective because many of the groups also attack broadleaf crops. Application of a starter fertilizer in the seed row has been documented to reduce damage to the seedlings and increase yield by placing the nutrients adjacent to seedling roots and by overcoming early nutrient deficiencies caused by root pruning.

Chemical

Several classes of seed treatments have been shown to improve seedling health at early stages. These include triazoles, fludioxonil, strobilurins, penflufen, and sedaxane, an SDHI. Under high inoculum levels, they can improve plant height, number of roots, etc. However, these chemicals are not systemic in the plant and cannot protect roots of older plants. In most cases, yield will not be increased significantly.

For more information on Rhizoctonia root rot, see Smiley et al. (2012) and Schroeder (2014).

Take-All Disease

Background, causal agents, and distribution

This disease is called take-all because it takes a major proportion of the yield under severe conditions. It attacks the roots, lower crown, and lower stem of wheat plants. It is found worldwide in wheat production areas with higher precipitation or irrigation and neutral-alkaline soils in temperate areas, where wheat is sown in the autumn. It is caused by the pathogen *Gaeumannomyces graminis* var. *tritici* (Ggt). Low populations are found across the dryland PNW region. Cereal hosts include wheat, barley, triticale, and, to a lesser extent, rye. Another subspecies, *G. graminis* var. *avenae* attacks oats, and *G. graminis* var. *graminis* attacks other grasses such as turfgrass and rice. Weed hosts of Ggt include brome grasses, wheatgrass, and quackgrass. However, Ggt populations do not cause disease on broadleaf rotation crops such as pea, lentil, and chickpea (Table 10-5).

The main diagnostic feature is black discoloration of roots, crowns, and lower stems caused by external mycelial growth on the surface of the root or stem. This is seen even after the roots are washed free of soil. In some cases, adhering soil can also cause blackening of the root, but this is not take-all. Black discoloration may also be seen on the subcrown internode and seminal and crown roots. Under severe conditions, the blackening can extend up to the first internode. Crown discoloration can also be seen in other diseases, but the symptoms are distinct. With Fusarium crown rot, the discoloration is brown in color, not black. In

Table 10-5. Take-all characteristics and management options for dryland cereal producers.

Take-all disease	
Background • Causal agents: *Gaeumannomyces graminis* var. *tritici* • Source: decaying roots and host residue in soil • Wide distribution: low levels found in most dryland wheat • Host range: wheat, barley, triticale, rye, grasses • High risk: winter wheat, wheat after wheat, irrigation, neutral-alkaline soils (pH > 6), infertile soils especially where Mn is deficient	**Economic impact** • 30% average annual losses **Management options** • Eliminate green bridge • Rotation with 1–2 years of a non-host broadleaf • Accelerate residue decomposition
Key diagnostics • Black discoloration of roots and lower stem • Dark runner hyphae on roots • Whiteheads • Plants easy to pull or breakage	**Ongoing research** • Screening for resistance (currently no resistant varieties)

common root rot, the discoloration is dark brown to black, but is seen more on the subcrown internode rather than the lowest internode. With eyespot or sharp eyespot, the discoloration is in a distinct, elongated, eye-shaped lesion with distinct margins. The other distinct symptom is whiteheads, seen after heading, when normal wheat heads should still be green. Whiteheads turn prematurely, and the grain is smaller in size. This is because severe infections in the lower stem cut off the flow of water and nutrients during grain filling. But other diseases or pathogens can cause whiteheads, including Fusarium crown rot, Cephalosporium stripe, and cereal cyst nematode.

Disease cycle and conditions that favor the pathogen

Disease is most severe when wheat is grown continuously for 2 or 3 years without rotation to a non-host such as a legume. It is also most severe under high precipitation conditions or irrigation including areas west of the Cascade Mountains. It is mostly a disease of winter wheat, grown in soil with **pH** > 6. It is also most severe in soils deficient in N, P, and especially manganese (Mn). The pathogen survives in infected roots, and colonizes the roots as runner hyphae on the outer parts of roots. Unlike other soilborne pathogens, it does not form a resistant spore that survives in the soil. This is why just one year of rotation to a non-host can reduce the disease. It can form a sexual spore called an ascospore over the winter on fruiting bodies on infected roots and leaf sheaths, but this is probably not important in the epidemiology of the pathogen in the PNW. The disease spreads by runner hyphae from infected roots to new roots.

Potential effects of climate change

If spring and summer precipitation becomes more frequent and total spring and summer precipitation increases, the incidence and severity of take-all could increase. However, the frequency of wheat in the crop rotation and soil pH will likely have a stronger influence than changes in climate.

Take-all management strategies

Prevention

Not feasible since the pathogen is already widely distributed across the dryland PNW and is not seed transmitted.

Avoidance

There is no resistance in any commercially available varieties. Crop rotation with a non-host such as a legume, other broadleaf crop, or fallow is a means of avoidance.

Monitoring

Monitoring field symptoms during the current growing season is critical to making informed management decisions for the following growing season. There are no effective control actions to suppress the disease once symptoms appear in a crop. The disease can easily be identified by looking at black discolored crowns and whiteheads, but by then the damage is done.

Suppression

Resistance

There is no resistance in any commercially available varieties.

Other cultural practices

Rotation with a broadleaf non-host, such as a legume, or cereal, such as oat or corn, can be effective; a 1 to 2-year break is sufficient to reduce inoculum. Eliminating the green bridge is important because the take-all pathogen can survive on volunteers and grassy weeds. Residue management to accelerate breakdown of residues can reduce disease. Optimum fertility and soil pH suppresses the disease; avoid N, P, or Mn deficiencies and limit liming to soil pH > 6.

Chemical

Some seed treatments have been shown to be effective in Europe and Australia, but none are registered in the US. The most commonly used seed treatments have no effect on take-all.

For more information on take-all disease, see Cook (2003).

Pythium Root Rot

Background, causal agents, and distribution

Pythium is a soilborne, parasitic, fungus-like organism that can attack the seeds and root systems of wheat and barley, pruning and rotting the roots and inhibiting their ability to take up water and nutrients. As a result, plants are stunted and show decreased yields. Once classified with fungi, these organisms have been separated out based on several key differences, but plant pathologists still consider their behavior in soil to be like fungi. This pathogen is widespread across all dryland cereal production areas (Table 10-6).

Key diagnostic features

Soilborne pathogens are often difficult to diagnose based on aboveground symptoms. In fact, *Pythium* is often called "the common cold of wheat,"

Table 10-6. Pythium root rot characteristics and management options for dryland cereal producers.

Pythium root rot	
Background • Causal agents: numerous *Pythium* species, but *P. ultimum* and *P. irregulare* group I and IV are most virulent • Source: infested soil, decaying roots • Wide distribution • Wide host range: cereals, grasses, rotation crops • Highest risk: cool, wet, spring conditions; lower, poorly drained areas of the field	**Economic impact** • 15–20% potential yield loss **Management options** • Seed treatment • Eliminate green bridge • Starter fertilizer below or with seed • Residue management to reduce load in no-till systems (use of chaff spreaders, straw choppers, or mowers)
Key diagnostics • Stunted plants, yellowing, reduced tillering • Reduced root system • Delayed heading and poor grain fill	**Ongoing research** • Distribution of species across eastern Washington

because it is ubiquitous and the symptoms are so non-descript. The impacts of *Pythium* were not known until the early 1980s when R.J. Cook did landmark experiments with the new fungicide metalaxyl, which was specific for Oomycetes. When plots treated with a soil drench were compared side-by-side with non-treated plots, the effect was dramatic: greener tissue, more canopy, greater plant height, and better yield. *Pythium* causes stunting and yellowing of seedlings, which is often seen in wetter, lower parts of the field. Unlike *Rhizoctonia*, it is hard to see *Pythium* symptoms on the roots because infected roots are quickly rotted away. However, overall, the root biomass is less, and there are fewer lateral roots (Figure 10-4). Root hairs are reduced, but this can only be seen under a microscope. The first leaf is often reduced in length because of early infection while still in the embryo stage. One of the best diagnostic features is the observation of resting spores, called oospores, inside rotted roots, but this also requires a microscope. However, researchers at the USDA-ARS have developed molecular methods of detecting and quantifying *Pythium* in the soil. These tests can be performed by commercial labs (e.g., Western Labs in Parma, Idaho).

Figure 10-4. Top: *Pythium* symptoms on untreated wheat (rows on the left) and metalaxyl-treated wheat (4 rows on the right). Bottom: Wheat grown in pasteurized soil shows healthy root growth (left) compared to roots grown in natural soil (right). (Photos by R. James Cook.)

Disease cycle and conditions that favor the pathogen

Moist soil and cool temperatures (50–60°F) in the spring favor *Pythium* for several reasons. First, many species produce motile swimming spores, called zoospores, that are attracted to roots and seeds and can initiate infections. These require free water in the soil to move, so they are only active in wet soils. However, many of the species in the dryland areas of the PNW do not produce zoospores. Also, cool conditions delay the emergence of seeds and seedlings, giving more time for the pathogen to attack succulent juvenile tissue. *Pythium* species are considered pioneer pathogens because they can grow to the seed or seedling and infect it in a matter of hours as well as rot the root ahead of other pathogens. The slower the emergence (due to cool temperatures), the more damage. However, *Pythium* can still continue to attack the growing roots as long as soil moisture is adequate. Once the root is infected, the pathogen will destroy the root and root hairs. Then, in a matter of days, it will reproduce by producing sporangia, which can produce more zoospores, or produce oospores, which are the thick-walled, resistant survival structures. These can survive in the soil under hot, dry, or cold conditions for many years before germinating to infect roots. This disease can be favored by direct tillage conditions because the heavy residue in the spring will keep the soils wetter and retards the heating of the soil from solar radiation.

Potential effects of climate change

Pythium would be favored by cool, wetter conditions in the spring.

Pythium management strategies

Prevention

Prevention is not feasible. The pathogen is already widely distributed, and is not seed transmitted. However, use of fresh, certified seed contributes to overall seedling vigor and health. This is especially important with *Pythium* because older seed takes longer to emerge and gives more time for *Pythium* to attack.

Avoidance

There are no resistant wheat or barley varieties at the present time, and crop rotation is not effective.

Monitoring

Monitoring field symptoms during the current growing season is critical to making informed management decisions for the following growing season. There are no effective control actions to suppress the disease once symptoms appear in a crop. Sampling for positive identification of *Pythium* can aid management in decisions. Historically, quantification of most of the soilborne pathogens has been very difficult. However, recently developed molecular methods of quantification (real-time quantitative PCR) are now available and the technology has been transferred to Western Labs in Parma, Idaho.

Suppression

Chemical

Most commercially available cereal seed treatments contain metalaxyl or mefenoxam. This chemical is very effective against *Pythium* to protect the seed and young seedling. It is especially important to treat seeds such as pea, chickpea, or lentil, which may not emerge without protection. However, these chemicals are not systemic in the plant, and cannot protect roots of older plants. In most cases, yield of cereals will not be increased significantly. Recently, *Pythium* species from chickpea have been identified with resistance to mefenoxam, but the impact on cereal crops is not known.

Resistance

There is no resistance in any commercially available varieties.

Other cultural practices

One of the most important strategies is elimination of the green bridge using appropriate herbicide timing, which can also be effective against other soilborne pathogens such as *Rhizoctonia*. *Pythium* can attack the roots of volunteer crops and grassy weeds in the fall and spring. When these plants are killed by herbicides such as glyphosate, the herbicide shuts down the defense pathways in the plant, enabling *Pythium* to more readily attack the roots, greatly increasing in population. If the new crop is planted soon after herbicide application, *Pythium* can attack new seedlings that contact these dying roots, causing extensive damage. However, if there is a suitable interval between spraying of weeds and planting, disease is reduced because natural microbial activity will reduce the inoculum. Research has shown the ideal interval for *Rhizoctonia* to be about 2–3 weeks, and this may also apply to *Pythium*. Residue management such as chaff spreaders, straw choppers, mowers, or harrows may reduce the residue in the spring under no-till conditions, allowing soil to warm and dry faster to reduce *Pythium* damage. Rotation has not been shown to be effective because many of the groups also attack broadleaf crops. Fallow may not be effective for all species because of the long survival period of oospores. Application of a starter fertilizer in the seed row has been documented to reduce damage to the seedlings and increase yield by placing the nutrients adjacent to seedling roots and by overcoming early nutrient deficiencies caused by root pruning.

For more information on Pythium root rot, see Smiley et al. (2012).

Fusarium Crown Rot

Background, causal agents, and distribution

This disease is known by a variety of names: Fusarium crown rot, dryland foot rot, and Fusarium root rot. A number of *Fusarium* spp. can colonize the roots, lower stem (crown), and leaf sheaths of wheat and barley, but the most virulent and important are *Fusarium culmorum* and *F. pseudograminearum*. This disease is distinguished from another crown disease, called common root and foot rot, caused by the pathogen *Bipolaris sorokiniana*, which is present in dryland wheat but has not been a major problem like *Fusarium*. To make things even more confusing, the eyespot disease caused by *Oculimacula yallundae* and *O. acuformis* is also commonly referred to as foot rot or strawbreaker foot rot. In dryland areas with little summer precipitation, *Fusarium* pathogens are mainly confined to the lower stem, but in areas with precipitation at flowering or in irrigated areas, these pathogens may also infect heads causing head blight, along with another species, *F. graminearum*.

Fusarium is a parasitic fungus; both species are pathogens of cereals and grasses, including grassy weeds. Cereal hosts include wheat, barley, oats, and rye. Weed hosts include wheatgrass, downy brome, and fescue.

Table 10-7. Fusarium crown rot characteristics and management options for dryland cereal producers.

Fusarium crown rot	
Background • Causal agents: *Fusarium pseudograminearum; F. culmorum* • Source: infested soil, residue, chlamydospores in soil • Wide distribution, in 95% of PNW fields • Host range: cereals, grasses, not broadleaf crops • High risk: drought, water stress, excess N fertilizer, highly susceptible varieties	**Economic impact** • 10% average yield loss, but as high as 30% **Management options** • N management: do not over fertilize: excess N can lead to drought stress • Avoid early planting of winter wheat • Avoid highly susceptible varieties • Residue management
Key diagnostics • Brown discoloration of lower stem, subcrown internode • Whiteheads	**Ongoing research** • Resistance/tolerance breeding

However, they do not cause disease on broadleaf rotation crops such as pea, lentil, and chickpea (Table 10-7).

Key diagnostic features

The key diagnostic feature of both *F. culmorum* and *F. pseudograminearum* is the chocolate brown discoloration that is found on the lower stem and internodes of a mature wheat plant. In the early part of the growing season, the outer leaf sheaths are also brown and the infection can extend into the main culm. Brown discoloration can also be seen on the subcrown internode and seminal and crown roots. Crown discoloration can also be seen in other diseases, but the symptoms are distinct. In take-all, the discoloration is black in color, not brown. In common root rot, the discoloration is dark brown to black, especially on the subcrown internode. With eyespot or sharp eyespot, the discoloration is in a distinct, elongated, eye-shaped lesion with distinct margins. The discoloration caused by *Fusarium* is more diffuse, present on the entire culm. The other distinct symptom is whiteheads, seen after heading, when normal wheat heads should still be green (Figure 10-5). Whiteheads turn prematurely and the grain is smaller in size. This is because severe infections in

Figure 10-5. Crown damage and whiteheads caused by Fusarium crown rot. (Photos by Richard Smiley, Oregon State University.)

the lower stem cut off the flow of water and nutrients during grain filling. *Fusarium* can easily be isolated from discolored stems, and pinkish mycelium is often seen inside infected culms.

Disease cycle and conditions that favor the pathogen

Disease is often most severe under water stressed conditions and where excess nitrogen has been applied. Drought stress predisposes the pathogen to move into the crown from previous latent infections. When too much nitrogen is applied under dryland conditions, plants produce luxurious vegetative growth, but then run out of water and go into drought stress. Early planting of winter wheat can also result in plants outstripping the water supply and allows for a greater period of time for *Fusarium* to infect in the fall. Both *Fusarium* pathogens can grow under very dry conditions, much drier than most fungi. Both pathogens survive in infected stubble. Poole et al. (2013) found that *F. pseudograminearum* occurred more frequently at lower elevations with higher temperatures than *F. culmorum*, which was found more often at higher, moister, cooler sites. The pathogens can infect roots in the fall (in the case of winter wheat) but can also infect the lower stem and crown at or below the soil line from contact with infected stubble. These infections can proceed into the outer leaf sheaths and into the culm, and later move up 2 or 3 internodes. Both species also produce thick-walled chlamydospores in the soil, which can survive for many years, especially in the case of *F. culmorum*, hence the ineffectiveness of short rotations away from cereals.

Potential effects of climate change

This disease may increase under more frequent drought conditions.

Fusarium management strategies

Prevention

The pathogen is already widely distributed across the dryland PNW, and is not seed transmitted.

Avoidance

There are no resistant wheat or barley varieties at the present time. Crop rotation is not an effective management strategy because the pathogen (especially *F. culmorum*) has the ability to survive for several years between the presence of a host crop. However, rotation may reduce inoculum.

Monitoring

Monitoring field symptoms during the current growing season is critical to making informed management decisions for the following growing season. There are no effective control actions to suppress the disease once symptoms appear in a crop. The disease can easily be identified by looking at discolored crowns and whiteheads, but by then the damage is done. But this can give the grower an idea of the susceptibility of varieties and where the disease is a problem in the field.

Suppression

Chemical

Most seed treatments contain triazoles, and claims have been made that seed treatments can reduce Fusarium crown rot, but good data is lacking. Part of the problem is that this is a crown disease, and most seed treatments do not provide long-term systemic protection to older plants at a time when infection may occur in roots or at the soil line. However, new seed treatment fungicides with different modes of action are currently being tested, but efficacy is unknown.

Resistance

There is no resistance in any commercially available varieties. However, there is a range of susceptibility and some varieties may be more tolerant than others. Growers should avoid highly susceptible varieties.

Other cultural practices

One of the most important suppression strategies is to manage nitrogen and water stress. The fertilizer rates should be based on realistic yield goals dependent on the stored soil moisture so that excessive vegetative growth and depletion of soil water is avoided. Later planting of winter wheat may also avoid outstripping the water supply and reduce the time in the fall when root infections can occur. Rotation with a broadleaf crop or fallow is only effective if there are two or more years out of cereals because of the long survival of the pathogen in the soil. Studies on residue management have been mixed. Burning has not been effective in reducing disease because the fungus can survive in the soil and lower crowns, which do not reach lethal temperature during burning. Stubble sizing and harrowing did not affect disease level probably because the inoculum is then spread around the field. Some studies from Australia have shown that row cleaners to remove stubble from the rows and precision placement of rows between the previous rows may reduce disease.

For more information about Fusarium crown rot, see Smiley et al. (2005; 2012).

Root-Lesion Nematode

Background, causal agents, and distribution

Root-lesion nematodes are tiny, worm-shaped, migratory endoparasites that live throughout the root zone, feed on living plant roots, and are well-adapted to survive between host crop growing seasons. Root-lesion nematodes have multiple hosts (cereals, oilseeds, grain legumes, and grasses) and are adapted to different cropping systems and agroclimatic conditions across the PNW. Management options are limited; inoculum can increase and spread rapidly once introduced in a field. Two species, *Pratylenchus thornei* and *P. neglectus*, are important economically in the dryland region; *P. neglectus* is more prevalent in the inland PNW, whereas *P. thornei* typically causes greater impact worldwide (Table 10-8). The

Table 10-8. Root-lesion nematode characteristics and management options for dryland cereal producers.

Root-lesion nematode	
Background • Causal agents: *Pratylenchus neglectus; P. thornei* • Source: soil, eggs, and host crop roots • Wide distribution across the PNW including driest zones • Wide host range: small grains, grasses, lentils, chickpea, peas, oilseeds, pasture legumes • High risk: continuous cropping	**Economic impact** • Average yield reduction of 5% but can reduce grain yield up to 50% in the PNW; $51 million annual impact **Management options** • Monitor populations and risk (soil test) • Fallow • Eliminate green bridge • Rotate with less susceptible crop such as barley
Key diagnostics • Decrease in number of lateral roots, reduced root mass; presence of root lesions • Stunted, yellowed plants • Can be confused with fungal root rots	**Ongoing research** • Rotation and tillage effects • Development of varieties with both resistance and tolerance to *P. neglectus* and *P. thornei*

damage caused by root-lesion nematodes reduces wheat profitability in the region by an estimated $51 million annually (Smiley 2015b). Root-lesion nematodes have been found in up to 90% of sampled fields in Washington, Oregon, and Idaho, including the driest grain-fallow areas, and reduce small grain yields by an estimated 5% annually in the tri-state region, although damage can be up to 50%. Oregon studies indicate that yields may be reduced when populations exceed a potential damage threshold of 1,000 nematodes per pound of soil. However, relationships between plant-parasitic nematode densities, yield response, and economic damage are difficult to generalize across regions because they are influenced by site-specific interactions with climate, soils, host crop tolerance, and other biotic factors.

Key diagnostic features

Damage caused by root-lesion nematodes is often not recognized and is underestimated. Aboveground symptoms are non-specific, including stunting, reduced tillering, and chlorosis. Moisture or nutrient deficiencies occur earlier than in adjacent healthy plants, limiting yield (Figure 10-6). Root penetration and feeding reduces the number of root hairs and the extent of root branching on intolerant crops, restricting water and nutrient uptake. Symptoms appear on roots when plants are 6–8 weeks old and can be confused with *Pythium* or *Rhizoctonia* symptoms. Lesions caused by nematode penetration of root tissues predispose crops to secondary infections by other root rot pathogens. Typically, root-lesion nematode distribution is variable within fields; crop canopies may show irregular height and growth stages.

Disease cycle and conditions that favor the pathogen

The root-lesion nematode completes its life cycle in 6 to 9 weeks and increases rapidly throughout the growing season. Adult females deposit up to 1 egg per day inside susceptible host roots or in moist soils when temperatures are favorable (68–77°F) and can remain active in cold, moist soil. Juveniles emerge from eggs at around one week; juveniles and adults feed both on and inside plant root tissue.

Figure 10-6. Reduced productivity of wheat grown in root-lesion nematode-infested soil (right, untreated) compared to nematicide-treated wheat (left). (Photo by Richard Smiley, Oregon State University.)

Distribution and population are influenced by many agronomic and environmental factors. Continuous cropping with susceptible host crops favors root-lesion nematodes; populations increase with the planting frequency of susceptible small grains, oilseeds, peas, lentils, and chickpeas. Spring wheat is more susceptible than winter wheat; barley is less susceptible than wheat. Volunteer host crops and weeds harbor inoculum between and during growing seasons. Soil texture does not appear to limit population density. Economic levels occur in both the grain-fallow and the higher precipitation annual crop areas. However, damage is typically greater in more limited soil moisture conditions. Root-lesion nematodes may move vertically in the soil profile to reach optimal soil moisture. Conservation tillage does not appear to favor populations. However, imazamox-resistant wheat varieties may be impacted by root-lesion nematodes migrating from dying weeds as a result of imazamox applications to control winter annual grassy weed infestations. Conservation Reserve Program (CRP) fields can support high populations.

Potential effects of climate change

The potential effect of climate change on root-lesion nematodes is unknown at this time. However, studies show that distribution and density are impacted by variability in temperature and in winter precipitation levels (Kandel et al. 2013).

Root-lesion nematode management strategies

Prevention

Field and equipment sanitation are the first lines of defense to prevent the introduction of inoculum from infested soil into clean fields. However, these strategies may be of limited utility due to the widespread distribution of these nematodes in the PNW.

Avoidance

Avoidance by host resistance or rotation is not currently an option. All locally adapted commercial wheat varieties that have been tested are susceptible to root-lesion nematodes and a wide range of hosts crops are also susceptible.

Monitoring

From a management perspective, it is highly useful to identify root-lesion nematodes to the species level; crop varieties vary widely in their response to *P. thornei* or *P. neglectus*. These species are very similar and difficult to distinguish. Recently developed DNA-based molecular testing can precisely identify and quantify individual species. Risk of economic damage increases as populations exceed 1,000 nematodes per pound of soil, at any soil depth.

Suppression

Risk can be reduced by decreasing populations; inoculum increases with the frequency of a susceptible crop.

Chemical

There are no foliar or seed-applied treatments to control root-lesion nematodes; no nematicides are currently registered for use in the PNW, and no commercial biological controls are available.

Resistance

No locally adapted commercial wheat varieties are resistant to either *P. neglectus* or *P. thornei*; several varieties show tolerance to *P. neglectus*. Spring wheat varieties with moderate tolerance to both root-lesion nematode species include 'Buck Pronto,' 'Tara 2002,' and 'Jerome.' Barley is less susceptible than wheat and can help reduce inoculum. Barley varieties respond variably to *Pratylenchus*; two-rowed feed barleys 'Camas' and 'Bob' have tolerance to both species and typically perform better than spring wheat varieties. Planting barley in fields transitioning from CRP can reduce risk to a subsequent wheat crop. Regional breeding program goals include developing wheat varieties with dual species resistance and tolerance to reduce damage and eliminate the need to identify root-lesion nematodes by species.

Other cultural practices

Rotation is not an effective standalone management practice. Where economical, a winter wheat-spring barley-summer fallow rotation can help reduce *Pratylenchus* populations in two phases: barley is less susceptible than wheat, and clean fallow controls host plants. Eliminating the green bridge and controlling volunteer and weed hosts during and between crop seasons also help reduce populations.

For further detail on root-lesion nematodes, see Smiley (2015a; 2015b), the primary sources for the information presented in this section.

Cereal Cyst Nematode

Background, causal agents, and distribution

Cereal cyst nematodes are sedentary endoparasites belonging to the *Heterodera avenae* cyst nematode group, which feed on and form egg-bearing cysts in living roots of small grain cereals and grasses. In contrast to root-lesion nematodes, cereal cyst nematodes do not infest broadleaf crops. Two *Heterodera* species, *H. avenae* and *H. filipjevi*, are important economically in the cereal production regions in Oregon, Washington, and Idaho, reducing average annual wheat profitability $3.4 million (Table 10-9). *H. avenae*, found in most major wheat production areas worldwide, was first reported in western Oregon in 1974, and in eastern Oregon and Washington fields in 1984. By 2005, surveys showed that *H. avenae* had become more widespread in the wheat production areas of all three states. *H. filipjevi*, a quarantine pest, was first identified in Union County, Oregon, fields in 2008 and, more recently, *H. filipjevi* was found at sites sampled in southeast Whitman County, Washington, in 2014. *H. filipjevi* will likely be detected in additional locations in the region using recently developed species-specific molecular testing techniques. Currently there is no evidence that *H. filipjevi* causes greater damage than *H. avenae*. The risk of quarantines being required is small since the pathogen is already established in the area.

Molecular identification and quantification data have helped increase our understanding of the distribution and epidemiology of plant parasitic nematodes. *H. avenae* and *H. filipjevi* are closely related; minor morphological differences distinguish them. However, recent studies show that spring wheat and barley cultivars differ in their response to the two species. For example, spring wheat cv. 'Louise' is susceptible to both species, cv. 'WB-Rockland' is resistant to *H. avenae* but susceptible to *H. filipjevi*, and cv. 'SY Steelhead' has resistance to *H. filipjevi* and is

Table 10-9. Cereal cyst nematode characteristics and management options for dryland cereal producers.

Cereal cyst nematode	
Background • Causal agents: *Heterodera avenae, H. filipjevi* • Source: soil, cysts, eggs, and host crop roots • Distribution: *H. avenae* is widespread in region; *H. filipjevi* has been identified in northeastern Oregon and eastern Washington • Hosts: small grains, grasses • High risk: annual cropping of susceptible host; spring wheat more susceptible than winter wheat	**Economic impact** • Yield reductions up to 50% on intolerant varieties; average reductions of 10% • *H. avenae* reduces annual wheat profitability >$3.4 million in the PNW **Management options** • Eliminate green bridge • Rotate with non-host broadleaf crops • Resistant + tolerant varieties and cultivars
Key diagnostics • Bushy proliferation of small, shallow roots at the points of nematode feeding • White females (pinhead sized) protruding from roots at heading • Patches of stunted, yellowed plants • Whiteheads	**Ongoing research** • Distribution surveys and species determination • Screening varieties for resistance plus tolerance • Resistance breeding with known *Cre* (wheat) and *Rha2* (barley) genes • Yield impact

susceptible to *H. avenae*. Races within a species can vary in reproductive capacity, or virulence, complicating risk assessment and resistance breeding. Recent studies indicate that the *H. avenae* and *H. filipjevi* races found in the PNW differ from those already described worldwide.

Key diagnostic features

Similar to root-lesion nematode, cereal cyst nematode symptoms on small grains are often not recognized and are confused with other causes. Growing an untreated susceptible variety next to a nematicide-treated crop helps researchers determine damage potential (Figure 10-7); no nematicides are registered for commercial use in the PNW. Symptoms of

H. avenae and *H. filipjevi* are indistinguishable. Aboveground symptoms are consistent with nutrient and water deficiency symptoms and mimic those caused by other root diseases or environmental stresses due to field variability. Populations may be randomly distributed resulting in irregular patterns or patches of pale, stunted plants across a field. The number, extent, and location of patches depend on the size of the population and their distribution. Root symptoms typically appear when plants are 6–8 weeks old. Tiny juveniles puncture and feed on young wheat and barley root tips causing root division and short, bushy root structure. At the cereal heading stage, white, pinhead-sized adult females become visible on roots. Once embedded in the root to feed, their bodies swell and protrude from the root surface. As infected roots die, females form a protective egg-filled cyst, dark brown (*H. avenae*) or lighter golden brown (*H. filipjevi*) in color. Root damage may predispose plants to secondary infestation by other root-infecting organisms.

Figure 10-7. *Heterodera avenae* symptoms on 'Alpowa' spring wheat (untreated, left) compared to nematicide treatment (right). (Photo by Richard Smiley, Oregon State University.)

Disease cycle and conditions that favor the pathogen

The cereal cyst nematode completes just one generation per cropping season and lives belowground for its entire life cycle. Cyst-encased eggs remain viable in the soil for many years, bridging growing seasons. Second stage juveniles emerge from overwintered cysts the following spring, as soils warm and moisture is favorable. These juveniles migrate to susceptible host crop roots, and puncture and feed on young meristematic tissue at root tips. Some eggs remain in the cyst for years to better ensure emergence of juveniles in an optimal environment. Emergence of *H. avenae* occurs from late February to late May in eastern Oregon, with the peak in mid-April. Emergence patterns of *H. filipjevi* are not yet well-understood; preliminary studies indicate *H. filipjevi* emergence peaks a few weeks ahead of *H. avenae*.

Continuous cropping with wheat or barley and 2-year grain-fallow sequences favor cereal cyst nematode populations; once infested, damage can spread across a field within 3–4 years. Spring wheat is more susceptible than winter wheat or spring barley, and late-planted winter grains are more susceptible than early-planted winter grains. Low fertility and deficient soil water intensify symptoms; plants benefit from adequate nutrition but do not respond to above-optimal rates. Cysts are sensitive to very dry soils. Conservation tillage does not appear to favor populations; no-till may reduce spread of inoculum throughout a field. While relative damage can be higher in sandy or droughty soils, cereal cyst nematodes are found in many soil types and are not restricted by soil texture. Risk of economic damage increases when *H. avenae* populations exceed 1,400 eggs + juveniles (from cysts and soil matrix) per pound of soil. These levels are commonly found in fields in this region. Reducing populations to fewer than 1,000 eggs + 2nd stage juveniles per pound of soil helps minimize damage (Smiley 2016). Yield is expected to decrease as cereal cyst nematode population increases. However, similar to root-lesion nematodes, relationships between population and yield response vary, as they are impacted by interactions of climate, host crop, and soil factors.

Potential effects of climate change

The potential effect of climate change on cereal cyst nematodes is unknown at this time.

Cereal cyst nematode management strategies

Prevention

Eradication is extremely difficult once cereal cyst nematodes become established. Avoid spreading infested soil to non-infested areas via equipment, animals, shoes, or crops. Infested soil may also be dispersed by wind or water.

Avoidance

Winter wheat sown during typical recommended planting dates will have less damage than spring wheat; plants can be well-established with healthy roots prior to peak emergence of juveniles in the spring. Late-planted winter grains are more susceptible than early-planted winter grains; spring grains are more susceptible than winter grains.

Monitoring

Risk of economic damage increases as *H. avenae* populations exceed 1,400 eggs + juveniles per pound soil. Identification of species is useful when a grower's primary control strategy is based on the selection of variety resistance or tolerance. Wheat, barley, and oat varieties may respond differently to *H. avenae* or *H. filipjevi*. DNA-based molecular testing can precisely identify and quantify individual species and is available through regional commercial and research labs.

Suppression

Chemical

There are no foliar or seed treatments to control cereal cyst nematodes, and no nematicides are registered for use in the PNW.

Biological

No commercial biological controls are available. However, existing fungal and bacterial parasites of *H. avenae* may offer potential for study of or development as a biocontrol in the future.

Resistance

Planting wheat and barley cultivars with moderate resistance and tolerance to cereal cyst nematodes can reduce risk. Ideally, a cultivar should be both resistant and tolerant to prevent buildup of inoculum and yield reduction. Breeding programs are focused on developing cultivars with dual resistance and tolerance to both *H. avenae* and *H. filipjevi*. Recent inland PNW trials identified spring wheat and barley varieties that showed resistance to or tolerance of cereal cyst nematodes (Marshall and Smiley 2016; Smiley 2016; Smiley et al. 2013). Response varied by cultivar, location, and species. For example, the hard red spring wheat cultivar 'WB-Rockland' showed both resistance and tolerance to *H. avenae* but was highly susceptible to *H. filipjevi*. Few wheat cultivars showed both resistance and tolerance to either species. Soft white spring wheat 'Louise' showed susceptibility to both species while 'Ouyen' was resistant to *H. avenae* but susceptible to *H. filipjevi*. Idaho studies identified 2-rowed and 6-rowed spring barley feed cultivars that showed moderate resistance plus moderate tolerance to *H. avenae*. Several barley malt cultivars also showed either resistance or tolerance; less is known about spring wheat and barley responses to *H. filipjevi*. 'SY Steelhead' spring wheat showed resistance to *H. filipjevi* but susceptibility to *H. avenae*. Variety resistance ratings are found in Smiley (2016).

Other cultural practices

Crop rotations which include resistant cereal varieties, non-host broadleaf crops, or fallow, with only 1 year of susceptible wheat, barley, or oats in a 3-year period can significantly reduce cereal cyst nematode numbers. Effective rotations include: (1) a 3-year sequence of winter wheat and two years of a non-host. The two non-host years could include two years of a broadleaf (oilseed or grain legume) crop, two years of clean fallow, or a single year of each; and (2) a 3-year sequence of winter wheat, spring wheat, and fallow or a broadleaf crop, where a resistant variety of spring wheat, winter wheat, or both are used. The traditional 2-year winter wheat-fallow rotation can be effective in the grain-fallow region if using a resistant winter wheat cultivar. Long rotations away from wheat are likely not going to be economical.

Eliminating the green bridge and controlling volunteer host crops and grass weeds during all phases of a rotation helps reduce inoculum. Maintaining optimal fertility levels supports crop vigor.

For further detail on cereal cyst nematodes, see Smiley (2015a; 2016), the primary sources for the information presented in this section.

Eyespot (Strawbreaker Foot Rot)

Background, causal agents, and distribution

The eyespot pathogens are capable of infecting wheat, barley, oats, rye, and several other grasses. However, winter wheat is the primary economic host, with spring wheat and barley only affected occasionally. Winter wheat losses can be up to 50% with severe infections (Table 10-10). The name eyespot comes from the characteristic eye-shaped lesions that occur on infected stems near the soil surface. This widespread disease in the PNW has been called strawbreaker foot rot locally since the mid-1900s, but the rest of the world knows it as eyespot. The name strawbreaker foot rot comes from the disease occurring near the base or foot of the stem

Table 10-10. Eyespot characteristics and management options for dryland cereal producers.

Rhizoctonia root rot and bare patch	
Background • Causal agents: *Oculimacula yallundae; O. acuformis* • Source: infested crop residue • Wide distribution across PNW; more common in the higher precipitation zones • Host range: mainly winter wheat, but some spring wheat and barley • High risk: early planting, 45–55°F with fall rains, open winter, susceptible variety	**Economic impact** • Up to 50% yield reduction when severe **Management options** • Resistant winter wheat varieties • Foliar fungicide in spring before stem elongation • Delayed fall seeding
Key diagnostics • Eye-shaped lesions on stem or leaf heath (honey-brown with dark centers) • Whiteheads • Multi-directional lodging	**Ongoing research** • New fungicide efficacy testing • Resistance screening of advanced winter wheat lines and wild wheat, determining if eyespot resistance genes are equally effective for both pathogen species

Adapted from Murray 2014a.

and the tendency of infected stems to break and fall over, resulting in widespread lodging (Figure 10-8). An older and now less commonly used name is Cercosporella foot rot, which is derived from the old name of the causal fungus, *Cercosporella herpotrichoides*.

Eyespot is now recognized as being caused by two closely related fungi, *Oculimacula yallundae* and *O. acuformis*. Until about 1989 when the sexual reproductive stage was discovered in South Australia, these fungi were grouped into the single species *Pseudocercosporella herpotrichoides* with varieties *herpotrichoides* and *acuformis*, respectively. At that time, variety *herpotrichoides* (*O. yallundae*) was the predominant eyespot pathogen in the PNW, but now variety *acuformis* (*O. acuformis*) is equally common.

Two other diseases that can be confused with eyespot are sharp eyespot, caused by *Rhizoctonia cerealis*, and Fusarium crown rot, caused by *Fusarium culmorum* or *F. pseudograminearum*. As the name suggests, sharp eyespot has lesions on stems that are eye-shaped with a distinct

Figure 10-8. Lodging of winter wheat caused by eyespot. (Photo by Tim Murray.)

margin, but in the PNW are more superficial and rarely serious enough to cause yield loss. Fusarium crown rot is widespread and potentially damaging, but is distinguished by infected roots, crowns, and stem bases as opposed to eyespot, which only infects stem bases.

Key diagnostic features

Eyespot is very difficult to detect and identify with certainty in the early stages of disease development, and there is no way to determine which of the eyespot fungi is present by looking at stem lesions. The key diagnostic feature of eyespot is the presence of honey-brown, elliptical lesions on the leaf sheaths and true stem (Figure 10-9). Eyespot lesions have a diffuse, dark-brown margin, lighter brown center, and often have dark-colored centers, which is composed of fungal hyphae. One or more lesions can be present on the same stem. As plants age, lesions on true stems may become sunken in the center with bending or breaking of the stem. Lesions can also coalesce into larger lesions when more than one occurs on a stem.

Figure 10-9. Characteristic early season stem lesions of eyespot. (Photo by Tim Murray.)

Dead standing stems known as whiteheads may appear during warm weather after grain begins to develop. Infected stems may also fall over or lodge after grain has begun to develop. In some instances, lodged stems may fall in different directions, a symptom known as "straggling," or they may fall in the same direction, which often occurs after a storm.

Disease cycle and conditions that favor the pathogen

The eyespot fungi survive in the residue of plants that were infected while they were alive and nowhere else. The length of time they can survive depends on the environmental conditions, but typically can survive 3 or more years under PNW conditions (longer under dry conditions). In the fall when temperatures are about 40–50°F and rain is common, these fungi begin producing millions of microscopic spores on the infested residue. Spores spread to nearby seedlings by splashing and blowing rain where they land on leaf sheaths, begin growing, and penetrate and infect the plant. The eyespot fungi grow slowly and colonize the outer leaf sheaths of the developing plant and remain there until the true stem develops in the spring. The true stem becomes infected when it grows up through the colonized leaf sheaths, giving the eyespot fungi an opportunity to penetrate it. Once in the stem, the eyespot fungi colonize it and destroy structural vascular tissues that can result in lodging. Yield can be reduced even when the crop does not lodge, although greatest damage occurs when the crop lodges.

Fall weather is most important for eyespot: cool temperatures from 40–50°F with frequent rain is important. Eyespot is likely to be more severe in years when winter conditions are mild with minimal snow cover because the pathogen can continue to spread and develop in infected plants. In contrast, cold winters with prolonged snow cover reduce the potential for eyespot because the pathogens do not spread, and they develop slowly inside infected plants at the low temperatures under snow.

Early seeding favors eyespot, likely because the larger plants have more susceptible leaf sheaths for infection and are more likely to be contacted when spores are splashing around than smaller plants. Although the worldwide literature on residue management practices is mixed, eyespot is less severe in no-till fields than in conventionally tilled fields in the PNW, and this is likely due to the later seeding dates associated with reduced or no-till and not the presence of residue per se. Short rotations, dense canopy, spring frost, and excess N status may also favor eyespot. Laboratory studies indicate that the rate of asexual sporulation is sensitive to temperature, light, water, and nutrient status, but it is not known how these factors may influence sporulation under field conditions. Increased understanding of population biology and epidemiology are needed to improve eyespot disease management.

Potential effects of climate change

Eyespot may become more severe in years when winter temperatures are warmer and there is less snow cover, as occurred in 2015–2016 because these conditions are more favorable for infection and disease development.

Eyespot management strategies

Prevention

The eyespot pathogens were first reported in the PNW over 100 years ago and are widely distributed, so sanitation practices such as cleaning equipment to prevent infestation of a field are not practical or effective.

Avoidance

Planting an eyespot-resistant variety is the primary recommendation for its control. Several winter wheat varieties with effective eyespot resistance are available, but no resistant spring wheat or barley varieties have eyespot resistance owing to the relative unimportance of this disease in spring cereals.

Monitoring

Fields planted to eyespot-susceptible varieties should be scouted in early spring before stem elongation to determine whether eyespot is present in sufficient amount to justify a foliar fungicide application. Eyespot is favored by moist soil conditions and, consequently, often found in low areas of fields such as draws, swales, and in the toeslope of hills. Collect enough plants from around the field to give 50 stems; wash those stems and determine if they have eyespot lesions. Consider a foliar fungicide when 10% or more of the stems have recognizable eyespot lesions before stem elongation begins.

Suppression

Chemical

There are no seed treatments that control eyespot. Fungicide application should be considered when 10% or more of stems have recognizable lesions before stem elongation begins (Zadok's growth stage 30). Several foliar fungicide treatments are registered for eyespot control: propiconazole + thiophanate-methyl; cyproconazole + thiophanate-methyl; fluxapyroxad + pyraclostrobin; and azoxystrobin + propiconazole. Resistance to the benzimidazole fungicides, thiabendazole, and thiophanate-methyl is present in Washington and Oregon. For this reason, use of a fungicide mixture containing more than one mode of action is recommended.

Resistance

See the Avoidance section.

Other cultural practices

Avoiding early seeding relative to the production area can be helpful in limiting eyespot development, but will not prevent it. Likewise, crop rotation may be useful for eyespot management by allowing time for infested residue to decompose, but again will not prevent eyespot. Planting spring wheat or barley instead of winter wheat in fields with history of severe incidence can suppress damage.

For more information on eyespot disease, see Murray (2006; 2014a).

Cephalosporium Stripe

Background, causal agents, and distribution

Cephalosporium stripe is a chronic disease of winter wheat in the inland PNW. It was first reported in Washington State in the early 1950s, having been described in Japan in 1934. It has since been described in several other areas of the US and Europe. Cephalosporium stripe is a vascular wilt-type disease because the pathogen infects and colonizes the water-conducting tissue (xylem) of the plant while it is alive and, in doing so, spreads throughout the plant.

Cephalosporium stripe is caused by the fungus *Cephalosporium gramineum*, which is an asexually reproducing fungus. Although this fungus has a wide host range among cereals and other grasses, the primary economic host is winter wheat. This disease can cause total loss of a wheat crop when environmental conditions are favorable and a susceptible variety is grown. The pathogen can be found in all precipitation zones in the inland PNW but is more common in the higher precipitation zones (Table 10-11).

Key diagnostic features

Cephalosporium stripe is easy to diagnose when characteristic yellow stripes develop in the leaves (Figure 10-10). Stripes run the length of the leaf blade and then extend down the leaf sheath. Symptoms may be present in late winter to early spring, depending on the susceptibility

Table 10-11. Cephalosporium stripe characteristics and management options for dryland cereal producers.

Cephalosporium stripe	
Background • Causal agent: *Cephalosporium gramineum* • Source: infested crop residue • Wide distribution across PNW; more common in the higher precipitation zones • Host range: cereals, especially winter wheat; fall annual grasses • High risk: early seeding, 45–55°F with fall rains, open winters with soil heaving, low soil pH, susceptible variety	**Economic impact** • Up to 100% yield loss on winter wheat when disease is severe **Management options** • Tolerant varieties • 3-year crop rotation out of winter wheat • Delayed seeding • Reduce residue (fragment or bale) • Liming to raise soil pH > 5.5
Key diagnostics • Vascular wilt, long yellow stripes in leaf blade extending down sheath; brown streaks in yellow stripes • Whiteheads, stunting, double canopy	**Ongoing research** • Resistance screening • Seed transmission • Molecular detection • Transfer of genetic resistance from wheatgrass • Genetic variation

Adapted from Murray 2014b.

Figure 10-10. Characteristic yellow striping of Cephalosporium stripe disease. (Photo by Tim Murray.)

Figure 10-11. Whiteheads (stunted, dead standing stems) on wheat caused by severe Cephalosporium stripe. (Photo by Tim Murray.)

of the variety. Eventually, small, brown, necrotic streaks develop in the center of the stripes and ultimately the entire width of the stripe may turn brown as the tissue dies. Depending on environmental conditions and susceptibility of the variety, stripes may appear in the flag leaf and eventually the head, resulting in dead standing stems known as whiteheads (Figure 10-11). Infected stems are often stunted, resulting in a "double canopy" with heads on healthy stems standing taller than heads on infected stems. Cephalosporium stripe is favored by moist soil conditions and, consequently, often found in low areas of fields such as draws, swales, and in the toeslope of hills.

Disease cycle and conditions that favor the pathogen

C. gramineum survives primarily in the residue of plants that were infected while they were alive. The length of time the pathogen can survive depends on the environmental conditions, but typically it can survive 3 or more years under PNW conditions (longer under dry conditions). *C. gramineum* is also seedborne in very low percentages, but this source of inoculum is not important under PNW conditions. The disease cycle is similar to that of eyespot: in the fall when temperatures are about 40–50°F and rain is common, *C. gramineum* produces millions of microscopic spores on infested residue that are washed into the soil near the roots and crowns of winter wheat plants. Spores germinate and penetrate the plant through wounds in stem bases and roots near the crown. Once inside the plant, the fungus grows into the young xylem tissue and begins producing more spores and toxic materials that result in formation of the yellow stripes. As the plant is dying, the fungus colonizes plant tissues outside the xylem and uses it as a food source for survival. Early seeding of winter wheat is favorable to Cephalosporium stripe because larger plants have larger root systems that are more susceptible to injury and subsequent infection. Open winters with multiple soil freeze-thaw events and short crop rotations (i.e., wheat-fallow) favor development of Cephalosporium stripe. Acid soil conditions can strongly influence development of this disease, with increased incidence and severity as soil pH drops below 5.2.

Potential effects of climate change

The impacts of climate change are difficult to predict with Cephalosporium stripe. In the near term, this disease may become more severe because open winters with frequent freeze-thaw events seem to favor disease development.

Cephalosporium stripe management strategies

Prevention

Cephalosporium stripe is widely distributed in the inland PNW, so sanitation practices such as cleaning equipment and tires to prevent infestation of fields are not practical.

Avoidance

Wheat varieties vary in their response to Cephalosporium stripe, ranging from tolerant to very susceptible; however, none have highly effective resistance.

Monitoring

Cephalosporium stripe is difficult to observe in fields before heading. Moreover, there are no management practices that will mitigate the impact of Cephalosporium after the crop has been planted. A molecular test is available to detect the pathogen in seed intended for export, but it has not been used commercially.

Suppression

Chemical

There are no effective seed-applied or foliar fungicides for the control of Cephalosporium stripe.

Other cultural practices

Avoiding early seeding and planting a tolerant variety can greatly reduce the development and impact of Cephalosporium stripe. Likewise, use of a 3-year crop rotation with winter wheat no more than once every three years can also reduce the impact of Cephalosporium stripe by allowing time for infested residue to decompose. Fragmenting infested residue to speed decomposition can help reduce the impact of Cephalosporium stripe. The literature on the effect of residue management practices on Cephalosporium stripe is mixed, with some reporting greater disease in reduced tillage systems than conventional and vice versa. In the PNW, Cephalosporium stripe is less severe in no-till fields than in conventionally tilled fields; as with eyespot, this response is likely due to the later seeding dates associated with reduced or no-till and not the presence of residue per se. However, no-till fields also have fewer freeze-thaw events and less soil heaving than conventionally tilled fields, which may contribute to reduced disease. Liming of soils to raise pH above 5.5 is beneficial in reducing the impact of Cephalosporium stripe where soil pH is low.

For more information on Cephalosporium stripe, see Murray (2014b), Quincke et al. (2014), and the Cephalosporium stripe page on the WSU Small Grains website.

Wheat Soilborne Mosaic

Background, causal agents, and distribution

Wheat soilborne mosaic (WSBM) disease is caused by the *Soilborne wheat mosaic furovirus* (SBWMV), which is transmitted by the fungus-like organism *Polymyxa graminis*. WSBM is a disease of winter wheat that was discovered in 1919 in Illinois and called "rosette" disease, but it wasn't until the 1960s that *Polymyxa* was identified as the vector. WSBM has been an important disease in the Great Plains, Midwestern, and Northeastern wheat-producing areas since its discovery.

WSBM was first identified in Washington in 2008, but was reported across the border in adjacent Umatilla County, Oregon, in 2005, and before that in the Willamette Valley of western Oregon in 1994. Whether the virus spread from western Oregon or how is not known. In Washington, it appears to be localized in the Walla Walla area. Because this disease is newly recognized in the region, breeding for resistance has not been a priority, and most winter wheat varieties are susceptible. Yield losses can be up to 75% with severe infection of highly susceptible varieties. Spring wheat and spring barley typically do not develop symptoms (Table 10-12).

Key diagnostic features

Symptoms of WSBM disease include a green to yellow mosaic on the leaves that appears in late winter or early spring as plants are beginning to grow. Depending on the virus strain and susceptibility of the variety,

Table 10-12. Wheat soilborne mosaic characteristics and management options for dryland cereal producers.

Wheat soilborne mosaic	
Background • Causal agent: *Soilborne wheat mosaic furovirus*, vectored by *Polymyxa graminis* • Source: infested soil • Limited distribution: Walla Walla, WA area and adjacent Umatilla County, OR • Host range: wheat, barley, rye, other grasses • High risk: cool, moist soil following seeding	**Economic impact** • Varies with cultivar susceptibility and degree of infestation, from minor to over 75% reduction in grain yield **Management options** • Field and equipment sanitation (prevent infected soil moving to clean fields) • Resistant varieties • Irrigation management following seeding
Key diagnostics • Green to yellow leaf mosaic • Stunting, chlorosis • Rosetting in very susceptible cultivars	**Ongoing research** • Screening PNW cultivars to identify resistance

plants may be severely stunted, a symptom referred to as a "rosette" because the stems don't elongate normally (Figures 10-12 and 10-13). Symptoms often appear in patches that range from small to large, occur in low areas or places where water moves, and may appear in patterns associated with tillage operations. The latter symptoms are associated with distribution of the vector, which is favored by high soil moisture and is moved with infested soil. Symptoms fade as temperatures warm and plant growth increases, leading to misdiagnosis as a nutrient problem associated with cold soil. Although symptoms may fade, plants remain damaged and yield is reduced in affected areas of the field. Yield loss ranges from minor to over 75% when the infestation is extensive and the variety is very susceptible.

Disease cycle and conditions that favor the pathogen

SBWMV survives only in association with its vector, *P. graminis*, in soil and is not seed transmitted. Following seeding of winter wheat, when soil is cool (~50°F) and moist, resting spores of *P. graminis* germinate, penetrating and infecting root hairs of young plants carrying SBWMV with

Figure 10-12. Green to yellow mosaic symptoms of wheat soilborne mosaic (WSBM) disease. (Photo by Tim Murray.)

Figure 10-13. Wheat variety trial showing a variety highly susceptible to wheat soilborne mosaic (WSBM) disease (left) next to a highly resistant variety (right). (Photo by Tim Murray.)

it. Once inside the plant, SBWMV replicates and spreads throughout the leaves in the fall and early winter, eventually resulting in the formation of symptoms. Soil conditions following seeding are critical to infection, with cool and moist soils favoring germination of *P. graminis*. Consequently, WSBM disease may develop with early or late seeding, depending on soil moisture and temperature conditions that occur afterwards. Infection declines when soil temperature is below 45°F.

Potential effects of climate change

It is difficult to predict the impact of near-term climate change on the frequency and severity of WSBM.

Wheat soilborne mosaic management strategies

Prevention

SBWMV appears to be a relatively recent introduction to the PNW and its distribution is limited. Consequently, sanitation practices that reduce or prevent movement of soil from infested to non-infested fields are effective in reducing the impact of this disease. Such practices include cleaning equipment, vehicle tires, and even shoes when traveling between fields.

Avoidance

Growing a resistant winter wheat variety is the primary method for managing WSBM. Wheat varieties adapted to the PNW vary in their response from very susceptible to very resistant (Figure 10-13).

Monitoring

Lab tests are available to detect and confirm the presence of SBWMV in symptomatic plants, but there are no post-infection treatments that can mitigate the damage.

Suppression

Chemical

There are no effective seed-applied or foliar treatments that will mitigate damage from SBWMV. Soil fumigation is partially effective in reducing resting spores of *P. graminis*, but does not eliminate the organism and is not cost-effective.

Other cultural practices

Crop rotation has little effect on SBWMV because the resting spores of *P. graminis* are capable of surviving for long periods of time in soil. For irrigated production, irrigation prior to seeding followed by no irrigation for several weeks after seeding may help reduce the impact of SBWMV.

For more information on WSBM disease, see Flowers et al. (2012) and Murray et al. (2009a).

Looking Ahead

Management strategies will continue to evolve in response to increased understanding of pathogen distribution and pathogen response to changing conservation cropping technologies, production practices, or environmental conditions. Continued climate change may affect pathogen distribution, virulence or aggressiveness, and host crop resistance, tolerance, and susceptibility. System-wide monitoring of crop response to current management strategies is an important tool to help determine if climate change or cropping practices reduce effectiveness of current management strategies. Future adaptations may include improved host resistance, altered planting schedules, new chemistries and adjusted timing and rates of application, or biological control methods.

Resources and Further Reading

Publications

Acid Soils: How Do They Interact with Root Diseases? Washington State University Extension Publication FS195E.

http://pubs.wpdev.cahnrs.wsu.edu/pubs/fs195e/?pub-pdf=true

Green Bridge Control Begins in the Fall. STEEP Conservation Tillage Handbook Chapter 4 No.18.

http://pnwsteep.wsu.edu/tillagehandbook/chapter4/041893.htm

Pacific Northwest Plant Disease Management Handbook

http://pnwhandbooks.org/plantdisease/

Small Grain Seed Treatment Guide. Montana State University Extension MT199608AG.

http://store.msuextension.org/publications/AgandNaturalResources/MT199608AG.pdf

2015–2016 Winter Wheat Breeder Variety Portfolio. Washington State University Extension Publication TB15E.

http://cru.cahe.wsu.edu/CEPublications/TB15/TB15.pdf

Websites

Oregon State University Wheat Research

http://cropandsoil.oregonstate.edu/group/wheat

Oregon State University Columbia Basin Agricultural Research Center (plant pathology research)

http://cbarc.aes.oregonstate.edu/plant-pathology/research-projects

Oregon State University Umatilla Co. Cereal Central (pests)

http://extension.oregonstate.edu/umatilla/pests

University of Idaho North Idaho Cereals (Publications: diseases and insect pests)

http://www.uidaho.edu/extension/cereals/north/publications

Washington State University Small Grains (disease resources)

http://smallgrains.wsu.edu/disease-resources/

Washington State Crop Improvement Association Seed Buyers Guide

http://washingtoncrop.com/

References

Bockus, W.W., R.L. Bowden, R.M. Hunger, W.L. Morrill, T.D. Murray, and R.W. Smiley, eds. 2010. Compendium of Wheat Diseases and Pests, 3rd ed. American Phytopathological Society Press, St. Paul, MN.

Chen, X.M. 2005. Epidemiology and Control of Stripe Rust [*Puccinia striiformis* f. sp. *tritici*] on Wheat. *Canadian Journal of Plant Pathology* 27: 314–337.

Chen, X.M. 2014. Integration of Cultivar Resistance and Fungicide Application for Control of Wheat Stripe Rust. *Canadian Journal of Plant Pathology* 36: 311–326.

Cook, R.J. 2003. Take-All of Wheat. *Physiological and Molecular Plant Pathology* 62(2): 73–86.

Cook, R.J., W.F. Schillinger, N.W. Christensen, R. Jirava, and H. Schafer. 2002. Occurrence of the Rhizoctonia "Bare Patch" Disease in Diverse Direct-Seed Spring Cropping Systems in a Low-Precipitation Zone. Pacific Northwest Conservation Tillage Handbook Series Chapter 4: 19. STEEP.

Kandel, S.L., R.W. Smiley, K. Garland-Campbell, A.A. Elling, J. Abatzoglou, D. Huggins, R. Rupp, and T.C. Paulitz. 2013. Relationship between Climatic Factors and Distribution of *Pratylenchus* spp. in the Dryland Wheat-Production Areas of Eastern Washington. *Plant Disease* 97: 1448–1456.

Ma, L.J., J.X. Qiao, X.Y. Kong, Y.P. Zou, X.M. Xu, X.M. Chen, and X.P. Hu. 2015. Effect of Low Temperature and Wheat Winter-Hardiness on Survival of *Puccinia striiformis* f. sp. *tritici*. *PLOS One* 10(6): e0130691.

Mahoney, A.K., E.M. Babiker, T.C. Paulitz, D. See, P.A. Okubara, and S.H. Hulbert. 2016. Characterizing and Mapping Resistance in Synthetic-Derived Wheat to Rhizoctonia Root Rot in a Green Bridge Environment. *Phytopathology* (posted online July 26 2016).

Marshall, J.M., and R.W. Smiley. 2016. Spring Barley Resistance and Tolerance to Cereal Cyst Nematode *Heterodera avenae. Plant Disease* 100: 396–407.

Matanguihan, J.B., K.M. Murphy, and S.S. Jones. 2011. Control of Common Bunt in Organic Wheat. *Plant Disease* 95(2): 92–103.

Mathre, D.E., ed. 1997. Compendium of Barley Diseases, 2nd ed. American Phytopathological Society Press, St. Paul, MN

Murray, T.D. 2006. Strawbreaker Foot Rot or Eyespot of Wheat. Washington State University Extension Publication EB1378. ***http://cru.cahe.wsu.edu/CEPublications/eb1378/EB1378.pdf***

Murray, T.D. 2014a. Strawbreaker Foot Rot Identification and Management. Washington State University Small Grains Diagnostics workshop video. ***https://www.youtube.com/watch?v=9kTWWUwShPU&list=PLajA3BBVyv1z8QXYCiWDCZHjxDwRcoUPY***

Murray, T.D. 2014b. Cephalosporium Stripe Identification and Management. Washington State University Small Grains Diagnostics workshop video. ***https://www.youtube.com/watch?v=z7ST-waq92C4&list=PLajA3BBVyv1z8QXYCiWDCZHjxDwRcoUPY&index=3***

Murray, T.D. 2016. Agronomic Management and Variety Selection for Conservation Tillage Systems: Diseases of Wheat. Pacific Northwest Direct Seed Association Cropping Systems Conference. PowerPoint presentation. January 12, 2016. ***http://www.directseed.org/files/4914/5339/7640/Agronomic_Management_and_Variety_Selection_for_Wheat_Diseases_Tim_Murray.pdf***

Murray, T.D., H.R. Pappu, and R.W. Smiley. 2009a. First Report of Soilborne Wheat Mosaic Virus on *Triticum aestivum* in Washington State. *Plant Health Progress.* ***http://www.plantmanagementnetwork.org/pub/php/brief/2009/sbwmv/***

Murray, T.E., D.W. Parry, and N.D. Cattlin. 2009b. Diseases of Small Grain Cereal Crops. A Colour Handbook. CRC Press.

NRCS (USDA Natural Resources Conservation Service). 2010. Integrated Pest Management Conservation Practice Standard 595-1. In Field Office Technical Guide. NRCS. ***http://www.nrcs.usda.gov/Internet/FSE_DOCUMENTS/stelprdb1044470.pdf***

Okubara, P.A., K.L. Schroeder, J.T. Abatzoglu, and T.C. Paulitz. 2014. Agroecological Factors Correlated to *Rhizoctonia* spp. in Dryland Wheat Production Zones of Washington State, USA. *Phytopathology* 104(7): 683–691.

Paulitz, T.C., and K. Adams. 2003. Composition and Distribution of *Pythium* Communities in Wheat Fields in Eastern Washington State. *Phytopathology* 93(7): 867–873.

Paulitz, T.C., P.A. Okubara, and K.L. Schroeder. 2009. Integrated Control of Soilborne Pathogens of Wheat. In Recent Developments in Management of Plant Diseases, U. Gisi, I. Chet, and M.L. Gullino, eds. Springer, Dordrecht, The Netherlands.

Poole, G.J., R.W. Smiley, C. Walker, D.R. Huggins, R. Rupp, J. Abatzoglou, K. Garland-Campbell, and T.C. Paulitz. 2013. Effect of Climate on the Distribution of *Fusarium* spp. Causing Crown Rot in the Pacific Northwest of the United States. *Phytopathology* 103: 1130–1140.

Quincke, M.C., T.D. Murray, C.J. Peterson, K.E. Sackett, and C.C. Mundt. 2014. Biology and Control of Cephalosporium Stripe. *Plant Pathology* 63: 1207–1217.

Schillinger, W.F., and T.C. Paulitz. 2014. Natural Suppression of Rhizoctonia Bare Patch in a Long-Term No-Till Cropping Systems Experiment. *Plant Disease* 98: 389–394.

Schroeder, K.L. 2014. Rhizoctonia Root Rot Identification and Management. Washington State University Small Grains Diagnostics workshop video. ***http://smallgrains.wsu.edu/video-rhizoctonia-root-rot-identification-and-management/***

Schroeder, K.L., and T.C. Paulitz. 2008. Effect of Inoculum Density and Soil Tillage on the Development and Severity of Rhizoctonia Root Rot. *Phytopathology* 98: 304–314.

Sharma-Poudyal, D., and X.M. Chen. 2011. Models for Predicting Potential Yield Loss of Wheat Caused by Stripe Rust in the US Pacific Northwest. *Phytopathology* 101: 544–554.

Smiley, R.W. 2015a. Plant-Parasitic Nematodes Affecting Small Grain Cereals in the Pacific Northwest. Pacific Northwest Extension Publication PNW674. Oregon State University. ***https://catalog.extension.oregonstate.edu/pnw674***

Smiley, R.W. 2015b. Root-Lesion Nematodes: Biology and Management in Pacific Northwest Wheat Cropping Systems. Pacific Northwest Extension Publication PNW617. Oregon State University. ***https://catalog.extension.oregonstate.edu/pnw617***

Smiley, R.W. 2016. Cereal Cyst Nematodes: Biology and Management in Pacific Northwest Wheat, Barley, and Oat Crops. Pacific Northwest Extension Publication PNW620. Oregon State University. ***https://catalog.extension.oregonstate.edu/sites/catalog/files/project/pdf/pnw620_2.pdf***

Smiley, R.W., J.A. Gourlie, S.A. Easley, L.M. Patterson, and R.G. Whittaker. 2005. Crop Damage Estimates for Crown Rot of Wheat and Barley in the Pacific Northwest. *Plant Disease* 89: 595–604.

Smiley, R.W., J.M. Marshall, J.A. Gourlie, T.C. Paulitz, S.L. Kandel, M.O. Pumphrey, K. Garland-Campbell, G.P. Yan, M.D. Anderson, M.D. Flowers, and C.A. Jackson. 2013. Spring Wheat Tolerance and Resistance to *Heterodera avenae* in the Pacific Northwest. *Plant Disease* 97: 590–600.

Smiley, R.W., T.C. Paulitz, and J. Marshall. 2012. Controlling Root and Crown Diseases of Small Grain Cereals. Pacific Northwest Extension Publication PNW639. Oregon State University. ***https://catalog.extension.oregonstate.edu/sites/catalog/files/project/pdf/pnw639.pdf***

Chapter 11
Insect Management Strategies

Sanford Eigenbrode, University of Idaho
Edward Bechinski, University of Idaho
Nilsa Bosque-Pérez, University of Idaho
David Crowder, Washington State University
Arash Rashed, University of Idaho
Silvia Rondon, Oregon State University
Bradley Stokes, University of Idaho

Abstract

This chapter provides an overview of the pests affecting wheat systems in the inland Pacific Northwest (PNW). The chapter begins by reviewing the principles of integrated pest management (IPM) and the challenges for insect pest management under projected climate change for the region, along with other potential changes such as biological invasions and the effects of changes in production technology. It then provides specific information about the most important of the region's pests of wheat including their life cycles, injurious stages, management, biological control, and potential responses to climate change. Each is accompanied by photographs and other information for pest identification. Key publications from scientific and Extension literature are provided at the end of the chapter for use by pest managers and others.

Key Points

- Changes in technology and production practices, as well as anticipated changes in climate, have implications for IPM in cereal production systems.

- Effective management of insect pests depends on managers having an understanding of each species distribution, life cycle, crop damage caused, and principles and practices for IPM specific to each pest.
- Evolution of cropping systems including changes in tillage regimes and rotational crops will have important implications for pest management.
- Though anticipated effects on insect pests vary by species, possible mechanisms by which climate change can impact insect pests include: changes in the timing of pest activity, shifts in the geographical range of pests, and shortened life cycle time (thus increasing the number of generations per year).
- The entire wheat system, including rotational crops and surrounding landscapes, influences pest abundance. Spatial and temporal variability in landscapes and agricultural production systems impact populations of insect pests.

Introduction and Background

Overview of Pests Affecting Inland PNW Cereal Production Systems

Cereal systems of the US are subject to economic injury from approximately 30 insect species. The PNW is home to more than 20 of these, including aphids, wireworms, Hessian fly, wheat midge, cereal leaf beetle, Haanchen barley mealybug, armyworms, and cutworms. Their collective potential to reduce yields across our region is substantial, but pressures vary across the region and among years so that relatively few are problematic at any one time or place. For this reason, however, their management requires that producers know these pests, can sample them effectively, and can make prudent decisions about treatments and management practices to minimize their impacts. This section of the handbook provides an overview of the most prevalent insect pests of the **inland PNW**, with information about their biology and life cycles, types of injury, and approaches for their management. It also considers how various farming practices can affect each pest. Finally, it takes a look forward to anticipate possible changes in pressure from these pests as the region's climate changes.

The Elements and Principles of Integrated Pest Management

Effective **integrated pest management** (IPM) depends on regularly scouting fields for the presence and abundance of **pests**. Pest presence indicates a need for vigilance to monitor pest densities to anticipate their population from reaching local economic **thresholds**. Correctly identifying pest species is critical, as many insects look alike but carry different risks and require different management approaches. This chapter provides guides for identifying common pests and citations to sources to help with identification. When in doubt, get the help of a specialist. Basic scouting principles include considering the timing of scouting (to focus on vulnerable stages of the crop and anticipated timing of infestations), scouting weekly during the vulnerable period, sampling randomly within the field to ensure accuracy, and using a sampling method that aligns with treatment thresholds when these are available. Appropriate methods must be used depending upon the goal, which might be to assess pest presence or to estimate pest densities as a basis for making treatment decisions.

Since each of the sections on individual pests discusses management options, here we provide a brief overview of the principles of IPM applicable to most pests. IPM combines nonchemical approaches with judicious use of pesticides to achieve economically viable pest control. A central principle in IPM programs is never to use pesticides as "just in case" insurance treatments or scheduled calendar applications. Rather, IPM producers look first to nonchemical controls and instead use pesticides "just in time," based on pest forecasts and economic thresholds. Nonchemical management tactics include cultural control, mechanical control, and biological control. Cultural controls (modifying the growing environment to reduce the prevalence of unwanted pests) include crop rotation, variety selection, altered planting date, fertilizer application, or fertilizer timing. Mechanical control tactics include physical removal, insect trapping, tillage, and other physically controlling management tactics. Biological controls include naturally occurring or augmented predators, parasitoids and pathogens that attack and kill pests, and any management practices designed to preserve or encourage these beneficial organisms. Chemical controls include so-called "least-toxic" biorational pesticides (that specifically target particular pests over beneficials and other nontarget species) and conventional broad-spectrum insecticides. Over the past decade several least-toxic biorational pesticides have been marketed for use in cereals. Efficacy can vary but these products, because of their narrow spectrum of activity, are environmentally safer than broad-spectrum conventional insecticides such as pyrethroids, carbamates, organophosphates, and neonicotinoids. At the same time, recent changes in application practices (seed treatments, lower rates of application, reduced frequency application) can minimize the effects of broad-spectrum conventional insecticides on beneficials and other nontarget organisms. These materials must be applied according to labels to comply with regulations and minimize environmental impacts, including disruption of biological control. While scouting for the presence of pests, natural enemies should also be noted to determine their relative abundance and assess their potential to control pest populations.

Insect Pest Management in a Diverse and Changing System

A Heterogeneous System: Eastern Idaho to Central Washington

Although inland PNW wheat production systems are united by similarities in climate, terrain, markets, and histories, they are also remarkably heterogeneous. The climate is generally Mediterranean-like, with cold, wet winters and warm to hot, dry summers, but there are significant gradients in average annual precipitation (from <7 to >25 inches) and mean annual temperatures (from 43°F to 55°F). Soils are dominated by Mollisol and Aridisol orders, but Alfisols are present in the wetter subregions. This edaphic and climatic heterogeneity can be delineated into **agroecological classes** in which specific cropping systems from wheat-fallow to continuous cropping with rotations predominate. Which cropping system is used in any parcel is also affected by many local factors. Across large parts of the region, precipitation is inadequate for crop production and irrigation is required (see Figure 1-3 in Chapter 1: Climate Considerations). The variable climate and production systems employed, in turn, can affect insect pests. The distributions of the several wireworm species in the region, for example, differ. As another example, the wheat midge is currently confined to a small portion of our region, while the wheat stem sawfly (*Cephus cinctus* Norton) is a serious problem in Montana wheat systems but with rare reports of injury in the Columbia Basin and Palouse regions.

As detailed in Chapter 1: Climate Considerations, the region's climate is dynamic, and is experiencing a warming trend accompanied by shifts in precipitation that include drier summers. Based on models, the changes will not necessarily be uniform, with warming and precipitation changes occurring at different rates and directions in different parts of the region. One motivation for this book is to anticipate the implications of a changing climate on inland PNW wheat systems and equip producers with scientific knowledge to cope with them. Climate change can affect insect pests of wheat and other crops (Eigenbrode and Macfadyen 2017; Lehmann et al. 2017). The effects can be directly on the pests, or they can be indirect, influencing biological control (Eigenbrode et al. 2015). Science-based projections of the implications of climate change for specific insect pests of inland PNW wheat systems, when available, are presented in this chapter.

Variability and Change in Technology

In addition to ongoing and anticipated changes in climate, cereal production is affected by changes in technology and production practices that have implications for IPM. Although reduced tillage methods have been adopted on much of PNW farmland (see Chapter 3: Conservation Tillage Systems), most of our wheat systems are still grown using **conventional tillage**. Adoption of reduced tillage, which is ongoing, can change pests and their management. Impending technology that could affect pest management includes remote sensing, which is just beginning to include capabilities for sensing biotic stresses in crops, like disease, weed, and insect infestations. During the useful life of this book, we anticipate these sorts of tools will become available, whether deployed via unmanned aerial vehicle (drones), tractor-mounted devices, or otherwise. For example, Russian wheat aphid (*Diuraphis noxia* Mordvilko), which causes distinctive changes in spectral reflectance of infested wheat plants, can be detected remotely based on normalized difference vegetation index imaging (Mirik et al. 2012). There has also been some success in detecting English grain aphid (*Sitobion avenae* F.) in experimental systems (Luo et al. 2013). Russian wheat aphid is of minor concern in our region and the two species differ in the economic injury they cause, necessitating a system that can discriminate between them but has not yet been investigated. Because reliable remote sensing would be such an enormous boon, allowing prudent, "just in time" pest management interventions without time-consuming sampling, its promise has been much discussed and its advent anticipated for more than 20 years. Development of successful applications is inevitable, but remains elusive.

Changing Cropping Systems

Wheat in the region is currently produced under annual crop, annual crop-fallow transition, and wheat-fallow production systems (as described in Chapter 1: Climate Considerations). In recent years there have been trends to increased incorporation of canola and legumes into crop rotation, which is facilitated in drier zones by the availability of fall-planted varieties of these commodities (see Chapter 5: Rotational Diversification and Intensification). There is also interest in other alternative crops or **cover crops**. Rotation out of wheat, which occurs in annual cropping systems, helps to break disease and pest cycles. The adoption of more diverse rotations will likely affect the abundance of insect pests and their natural enemies with implications for pest management.

Invading Pests

Most insect pests of the inland PNW cereals are non-native invaders. That is, their native ranges coincide with the origins of cereal crops and people have accidentally spread them throughout the globe. At intervals, since wheat production began here, new members of this pool of potential pests have arrived to join the inland PNW pest complex. Key aphid species—bird cherry-oat (*Rhopalosiphum padi* L.), English grain, and rose-grass (*Metopolophium dirhodum* Walker)—probably arrived with the first wheat crops grown in western Oregon in the mid-19th century. Bird cherry-oat aphid feeds on many grasses and may have been distributed globally even before European colonization. As wheat moved east into Washington, Oregon, and Idaho, aphids could readily colonize these crops on prevailing westerly winds. The Russian wheat aphid arrived in the US in 1986 and spread rapidly, reaching Washington by 1988. In 2011 an aphid new to

North America (native to the UK), *Metopolophium festucae cerealium* Stroyan ('wheat and grass' aphid in this book) was found to be abundant and widespread throughout our region. Another relatively recent invader is the cereal leaf beetle (*Oulema melanopus* L.), first detected in Idaho in 1992 and Washington in 1999. Haanchen barley mealybug (*Trionymus haancheni* McKenzie) was first detected in 2003. Hessian fly (*Mayetiola destructor* Say) was first recorded in western Washington in the 1930s and in the semiarid regions of eastern Washington in the 1960s. The yellow underwing noctuid (*Noctua pronuba* L.) has been absent from southern Idaho until very recently but is beginning to appear there. This invasion process is certain to continue as pests move throughout the world, presenting new challenges to production systems. Very rarely do pests disappear from a region, so the process is cumulative. An analogous process occurs for invasive weeds and pathogens affecting cereal crops. Insect pests that are present in the US but not yet present or that have minor pest status in the PNW include wheat stem sawfly (*Cephus cinctus* Norton), wheat stem maggot (*Memoryze americana* Fitch), and white grubs (various species).

Principal Insect Pests of PNW Cereal Production Systems

In the following sections, we provide an overview of the principal insect pests affecting cereal systems in the PNW. For each we describe its distribution, a description of the insect and its life cycle, the damage it causes, principles and practices for its management, where more information is available, and its projected response to climate change in the region. Although these sections mention insecticides and their use, please refer to the PNW Insect Management Handbook for more information: ***http://insect.pnwhandbooks.org***.

Aphids

Pest status & distribution

Although as many as 12 species of aphids (Hemiptera: Aphididae) can be found in PNW wheat production systems, six species predominate. Listed here roughly in their order of relative abundance in recent surveys

(2011–2014): English grain aphid, the newly invasive 'wheat and grass' aphid, rose-grass aphid, Russian wheat aphid, bird cherry-oat aphid, and greenbug (*Schizaphis graminum* Rondani). All are equally prevalent throughout the region except Russian wheat aphid, which is more abundant in northern Oregon. With the exception of Russian wheat aphid and 'wheat and grass' aphid, the pests have been part of wheat production systems throughout their history in the PNW. Russian wheat aphid rapidly invaded the PNW in 1988, soon after its first occurrence in North America. The 'wheat and grass' aphid has been detected in large densities throughout central Washington, northern Idaho, and Oregon since its detection in surveys in 2011, but may have been in the region since the 1990s (Halbert et al. 2013). This species has not been detected in southern Idaho or Oregon. It feeds on wheat and other grasses in the region (Davis et al. 2014a)

Pest description

Aphids are small soft-bodied, oval or teardrop-shaped insects. They can be distinguished from similar insects by the presence of a pair of cornicles, backward-projecting organs that look like "tail pipes" extending from the abdomen that extrude a defensive fluid (Figure 11-1). The species affecting wheat differ in the type and level of damage they can inflict, and

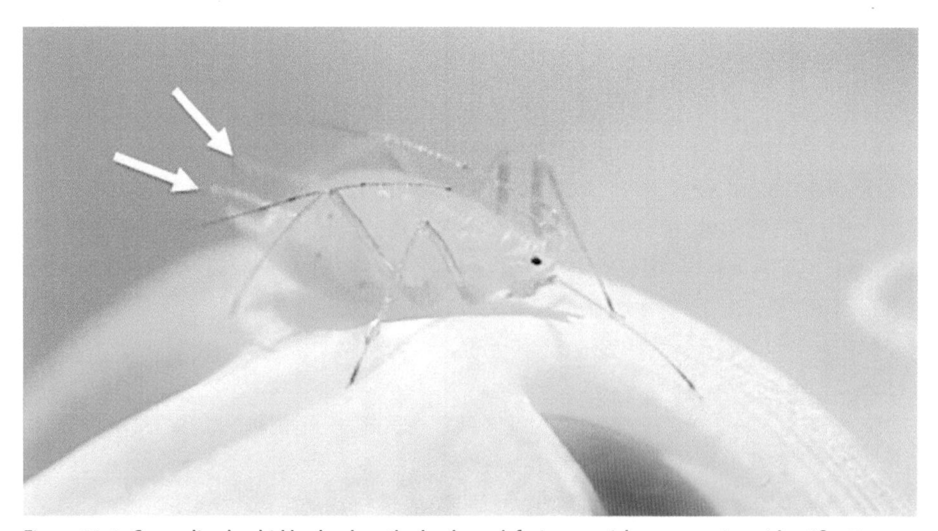

Figure 11-1. Generalized aphid body plan, the backward-facing cornicles are a unique identification characteristic only present in aphids (arrows). (Photo: Brad Stokes, University of Idaho.)

Table 11-1. Identifying characteristics of the principal aphid pests of PNW cereal production systems.

Common Name	Shape & Size	Cornicles	Antennae	Abdomen
Russian wheat aphid	Spindle; ~2 mm (0.08 inch)	Very short; not longer than wide	Shorter than length of the entire body	Little to no pigmentation with two-tailed appearance
Rose-grass aphid	Spindle; 1.6–2.9 mm (0.06–0.12 inch)	Long, pale & cylindrical	3/4 length of the entire body, dark between segments	Green with light pale stripe down midline
'Wheat and grass' aphid	Spindle; ~2 mm (0.08 inch)	Long, pale & cylindrical	3/4 length of the entire body, becoming darker distally	Uniformly green with no pale stripe down midline
English grain aphid	Spindle; 1.8 mm (0.1 inch)	Short, dark & cylindrical	3/4 length of the entire body, uniformly dark	Little to no pigmentation (yellow-green to reddish brown)
Bird cherry-oat aphid	Oval; ~2 mm (0.08 inch)	Short, swollen & flanged (red patches around base)	3/4 length of the entire body, uniformly dark	Dark pigmentation (olive-green to greenish-black appearance)
Greenbug	Spindle; 1.3–2.1 mm (0.05–0.08 inch)	Short, pale & cylindrical	3/4 length of the entire body, pale joints between segments	Green with dark stripe down midline

Figure 11-2. Representative photos of adult, wingless aphids. Bird cherry-oat aphid (top left), 'wheat and grass' aphid (top middle), Russian wheat aphid (top right), English grain aphid (bottom left), rose-grass aphid (bottom middle), and greenbug (bottom right). (Photos: top left, top middle, top right, and bottom left, Brad Stokes, University of Idaho; bottom middle, Claude Pilon Les Pucerons du Québec; bottom right, Kansas Department of Agriculture.)

in their distribution in the PNW, so producers should be familiar with the species as a foundation for pest management. Both winged and wingless forms occur in the same species, but during the summer months nearly the entire population on infested plants will be wingless. These wingless forms can be distinguished based on morphology and distribution within the plant, although a hand lens is often required (Table 11-1; Figure 11-2).

Life cycle

All aphids undergo periods of asexual, parthenogenetic viviparous reproduction. This means that summer populations entirely consist of females that do not lay eggs but instead give birth without mating to many dozen live nymphs; all these offspring are female and can mature into reproductive adults within a week. As a result, aphid infestations increase exponentially during the summer. Most aphid life cycles also include sexual reproduction in which males and nonparthenogenetic females are produced, mate, and lay eggs, which are winter hardy. Two PNW species, bird cherry-oat aphid and rose-grass aphid, are host-alternating (as their dual-host plant names suggest), which means the winged sexual forms migrate in the fall from cereals to a woody host where eggs are laid and overwintering occurs. All the other PNW cereal aphid species overwinter in grassy habitats including winter wheat fields, either undergoing a sexual phase and laying eggs there or as hardy asexual forms. In all cases, seasonal migrations of winged forms from overwintering locations back into the wheat crops establish the pest populations each year. These fall and spring migratory movements have been tracked over years and their timing and size vary in response to weather patterns. Due to the timing of these movements and the production cycle in the region, spring-planted cereals in the PNW are at greater risk of aphid-induced injury than fall-seeded cereals.

Host plants & damage

In the PNW, aphids feed on every commercially produced cereal crop: wheat, barley, oats, and rye. They feed on plants by inserting their mouthparts (stylet) into the phloem of the plant, extracting the nutritious phloem sap. Sufficiently high densities of aphids can deplete plant resources and reduce plant growth or kill plants outright. In addition, some aphids have toxic saliva that can injure plants and reduce yield. Russian wheat aphid, 'wheat and grass' aphid, and greenbug can cause this additional type of injury (Figure 11-3). Russian wheat aphid and 'wheat and grass'

Figure 11-3. Damage on wheat from feeding by the 'wheat and grass' aphid is evidently caused by salivary toxins. Other aphids affecting wheat with salivary toxins that cause plant injury are greenbug and Russian wheat aphid. (Photo: Brad Stokes, University of Idaho.)

aphid can cause considerably more injury per individual aphid than bird cherry-oat aphid, which is not known to introduce a toxin when it feeds.

Importantly, aphids can carry plant viruses that potentially cause much more severe plant injury than direct aphid feeding. A single aphid can transmit a pathogenic virus to the plant, which severely reduces yield. In the PNW, several different species of *Barley yellow dwarf virus* (BYDV) are the primary aphid-borne pathogens affecting cereals. Bird cherry-oat aphid, English grain aphid, rose-grass aphid, and greenbug are capable of transmitting BYDV. Recent research indicates that the 'wheat and grass' aphid is not a vector for BYDV (Sadeghi et al. 2016). Symptoms of a BYDV infection in wheat include leaf chlorosis (sometimes reddening) (Figure 11-4), leaf roll, stunted plants, small irregularly shaped seed heads, and reduced seed size after maturity. BYDV is an obligate pathogen of many grass species and overwinters exclusively in wild and cultivated grasses and volunteer cereal/corn and weedy grasses, including the newly invasive African wiregrass (*Ventenata dubia*) (Ingwell and Bosque-Pérez

Figure 11-4. Wheat with barley yellow dwarf disease caused by *Barley yellow dwarf virus* (BYDV) in southern Idaho. (Photo: Juliet Marshall, University of Idaho.)

2015). BYDV infections frequently result in the plant acquiring barley yellow dwarf disease, causing additional yield damage to wheat crops in the inland PNW.

Integrated pest management

Monitoring and thresholds

Scout for cereal aphids weekly from emergence until crop maturity since aphid populations can build rapidly. Use a standard 15-inch sweep net to detect early infestations when densities are low. When infestations are detected, frequent monitoring is advisable. Count aphids per stem to determine if the nominal thresholds for treatment have been exceeded. Both scouting for the presence of pests and monitoring for pest abundance should be done at multiple sites in field margins and interior since aphids are usually highly aggregated, especially during early infestations. Precise economic thresholds do not exist for cereal aphids in our region, but the literature provides some guidelines for the use of chemical treatments. Rules of thumb recommend treatment when aphids (regardless of species) reach two to ten per tiller, per stem, or per head, prior to dough stage. After dough stage, there are no benefits from treating aphids, as they are not damaging to yield after this growth stage. When the risk of virus is high, thresholds are not useful since a single aphid can transmit the virus. At this time, virus risk monitoring systems do not exist. In recent years, virus infection has been negligible in northern Idaho, but more prevalent in central Washington and eastern Idaho, where significant virus outbreaks have occurred.

Biological control

In our region, aphids are generally held below nominal thresholds for chemical control by generalist predators and parasitoids. Well-known aphid predators that can readily be observed at work in cereal crops include several species of lady beetles (Coccinellidae), fly larvae that specialize on aphids (Syrphidae), lacewing larvae and some adults (Chrysopidae and Hemerobiidae), big-eyed bugs (Geocoridae), assassin bugs (Reduviidae), minute pirate bugs (Anthocoridae), and rove beetles (Staphylinidae). In addition, PNW aphids are attacked by at least eight different species of parasitic wasps (Hymenoptera) (Bosque-Pérez et al. 2002). Aphids also are **susceptible** to some specific entomopathogenic fungi, but these rarely have significant impacts in dryland systems because they require persistent humid conditions to create epidemics.

Cultural control

A well-known and often used practice to reduce the risk of aphids and their associated viruses is to plant spring wheat as early in the growing season as possible, reducing the amount of time for aphids to feed and/or transmit viruses to the plant. Eliminating the **green bridge** for BYDV (wild, volunteer cereals and weedy hosts) may also reduce the number of primary infections in a given field. Aphids can also fly into the fall months, so early planting of winter wheat potentially places the crop at greater risk of virus infection.

Resistant varieties

Although sources of host plant resistance to several cereal aphid species are known, no PNW varieties carry deliberately developed resistance to any common aphid pests. This mirrors the situation globally. Research continues to improve understanding of the genetics and mechanisms of resistance to aphids, but few varieties have been released. Eventually, this knowledge, coupled with demand, may lead to adapted **resistant** varieties for our region.

Chemical control

Neonicotinoid seed treatments, often used for wireworm control, can also have some efficacy against aphids, and there is evidence that these treatments can also limit spread of BYDV for which some aphid species serve as vectors. Tighter regulations on these materials may limit their utility before long. Neonicotinoids are also available as foliar sprays for aphids in cereals, as are many pyrethroid products and a few organophosphate and carbamate products. Foliar applications especially run the risk of reducing natural enemy populations that are important for keeping aphids and other pests in check in cereals under most situations in our region. Please refer to the PNW Insect Management Handbook for current insecticide recommendations.

Climate change

Aphids potentially respond to climate variability through changes in their geographic ranges and the timing and abundance of their annual migrations, which can affect their arrival into fall- and spring-planted crops, with implications for direct injury and viral disease epidemiology. Given the relatively abundant historic data and importance of aphids, they have received considerable attention in the context of climate change. Different species respond differently to climatic drivers. In the PNW, 20-year suction trap records indicate that bird cherry-oat aphid, rose-grass aphid, and Russian wheat aphid each responded differently to climate (Davis et al. 2014b). Russian wheat aphid abundances were negatively correlated with increasing temperatures, rose-grass aphid abundance was positively correlated with increasing cumulative precipitation, and bird cherry-oat aphid abundances were unrelated to any climate variables. This heterogeneity is similar to studies of aphids and climate around the world. At this juncture, no clear projections can be offered.

There are numerous ways climate change can potentially affect the bird cherry-oat aphid and hence BYDV (Finlay and Luck 2011), but there is scant research on the topic. Only two studies exist examining effects of climatic factors on BYDV. In separate controlled studies, elevated carbon dioxide and sharply elevated temperature (+5°C; 41°F) increased virus titer (abundance of virus particles) in infected plants, leading the authors to suggest that virus spread could be enhanced under future projected climate change conditions (Trebicki et al. 2015; 2016). In a surprising twist, recent work shows that BYDV-infected wheat plants tolerate drought stress better than non-infected controls, suggesting the system-wide response to drought could be complex (Davis et al. 2015).

Hessian Fly

Pest status & distribution

The Hessian fly has been a pest of US wheat since its accidental introduction into the country over 200 years ago (Bosque-Pérez 2010). It has been present in parts of the inland PNW since the 1930s. Damaging infestations of Hessian fly have only occurred in the inland PNW over the last two decades where climatic conditions are suitable for fly survival and development. In 2015, the fly was detected in southern Idaho for the first time. Wheat is the preferred host for Hessian fly, with spring wheat more commonly damaged than winter wheat.

Pest description

Adult Hessian flies are small (1/8") dark brown-reddish colored midges (Figure 11-5). They do not feed, and die a few days after emergence as adults. Hessian fly has four life stages: egg, larva, pupa, and adult. After mating, female flies lay 200 to 300 eggs on the upper surface of wheat leaves (Figure 11-5).

Life cycle

Eggs develop in between five to ten days depending on temperature. Larvae emerge as very small, bright red, legless maggots, which migrate to a node or the crown of the plant where they begin feeding. When larvae are fully developed, their cuticle hardens and darkens to form puparia that are attached to stems under leaf sheaths (Figure 11-6). Puparia are protected from the elements during periods of unfavorable conditions. Puparia are often referred to as "flaxseeds," which they resemble. Adults emerge from puparia during favorable environmental conditions in the spring. In the inland PNW, one or two generations of the fly occur per year (Castle del Conte et al. 2005).

Host plants & damage

Damage to the plant is caused solely by larval feeding. Larvae feed on the stem under the leaf sheaths, and high infestations can result in stunted plants or plant death. Feeding also causes reductions in grain quantity and quality, and weakened stems result in lodging and decreased yields. Yield reductions due to Hessian fly infestation of spring wheat without resistance range from 11–24% (Smiley et al. 2004). If infestations are severe, primary tillers may die, but sometimes plants develop new tillers (Schotzko and Bosque-Pérez 2002). Although no visible injury to the plant may be noticeable at the feeding site, infested plants might be stunted, lodged, exhibit erratic head heights in the field, or, in some genotypes, show erect dark green leaves.

Figure 11-5. Adult Hessian fly with single egg on wheat. (Photo: Scott Bauer, Bugwood.org.)

Figure 11-6. Hessian fly larva and puparia on wheat. (Photo: Dennis Schotzko, University of Idaho.)

Integrated pest management

The occurrence of fly infestations is difficult to predict. Therefore, control methods are mostly preventive, with the most common being resistant varieties and earlier seeding to escape infestation (Bosque-Pérez 2010). Additionally, crop rotation and destruction of volunteer wheat are important management tools. Fly parasitoids also provide some control. Parasitism levels vary widely depending on location and year, ranging from 10% to 85% (Bullock et al. 2004). Eight parasitoids are known to attack Hessian fly in the inland PNW (Bullock et al. 2004). Although the retention of wheat residue is known to increase fly survival (Clement et al. 2003), reduced tillage practices do not increase within-field abundance of the fly and have no consistent effect on spring wheat yield (Castle del Conte et al. 2005).

Fly biotypes (or genetic variants) that attack resistant varieties are known to exist in many parts of the US including the inland PNW (Ratcliffe et al. 2000). Such virulent biotypes pose a potential risk to the durability of resistant varieties. Utilization of multiple resistance genes and control via parasitoids will increase durability of resistance. Screening and breeding for resistance to Hessian fly is a continuous effort in the wheat breeding and host plant resistance programs in Idaho and Washington, and numerous resistant varieties are available for growers. Current wheat varieties include: 'Jefferson,' 'Jerome,' 'Cataldo,' 'Diva,' 'Louise,' 'Kelse,' 'Babe,' 'Whit,' 'Hollis,' and 'JD'. Spring barley 'Baronesse' is resistant to a predominant Hessian fly biotype. Additional varieties are in the pipeline and will be released in the future. In areas of heavy Hessian fly infestations such as northern Idaho and eastern Washington, growers are encouraged to always plant resistant spring wheat varieties to avoid economic losses.

Climate change

In the inland PNW, resistant varieties are used as the primary IPM tactic to manage Hessian fly (Ratcliffe et al. 2000; Schotzko and Bosque-Pérez 2002). Although warming climates can change the timing of fly activity and even increase the number of generations per year, resistance technology should remain effective. However, currently all resistant varieties deployed are spring wheats. There is an indication that warmer falls and wetter springs associated with climate change might result in higher incidence of Hessian fly in winter wheat. This will necessitate the development of Hessian fly-resistant winter wheat varieties adapted to the inland PNW region. There is one report of a Hessian fly-resistant gene that loses its efficacy under elevated temperatures, but that gene is not important for continuing resistance to the fly in inland PNW spring wheat varieties, which currently rely on a different set of Hessian fly-resistant genes.

Cereal Leaf Beetle

Pest status & distribution

The cereal leaf beetle (CLB) is an invasive pest of cereal crops. It was accidentally introduced into Michigan in the 1960s and has since expanded its distribution westward to all of the wheat/barley/oat/rye growing counties of Idaho, Washington, and Oregon. In the US, at least 30 states have confirmed CLB populations although they may be present in other regions where cereal crops are produced (Philips et al. 2011). Crop damage by this insect may result in significant yield quantity and quality reduction and reduced economic returns to producers. In some parts of its range in the US, CLB can cause up to 75% yield loss in cereal crops (Buntin et al. 2004).

Pest description

CLB adults are small (1/4"), oval-shaped beetles with blue-to-green metallic forewings (elytra) and a brightly colored red head, thorax, and legs (Figure 11-7). CLB eggs are very small and bright yellow as they are laid by the female, and then turn to a darker brown as they near maturity. CLB larvae have a brown head with a yellowish body; often they may appear darker because larvae use their own feces to coat themselves, and it is thought this behavior helps deter natural predators and parasitoids. CLB should not be confused with the similar-looking Collops beetles (*Collops vittatus* Say), which are beneficial insects that occur in cereal fields. The Collops beetle, or "red cross beetle," has a red thorax, but not a red head, and elytra that are metallic blue and red. Males have distinct swellings at the base of their antennae; a female is shown in Figure 11-8.

Figure 11-7. Adult cereal leaf beetle in a wheat field. (Photo: Nate Foote, University of Idaho.)

Figure 11-8. Adult female Collops beetle resting on a wheat head. (Photo: Brad Stokes, University of Idaho.)

Life cycle

CLB completes only one life cycle per season in the inland PNW. They overwinter as diapausing, non-feeding/inactive adults and start emerging in mid-April to early May depending upon climatic conditions. Under a "typical" winter in the inland PNW, the adults emerge and move onto winter wheat where they usually cause little damage; they will then move onto spring wheat where they start laying eggs. Females lay eggs on the surface of wheat (or other hosts) singly or in clusters of two to three, primarily during the latter part of May. Females deposit at least 300 eggs on the top of the leaves or margins close to the leaf-mid-rib. Within four to 23 days after oviposition, larvae begin emerging and can be active until July. There are four larval **instars** (growth stages) for this particular pest before pupation begins. Pupation occurs in the soil at a depth of one to two inches during June and July, after which adults emerge in three weeks. Adults are active until late fall when they begin to search out protected overwintering sites, often in close proximity to the infested field.

Host plants & damage

CLB has a wide range of cultivated grass host crops (wheat, barley, oats, rye, corn, sorghum, and sudangrass), and many other wild and native grasses may be acceptable as hosts. The larval and adult life stages are the only economically damaging life stages for this cereal crop pest. The fourth instar CLB larva, rather than the other life stages, causes most of the crop damage. CLB larvae feed on the upper mesophyll part of the leaves, typically between the veins of leaves, causing a characteristic "window pane" look to the infested host plant (Figure 11-9). Adult CLB chew through the entire leaf, making small slits between leaf veins.

Integrated pest management

Monitoring and thresholds

Adults should be monitored using a standard 15-inch insect sweep net during the later parts of the spring months, after they have emerged from their nearby overwintering sites. Grasses close to field margins may also be monitored during this time to get a sense of the local population. Larvae are more difficult to scout for by visually examining a number of random plants in the suspected infested field as they tend to be hidden in the field.

Figure 11-9. Cereal leaf beetle larva feeding on wheat. The "window pane" injury is characteristic of feeding by adults and larvae. (Photo: Brad Stokes, University of Idaho.)

The economic threshold for CLB on wheat in the inland PNW is three eggs or three larvae per tiller before the booting stage, and subsequently one larva per flag leaf during the remaining growth stages. Populations should be scouted after plant emergence prior to the booting stage in the inland PNW, and a control action may be justified if the population is approaching the economic threshold. Degree day models for CLB are reliable and can be used to time scouting efforts: ***http://uspest.org/cgi-bin/ddmodel.us?spp=clb&uco=1***. When determining whether to spray, it is prudent to assess potential impacts of natural enemies and to avoid spraying when these are abundant.

Biological control

Classical biological control via the introduction of parasitoids from the CLB's native range has been relatively successful. A pinhead-sized wasp, *Tetrastichus julis* Walker (Eulophidae), is known to parasitize all larval instars and is well established in the inland PNW (Figure 11-10) (Roberts et al. 2012). In the early 2000s, university and federal scientists released *T. julis* in key areas of the inland PNW to augment this parasitoid population.

Figure 11-10. Cereal leaf beetle larva (left) and its introduced natural enemy (right), the parasitoid wasp *Tetrastichus julis*. (Photo: Brad Stokes, University of Idaho.)

Several other parasitoids exist, *Diaparsis carinifer*, *Lemophagus curtus* (Ichneumonidae), and the egg parasitoid *Anaphes flavipes* (Mymaridae), though they are not as well established as *T. julis* for CLB control. Lady beetles (Coccinellidae) consume CLB larvae. Other generalist predators may also be effective, including ground beetles (Carabidae), soft-winged flower beetles (Melyridae), notably the Collops beetle, which resembles the CLB, assassin bugs (Reduviidae), damsel bugs (Nabidae), minute pirate bugs (Anthocoridae), and green lacewings (Chrysopidae). In the past, CLB control has relied heavily on insecticides and that will likely continue. Least-toxic biorational insecticides may eventually have a place in CLB pest management, but currently these are too costly for use in production agriculture.

Cultural control

Factors such as late planting, lack of nitrogen fertilization, or poor soil health can reduce field populations of the CLB; lower seeding rate in oats, mixed cropping of oats and barley, and combinations of nitrogen and potassium fertilizers have a limited effect on the CLB. Growing a strip of oats between spring wheat and winter wheat may provide a viable option of trap cropping for CLB infestations. These effects have been observed in limited studies in other parts of CLB range and there are no definitive guidelines on how to implement them. An aggregation pheromone has been identified that has potential for use as a monitoring tool for the beetle (Rao et al. 2003).

Resistant varieties

Host plant resistance is known from varieties of wheat that have higher than normal silica-rich trichomes, narrow-leaved varieties, and others that produce the secondary compound DIMBOA (2,4-dihydroxy-7-methoxy-1,4-benzaxazin-3-one) which has an antibiotic effect on the larvae. No inland PNW varieties are CLB resistant.

Insecticides

Numerous broad-spectrum, commercially available insecticides are registered and approved for use in wheat or other cereal crops for use in controlling CLB populations. Synthetic pyrethroids such as permethrin, cypermethrin, and fenvalerate have been effective in controlling leaf beetles; however, these compounds are lethal to natural enemies and should be applied judiciously. Please refer to the PNW Insect Management Handbook for current insecticide recommendations.

Climate change

The CLB has been studied for its potential response to climate change using bioclimatic models. Based on current models, CLB range in North America is expected to continue to expand northward (Olfert et al. 2004; Olfert and Weiss 2006). In much of the inland PNW, models indicate that the climate will become, in general, more hospitable for the beetle, making it a more serious pest. Furthermore, based on data from Utah, as the inland PNW climates warm, parasitism by *T. julis* is expected to be reduced, potentially releasing CLB from this very successful biological control agent and increasing its population to damaging levels (Evans et al. 2012).

Cutworms

Pest status & descriptions

The term cutworm refers to immature stages of multiple species of moths belonging to the family Noctuidae. They are relatively large and soft-bodied insects as larvae, and up to 2 inches long at their later stages of development. Cutworms are mostly active during the night and take refuge just below the soil surface during the day. Among several species of cutworms present in the inland PNW, black cutworms and variegated cutworms have been reported to emerge more frequently in numbers that may result in significant yield loss.

Black cutworm

Black cutworm (*Agrotis ipsilon* Hufnagel) adults are brownish-gray with a light silver-colored band (Berry 1998a) and "dagger-shaped" patterns (Cook et al. 2003) on the forewing (Figure 11-11). Larvae are dark gray with a dark brown or black head capsule; a lighter stripe runs along the backside of the body (Figure 11-12). In their final, largest instar, these larvae can be 1.5 inches long (Berry 1998a).

Figure 11-11. Black cutworm adult. (Photo: John Capinera, University of Florida, Bugwood.org.)

Figure 11-12. Black cutworm larva. (Photo: John Capinera, University of Florida, Bugwood.org.)

Variegated cutworm

Variegated cutworm (*Peridroma saucia* Hübner) adult moths are brown to reddish brown. They possess darker markings on the forewings as well as a kidney-shaped spot (Figure 11-13). Full-grown larvae can be up to 2 inches long. Their body color may range from light gray to dull brown, with a row of yellow dots along their back (dorsal) (Berry 1998b) (Figure 11-14).

Large yellow underwing

Large (greater) yellow underwing (*Noctua pronuba* L.) adult moths are large with brown forewings and brightly colored yellow-orange hindwings with a broad black band around the margin, though ten different color variants have been reported from Europe (Figure 11-15). Larvae are usually olive brown, though some may have a distinct reddish tinge. Larvae are also marked with a bold black and cream dash on each side of the midline; the overall appearance is a series of dark broken dashes that run the length of the body (Figure 11-16). The final, largest instar can be 1.5 inches long. This species is a recent but relatively minor invasive pest of cereal crops in the inland PNW, first reported from Oregon in 2001.

Figure 11-13. Variegated cutworm adult. (Photo: Pests and Diseases Image Library, Bugwood.org.)

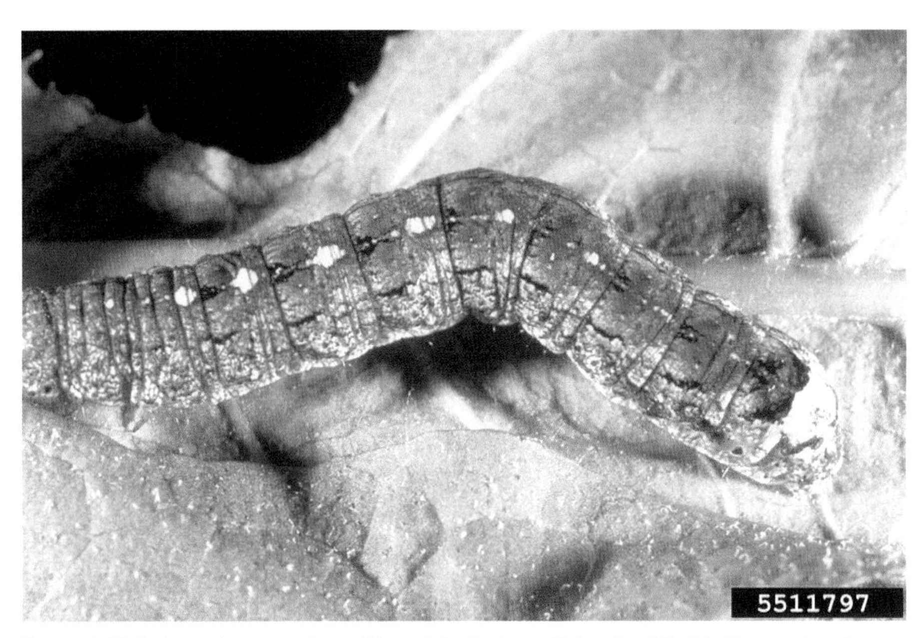

Figure 11-14. Variegated cutworm larva. (Photo: John Capinera, University of Florida, Bugwood.org.)

Figure 11-15. *Noctua pronuba* adult. Upper specimen is male; lower specimen is female. (Photo: Edward Bechinski, University of Idaho.)

Life cycles

Black cutworm is present across the inland PNW. They overwinter as pupae in areas with mild winters and emerge as adults in the spring. Adults could also migrate and disperse into areas with harsher winters, where survival of the overwintering larvae might have been jeopardized (Berry 1998a). Females continue to lay eggs through June. Eggs hatch within a week. Young larvae initially utilize foliage, but older larvae remain at, or

Figure 11-16. *Noctua pronuba* larva in Nez Perce County, Idaho. (Photo: Edward Bechinski, University of Idaho.)

just below, the soil surface for about a month before pupating. Adults will emerge from the pupae within two weeks and lay eggs to start another generation that may not successfully overwinter in some localities (Berry 1998a).

Variegated cutworms may overwinter at different developmental stages, primarily as late instar larvae in the soil. Adults emerge in late spring and early summer, laying eggs in clusters of up to several hundred on the underside of leaves. Eggs will hatch within a week and larvae continue to feed for up to 6 weeks (Berry 1998b). There are two generations in the inland PNW, with the second generation forming in the overwintering stage.

Large yellow underwing cutworms overwinter as partially grown to almost full-grown larvae under plant residue. It is believed that this insect only has one generation per year in the inland PNW, though it may have two generations. This cutworm has six larval instars, each larger than the previous. Adults emerge in mid-summer, they mate, and females begin laying egg masses (up to 2,000 eggs per female) until late September before dying off. During the winter and early spring when temperatures rise above 40°F, early to mid-instar larvae begin sporadically feeding on fall- and spring-planted crops throughout the night (nocturnal feeding) (Bechinski et al. 2009). Crop damage occurs from mid-September until early May. Damage is typical of other cutworms and includes stem girdling, crown feeding, leaf feeding, and leaf clipping (plants appear to be clipped with scissors) of various plants (Bechinski et al. 2009).

Host plants & damage

Feeding damage on the foliar tissue by black cutworm could occur during the early larval instars (Figure 11-17). However, later instars of the black cutworm feed on crowns and roots during the night. While young seedlings may be cut and eliminated, damage on the older plants may present as wilting plant leaves (Berry 1998a). Damage by the variegated cutworm can be devastating when they appear in large numbers as they may defoliate the crop (Berry 1998b) and cut seedlings off at the soil level. Large yellow underwing damage varies, though currently it is a minor pest with only a few isolated infestations in the inland PNW.

Figure 11-17. Typical cutworm damage on wheat. (Photo: Edward Bechinski, University of Idaho.)

Sampling, monitoring, and thresholds

To achieve an effective control, cutworm presence must be detected early in the season when larvae are young and plants are still at the seedling stage (see the Insecticides section). Monitoring could be done on a regular basis by observing damaged plants, sweep netting for larvae (variegated cutworms), and inspecting the soil surface to find larvae. Field inspections for larvae are most effective in the dark, when larvae are most active, by using a flashlight or headlamp. No detailed studies are available on economic thresholds; however, degree days can be used to estimate time for scouting and to predict timing for the most damaging larval stages (Cook et al. 2003). For black cutworm infestations, chemical treatment is suggested where the presence of two or three damaged plants in a 10-foot row section is observed in multiple spots (Berry 1998a). In addition to field scouting, adult flight monitoring may also help with early detection of potential outbreaks (Bechinski et al. 2009).

Biological control

There are several predaceous (e.g., ground beetles, spiders, centipedes) and parasitoid (e.g., wasps and flies) arthropods, pathogens, and birds that can reduce cutworm numbers. Large populations of cutworms are highly susceptible to fungal diseases, especially under moist conditions.

Cultural control

Removal of volunteer and grassy weeds with cultivation or herbicide applications will eliminate food sources available to cutworms, prior to spring crop emergence.

Insecticides

Early detection of cutworms would also increase the likelihood of foliar insecticide application success. This is because it would facilitate targeting early instars before inflicting damage to the susceptible seedlings, and before they move down on the foliar tissue and take cover in dense crowns or under the soil surface. Insecticide applications need to be conducted after sunset or before sunrise to maximize chances of targeting active larvae and minimizing the negative impact on beneficial insects and bees. Numerous broad-spectrum, nerve-poisoning insecticides containing the active ingredients beta-cyfluthrin, carbaryl, chlorpyrifos, gamma-cyhalothrin, cyfluthrin, lambda-cyhalothrin, and zeta-cypermethrin are labeled for cutworm control in wheat. Please refer to the PNW Insect Management Handbook for current insecticide recommendations.

Resistant varieties

None are currently available.

Climate change

There are no studies we know of that suggest significant changes in pressure from these pests due to climate change. One study from wheat production systems in China detected no trend in injury by noctuid pests over a 25-year period, despite a warming trend associated with changes in aphid pressure, suggesting the cutworms may not be affected by climate change elsewhere including the inland PNW.

Wheat Head Armyworm

Pest status & distribution

The genus *Dargida* (Lepidoptera: Noctuidae) consists of eight species, which includes *D. procinctus*, *D. gramminivora*, *D. quadrannulata*, *D. diffusa* (known in the PNW as the wheat head armyworm), *D. terrapictalis* (known as the false wheat head armyworm), *D. tetera*, *D. rubripennis*, and *D. aleada*. All of these species are found north of Mexico (Michaud et al. 2007). The most prominent species in the PNW are the wheat head armyworm, the false wheat head armyworm, and more recently the olive-green cutworm, *D. procinctus*. Because of the similarity of larvae and adults within this genus, definitive identification should be left to a taxonomic expert. Nonetheless, all *Dargida* larvae feed on wild grasses, grains, wheat, or other cereal crops, and the insects matching the descriptions below should be considered potential pests of wheat.

Pest descriptions

The adult moths are yellow-brown with a brown stripe running down the length of each of the forewings. This coloration provides camouflage from predators in cereal crop fields that are drying down near the end of the season; hindwings are darker in *D. terrapictalis* compared to *D. diffusa* (Figure 11-18). The larvae vary in color but have been noted as gray, cream, or green with distinct yellow, white, and brown strips along the length of the body (Figure 11-19).

Figure 11-18. *Dargida diffusa*, the true wheat head armyworm, adult. (Photo: Luc Leblanc, University of Idaho.)

Figure 11-19. *Dargida terrapictalis*, the false wheat head armyworm, adult. (Photo: Luc Leblanc, University of Idaho.)

All three inland PNW species, *D. diffusa*, *D. terrapictalis*, and *D. procinctus* have four distinct life stages: egg, larva, pupa, and adult. Larvae go through five instars. During the winter months, the larvae pupate in the soil. When spring arrives, moths emerge and, within a few days, the moths lay eggs on wheat or barley crops. Larvae that develop from eggs feed on wheat as early as late May, with increasing numbers into mid-June. This late-spring timing coincides with wheat flag leaf development. Larvae feed on wheat heads, primarily at night, when ambient temperatures are cooler. They crawl toward the base of stalks during hot days. *D. diffusa* larvae and moths are typically active only at night. In the western US, armyworms are considered sporadic pests. They can have up to two generations per year with a second generation developing on warm season grasses in the fall, after wheat has been harvested. The appearance of a typical later instar larva is shown in Figure 11-20.

Host plants & damage

All larvae of the genus *Dargida* feed on wheat and various other grain and grass crops in the PNW. Damage is caused exclusively by the larval stages

Figure 11-20. Typical *Dargida* spp. larva. (Photo: Frank Peairs, Colorado State University, Bugwood.org.)

of these pest insects, as they will often eat into part of the wheat/barley/ oat kernel causing direct damage to the product. Damage will often go unnoticed until screening at grain elevators. Damage takes the form of a small hole bored into the base of the floret. The pests are more likely to be found along field margins.

Integrated pest management

Sampling

Sampling for larvae and moths may be done with a sweep net. Focus on field margins for detection only. Once detected, sampling in the interior should be conducted if the aim is to assess field scale levels of infestation. A sex-attractant may be used to lure moths to the trap. Traps should be left in or adjacent to the field throughout the crop season and should be checked at least once per week. While no insecticides are specifically labeled for control of wheat head armyworms in the inland PNW, studies suggest pyrethroids may work well. Products specifically labelled for cereals can be legally used for *Dargida* control even though the pest is not cited on the label.

Biological control

There are confirmed accounts of parasitism of *Dargida* spp. by small wasps, but the species have not been identified. Similar to other armyworm species, wheat head armyworms are vulnerable to predation by ground beetles, spiders, birds, and rodents.

Cultural and chemical control

There are no established management plans or economic thresholds for this pest since they are sporadic. Infestations are usually concentrated around field margins, so it is recommended that scouting efforts focus on this area. There are no insecticides specifically labeled for this pest, but materials registered for other armyworms in wheat would likely provide control if applied sufficiently early. Larvae arriving with harvested wheat either die or emerge as moths, potentially surviving in storage. Please refer to the PNW Insect Management Handbook for current insecticide recommendations.

Resistant varieties

No resistant varieties are available.

Climate change

To date, there is no scientific study or associated data that would suggest a correlation with wheat head armyworms and climate change, though we could speculate based on biology alone that the number of generations per year, growth rate, or their known range may increase due to warmer overall temperatures in the inland PNW.

Wheat Midge

Pest status & distribution

Wheat midge, *Sitodiplosis mosellana* (Gehin), is a European species first detected in the inland PNW in 1991 in Boundary County, Idaho. Pest distribution in Idaho has since expanded to Benewah and Kootenai counties. Surveys in Washington confirmed wheat midge at low to potentially damaging levels in Garfield, Lincoln, Spokane, and Stevens Counties.

Pest description

Adults resemble mosquitoes but are smaller (1/8" long) and with bright orange bodies (Figure 11-21). They are most often seen at dusk resting on wheat heads during plant flowering, hence their colloquial name "orange wheat blossom midge." Mature larvae are 1/8 inches long, bright yellow-orange, legless maggots.

Life cycle

Wheat midge develops through a single generation annually. Mature larvae overwinter 2 to 4 inches in the soil, pupate during May, and begin to emerge by late-June as adult flies. Adults remain close to the soil surface during the day but fly to flowering wheat heads on warm, calm evenings when air temperature is at least 60°F and wind speed is less than 8 mph. Females lay eggs under the glumes and palea. Larvae feed on the developing kernels for 2 or 3 weeks, and then remain inactive on the head until rain or dew causes them to drop to the soil where they overwinter

Figure 11-21. Adult wheat midge resting on flowering wheat head (left) and larvae from dissected wheat head (right). (Photos: Robert Lamb, Ag and Food Canada; Diana Roberts, Washington State University.)

(Knodel and Ganehiarachchi 2008).

Host plants & damage

Wheat midge is primarily a pest of wheat. Non-economic infestations sometimes occur in barley, rye, and intermediate wheat grass.

Larvae are the sole damaging stage. They feed externally on the developing kernels but are hidden under the bracts that surround the seed and cannot be seen without dissecting the wheat plant head. Damaged wheat heads may lose their green coloration early in the field maturation, thus appearing unhealthy relative to the rest of the field. Injury ranges from shriveled, cracked, and underweight kernels to complete abortion of the seed (Figure 11-22).

Wheat is only susceptible to larval injury when eggs are laid on flowering heads from emergence to full flowering; larvae cannot complete

Figure 11-22. Wheat midge larval feeding injury to spring wheat as seen at crop harvest. (Photo: Dennis Schotzko, University of Idaho.)

development if oviposition occurs earlier or later than crop flowering. Infestations seldom develop in winter wheat because the crop flowers before midge ovipositional flights during late June and early July. In contrast, severe infestations can develop in spring wheat because flowering more likely coincides with midge oviposition.

Integrated pest management

Adult monitoring and thresholds

Larval management with foliar insecticides depends on field scouting for adult midges because applications must be timed to kill females before they lay eggs on flowering wheat heads.

Field studies by the University of Idaho showed that foliar insecticide applications are justified if sweep net sampling in flowering wheat fields at twilight on warm, calm evenings detects an average of 1 to 4 adult midges per five sweeps. This threshold is conservative and has a low probability of failing to treat an economic infestation, but a 1-in-3 chance of needlessly treating a non-economic infestation. Sweep net sampling during daylight hours has no value for IPM decisions because midge adults remain close to the soil surface during the day and so escape collection.

Commercial sticky traps baited with the female wheat midge sex pheromone are highly attractive to adult males. North Dakota State University tentatively recommended insecticide treatment if cumulative midge captures with pheromone traps exceed 10 per trap at 3 days after heading, but those thresholds have not been validated in our area.

Forecasting the timing of midge ovipositional flights

Research by the University of Idaho showed that 80% of seasonal midge flight activity occurs between 735 and 915 cumulative degree days above 5°C (41°F; $DD_{5°C}$) since January 1, with seasonal maximum flight at 820 $DD_{5°C}$. Hence, it is not necessary to monitor spring wheat fields before 735 $DD_{5°C}$ or after 915 $DD_{5°C}$ because oviposition does not occur before or after those periods.

Biological control

Carabid ground beetles can be important natural predators of midge larvae in the soil. Surveys have failed to detect any parasitoids of the wheat midge in Idaho and Washington.

Cultural control

Rotate crops to reduce likelihood of midge damage to spring wheat. The most effective rotation is spring wheat planted after canola or some other non-cereal crop; spring wheat after barley poses minimal risk, and spring wheat after winter wheat poses some but overall low risk. The worst cropping sequence is continuous spring wheat.

Plant spring wheat varieties as early as agronomically possible so that crop flowering does not coincide with peak seasonal midge oviposition during July. Midge-free seeding dates that allow spring wheat crops to escape egg-laying are given in Table 11-2 for Boundary County, Idaho. During "average years" (i.e., temperatures that occur 4 years in 5), seeding earlier than 11–20 of April allows the wheat crop to grow beyond the susceptible flowering stage before midge activity reaches seasonal peaks.

Table 11-2. Relative risk of wheat midge infestation as a function of spring wheat seeding date, Boundary County, Idaho.

infestation risk	temperature scenario (1 Jan - 31 Jul)		
	colder-than-normal	average year	warmer-than-normal
LOW: seed before	20 April	11 April	23 March
HIGH: seed during	2 May	25 April	5 April
LOW: seed after	13 May	7 May	17 April

LOW RISK	flowers before 10% seasonal midge flight (735 $DD_{5°C}$) or after 90% seasonal midge flight (915 $DD_{5°C}$)
HIGH RISK	flowers during maximum (50%) seasonal midge flight (820 $DD_{5°C}$)

Resistant varieties

Montana State University released a midge-resistant spring wheat variety in 2016, 'Egan,' which incorporates the *Sm 1* gene and causes plants to respond to larval feeding injury with elevated levels of phenolic acids that halt larval feeding. Midge-resistant *Sm 1* red spring and hard red spring wheat varieties have been commercially available in western Canada since 2010.

Insecticides

The following insecticides are labelled as foliar sprays applied to wheat for control of wheat midge or orange blossom wheat midge or orange wheat blossom midge: chlorpyrifos, dimethoate, gamma-cyhalothrin, lambda-cyhalothrin, and malathion. All are broad-spectrum insecticides that potentially disrupt biological control of aphids and CLB. Please refer to the PNW Insect Management Handbook for current insecticide recommendations.

Climate change

Bioclimatic models (Olfert et al. 2016) suggest that the present climate of the inland PNW is marginally conducive to wheat midge outbreaks and that pest expansion is unlikely beyond currently known distribution in the Idaho panhandle and adjoining eastern Washington. Based on 2030 and 2070 climate change projections, these same models predict that the future climate of the inland Northwest will remain marginal with no pest expansion westward or southward into adjoining inland PNW wheat production regions.

Mites

Pest status & distribution

Mites, also called spider mites, are not technically insects but instead are classified as arachnids. Some mites feed exclusively on plants and can impact yield in agricultural crops, such as spring and winter wheat, barley, oats, or Timothy hay. Mite distribution is widespread, occurring across all the counties of Idaho, Washington, and Oregon. Several mite species of economic agricultural importance occur in these states: brown wheat mite (*Petrobia latens* Müller), Banks grass mite (*Oligonychus pratensis* Banks), winter grain mite (*Penthaleus major* Duges), and the wheat curl mite (*Aceria tosichella* Keifer). In addition to feeding on plant material, the wheat curl mite successfully vectors *Wheat streak mosaic virus* (WSMV), causing even more yield damage.

Pest descriptions

Mites are minute creatures. The use of a hand lens is essential for assessing a suspected mite infestation and attempting to identify the species involved.

Brown wheat mite

Brown wheat mite (*Petrobia latens*) adults have a dark brown ovoid body with yellow-orange to slightly reddish legs and are 1/50 inches in total length (Figure 11-23) (Blodgett and Johnson 2002). They have a lighter stripe that extends from the head (cephalothorax) to the end of the body (abdomen). Larvae and nymphs resemble adults. As with most mite species this is difficult to see unless there is at least some magnification available.

Banks grass mite

Banks grass mite (*Oligonychus pratensis*) adults are dark green to a

Figure 11-23. Brown wheat mite adult. (Photo: Phil Sloderbeck, Kansas State University, Bugwood.org.)

Figure 11-24. Banks grass mite adult. (Photo: F.C. Schweissing, Bugwood.org).

darker brown color and are 1/32 inches in total length (Figure 11-24) (Brewer 1995). They have a row of spots on each side of their abdomen that distinguishes them from other mites that feed on wheat. Larvae and nymphs resemble adults.

Winter grain mite

Winter grain mite (*Penthaleus major*) adults are iridescent black in color (cephalothorax and abdomen) with yellow, orange or more often red-colored legs. They often have a red stripe, and unusually have an anal pore on the upper side of the abdomen (Bauernfeind 2005). This pest species only has two generations per year in Idaho, Washington, and Oregon. The first starting in late fall, September or October, with population and economic peaks in December or January. The second generation reaches high populations in the field during the months of March and April. This species excels in low temperature environments; females lay oversummering eggs as temperatures exceed their developmental limit. Larvae hatch later in the season and begin feeding on leaf tissue near the ground, wandering up the plant during cooler nights.

Wheat curl mite

Wheat curl mite (*Aceria tosichella*) adults are nearly microscopic white and 1/100 inches in total length. This species has a cigar-shaped body with only four legs (as opposed to eight in the other mite species listed above) pointed forward and a fleshly lobe located posterior. Even under a hand lens this species may be unrecognizable in the field. The wheat curl mite is unlike other mites because its main method of dispersing is wind. The most economically important factor with this species of mite is the ability of it to vector WSMV. WSMV was first detected in Kansas in 1987, with more infections found during the subsequent year (Townsend et al. 1996). It has since moved east into Montana, North and South Dakota, Idaho, Washington and Oregon. Plants infected with WSMV appear to have bright yellow or orange streaking, often most severely near the tip of the leaf (Figure 11-25). WSMV also facilitates injury by seed-borne *Wheat mosaic virus* and mechanically transmitted *Triticum mosaic virus*; severity and yield losses in individual fields is greater when all three viruses are present. Symptoms of each virus are nearly identical, making proper identification difficult at best. The wheat curl mite is the only

Figure 11-25. *Wheat streak mosaic virus* (WSMV). (Photo: Mary Burrows, Montana State University, Bugwood.org.)

known vector of WSMV, the mite remains infective for approximately a week after it obtains the virus.

Life cycles

Mites have a total of five life stages: the egg, larva, nymph (two nymphal stages), and the adult. Eggs are very small (1/200"), oval, and translucent on the plant, they are laid on the underside of leaves. Larvae are small (1/100"), green, yellow, or pale in color and have a total of six legs. Nymphs are slightly darker in color than the larvae and have a total of eight legs; there are two quiescent stages: one after each of the nymphal stages. Adults are small (1/50"), oval, and darker than the larval and nymphal stages. Eggs hatch in 3 to 10 days after they are laid, and numerous generations occur throughout the year; a complete generation may only take a total of 10 days during the summer.

Host plants & damage

Mites are often polyphagous, feeding on numerous different host plants. All commercially grown cereal crops (oats, barley, spring wheat, winter wheat, and Timothy hay) in Idaho, Washington, and Oregon are susceptible to each species of mites. Fields infested with mites typically have a silver or gray coloration. Individual plants may have chlorotic lesions (yellow speckling) from feeding damage or have silk strung between leaves, a sure sign of a mite infestation. Feeding damage causes reduced photosynthetic potential and reduced yield. WSMV vectored by the wheat curl mite is an additional factor in host plant damage. Younger plants are more susceptible to mite damage and can result in stunted plants with reduced foliage and smaller yield potential.

Integrated pest management

Management of each mite infestation is dependent upon which species is present in the field as each mite has a slightly different biology/ecology/behavior.

Adult monitoring and thresholds

Monitoring for mites should be done throughout the season depending upon your location and its history of mite infestation. Winter grain mite scouting should begin in October and go until April, when small populations start infesting plants. Brown wheat, Banks grass, and wheat curl mites should be scouted from April throughout mid-summer, especially if local climatic conditions are hot and dry. Plants should be visually inspected early in their development when they are more susceptible to mite infestation damage. Scouting should take place in several parts of the field with multiple replications; often mite infestations start near the margin of the field where they are coming from adjacent host plants. Examining the newest leaf tissue, the base of the plant, as well as the soil surface is advisable for examining plants for mites. Visible silk (webbing) on plants is typically found in moderate to severe infestations. Plants without silk should be looked over carefully for the beginnings of an infesting population. Plant samples may also be taken by shaking leaves over a white piece of paper to inspect for mite activity. Chlorotic plants that may show a silvery appearance indicate a mite infestation. A 10× or 20× hand lens is very useful for examining mite infestations, as these pests are very small. Specific thresholds (economic injury levels) are sparse for mite pest species in cereal crops; though if populations are high early in the growing season (October → winter grain mite; April → brown wheat, Banks grass, and wheat curl), an acaricide treatment may be justifiable.

Biological control

Numerous predatory arthropods attack mites and can readily keep mite infestations below economic importance in agricultural fields. Some species of predatory mites (*Phytoseiulus* spp. and *Neoseiulus* spp.) and predatory thrips (*Scolothrips* spp.) are typically present and prefer to feed specifically on pest mite species. Additionally, several insects also predate on these mites, including damsel bugs, minute pirate bugs, big-eyed bugs, green/brown lacewing larvae, and some small black lady beetle species (*Stethorus* spp.). The listed arthropod predators all feed on each and every life stage of pest mite species.

Cultural control

Proper crop rotation with non-host crops should reduce or eliminate previously infested fields. Destroying volunteer wheat or other green bridge host plants in early spring will reduce food availability for developing mite larvae and nymphs. Planting winter wheat later in the fall, as well as reducing any potential green bridge plant, would reduce the population of winter grain mite in certain fields and the amount of time they spend there.

Resistant varieties

Many varieties of spring and winter wheat may have partial resistance to WSMV and/or wheat curl mite.

Acaricides

Several commercially available acaricides are registered for use in treating mite infestations in cereal crops. Mite feeding is often on the underside of leaves, making acaricides difficult to directly apply to the pest in question. Please refer to the PNW Insect Management Handbook for current pesticide recommendations.

Climate change

With increasing temperatures and hotter/drier summer days in the not-so-distant future expect some mite species to become increasingly more abundant and economically important for cereal crop producers. Life cycles of all the pest species in our area may be sped up from increasing temperatures, hence lowering the total development time (from egg to adult) and increasing the number of generations per year.

Wireworms

Pest status & distribution

Wireworms are the larval stage of click beetles and common pests of field and row crops across the contiguous US and Canada, including inland PNW and Intermountain West cereal crops (Andrews et al. 2008; Milosavljević et al. 2016b). Surveys in the inland PNW found that the distribution and abundance of individual wireworm species varies across the region (Rashed et al. 2015; Milosavljević et al. 2016b). Wireworms are usually found in the greatest abundance in fields that have been planted to grasses, grains, or sod for several years (Andrews et al. 2008; Milosavljević et al. 2016b).

Pest descriptions

Wireworm adults are elongated, parallel-sided, brown, reddish-brown, or black beetles with serrate (saw-like) antennae (Figure 11-26) (Milosavljević et al. 2015; Rashed et al. 2015). Wireworm larvae are cylindrical, slender, flattened, and often elongated light yellow to dark brown and resemble mealworms (Figure 11-27). Wireworm larvae have fixed urogomphi (tails at the tip of the abdomen). The sugar beet wireworm [*Limonius californicus* (Mannerhein)], the western field wireworm (*L. infuscatus* Motschulsky), the Pacific Coast wireworm (*L. canus* Le Conte), the Great Basin wireworm [*Selatosomus pruininus* (Horn)], and the green wireworm [*S. aeripennis* (Kirby)] are some of the most commonly found wireworm species in the inland PNW (Milosavljević et al. 2015; 2016b; Rashed et al. 2015). Other genera in the region include *Agriotes* and *Melanotus*.

Life cycles

Overwintering adult click beetles generally emerge during spring and early summer and lay eggs on the surface or in the soil (Andrews et al. 2008). Soon after hatching, larval wireworms move within the soil until

Figure 11-26. Wireworm adult (click beetle). (Photo: Arash Rashed, University of Idaho.)

Figure 11-27. Wireworm larva feeding at the base of a wheat plant. (Photo: Arash Rashed, University of Idaho.)

they orient themselves by detecting volatiles and carbon dioxide released from sprouting seeds and root tissue. Wireworms may persist as larvae in the soil for 1 to 11 years, depending upon nutrition, host plant quality, and climatic conditions (Andrews et al. 2008).

Host plants & damage

Wireworms cause considerable damage to cereal crops by feeding on germinating grains, roots, and stems (Figure 11-28) (Andrews et al. 2008; Higginbotham et al. 2014; Esser et al. 2015). Crops attacked have poor stands that deteriorate over time because wireworms bore into underground portions of the stem. Early signs of damage may be characterized by the presence of a dead central leaf in developing seedlings (Esser 2012). Depending on the growth stage, this injury can cause eventual plant death. Plants affected at later stages of development would suffer from delayed growth/maturity (Andrews et al. 2008).

Factors affecting wireworms

The seasonal feeding activity of wireworms varies considerably across species, mediated by the crop and environmental conditions (Andrews et al. 2008; Milosavljević et al. 2016a; 2016b). For instance, in the inland PNW, while the sugar beet wireworm (*Limonius californicus*) remains active

Figure 11-28. Wireworm damage to wheat. (Photo: Arash Rashed, University of Idaho.)

throughout the season, other species such as *L. infuscatus* and *L. canus* show peak activities earlier in the season and cause significant damage to the planted seeds and/or young roots (Milosavljević et al. 2016a). Ambient temperature drives wireworm development. Larger larvae can withstand higher temperatures during the summer if there is plenty of moisture, whilst younger larvae tend to burrow downward into the soil if it becomes too warm (Andrews et al. 2008). The responses of wireworms to soil temperature are species-specific. For example, in Washington, optimal temperatures for *L. canus* and *L. californicus* are 70–74°F. *Selatosomus pruininus*, *S. aeripennis*, *S. destructor*, and *Hadromorphus glaucus* can withstand lower soil temperatures, and thus attack crops early in the season. Wireworm species in the inland PNW also differ in their moisture needs; *L. californicus* inhabits mostly damp soil, while *S. pruininus* is an obligate dryland species (Andrews et al. 2008; Milosavljević et al. 2016a).

The tolerance of wireworms to soil pH also varies among and within genera (Milosavljević et al. 2016a). Previous studies have suggested that *Limonius* larvae cause most injury in more alkaline soils, whereas *Agriotes* and *Melanotus* species usually prefer acidic soils. *Limonius* species can survive a considerable range of soil pH and can thus cause significant damage to the crops on both alkaline and acidic soils if other factors are favorable (Milosavljević et al. 2016a).

Integrated pest management

Adult and larval monitoring and thresholds

Random soil sampling and solar bait traps have been the two approaches used for monitoring larval presence (Esser 2012; Rashed et al. 2015). Establishing a solar bait trap consists of burying a mixture of germinating/soaked cereal and corn seeds 6 inches deep in the soil, which is then covered by a dark plastic (Esser 2012). As the dark plastic cover absorbs heat from the sun and keeps carbon dioxide, moisture, and volatiles localized, it provides an environment that would attract wireworms to the bait. The number of wireworms collected in the trap could be counted in about 10 to 14 days after placement. Ideally one to two solar bait traps per acre would provide a good assessment of wireworm situation. Trapping is most effective early in the season when soil temperatures reach 45°F (Esser 2012; Rashed et al. 2015). Using this method, an average of 1–2 wireworms per trap indicates insecticide treatments are merited (Esser 2012). Consult the PNW Insect Management Handbook for materials and rates.

Biological control

Biological control of wireworms has been a subject of very few studies, and more work is needed to evaluate the effectiveness of various biocontrol agents in reducing wireworm populations (Ansari et al. 2009; Reddy et al. 2014). While entomopathogenic nematodes have been isolated from wireworms, recent studies have indicated several species of entomopathogenic fungi can be effective in reducing wireworm populations and increasing stand counts in spring wheat (Reddy et al. 2014), although further studies are needed. Ground-foraging beetles like carabids and some birds are predators that might provide some biological control of wireworms (Andrews 2008).

Cultural control

To date, multiple studies have been conducted to examine effects of crop rotations in wireworm management; however, results have been context-dependent and differ based on the wireworm species present, cropping system, and region (Esser et al. 2015). Winter wheat-fallow rotations in the inland PNW have been shown to reduce wireworm populations by 50% compared to continuous spring wheat systems, suggesting that incorporating

fallow into rotations will provide benefits for wireworm control (Esser et al. 2015). Studies have also indicated that dense soil would have a negative impact on wireworms. Thus, proper seedbed preparation, in which the soil is well packed, will not only support healthy and vigorous plant growth but will also limit wireworm movement and reduce feeding damage. Repeated years of **no-till** planting may cause an increase of wireworm damage, creating a central linear furrow in fields where wireworms may concentrate and cause even more damage than usual.

Results of several studies have also shown that wireworm damage is not uniform across crops. Oats are highly **tolerant** of wireworms, with no insecticides needed for wireworm control in this crop (Higginbotham et al. 2014). Barley seems to be fairly tolerant of wireworms as well, although insecticides can increase yield compared to controls. In contrast, wheat is highly susceptible to wireworms, and insecticides will provide significant economic benefits for multiple wireworm species present (Higginbotham et al. 2014; Esser et al. 2015).

Resistant varieties

Although variation in susceptibility to wireworm damage has been documented among wheat genotypes, no resistant varieties are currently available. One study that evaluated 163 wheat genotypes found some genotypes were consistently tolerant (performed well in the presence of wireworms), but the mechanisms or whether the effect was genetically based are not known (Higginbotham et al. 2014).

Insecticides

Seed treated with neonicotinoids can provide stand and yield protection from certain wireworm species. In the inland PNW, applying thiamethoxam and imidacloprid as seed treatments can reduce wireworm populations and increase yields and economic returns in areas with *Limonius* spp. (Esser et al. 2015). However, it is also known that not all species respond similarly to insecticidal treatments. Higher rates of neonicotinoids are more effective against *L. californicus*, possibly because this species has a higher susceptibility to these compounds, as compared with other wireworm species (Esser et al. 2015). Please refer to the PNW Insect Management Handbook for current insecticide recommendations.

While insecticides can be effective tools for managing wireworms in cereals, neonicotinoids do not eliminate, but may reduce, populations to non-economically important infestations from fields (Esser et al. 2015). IPM strategies are likely to be most effective when cultural management practices are combined with insecticides.

Climate change

To date there is no scientific study or associated data that would suggest a correlation with wireworms and climate change, though we could speculate based on biology alone that number of generations per year, developmental time, life cycle, growth rate, or their known range may increase due to warmer overall temperatures in the inland PNW.

Haanchen Barley Mealybug

Pest status & distribution

Haanchen barley mealybugs (*Trionymus haancheni* McKenzie) were officially reported from California in the 1960s (McKenzie 1962). They were later found in other states including Montana, Wyoming, and the PNW states of Idaho and Washington (e.g., Garfield County). In Idaho, their distribution has been mainly limited to dryland production in eastern (e.g., Bonneville and Madison Counties) and southeastern (e.g., Caribou County) parts of the state. Haanchen barley mealybugs feed on a wide variety of grass crops, including barley, wheat, rye, and oats, but in our region have become pests on barley.

Pest description

Adult females are oval-shaped, 1/5 inches long, and may be covered with white powdery secretions (Figure 11-29), forming hair-like filaments along their body outline. Eggs are pinkish-red, microscopic, and protected in cottony wax secretions, also known as an ovisac. Immature stages of the mealybugs are named crawlers. There are several nymphal instars. Overall they resemble adults, as they are also oval-shaped and have three pairs of legs. Their presence, however, is hard to spot with the naked eye due to their very small size. Crawlers are also the most mobile stage of the mealybug life cycle; they may disperse short distances to nearby plants by crawling, or they may travel long distances by wind (Alvarez 2003).

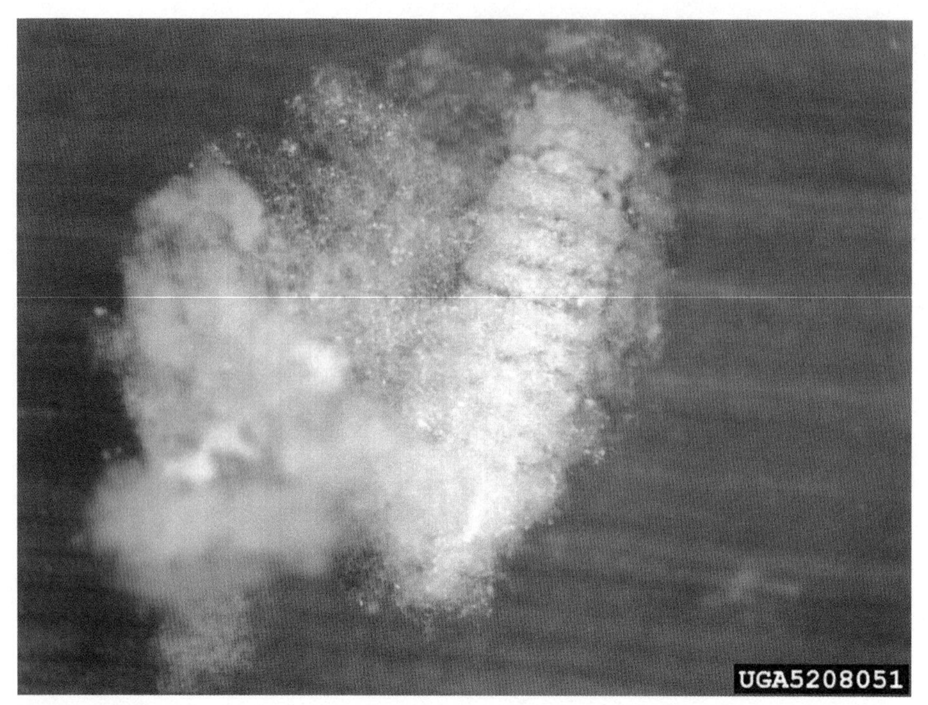

Figure 11-29. Haanchen barley mealybug adult. (Photo: Juan Manuel Alvarez, University of Idaho, Bugwood.org.)

Life cycle

The Haanchen barley mealybug life cycle is yet to be described in detail. To date, observations indicate that they may overwinter as eggs, protected by soil and plant residue, in eastern Idaho (Alvarez 2003). The presence of adults and cottony secretions are detectable later in the spring and early summer at the very base of the infested plants. As plants grow, insects may move up the stem to feed on fresh leaf tissues. Their presence in late summer, under the upper leaf sheaths, can be spotted in the form of visible brownish water stains.

Females lay eggs in protected areas of the plant close to the stem bases and roots, as well as leaf sheaths. Females are capable of ovipositing up to several hundred eggs in a relatively short period of time; the presence of males does not appear to be necessary for reproduction. Following egg hatch, crawlers disperse via crawling, by wind, and/or by assistance from animals and human traffic passing through infested fields. Haanchen mealybug outbreaks seem to be associated with mild winter conditions and dry summer months (Alvarez 2003). The exact number of generations is yet to be determined.

Host plants & damage

Although Haanchen barley mealybugs are known to be primarily damaging in barley, they also feed on wheat and other grasses. Later into the spring and early in the summer, the presence of the mealybug can be detected by the formation of white cottony elements at the very base of stems right around the soil surface (Figure 11-30). Adults can also be seen at the base of a leaf sheaf (Figure 11-31). Direct damage, in forms of excessive yellowing, may be caused by both adult and immature stages of the mealybug, as they use their piercing sucking mouthparts to utilize phloem sap (Mani 2016). Damage appears in fields as irregular bare patches and/or patches of weak plants. Although the exact cause of the reduction in chlorophyll content is yet to be determined, toxic saliva compounds injected into the plant tissue during the feeding process may be a contributing factor (Alvarez 2003;

Figure 11-30. Mealybugs on the plant, showing the white cottony elements. (Photo: Juan Manuel Alvarez, University of Idaho, Bugwood.org.)

Figure 11-31. Mealybug at the base of a barley leaf sheath. (Photo: Juan Manuel Alvarez, University of Idaho, Bugwood.org.)

Mani 2016). Indirect damage can be caused through honeydew production by the insect during feeding. Excessive honeydew (sugar-laden excretions) left on the plant tissue can reduce grain quality, interfere with harvest (Alvarez 2003), and facilitate fungal infections.

Integrated pest management

Currently, there are no established economic thresholds, registered pesticides, or effective integrated management programs for the Haanchen barley mealybug. Although variations in the degree of susceptibility have been reported among various barley varieties, studies are yet to screen for resistance to Haanchen barley mealybugs.

Biological control

Parasitoids and predators have been shown to provide the most effective, and relatively more sustainable, management option with other species of mealybugs. *Rhizopus* spp. (Encyrtidae) has been the predominant parasitoid species found on Haanchen barley mealybugs in Idaho (Blodgett 2009; Mani 2016). Generalist predators, such as lady beetles, have been frequently encountered in infested field plots and may provide some control of the mealybug.

Given current information on the life history traits of Haanchen barley mealybugs, proper rotation with non-cereal crops and proper seeding bed preparation (e.g., cultivation) is expected to interrupt the continuity of the insect life cycle in affected fields.

Chemical control

Currently there are no recommended foliar insecticides for managing Haanchen barley mealybugs in cereals. Having a concealed feeding habit, protective waxy cover, likely asexual reproduction capability, short generation time, and high dispersal potential, chemical control of Haanchen barley mealybugs is greatly challenging. In other crops, targeting crawlers with foliar insecticides, timed approximately a week after egg-laying, may offer a relatively greater chance of success in effectively reducing populations (Alvarez 2003). While seed treatments, foliar spray applications (with a surfactant), and systemic chemistries may help to reduce mealybug populations (Mani 2016), they may not be cost-effective especially under dryland production systems. In addition, excess use of broad-spectrum pesticides can potentially lead to subsequent outbreaks due to the elimination of the natural enemies (Alvarez 2003). Please refer to the PNW Insect Management Handbook for current insecticide recommendations.

Climate change

There are no scientific studies or indications of potential response to climate change by the Haanchen barley mealybug. Its range as a pest of wheat is very limited and it has not been well studied.

A Bigger Picture

Space – Production Landscapes

Cereal crops are grown in heterogeneous landscapes. Even in the most wheat-intensive portions of the inland PNW, wheat fields occur within matrices of other habitat types including fallow or rotational crops with wheat, perennial grasses (pasture or Conservation Reserve Program), and a variety of other crops especially in irrigated areas, scablands, and forests. Insect pests and beneficial organisms live within varied landscapes that include agricultural fields and other habitats. They move among these while foraging and during their annual life cycles. Aphid species, like bird cherry-oat aphid, must overwinter on woody hosts while other aphids overwinter in grassy habitats, and both types then recolonize wheat. Similarly, predators and parasitoids that attack aphids and other pests move into wheat fields out of this larger landscape. The benefits of this free and natural pest control are enormous. A simple experiment in which cereal aphids are protected from natural enemies by exclusion cages reveals that the unprotected populations can exponentially increase and kill wheat plants in just a few weeks. When feasible, natural or perennial habitats can be conserved to help sustain these benefits on or near production fields. The viability of inland PNW cereal systems in future decades will continue to depend upon these biological services.

Complexity – The Wheat Agroecosystem

Just as individual fields are part of a larger landscape-scale system, within each field, the crop, soil organisms, weeds, pathogens, and insects constitute an interactive system. The interactions among the components of this system contribute to its net productivity. A schematic of the continuous winter wheat systems in the inland PNW (Figure 11-31) illustrates the direct effects and interactions among its components. Some of these interactions are characterized well enough to be managed, like biological control of pests by parasitoids and predators. Others remain less well understood but potentially important. For example, evidence is accumulating that soil organisms and the conditions they promote can have emergent effects on pests and diseases that affect the aerial portions of plants. Disease-causing agents, like plant viruses, can affect the insects that are vectors of these pathogens or even the responses of plants to environmental stressors like drought. Going forward, we expect to gain a better understanding of these interactions and to find management approaches that exploit this knowledge to improve plant protection and productivity.

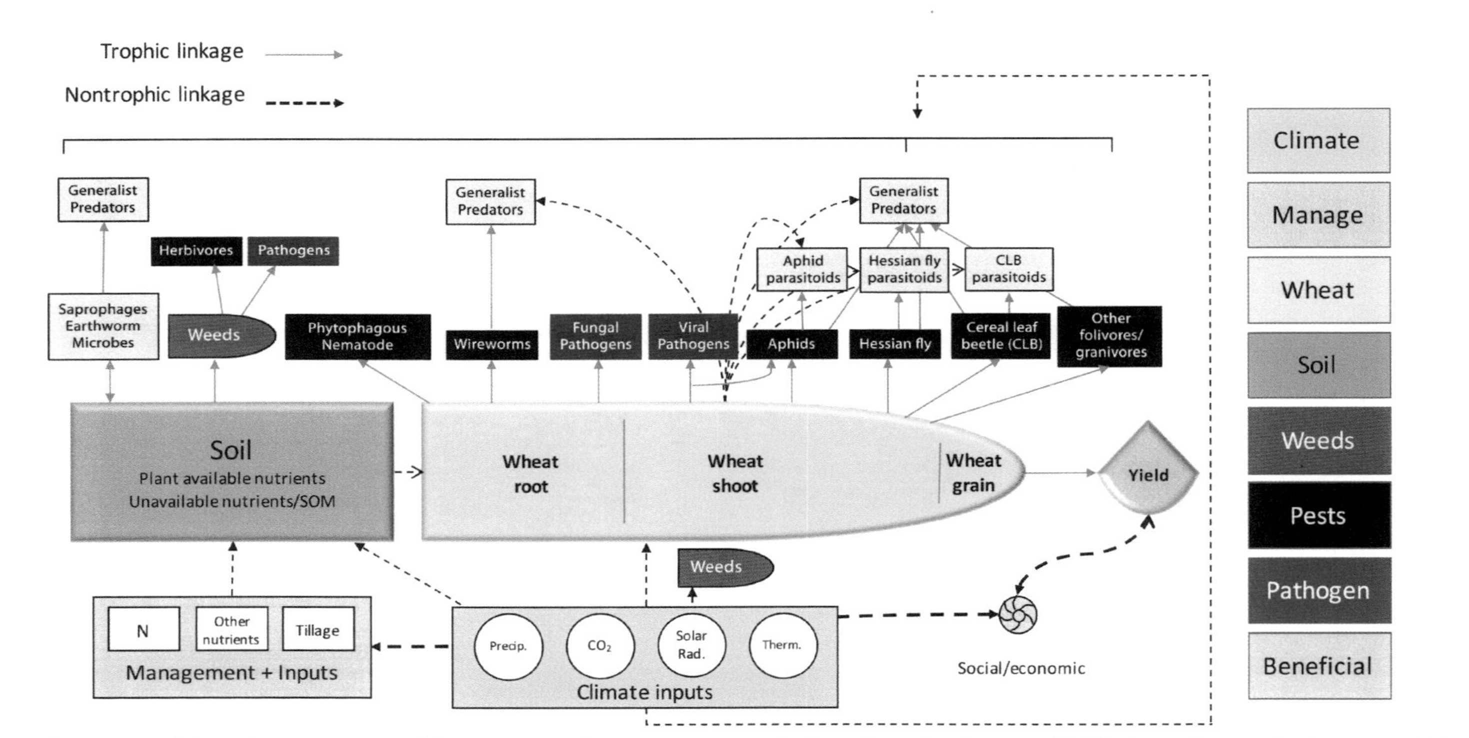

Figure 11-32. Schematic representation of the community of organisms associated with a wheat plant in a typical PNW winter wheat production system. Linkages among organisms are indicated with arrows. Trophic or feeding linkages are shown with solid arrows pointing towards the consumer. Nontrophic linkages, which are various sorts of indirect effects, are shown with broken-line arrows pointing towards the affected group. The portal to social/economic factors indicates that inputs and economic yield from the system are mediated by the human systems in which the production system isolated here exists. Also not shown here are potential biotic interactions between the focal crop (wheat) and conventional or alternative rotational crops occurring in the same landscape or in different years within the same field.

Time – Changing Conditions

Production conditions are always in flux. As climate, insect pests, weeds, diseases, technology, markets, and cultures change, agricultural systems must adjust to the new conditions. The scales of operation, extent of the influence of global markets, and types of technology in use on today's farms would have been difficult or impossible to imagine a few generations ago. We can be certain that the same will be true a few generations hence. To the credit of our farmers and partnerships between industry and agricultural universities, cereal production has continued to thrive in the US and the inland PNW. These dynamics may present significant challenges in the coming decades, but building on traditions, healthy partnerships, and new science should enable resilience and continued productivity.

Resources and Further Reading

Hessian fly

Bosque-Pérez, N.A. 2010. Hessian Fly. In Compendium of Wheat Diseases and Insects 126–128, 3rd ed. W.W. Bockus, R.L. Bowden, R.M. Hunger, W.L. Morrill, T.D. Murray, and R.W. Smiley, eds. The American Phytopathological Society Press, St Paul, MN.

Wheat midge

Knodel, J., and M. Ganehiarachchi. 2008. Integrated Pest Management of the Wheat Midge in North Dakota. North Dakota State University Extension Service.

Cereal Leaf Beetle

Kher, S.V., L.M. Dosdall, and H. Cárcamo. 2011. The Cereal Leaf Beetle: Biology, Distribution and Prospects for Control. *Prairie Soils and Crops Journal* 4: 32–41.

Armyworms, cutworms, and headworms

Bechinski, E.J., Smith, L.J., Merickel, F.W. 2009. Large Yellow Underwing, a New Cutworm in Idaho. University of Idaho Current Information Series 1172.

Berry, R.E. 1998a. Black Cutworm Factsheet. Modified from Insects and Mites of Economic Importance in the Northwest, 2nd ed. p. 221. ***http://mint.ippc.orst.edu/blackcutfact.pdf***.

Berry, R.E. 1998b. Variegated Cutworm. Modified from Insects and Mites of Economic Importance in the Northwest. 2nd Ed. p. 221. ***http://mint.ippc.orst.edu/vcfact.pdf***.

Cook, K.A., S.T. Ratcliffe, and M.E. Gray. 2003. Black Cutworm (*Agrotis ipsilon* Hufnagel). Insect Fact Sheet, University of Illinois IPM. ***https://ipm.illinois.edu/fieldcrops/insects/black_cutworm.pdf***.

Peairs, F.B. 2006. Crops: Caterpillars in Small Grains. Insect Series 5.577. Fort Collins, CO: Colorado State University Cooperative Extension.

Royer, T. 2007. Armyworms and Wheat Head Armyworms: How to Tell the Difference. Oklahoma State University Plant Disease and Insect Advisory 6(15).

Wireworms

Esser, A.D. 2012. Wireworm Scouting: The Shovel Method and the Modified Wireworm Solar Bait Trap. Washington State University Extension Fact Sheet FS059E.

Milosavljevic, I., A.D. Esser, and D.W. Crowder. 2015. Identifying Wireworms in Cereal Crops. Washington State University Extension Bulletin F5175E.

Rashed, A., F. Etzler, C.W. Rogers, and J.M. Marshall. 2015. Wireworms in Idaho Cereals: A Guide to Monitor Numbers and Identify Predominant Species in the Intermountain Region. University of Idaho Extension Bulletin 898.

Haanchen barley mealybug

Alvarez, J.M. 2003 Haanchen Barley Mealybug, A New Pest of Barley in Idaho. University of Idaho Current Information Series 1109.

Blodgett, S. 2009. High Plains Integrated Pest Management Guide. ***http://wiki.bugwood.org/HPIPM:Haanchen_Mealybug***.

Mani, M. 2016. Management of Mealybugs in Agricultural and Horticultural Crops, Barley. In Mealybugs and their Management in Agricultural and Horticultural crops, p. 249. M. Mani and C. Shivaraju, eds. Springer India.

Mites

Bauernfeind, R. 2005. Winter Grain Mites. Kansas State University. Extension Publication MF-2073.

Brewer, M.J. 1995. Banks Grass Mite *Oligonychus pratensis*. University of Wyoming Extension Publication B-1013.

Blodgett, S., and G.D. Johnson. 2002. Brown Wheat Mite. Montana State University Extension Publication MT200212 AG 12/2002.

References

Alvarez, J.M. 2003. Haanchen Barley Mealybug, A New Pest of Barley in Idaho. University of Idaho Current Information Series 1109.

Andrews, N., M. Ambrosino, G. Fisher, and S. Rondon. 2008. Wireworm Biology and Nonchemical Management in Potatoes in the Pacific Northwest. Pacific Northwest Extension Publication 607. Oregon State University.

Ansari, M.A., M. Evans, and T.M. Butt. 2009. Identification of Pathogenic Strains of Entomopathogenic Nematodes and Fungi for Wireworm Control. *Crop Protection* 28: 269–272.

Bauernfeind, R. 2005. Winter Grain Mites. Kansas State University Extension Publication MF-2073.

Bechinski, E.J., L.J. Smith, and F.W. Merickel. 2009. Large Yellow Underwing, A New Cutworm in Idaho. University of Idaho Current Information Series 1172.

Berry, R.E. 1998a. Black Cutworm. Modified from Insects and Mites of Economic Importance in the Northwest, 2nd ed., p. 221. ***http://mint.ippc.orst.edu/blackcutfact.pdf.***

Berry, R.E. 1998b. Variegated Cutworm. Modified from Insects and Mites of Economic Importance in the Northwest, 2nd ed., p. 221. ***http://mint.ippc.orst.edu/vcfact.pdf***.

Blodgett, S. 2009. High Plains Integrated Pest Management Guide. ***http://wiki.bugwood.org/HPIPM:Haanchen_Mealybug***.

Blodgett, S., and G.D. Johnson. 2002. Brown Wheat Mite. Montana State University Extension Publication MT200212 AG 12/2002.

Bosque-Pérez, N.A. 2010. Hessian Fly. In Compendium of Wheat Diseases and Insects, 3rd ed., p. 126–128. W.W. Bockus, R.L. Bowden, R.M. Hunger, W.L. Morrill, T.D. Murray, and R.W. Smiley, eds. The American Phytopathological Society Press, St Paul, MN.

Bosque-Pérez, N.A., J.B. Johnson, D.J. Schotzko, and L. Unger. 2002. Species Diversity, Abundance, and Phenology of Aphid Natural Enemies on Spring Wheats Resistant and Susceptible to Russian Wheat Aphid. *BioControl* 47: 667–684.

Brewer, M.J. 1995. Banks Grass Mite *Oligonychus pratensis*. University of Wyoming Extension Publication B-1013.

Bullock, D.G., N.A. Bosque-Pérez, J.B. Johnson, and F.W. Merickel. 2004. Species Composition and Distribution of Hessian Fly (Diptera: Cecidomyiidae) Parasitoids in Northern Idaho. *Journal of the Kansas Entomological Society* 77: 174–180.

Buntin, G.D., K.L. Flanders, R.W. Slaughter, and Z.D. DeLamar. 2004. Damage Loss Assessment and Control of the Cereal Leaf Beetle (Coleoptera: Chrysomelidae) in Winter Wheat. *Journal of Economic Entomology* 97: 374–382.

Castle del Conte, S.C., N.A. Bosque-Pérez, D.J. Schotzko, and S.O. Guy. 2005. Impact of Tillage Practices on Hessian Fly-Susceptible and Resistant Spring Wheat Cultivars. *Journal of Economic Entomology* 98: 805–813.

Clement, S.L., L.R. Elberson, F.L. Young, J.R. Alldredge, R.H. Ratcliffe, and C. Hennings. 2003. Variable Hessian Fly (Diptera: Cecidomyiidae) Populations in Cereal Production Systems in Eastern Washington. *Journal of the Kansas Entomological Society* 76: 567–577.

Cook, K.A., S.T. Ratcliffe, and M.E. Gray. 2003. Black Cutworm (*Agrotis ipsilon* Hufnagel). Insect Fact Sheet, University of Illinois IPM. ***https://ipm.illinois.edu/fieldcrops/insects/black_cutworm.pdf.***

Davis, T.S., Y. Wu, and S.D. Eigenbrode. 2014a. Host Settling Behavior, Reproductive Performance, and Effects on Plant Growth of an Exotic Cereal Aphid, *Metopolophium festucae* subsp. *cerealium* (Hemiptera: Aphididae). *Environmental Entomology* 43: 682–688.

Davis, T.S., J. Abatzoglou, N.A. Bosque-Pérez, S.E. Halbert, K. Pike, and S.D. Eigenbrode. 2014b. Differing Contributions of Density Dependence and Climate to the Population Dynamics of Three Eruptive Herbivores. *Ecological Entomology.*

Davis, T.S., N.A. Bosque-Pérez, T. Magney, N.E. Foote, and S.D. Eigenbrode. 2015. Environmentally Dependent Host-Pathogen and Vector-Pathogen Interactions in the *Barley Yellow Dwarf Virus* Pathosystem. *Journal of Applied Ecology.*

Eigenbrode, S.D., and S. Macfadyen. 2016. Insect Pests of Wheat Under Changing Climates – An Assessment of the State of Knowledge and Needs for Research. In Achieving Sustainable Wheat Cultivation. P. Langridge, ed. Burleigh Dodds, Cambridge.

Eigenbrode, S.D., T.S. Davis, and D.W. Crowder. 2015. Climate Change and Biological Control in Agricultural Systems: Principles and Examples from North America. In Climate Change and Insect Pests, p. 119–135. C. Bjorkman and P. Niemela, eds. CABI, Oxon.

Esser, A.D., I. Milosavljević, and D.W. Crowder. 2015. Effects of Neonicotinoids and Crop Rotation for Managing Wireworms in Wheat Crops. *Journal of Economic Entomology* 108: 1786–1794.

Esser, A.D. 2012. Wireworm Scouting: The Shovel Method and the Modified Wireworm Solar Bait Trap. Washington State University Extension Publication FS059E.

Evans, E.W., N.R. Carlile, M.B. Innes, and N. Pitigala. 2012. Warm Springs Reduce Parasitism of the Cereal Leaf Beetle through Phenological Mismatch. *Journal of Applied Entomology* 137: 321–400.

Finlay, K.J., and J.E. Luck. 2011. Response of the Bird Cherry-Oat Aphid (*Rhopalosiphum padi*) to Climate Change in Relation to its Pest Status, Vectoring Potential and Function in a Crop-Vector-Virus Pathosystem. *Agriculture, Ecosystems & Environment* 144: 405–421.

Halbert, S.E., Y. Wu, and S.D. Eigenbrode. 2013. *Metopolophium festucae cerealium* (Hemiptera: Aphididae): A New Addition to the Aphid Fauna of North America. *Insecta Mundi* 0301: 1–6.

Higginbotham, R.W., P.S. Froese, and A.H. Carter. 2014. Tolerance of Wheat (Poales: Poaceae) Seedlings to Wireworm (Coleoptera: Elateridae). *Journal of Economic Entomology* 107: 833–837.

Knodel, J., and M. Ganehiarachchi. 2008. Integrated Pest Management of the Wheat Midge in North Dakota. North Dakota State University Extension Service.

Mani, M. 2016. Management of Mealybugs in Agricultural and Horticultural Crops, Barley. In Mealybugs and their Management in Agricultural and Horticultural Crops, p. 249. M. Mani and C. Shivaraju, eds. Springer India.

McKenzie, H.L. 1962. Third Taxonomic Study of California Mealybugs, Including Additional Species from North and South America. *Hilgardia* 32: 637–687.

Michaud, J.P., P.E. Sloderbeck, and R.J. Whitworth. 2007. Wheat Insect Management. Kansas State University Research and Extension.

Milosavljević, I., A.D. Esser, and D.W. Crowder. 2016a. Seasonal Population Dynamics of Wireworms in Wheat Crops in the Pacific Northwestern United States. *Pest Science*.

Milosavljević, I., A.D. Esser, and D.W. Crowder. 2016b. Effects of Environmental and Agronomic Factors on Soil-Dwelling Pest Communities in Cereal Crops. *Agriculture, Ecosystems & Environment* 225: 192–198.

Milosavljević, I., A.D. Esser, and D.W. Crowder. 2015. Identifying Wireworms in Cereal Crops. Washington State University Extension Publication F5175E.

Olfert, O., and R.M. Weiss. 2006. Impact of Climate Change on Potential Distributions and Relative Abundances of *Oulema melanopus*, *Meligethes viridescens* and *Ceutorhynchus obstrictus* in Canada. *Agriculture, Ecosystems & Environment* 113: 295–301.

Olfert, O., R.M. Weiss, and R.H. Elliot. 2016. Bioclimatic Approach to Assessing the Potential of Climate Change on Wheat Midge in North America. *Canadian Entomologist* 148: 52–67.

Olfert, O., R.M. Weiss, S. Woods, H. Philip, and L. Dosdall. 2004. Potential Distribution and Relative Abundance of an Invasive Cereal Crop Pest, *Oulema melanopus* (Coleoptera: Chrysomelidae), in Canada. *Canadian Entomologist* 136: 277–287.

Philips, C.R., D.A. Herbert, T.P. Kuhar, D.D. Reisig, W.E. Thomason, and S. Malone. 2011. Fifty Years of Cereal Leaf Beetle in the U.S.: An Update on its Biology, Management, and Current Research. *Journal of Integrated Pest Management* 2: 1–5.

Rao, S., A. Cossé, B. Zilkowski, and R. Bartelt. 2003. Aggregation Pheromone of the Cereal Leaf Beetle: Field Evaluation and Emission from Males in the Laboratory. *Journal of Chemical Ecology* 29: 2165–2175.

Rashed, A., F. Etzler, C.W. Rogers, J.M. Marshall. 2015. Wireworms in Idaho Cereals: A Guide to Monitor Numbers and Identify Predominant Species in the Intermountain Region. University of Idaho Extension Bulletin 898.

Ratcliffe, R.H., S.E. Cambron, K.L. Flanders, N.A. Bosque-Pérez, S.L. Clement, and H.W. Ohm. 2000. Biotype Composition of Hessian Fly (Diptera: Cecidomyiidae) Populations from the Southeastern, Mid-Western, and Northwestern United States and Virulence to Resistance Genes in Wheat. *Journal of Economic Entomology* 93: 1319–1328.

Reddy, G.V.P., K. Tangtrakulwanich, S. Wu, J.H. Miller, V.L. Ophus, J. Prewett, and S.T. Jaronski. 2014. Evaluation of the Effectiveness of Entomopathogens for the Management of Wireworms (Coleoptera: Elateridae) on Spring Wheat. *Journal of Invertebrate Pathology* 120: 43–49.

Roberts, D. 2012. Extension Leads Multi-Agency Team in Suppressing a Pest in the West. *Journal of Extension* 50. ***https://www.joe.org/joe/2012april/a10.php***.

Sadeghi, S.E., J. Bjur, L. Ingwell, L. Unger, N.A. Bosque-Pérez, and S.D. Eigenbrode. 2016. Interactions between *Metopolophium festucae cerealium* (Hemiptera: Aphididae) and *Barley yellow dwarf virus* (BYDV-PAV). *Journal of Insect Science* 16: 1–6.

Schotzko, D.J., and N.A. Bosque-Pérez. 2002. Relationship between Hessian fly Infestation Density and Early Seedling Growth of Resistant and Susceptible Wheat. *Journal of Agricultural and Urban Entomology* 19: 95–107.

Smiley, R.W., J.A. Gourlie, R.G. Whittaker, S.A. Easley, and K.K. Kidwell. 2004. Economic Impact of Hessian Fly (Diptera: Cecidomyiidae) on Spring Wheat in Oregon and Additive Yield Losses with Fusarium Crown Rot and Lesion Nematode. *Journal of Economic Entomology* 97: 397–408.

Townsend, L., D. Johnson, and D. Hershman. 1996. Wheat Streak Mosaic Virus and the Wheat Curl Mite. University of Kentucky Extension Publication ENTFACT-117.

Trebicki, P., N. Nancarrow, E. Cole, N.A. Bosque-Pérez, F.E. Constable, A.J. Freeman, B. Rodoni, A.L. Yen, J.E. Luck, and G.J. Fitzgerald. 2015. Virus Disease in Wheat Predicted to Increase with a Changing Climate. *Global Change Biology* 21: 3511–3519.

Trebicki, P., R.K. Vandegeer, N.A. Bosque-Pérez, K.S. Powell, B. Dader, A.J. Freeman, A.L. Yen, G.J. Fitzgerald, and J.E. Luck. 2016. Virus Infection Mediates the Effects of Elevated CO_2 on Plants and Vectors. *Nature Scientific Reports* 6: 22785.

Chapter 12

Farm Policies and the Role for Decision Support Tools

Laurie Houston, Oregon State University
Clark Seavert, Oregon State University
Susan Capalbo, Oregon State University
John Antle, Oregon State University

Abstract

This chapter provides an overview of farm policies up to the most recent 2014 Farm Bill, the expected focus of future policies, and the potential role decision support and precision agriculture tools can play in both developing and analyzing farm policies. Farm policies have often been designed to manage risk and incentivize desired management practices and will continue to do so. The defining characteristic of future policy is expected to reside around the use of spatially explicit data and decision support tools that will inform policymakers regarding the design of future farm policies and inform farmers regarding the effect farm policies will have on net returns. The same spatially explicit data will also help farmers optimize the use of inputs, reduce costs, and improve environmental outcomes. In addition, this chapter will provide a sampling of decision support tools as they relate to agricultural policies and a reference guide for additional sources for tools and resources. We then provide an example of how spatially explicit data and a decision support tool such as AgBiz Logic can be used by growers and policymakers to examine the impact on net returns of a targeted conservation policy.

Key Points

- Policy will play a key role in influencing management practices in inland Pacific Northwest grain production systems including: conservation cropping, residue and soil water management, crop rotations, and pest management.
- The influence of policy will occur through the development of risk management options, management recommendations and incentives, and the adoption of agricultural technologies.
- The use of spatially explicit data and regional impact models will likely play a larger role in the design and implementation of future farm policies.
- Precision farming tools allow for spatially explicit management and have the potential to improve sustainability of management practices.
- Decision support tools, such as AgBiz Logic, can be used by growers and policymakers to assess potential impacts a variety of agricultural policies may have on farm level net returns and profitability.

Brief History of Farm Policy (up to the 2014 Farm Bill)

Prior to 1933, the policy of the United States Department of Agriculture (USDA) was primarily directed toward on-farm support and services. Services included agricultural research projects at land grant institutions, marketing services, and Extension programs. The Great Depression and the Dust Bowl changed farm policy. During this time, farm households accounted for nearly a quarter of the US workforce and 8 percent of gross domestic product, or GDP (Dimitri et al. 2005). Farm prices were falling and farm foreclosures were on the rise. In an attempt to increase the welfare of many rural Americans, Congress passed the Agricultural Adjustment Act of 1933 in order to raise the value of crops. The act created the first income-support subsidies and production controls for basic commodities, which at that time consisted mostly of corn, wheat, cotton, rice, and dairy products. Shortly thereafter, the Agricultural Adjustment Act of 1938 created a more permanent farm bill with a built-in requirement to update it every five years and the Federal Crop

Insurance Corporation was also created. The act was made permanent in 1949, but subsequent farm bills have regularly amended its provisions roughly every five years since then.

The various farm bills since 1949 have represented an evolution of farm policies in response to various market factors and political and social pressures of the time. This evolution is evident in the summary of the major changes in farm policies presented in Table 12-1. For example, the 1985 Farm Bill introduced a major new environmental provision: the Conservation Reserve Program. Changes in the 1996 Farm Bill eliminated deficiency payments and replaced them with production flexibility contract payments. These changes were precipitated by a gradual decrease in the reliance on Commodity Credit Corporation storage programs and more on direct payments to support commodity prices and farm incomes as well as the rising popularity of deregulation and less governmental interference (Ray 2001). The 2002 Farm Bill reinstated deficiency payments in the form of counter-cyclical payments. The 2008 bill created a new revenue support program called Average Crop Revenue Election (ACRE). The most recent 2014 Farm Bill brought about several more changes by eliminating direct payments, counter-cyclical payments, and ACRE, and replaced them with two new commodity programs aimed at risk management called Agriculture Risk Coverage (ARC) and Price Loss Coverage (Effland et al. 2014). The 2014 bill also consolidated environmental programs, expanded the crop insurance program, and tied crop insurance closer to commodity programs and conservation programs (Chite 2014; ERS n.d.). The Wildlife Habitat Incentive Program (WHIP) was combined with the Environmental Quality Incentives Program (EQIP); The Grassland Reserve Program, Wetlands Reserve Program, and Farm Ranchland Protection Program were combined into one easement program titled Agricultural Conservation Easement Program. The 2014 Farm Bill also strengthened and expanded the Federal Crop Insurance (FCI) program to include a whole farm policy, and required conservation compliance in order to receive premium subsidies. Although the features of the farm bill have changed over time, the idea of supporting income and minimizing risk for farmers and food production has remained constant.

The new farm bill also slightly increased funding for Research, Extension, and Related Matters: Title VII. Part of this funding went to establish the

Table 12-1. Summary of key agricultural policies, 1933–2015.

Program Title	Policy Enacted
Agricultural Adjustment Act of 1933	Introduced price and income-support programs, created the Commodity Credit Corporation (CCC), and made price-support loans for designated basic storable commodities (corn, wheat, and cotton). The government also agreed to buy excess grain from farmers, which could be released in later years when bad weather affected yields.
Soil Conservation and Domestic Allotment Act of 1936	Provided for soil conservation and soil-building payments to participating farmers. First link between soil conservation and commodity programs.
Agricultural Adjustment Act of 1938	Requirement to update the farm bill every five years. Created mandatory price supports for corn, cotton, and wheat.
Federal Crop Insurance Act of 1938	Also established the Federal Crop Insurance Corporation.
Agricultural Act of 1949	Established high, fixed-price supports and acreage allotments as permanent farm policy. Programs revert to the 1949 provisions anytime a new farm bill fails to pass.
Agricultural Act of 1954	Introduced flexible price supports for basic commodities, and authorized a CCC reserve for foreign and domestic relief.
Agricultural Act of 1956.	Created the Conservation Reserve Program (CRP), and the Acreage Reserve Program (ARP) for wheat, corn, rice, cotton, peanuts, and tobacco.
Agricultural Act of 1965.	Introduced new income-support payments in combination with reduced price supports and continued supply controls.
Agriculture and Consumer Protection Act of 1973	Established target prices and deficiency payments to replace former price-support payments, and authorized disaster payments and disaster reserve inventories to alleviate distress caused by natural disaster.

Program Title	Policy Enacted
Agricultural Adjustment Act of 1980	Amended the Food and Agriculture Act of 1977 primarily to raise the target prices of wheat and corn.
Federal Crop Insurance Act of 1980	Expanded crop insurance into a national program with the authority to cover the majority of crops.
Federal Crop Insurance Reform Act of 1994	Made participation in the crop insurance program mandatory for farmers to be eligible for deficiency payments under price-support programs, certain loans, and other benefits.
The Food Security Act of 1985	Created a Conservation Reserve Program under which the Federal Government entered into long-term retirement contracts on qualifying land.
Federal Agricultural Improvement and Reform Act of 1996	Introduced a 7-year phase-out of government income-support payments by replacing price support and supply control programs with direct payments. Repealed the mandatory participation requirement for crop insurance, but required farmers who accepted other benefits to purchase crop insurance or waive their eligibility for disaster benefits.
Farm Security and Rural Improvement Act of 2002	Brought back price supports with the Direct and Counter-cyclical Payments program (DCP). Introduced working-lands conservation payments through the Conservation Security Program (CSP).
Food, Conservation, and Energy Act of 2008	Created Supplemental Revenue Assurance (SURE) – insures against crop revenue losses and Average Crop Revenue Election (ACRE), an alternative to counter-cyclical payments.
Agricultural Act of 2014	Replaced Direct and Counter-cyclical Program, and ACRE with Price Loss Coverage (PLC) and Agriculture Risk Coverage (ARC)

For a complete description of these agricultural policies and others, see Becker (2002), Bowers et al. (1984), Dimitri et al. (2005), Limpton and Pollack (1996), Mercier (2012), O'Donoghue (2016), or Womach (2005).

Foundation for Food and Agriculture Research, which is designed to encourage public-private partnerships in research by requiring private matching funds. This slightly reduced the downward trend in public funding of research and development, but also encouraged the shift from public funding to private funding. This may increase total funding for research and development, but it also changes the research focus more towards post farm research and development and less towards productivity, efficiency, and conservation issues (Pardey et al. 2015).

Policy Impacts in the Pacific Northwest

According to the 2007 Census of Agriculture, 38% of US farms received some form of government payment. Using individual responses to the 2007 Agricultural Census, Antle and Houston (2013a; 2013b) examined the distribution of the major types of farm program payments for the wheat region of the Pacific Northwest (PNW), along with three other regions of the country. These farm programs included Direct Payments; Conservation Programs, such as the Conservation Reserve Program (CRP), Wetlands Reserve Program (WRP), Conservation Security Program (CSP), and EQIP; Market Assistance Loans; and Loan Deficiency Payments. The census does not provide information on crop insurance premium subsidies or insurance indemnity payments so these types of payments were not included in the analysis. About 44% of farms in this region received government payments from these programs. Direct payment subsidies represented about 40% of total mean government payments for most farms, and as much as 83% for some large farms (Antle and Houston 2013a; 2013b). The recent elimination of direct payments in the 2014 Farm Bill, along with current and upcoming changes in crop insurance, may significantly change the amount and distribution of farm payments. It may also play a role in the ability of farmers in the region to adopt new technologies, such as **precision agriculture** technologies, that have the potential to both improve net returns and environmental outcomes. The magnitude of the impact will depend on the extent to which changes in crop insurance subsidies compensate for the losses from direct payments (Antle and Houston 2013b).

The Center for Agricultural and Environmental Policy at Oregon State University has provided a preliminary analysis of the impacts of the most recent Farm Bill on California, Oregon, and Washington (Olen and Wu 2014). Most notably, for the major commodities (including small grains) and the dairy/livestock sectors, the 2014 Farm Bill ends direct payments, counter-cyclical payments, and crop revenue election programs; establishes price loss coverage and risk coverage programs; establishes margin protection programs for dairy; establishes supplemental agricultural disaster assistance programs for livestock; establishes payment limits and income caps for payments; and provides weather-related coverage for commodities not included in crop insurance policies. Farmers in the West, including wheat farmers in Oregon and Washington, will benefit from the expanded crop insurance program known as the Supplemental Coverage Option (SCO) (Olen and Wu 2014). According to the Risk Management Agency (RMA n.d.) state profiles, 77% of the wheat acreage in Oregon and 90% of the wheat acreage in Washington was insured in 2013. This is expected to increase in 2015 and 2016 as a result of the SCO. The flexibility to choose different levels of protection for irrigated and non-irrigated crops may also boost insurance usage for many crops (Olen and Wu 2014).

Agriculture also depends on support from the research programs and Extension efforts of land-grant universities. These institutions provide essential research regarding many aspects of agricultural production, such as improvements in crop varieties, food safety, environmental conservation, effectiveness of cropping systems, and short-term and long-term climate projections and their effects on crop yields and soil quality, just to name a few. The agriculture industry also depends heavily on outreach and Extension efforts such as field days and a variety of special workshops.

Better prediction of the impacts of farm policies will be possible as the analyses of farm programs are spatially downscaled. The use of farm-level decision support tools coupled with regional impacts models is needed for fine-tuning the effects on individual growers and regions, and for predicting overall participation rates in numerous federal programs. Big data initiatives will help to provide some of the data necessary to accurately assess the impact of current and future farm programs on farm welfare, sustainability, food security, and environmental outcomes. Big data refers to extremely large data sets that may be analyzed computationally to reveal patterns, trends, and associations. The Agriculture and Technology department of The State University of New York is currently in the process of building a cloud-based big data clearinghouse for agriculture they call BRAG cloud (short for Broadband Rural Agriculture cloud). Their goal is to help farmers and others in the food industry make use of the large amount of data that is being generated from the increasing use of precision agriculture tools and strategies (Desmond 2016).

Prospects for Agricultural Policy in the Future

United States farm policies cover a wide range of objectives such as stabilizing farm income, assuring adequate nutrition, food security, and safety, and protecting the environmental. Among all these objectives, there are several possible directions for agricultural policy. It is fairly certain however, that these basic priorities will remain and that the projected expansion of the world's population along with climate change will shape future policy. With the world's population expected to grow by more than 2 billion people by 2050 (UN 2015), the stagnated yields of the world's major cereal crops (Ray et al. 2012), and the projections that climate change will have a detrimental impact on crop yields in the future (Hatfield et al. 2014), agricultural policies will need to focus on reducing waste, improving the equitable distribution of food, and increasing production efficiencies, food quality, and nutrition while managing risk and environmental outcomes. In order to feed more than 2 billion additional people under these conditions, we will need to make more efficient use of our resources.

Future agricultural policies will likely focus more on sustainable management of agricultural landscapes with an aim to maintain and improve food availability and quality while also maintaining and enhancing the natural resource base and risk management. Sustainable management will target soil quality, water quality, nitrogen (N) cycles, and greenhouse gas emission reductions including nitrous oxide (N_2O), carbon dioxide (CO_2) and methane (CH_4). These goals are evident in the current calls for climate-smart agriculture, sustainable intensification, managing agro-ecosystems to enhance ecosystem services, and land-use policies calling for land sparing or land sharing (Power 2010; Phalan et al. 2011; Garnett et al. 2013; The World Bank 2011; 2014). The current Strategic Plan for the USDA encourages voluntary practices such as **conservation tillage**, manure and nutrient management, fertilizer efficiency, increasing energy efficiency, and developing renewable sources of energy (USDA 2014). The USDA has also established seven Regional Climate Hubs in order to deliver science-based knowledge and practical information to farmers. The information and guidelines within previous chapters are also geared toward understanding how to more effectively meet these goals and provide timely information to growers and land managers.

One way to address these concerns about sustainable management is through policies and programs that address plant breeding to increase resiliency, yields, and nutritional qualities of crops while also reducing input requirements. Genetically modifying organisms has the potential to increase crop yields, reduce herbicide and pesticide use (Klümper and Qaim 2014), as well as increase shelf life, vitamin content, and resistance to diseases (NAS 2016). The debate over genetically modified organisms (GMOs) and the strict regulations in some countries concerning them, may hinder this mode. However, there are new plant breeding techniques that use genome editing which may potentially be more acceptable to consumers (Hartung and Schiemann 2014). Genome editing allows breeders to determine if a plant will have the desired characteristics before the plant is fully mature. It is a much quicker process that imitates the traditional mutation process of conventional breeding. Unlike GMO techniques that introduce genes that do not arise naturally in the species into a plant's DNA, genome editing allows breeders to develop plants that do not differ in any way from a plant whose genome was altered through breeding (Rosch 2016). Hartung and Schiemann (2014) and the National Academies of Sciences (2016) argue that new plant varieties should be regulated based on novel characteristics and hazards rather than the technique used to create it. Current policy in the US is product-based in theory, but the USDA and the Environmental Protection Agency determine which plants to regulate at least partially on the process by which they are developed (NAS 2016). We view this as an emerging field that agricultural policy will grapple with in the near future as more varieties are developed using genome editing.

Another way to meet these goals is by closely linking commodities, crop insurance, and environmental conservation programs and by designing targeted agricultural policies that will strengthen the farm financial safety net. The next farm bill may introduce hybrid conservation-risk policies that incorporate counter-cyclical and risk components of current farm programs and crop insurance (Coppess 2016). Currently, crop insurance has links to commodity programs, but these two programs work mostly independent of conservation programs. By linking these policies, there should be less adverse consequences from policies that target only one goal. As a move in this direction, the USDA plans to expand crop insurance availability and product coverage, and to use geographical information systems, remote sensing, precision agriculture and data mining, to improve crop insurance products and rapidly assess damage (USDA 2014). A Working Landscapes Initiative, regarding conservation and crop insurance advocated by the Meridian Institute, is also supporting a policy move in this direction (AGree 2016). This initiative supports several research and advocacy efforts to assess the correlation between soil type and yield risk; update the USDA's data collection system to increase the efficiency of data collection and integration of data and reduce respondent burden; and initiate changes in crop insurance that will support innovation and conservation. These initiatives will require gathering spatially explicit data. Until recently, designing targeted agricultural policies was virtually impossible on a large scale due to lack of data and the cost of obtaining it. However, advances in technology and data management have begun to make processing large quantities of spatially explicit climatic, geographic, and economic data possible and affordable. (For more detail about data and precision agriculture, see Chapter 8: Precision Agriculture.) Privacy and confidentiality concerns regarding such data gathering and sharing have been a concern of many farmers and commodity organizations (AGree 2014). In response, the American Farm Bureau together with a consortium of farmer organizations and agriculture data technology providers, is developing data privacy and security principles in order to ensure that data not be misused (Plume 2014). For a larger discussion on the use of big data for agro-environmental policies, see Antle et al. (2015).

Climate initiatives will also frame future agricultural policies. The USDA Building Blocks for Climate-Smart Agriculture and Forestry: Implementation Plan and Progress Report (USDA 2016) provides details regarding the USDA's framework for helping farmers, ranchers, and forestland owners respond to climate change. The framework will work within many existing conservation programs such as EQIP and CRP, and make them more flexible to unique conditions on farms in different parts of the country. The plan also focuses on soil quality to increase organic matter and improve microbial activity. This will sequester more carbon, which will have several benefits beyond the reduction of greenhouse gases, such as improved water management, improved wildlife and pollinator habitat, as well as improved yields. The USDA also established a Soil Health Division at the Natural Resources Conservation Service (NRCS) in 2015, to focus on providing financial and technical assistance to farmers to implement conservation practices such as tillage management, **cover crops**, and grassed waterways. Additionally, 'sensitive lands' will be identified and NRCS will target owners of this land and encourage the adoption of conservation systems using financial and technical assistance incentives.

Future policies will likely take advantage of 'big data' to design spatially explicit policies to enhance the efficiency and reduce the cost of farm programs such as crop insurance subsidies, which are expected to increase substantially in the next ten years and to outpace spending on traditional commodity programs by about one-third (Shields 2010). An example of a policy that could benefit from a more targeted approach would be the requirements in the 2014 Farm Bill which states that farmers must practice soil and water conservation measures on vulnerable lands in exchange for receiving subsidies for crop insurance premiums. Having spatial information about the effectiveness of various practices at specific locations would allow policymakers to target highly sensitive parcels and provide landowners with information necessary to make informed decisions at the farm level to maximize net returns and minimize environmental impacts. Landowners with highly sensitive parcels of land could then be incentivized to adopt specific conservation practices on these lands by offering larger subsidies or lower crop insurance premiums for farmers with the most effective conservation outcomes, sometimes referred to as precision conservation. This would increase the efficiency by increasing enrollment of highly sensitive lands, and possibly reduce the subsidy payments for less sensitive land, improve the environmental benefits, and reduce the social cost of achieving the environmental benefits. Thus, we foresee a movement toward a more spatially oriented form of policy that is more efficient and better able to meet both the needs of growers and policymakers by providing risk management while also emphasizing sustainability. We also note, however, that better data and targeted approaches that are parcel-specific may not be applicable for all conservation strategies, especially ones that require conservation efforts across large areas of land encompassing many different landowners and land uses.

Additional Sources of Information for Farm Policies

Farm policies are constantly changing and it is often difficult to find current information regarding farm policies and what they mean at the farm level. This section does not provide specific information on current farm polices due to the changing nature of policies. However, we provide several sources for obtaining information about current policies as well as sources that analyze current or proposed policies.

United States Department of Agriculture

http://www.usda.gov/wps/portal/usda/usdahome?navid=farmbill

The most current information about the current farm bill can be obtained from the United States Department of Agriculture's (USDA) website for the farm bill. This page also contains links to the latest farm bill news and blogs.

National Agricultural Law Center

http://nationalaglawcenter.org/farmbills/

The National Agricultural Law Center provides a complete list of web links to both current and historical farm bills as well as Congressional Research Service reports related to farm bills and agricultural programs.

Farm Services Agency

http://www.fsa.usda.gov/programs-and-services/farm-bill/index

http://www.fsa.usda.gov/programs-and-services/index

The Farm Services Agency (FSA) within the USDA also provides information about the farm bill. This site provides highlights of the farm bill as well as specific information on various policies. Individuals can sign up to receive email updates on a regular basis. A listing of all Programs and Services offered by the Farm Service Agency such as Farm Loan Programs and Price-Support Programs is also provided on this page.

Economic Research Service

http://www.ers.usda.gov/topics/farm-economy/farm-commodity-policy.aspx

The Economic Research Service (ERS), which is also part of USDA, has a webpage for Farm & Commodity Policy. This page covers evolving farm and commodity policies. Often new farm bills extend, revise, and replace provisions of previous farm bills. In other cases, provisions of a new farm bill extend, revise, and replace language in laws regulating areas that overlap farm bill authorities, including food and nutrition, food safety, trade, credit, research and Extension, forestry, food safety, organic production, pesticides, and crop insurance. Details on farm bill provisions and related legislation are available at this site as well as reports and articles that analyze the impacts and implications of these policies.

National Association of Wheat Growers

http://www.wheatworld.org/

National Association of Wheat Growers provides weekly news updates on activities and policies that directly impact wheat producers.

OreCal

http://oregonstate.edu/caep/

http://aic.ucdavis.edu/

OreCal publishes briefs as a collaboration between the Center for Agricultural & Environmental Policy at Oregon State University and the University of California Agricultural Issues Center at UC Davis. Their mission is to improve public and private decision making by providing objective economic analysis of critical public policy issues concerning agriculture, natural resources, food systems, and the environment, with an emphasis on the western United States.

Center on Budget and Policy Priorities

http://www.cbpp.org/research

The Center on Budget and Policy Priorities is a research and policy institute that analyzes federal budget priorities.

Farm Policy Facts

http://www.farmpolicyfacts.org/about-farm-policy-facts/

Farm Policy Facts is a coalition of farmers and commodity groups created to educate Congress about the importance of agriculture and to ensure farmers have a voice in the legislative process.

Decision Tools for Sustainable or Climate-Smart Agriculture

The projected impacts of alternative farm policies are dependent upon who adopts the policy changes or who agrees to participate in the new or revised policies. For example, if only 10% of eligible farms are enrolled in a given policy in a specific area, the impacts will be quite different than if there is 90% participation. Often times, policy analysts make assumptions about the level of participation or adoption that may not reflect the actual behavior of farmers in the area. These projections may be significantly off base, especially regarding policies that require substantial changes to existing practices. Without a tool that farmers can use to project or explore the advantages of changing their management practices, and without some means to communicate the resulting changes in net returns to research and policy community, there is little information to guide adoption decisions and little information to make informed ex-ante participation rates projections for proposed policy or policy changes. For example, recent changes in the 2014 Farm Bill have expanded crop insurance and eliminated direct payments, which could lead to marginal lands being brought into production due to the absence of risk. Decision support tools that examine effects of these programs on farm-level decisions and net returns could be used to examine the extent to which this would be true, and to target or adjust problematic areas of the policy.

Being able to make informed decisions at the farm scale is essential to enhancing and expanding sustainable and climate-smart agriculture. Without information that is readily accessible to farmers on the farm-level economic costs and returns of taking certain actions farmers may be reluctant to take those actions (GAO 2014). With this in mind, the USDA is working to provide this type of information to farmers, and the regional climate hubs will be a clearing house for supporting, collecting, and disseminating this type of information.

The advent of mobile computing and communication devices has enhanced our ability to make these informed decisions (Antle et al. 2015). Therefore, it is critical that farmers and land managers have access to decision support tools that will allow them to make more efficient use of inputs and capital and better analyze outcomes and tradeoffs of alternative management pathways. Many farm-level data and decision tools from private and public sources are currently in use and are developing rapidly (Antle et al. 2015). These tools can help farmers better understand the likely impacts external factors, such as changing weather patterns, long-term climate projections, and agricultural policies, could have on the sustainability of their operations. They can also help farmers track and understand the spatial variability on their farms, which can lead to greater efficient use of inputs and enhance the economic and environmental sustainability of their farm.

Presented here is a small sampling of tools and resources that can aid farmers in informed decision making regarding the economic implications of various farm bill policies, daily and long-term management options, as well as enhancing the ability to track and monitor spatial variations in fields which is necessary for efficient allocation of inputs. The reader should be aware that this is not a comprehensive list of tools, nor is it an endorsement of these tools, rather it is a compilation of resources that are available, and some may be more appropriate for individual applications than others. The reader should be aware of that. The decision tools discussed in this section provide users with likely changes in directions and relative magnitudes of key outcomes and indicators. Thus, many of these tools are most useful from a comparative or relative perspective and are not intended as prescriptive recommendations.

Policy Tools and Resources

National Association of Agricultural and Food Policy

https://usda.afpc.tamu.edu/

Several departments within the USDA provide tools and resource aids to farmers. For example, the National Association of Agricultural and Food Policy has developed several tools to better understand the economic implications of choices under the 2014 Farm Bill. The suite of integrated tools is designed to help farmers make the choices required for participation in the 2014 Farm Bill and for choices available under crop insurance.

Natural Resource Conservation Service

http://www.nrcs.usda.gov/wps/portal/nrcs/main/national/technical/tools/

The NRCS also has a suite of tools and resources available to farmers. This site offers links to information, tools, and technical assistance regarding climate change, energy use, cover crops, nutrient management, and many other resources. In the economic tools section for example, there are links to tools in information about general conservation planning, watershed protection, general economic planning, financial functions, investments and retirement, etc. Each link also has a contact person so individuals can quickly email knowledgeable personnel regarding questions they may encounter while researching the tools and resources provided.

Cover Crop Economics Tool

http://www.conservationwebinars.net/webinars/cover-crop-economics-decision-support-tool

An example of one of the tools offered on the NRCS website that would be particularly useful tool for wheat growers in the PNW considering incorporating cover crops into their rotation is the Cover Crop Economics Tool. This is a spreadsheet that measures direct nutrient credits, input reductions, yield increases and decreases, seed and establishment costs, erosion reductions, grazing opportunities, overall soil fertility levels, and water storage and **infiltration** improvements. The tool focuses on monetary benefits and costs. Though there are many benefits to cover crops, they may not always be cost effective in the short run. This tool helps the farmer determine if a cover crop makes sense for their individual situation.

Web Soil Survey (3.1) Tool.

http://websoilsurvey.sc.egov.usda.gov

Another useful tool for planning is the Web Soil Survey (3.1) tool. This tool can be used for general farm, local, and wider area planning by providing maps and tables of soil property data and information. The information is downloadable for use in a local geographic information system. It will also generate a custom soil survey report for any selected area.

Energy Estimators

Other useful tools that estimate economic gains or losses from farm activities include:

Energy Estimator for Irrigation

http://ipat.sc.egov.usda.gov

The Energy Estimator for Irrigation which estimates potential energy savings associated with pumping water for irrigation.

Energy Estimator for Nitrogen

http://nfat.sc.egov.usda.gov

The Energy Estimator for Nitrogen which calculates the potential cost-savings related to N use on a farm or ranch.

Energy Estimator for Tillage

http://ecat.sc.egov.usda.gov

The Energy Estimator for Tillage which estimates diesel fuel use and costs in the production of key crops in a region and compares potential energy savings between conventional tillage and alternative tillage systems.

Other commonly used NRCS tools can be accessed at:
http://www.nrcs.usda.gov/wps/portal/nrcs/detail/national/programs/?cid=STELPRDB1261051

Decision Support Tools Developed and Assembled by the Regional Approaches to Climate Change Project

https://www.reacchpna.org/tools

The decision support tools assembled on the Regional Approached to Climate Change (REACCH) website have been developed based on specific needs faced by dryland cereal farmers in the PNW. Below is a sampling of some of the tools currently available.

- Wireworm Identification Tool—this tool helps identify the species of wireworms in your field.
- Aphid Tracker Map and Aphid Tracker Calculators—interactive maps and aphid calculators are designed to help manage for aphids and aphid viruses in cool season legumes.
- Weather and Winter Wheat Yield—this tool allows the user to visualize year-to-year variability in winter wheat yields for Oregon, Washington, and Idaho over the past three decades and its relationship to variability in climate.
- A variety of climate and weather tools allow users to explore past climate, short-term forecasts and long-term projections for specific locations, these include:
    - Climate Dashboard
    - Climate Projections
    - Seasonal Climate Forecasts
    - Climate Historical Averages
    - USDA Plant Hardiness Zone Maps

AgWeatherNet

http://www.weather.wsu.edu/

This website provides access to current and historical weather data from Washington State University's automated weather station network along with a range of models and decision aids. Weather variables include air temperature, relative humidity, dew point temperature, soil temperature at 8 inches, precipitation, wind speed, wind direction, solar radiation, and leaf wetness. Some stations also measure atmospheric pressure. These variables are recorded every 5 seconds and summarized every 15 minutes by a data logger.

Precision Farming Tools

With the recent proliferation of spatially oriented technologies and data processing, the use of precision agriculture technologies is on the rise, but not all farming locations can take full advantage of these technologies due to lack of broadband data services. The Palouse area, for example, has historically been a difficult region to secure effective broadband coverage due to the terrain. This is partly why President Obama signed an Executive Order in 2012 to make broadband construction along federal roadways and properties up to 90% cheaper and more efficient. Additionally, the Precision Farming Act of 2016 was introduced to help expand the adoption of precision agriculture by providing funding for the expansion of broadband infrastructure and services to cover rural areas. If this act or some variation of this act is passed, more farmers in rural areas will be able to take advantage of precision agriculture technologies that require the use of broadband services to guide tractors, manage data, and access mobile apps.

Precision farming techniques will also help farmers adapt to policies that will likely result from recent initiatives focused on climate-smart agriculture, sustainability, spatially explicit management such as precision conservation, risk management, and environmental policies. Tools such as variable rate technology, unmanned aerial vehicles, and global positioning systems (GPS) will allow farmers to be more precise in their input use, thus benefitting quality and yield outcomes and environmental impacts. For example, variable rate technology can automatically adjust seeding, spraying, and spreading based on such variables as slope, soil texture and moisture content. It can be either map-based or sensor-based; unmanned aerial vehicles can fly over fields and inform farmers about the spatial differences in their fields regarding N needs on specific areas of their fields, or the extent of weeds or insect damage in certain zones; and GPS can enable farmers to geo-reference trouble spots in fields or to guide machinery for more precise tillage and application of seeds and sprays.

Large agricultural and chemical companies such as Monsanto, DuPont, and John Deere have been investing heavily in precision agriculture technologies. For example, Monsanto has developed a system which provides field-by-field recommendations for ways to increase yield, optimize inputs and enhance sustainability, as well as an iPad app that combines historical yield data, satellite imagery, field information about soil and moisture, and plant varieties to make customized variable rate seeding prescriptions for individual fields in order to maximize yield potential on a field-by-field basis. Similarly, DuPont has also developed software to help growers make informed management decisions. Their software combines current and historical field data with real-time agronomic information. They have also developed an app for taking field notes and photos with GPS tags to track field agronomic status. Likewise, John Deere has also invested in the development of GPS guidance and variable rate application systems that help control input costs.

Smartphone Apps for Agriculture

There are numerous smartphone apps for agriculture that range from storm and frost alarm weather apps to weed identification apps. A few sources for farm apps are AgWebAppFinder, Farms.com, and Croplife.com. These sites have application directories that allow users to search for apps by key word. The AgWebAppFinder site also has an Editor's Best and Highest Rated tab to help you find useful apps. Croplife.com and AgWebAppFinder also post periodic reviews of farm apps.

AgWebAppFinder

http://www.AgWebAppFinder.com

Farms.com

http://www.farms.com/agriculture-apps/

Croplife.com

http://www.croplife.com

Fertilizer and Spray Application Decision Tools

Washington State University Fertilizer Application Tools

http://wheattools.wsu.edu/Applications/Fertilizer%20Use%20 Calculator/

Nitrogen fertilizer represents a significant portion of input costs. Overuse of N also contributes to air and water pollution. Therefore, it is important to be able to accurately evaluate the need for fertilizer on a variety of different terrains and soil types. Washington State University has two fertilizer application tools, one is an N calculator and the other is a post-harvest calculator.

Nitrogen Index App

https://play.google.com/store/apps/details?id=gov.usda.ars.spnr. driver&hl=en

This tool can be used to assess the potential risk of N losses associated with a given set of management practices. Users can evaluate how changes in management can reduce the potential risk of nitrate leaching.

The Nitrogen Index for desktop and laptop computers is also available for download at the USDA-ARS Soil Plant Nutrient Research Unit website: ***http://www.ars.usda.gov/npa/spnr/nitrogentools***.

Aphid Calculator

See the section on REACCH decision support tools above.

Greenhouse Gas Tools

Agriculture in the United States produces about 9% of total US greenhouse gas emissions (EPA n.d.). With growing concerns about climate change and the push for either a carbon market or a carbon tax, carbon calculators will be useful tools for analyzing sources of on-farm emissions that will allow producers to determine sources of emissions on their farm as well as possible sequestration options. NRCS has developed three carbon planning tools: COMET-Farm, COMET-Planner and COMET-Energy.

COMET-Farm

http://www.comet-farm.com

COMET-Farm calculates farm-scale greenhouse gas emissions and carbon sequestration associated with farm management practices and strategies. It allows producers to evaluate alternative management strategies and the associated impact on greenhouse gas emissions and carbon sequestration. The user inputs site-specific management data, and then site-specific soil and climate data is used to generate reports that compare current greenhouse gas emissions with emissions from alternative management scenarios that are accurate estimates tailored to an individual's specific situation. It is applicable to all agricultural lands in the lower 48 states.

COMET-Planner and COMET Energy

COMET-Planner and COMET Energy evaluate potential carbon sequestration and greenhouse gas reductions from adopting NRCS conservation practices and reductions in greenhouse gas emissions based on anticipated fuel savings.

For an overview and comparison of more than 30 publicly accessible carbon calculator tools for the agriculture and forestry sectors, see Denef et al. (2012).

Crop Simulation/Yield Tools

The crop simulation and yield tools consist primarily of agronomic relationships among key biophysical factors and management inputs. For the most part, they do not explicitly incorporate economic variables or optimization algorithms and approaches. These tools are essential for building economic tools that require changes in projected yields associated with policy, weather, and management changes.

Decision Support for Agro-Technology Transfer

http://dssat.net/

Decision Support for Agro-Technology Transfer (DSSAT) is a crop simulation model that has been in use for over 20 years, developed through a collaboration of scientists at the University of Florida, University of Georgia, University of Guelph, University of Hawaii, the International Center for Soil Fertility and Agricultural Development, USDA-Agricultural Research Service, Universidad Politecnica de Madrid, Washington State University, and other scientists associated with the International Consortium for Agricultural Systems Applications. The simulated yields are based on site-specific daily weather data, soil characteristics, and crop management activities. The tool is used to evaluate how changes in crop characteristics, management and environmental conditions may impact crop yields.

Kansas Wheat Yield Calculator

https://itunes.apple.com/us/app/kansas-wheat-yield-calculator/id925624508?mt=8

The Kansas Wheat Yield Calculator is a phone app developed by the Kansas Wheat Commission and allows growers to estimate potential wheat yields by collecting information about wheat fields to assess potential yield ahead of harvest.

CropSyst

http://modeling.bsyse.wsu.edu/CS_Suite/CropSyst/index.html

CropSyst is a user-friendly, multi-year, multi-crop daily time step simulation model. The model simulates the soil water budget, soil-plant N budget, crop canopy and root growth, dry matter production, yield, residue production and decomposition, and erosion. Management options include cultivar selection, crop rotation (including fallow years), irrigation, N fertilization, tillage operations (over 80 options), and residue management. The model is designed to be an analytic tool to study the effect of cropping systems management on productivity and the environment.

Whole Farm Assessment Tools

Whole farm assessment tools address impacts and tradeoffs at the farm scale rather than on a field-by-field basis. These are more complex tools that capture the scale and complementarities inherent in farm-level planning, and, for the most part, allow for tradeoffs among operations. For example, we describe two whole farm assessment tools here: AgBalance™ and AgBiz Logic™.

AgBalance

AgBalance is an assessment tool designed by BASF Corporation to analyze agricultural practices on the farm and throughout the chain of production. It is based on environmental, economic, and social indicators that are aimed at helping producers balance demand with sustainable production. It can be used to assess current agricultural practices, identify areas for potential improvements, assess the impact of regulations on products and farming practices, and demonstrate the relationship between farming practices and biodiversity or resource consumption. Findings from this process can be presented to policymakers and partners throughout the food production chain.

AgBiz Logic

AgBiz Logic developed at Oregon State University (with support from REACCH, the Northwest Climate Hub, and a USDA-SCRI grant) consists of several modules: AgBizProfit, AgBizLease, AgBizFinance, AgBizClimate, and AgBizEnvironment (Figure 12-1). The modules are designed help growers assess operational investment choices to make profitable decisions. AgBizProfit can help make short-, medium-, and long-run investment decisions based on profitability. AgBizLease can help establish equitable short- and long-run crop share and cash rent payment leases based on each party's contributions to the lease. AgBizFinance can assist in making long-run decisions on a whole farm and ranch basis based on financial ratios and performance measures. AgBizClimate and AgBizEnvironment (both under development) will incorporate climatic projections and environmental considerations into the decision-making process.

Using a Whole Farm Assessment Tool to Evaluate Tradeoffs

In this section we use AgBiz Logic to illustrate how whole farm decision support tools can be used to evaluate farm-level decisions related to

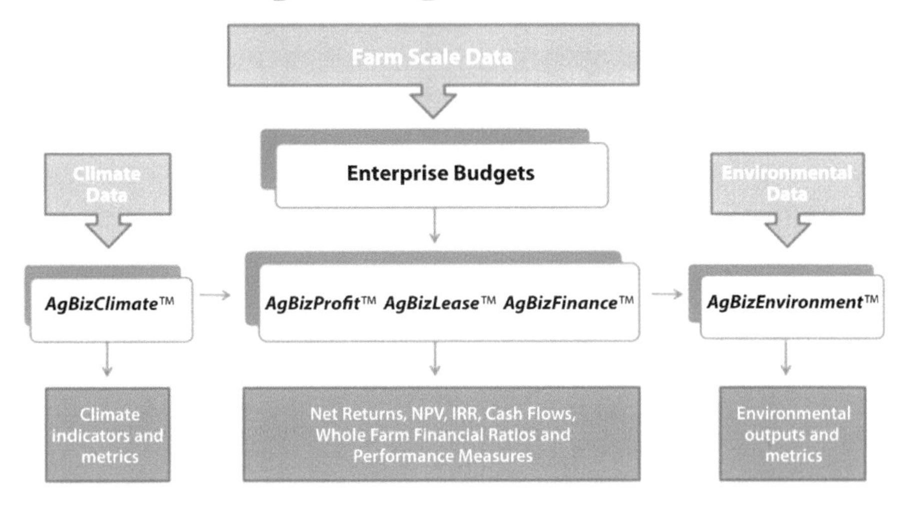

Figure 12-1. AgBiz Logic platform. (NPV = net present value; IRR = internal rate of return.)

agricultural policies and related investment decisions. As mentioned earlier, requirements in the 2014 Farm Bill state that farmers must practice soil and water conservation measures on vulnerable lands in exchange for receiving subsidies for crop insurance premiums. There is also a movement towards more targeted farm policies. Let's assume, then, that in the future, NRCS will be able to offer farms with highly vulnerable land a higher subsidy for crop insurance premiums if they enroll those lands in CRP or some other conservation program. We will illustrate how farmers can use AgBiz Logic to determine if it makes economic sense for their operation to target their conservation efforts.

What follows is an example of a hypothetical case study using the mid-Columbia region of Umatilla County, Oregon, to examine the impacts on profitability and feasibility of alternative cropping systems on leased land when considering precision conservation practices. (In other words, can a grower make money in an investment and does the grower have the financial resources to implement the decision?) We will first present an example of how the AgBizProfit module can be used to evaluate the profitability of various cropping systems using precision conservation. Then we show how AgBizLease can determine if current lease arrangements are equitable given the changes in cropping systems

AgBiz Logic Platform

http://www.agbizlogic.com/

AgBiz Logic consists of the following economic and financial calculators:

AgBizProfit is a capital investment tool that evaluates an array of short-, medium-, and long-term investments. The module uses the economic concepts of net present value, annual equivalence, and internal rate of return to analyze the potential profitability of a given investment.

AgBizLease is designed to help agricultural producers establish equitable short- and long-run crop, livestock and other capital investment leases. The module uses the economic concept of net present value to analyze an equitable crop share or cash rent lease for a tenant and landowner.

AgBizFinance is designed to help agricultural producers make investment decisions based on financial liquidity, solvency, profitability, and efficiency of the farm or ranch business. After an AgBizFinance analysis has been created, investments in technology, conservation practices, value-added processes, or changes to cropping systems or livestock enterprises can be added to or deleted from the current farm and ranch operation. Changes to a business's financial ratios and performance measures are also calculated.

AgBizClimate (under development at the time of this publication) delivers essential information about climate change to farmers and land managers that can be incorporated into projections about future net returns, via changes in expected yields. By using data unique to their specific farming operations, growers can develop management pathways that best fit their operations and increase net returns under alternative climate scenarios.

AgBizEnvironment (under development at the time of this publication) uses environmental models and other ecological accounting to quantify changes in environmental outcomes such as erosion, soil loss, soil carbon sequestration, and greenhouse gas emissions resulting in the ability to incorporate on-farm and off-farm environmental outcomes into the decision support software and platform.

and, if not, what would make a lease equitable and how AgBizFinance can be used to evaluate the feasibility of working capital and solvency when investing in the new equipment and technologies. We then examine the tradeoffs that the farm manager must weigh when making precision agriculture and precision conservation decisions.

Initial setup

In this hypothetical example the farm operation is a representative 2,000-acre dryland farm growing annual crops in a region that receives between 18 to 24 inches of precipitation annually. In keeping with common practice, the producer uses a winter wheat, spring barley, and dry pea crop rotation that includes **direct seeding** to conserve soil moisture, increase yields, reduce soil erosion, and reduce fuel usage. The farm's average yields for winter wheat are 104.5 bushels of winter wheat, 1.8 tons of spring barley, and 1,900 pounds of dry peas per acre. Approximately 867 acres are leased and the farm operator owns the remaining acres. The land lease is based on the landowner receiving 1/4 of the crop and paying 1/4 of the weed control, fertilizer, and crop insurance costs (hail, fire, and crop revenue coverage) and 100% of the property insurance and taxes.

On the 867 acres of leased land the farmer has 307 acres that are identified as highly vulnerable and qualifies the farmer and landowner to receive the higher crop insurance premium subsidy offered for highly vulnerable land areas. The soil on much of this acreage is very cobbly loam, consisting of 45% sand, 43% silt, and 12% clay with 1–7% slope; other vulnerable soils are heavy clays that are extremely saturated in years with above average moisture, which greatly reduces yields (Figure 12-2). Note that this type of soil information can be obtained for any parcel of land from the Web Soil Survey Tool developed by NRCS. Due to the soil quality on these acres, crop yields are generally about 15% lower than average yields for the rest of the farm.

Before the farmers decide to enroll these acres in the conservation program, they will want to examine the cost of investing in precision agriculture equipment that will be necessary for this targeted conservation as well as the impact on net returns. In this example, the initial costs to the tenant for the GPS technologies is assumed to be $25,000, with a 5-year life and a maintenance cost of 2% of the initial cost, resulting in $2.75 per acre

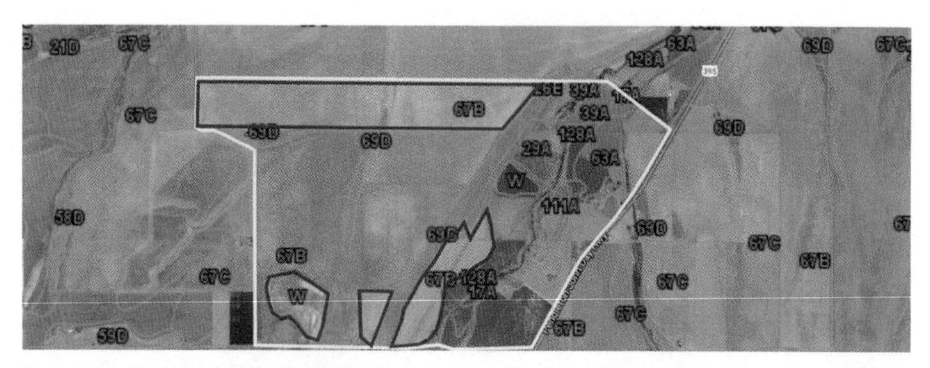

Figure 12-2. Aerial view of the rented land, within the yellow boundary line. Thirty-five percent (or 307 acres) of the leased land that qualifies as highly vulnerable land, within each of the four red boundary lines. The majority of the vulnerable soils are very cobbly loam, consisting of 45% sand, 43% silt, and 12% clay with 1–7% slope. The other vulnerable soils are heavy clays that are extremely saturated in years with above average moisture, which greatly reduces yields.

for all acres on the farm. The landowner has agreed to pay $1,700 for the upfront costs to establish the conservation practices, and the tenant pays the annual costs to maintain the land of $52.

To make a fully informed decision, the landowner and the tenant will want to examine the impacts on the lease arrangement to determine if the current 1/4 crop share lease to the landowner is equitable if the additional land is taken out of production, as well as examine financial feasibility of investing in the new equipment necessary for each scenario. With this information, they can then determine how much of a premium would be needed to compensate for taking land out of production. To do this, the farm operator would initially use the AgBizProfit module to analyze this precision conservation investment strategy based on three options.

The first option is to continue with the current situation of winter wheat, spring barley, and dry peas. The other two options require the farmer to purchase precision agriculture technology such as a GPS guidance and yield mapping systems. Option two is to plant less valuable crops such as canola and camelina on the vulnerable acreage that has lower yielding soils. These crops generally result in lower returns but also have less input costs and have the added benefit of diversifying the crop mix to include bioenergy crops. The third option is to implement a long-term conservation plan of a CRP-type planting. Options two and three require the farmer to purchase precision agriculture technology such as

GPS guidance and yield mapping systems. The next step would be to use AgBizLease to analyze the most profitable cropping strategy to determine how the lease arrangement might change based on the costs of new technology and alternative crops. Then an AgBizFinance analysis can be used to determine if required investments for options two or three can be implemented and funded from annual cash flows or must be financed with a loan. Finally, the results from the analysis can be used by the farmer to examine the tradeoffs involved in each decision and make a fully informed decision.

Data needs

Data needs for AgBizProfit: To use precision farming techniques effectively, it is essential that the producer generate profit and loss maps on a field-by-field basis. This would include identifying the higher and lower yielding areas of a field as well as any variation in production input costs associated with the fluctuating yields. Profit and loss maps generally provide net returns over and above fixed costs. Overlaying the maps of net returns to NRCS soil maps could also validate the variation in yields and profits by soil characteristics, thus illustrating the potential economic benefits of changing cropping systems and purchasing precision farming technologies.

The next set of data includes establishing projected yields for winter wheat, spring barley, and dry peas planted on the higher productive areas of the field as well as the canola and camelina planted on marginally productive soils and any associated production input cost differences. The final set of data required for an AgBizProfit analysis are the costs to establish and maintain the conservation planting on the 307 acres identified as highly vulnerable.

Data Needs for AgBizLease: Equitable land leases are usually calculated based on the cost contributions of the tenant and landowner to the lease. As with the current crop share lease, it is assumed that the tenant provides 3/4 of the costs (labor, machinery, production inputs, etc.) and the landowner 1/4, which includes any production inputs, a reasonable return to the market value of the land, property taxes, insurance, etc. The data required for this analysis are any changes in costs by the tenant and landowner when considering either option two or three.

Data Needs for AgBizFinance: The requirements to conduct a whole farm analysis include balance sheet information, principal and interest payments on current loans (short-, medium-, and long-term), capital leases, and family withdrawals from the business. The outcomes from a whole farm analysis varies greatly by the amount of working capital in the year of the decision, how much farm equity is the owner's versus the lender's, and the anticipated net income from the future investment.

The Informed Decision

Output regarding changes in net returns, the tenant and landowner contributions to total costs, as well as debt-to asset ratios and working capital generated from the AgBiz Logic tool can now be used to make a fully informed financial decision regarding these options. Output regarding differences in net returns as well as the contributions to total costs by the landowner and tenant from our hypothetical example are presented in Table 12-2.

The current lease for Option 2 may not need to be adjusted. However, due to the considerable shift in the landowner's and tenant's contributions of total costs, adjustments to the lease arrangement may be desired in Option 3. An issue with changing lease arrangements is the willingness of the landowner and tenant to modify the current lease. In order to agree to changes in the agreement, there is an education effort that needs to take place so that each party understands and feels comfortable with the reasons behind the changes in input costs and crop returns.

Once it is determined that a particular option such as Option 2 or Option 3 is profitable, the next consideration is to determine if it is financially feasible. The AgBizFinance module is designed to make comparisons of the changes to working capital and debt-to-asset ratios for each option which can then provide the necessary information for a financial decision. Even without examining a specific financial scenario for this hypothetical scenario, it is likely that the tenant could be able to pay for the additional GPS technology from annual cash flows from the entire farm under either option. Therefore, the deciding factor would be determined by the size of the premium subsidy.

Table 12-2. Pertinent Information from the AgBiz Logic modules (AgBizProfit and AgBizLease).

	Option 1 Current	Option 2 Add Canola + Camelina	Option 3 Add Conservation
Crop Planted		**Acres Planted**	
Wheat	434	280.0	280
Dry Peas	217	140.0	140
Spring Barley	217	140.0	140
Canola	—	153.5	—
Camelina	—	153.5	—
Total Acres Planted	867	867	560
Tenants Net Returns	$102,280	$106,286	$75,097
NR (Opt 2-Opt 1)		$4,006	
NR (Opt 3-Opt 1)			-$27,183
NR (Opt 3-Opt 2)			-$31,189
Lease Information			
Tenant's Costs	$190,659	$174,007	$118,744
Landowner's Share of Costs	25%	26%	34%
Tenant's Share of Costs	75%	74%	66%

Conclusion

Decision support tools, such as AgBiz Logic, and other tools that utilize spatially explicit data will likely help shape future agricultural policies. Such tools will aid in the design of policies that address risk management as well as encourage adoption of technology and management options that enhance sustainable climate smart agriculture. Many of these tools can be used by both growers and policymakers to assess potential impacts at the farm and regional landscape scales for a variety of agricultural policies. These policies may influence management practices in inland PNW grain production systems such as conservation cropping, residue and soil water management, crop rotations, and pest management—all of which have the potential to improve crop production and environmental outcomes.

References

AgBiz Logic. *http://www.agbizlogic.com/*

AGree. 2014. Farm Bureau to Discuss 'Big Data' – DTN The Progressive Farmer. *http://foodandagpolicy.org/news/story/farm-bureau-discuss-big-data-dtn-progressive-farmer*

AGree. 2016. Working Landscapes Initiative: Conservation and Crop Insurance. *http://foodandagpolicy.org/sites/default/files/160128%20CCITF%20Handout.pdf*

Antle, J., and L. Houston. 2013a. A Regional Look at the Distribution of Farm Program Payments and How It May Change with a New Farm Bill. Choices. Quarter 4. *http://www.choicesmagazine.org/UserFiles/file/cmsarticle_339.pdf*

Antle, J., and L. Houston. 2013b. Got Subsidies? Who Benefits from Farm Programs and How that Could Change with a New Farm Bill. OreCal Issues Brief #003. *http://static1.1.sqspcdn.com/static/f/2028237/22525527/1384721702237/OreCal+Issues+Brief+003.pdf?token=AYh%2F%2BxkEyIYNyFxJS3D6FmIxoM0%3D*

Antle, J., S. Capalbo, and L. Houston. 2015. Using Big Data to Evaluate Agro-Environmental Policies. Choices. Quarter 3. *http://choicesmagazine.org/choices-magazine/submitted-articles/using-big-data-to-evaluate-agro-environmental-policies*

Becker, G.S. 2002. Congressional Research Service Report for Congress. Farm Commodity Legislation: Chronology, 1933–2002, Order Code 96-900 ENR. *http://nationalaglawcenter.org/wp-content/uploads/assets/crs/96-900.pdf*

Bowers, D.E., W.D. Rasmussen, and G.L. Baker. 1984. History of Agricultural Price-Support and Adjustment Programs, 1933-84. USDA, ERS, Agriculture Information Bulletin Number 485. December 1984. *https://www.ers.usda.gov/publications/pub-details/?pubid=41994*

Chite, R.M. 2014. The 2014 Farm Bill (P.L. 113-79): Summary and Side by Side. Congressional Research Service R43076. ***http://nationalaglawcenter.org/wp-content/uploads/2014/02/R43076.pdf***

Coppess, J. 2016. The Next Farm Bill May Present Opportunities for Hybrid Farm-Conservation Policies. *Choices* Quarter 4. ***http://www.choicesmagazine.org/choices-magazine/theme-articles/looking-ahead-to-the-next-farm-bill/the-next-farm-bill-may-present-opportunities-for-hybrid-farm-conservation-policies***

Denef, K., K. Paustian, S. Archibeque, S. Biggar, and D. Paupe. 2012. Report of Greenhouse Gas Accounting Tools for Agriculture and Forestry Sectors. Interim Report to the USDA under Contract No. GS-23F-8182H. ***https://www.usda.gov/oce/climate_change/techguide/Denef_et_al_2012_GHG_Accounting_Tools_v1.pdf***

Desmond, P. 2016. Big Data and IoT Promise Big Changes to Agriculture. The Enterprisers Project. ***https://enterprisersproject.com/article/2016/8/big-data-and-iot-promise-big-changes-agriculture***

Dimitri, C., A. Effland, and N. Conklin. 2005. The 20th Century Transformation of US Agriculture and Farm Policy. USDA-ERS, Economic Information Bulletin Number 3. ***https://www.ers.usda.gov/webdocs/publications/eib3/13566_eib3_1_.pdf***

Effland, A., J. Cooper, and E. O'Donoghue. 2014. 2014 Farm Act Shifts Crop Commodity Programs Away from Fixed Payments and Expands Program Choices. Amber Waves, July 24, 2014. ***https://www.ers.usda.gov/amber-waves/2014/july/2014-farm-act-shifts-crop-commodity-programs-away-from-fixed-payments-and-expands-program-choices/***

EPA (Environmental Protection Agency). n.d. Sources of Greenhouse Gas Emissions. ***https://www.epa.gov/ghgemissions/sources-greenhouse-gas-emissions#agriculture***

ERS (Economic Research Service). n.d. Farm & Commodity Policy. ***http://www.ers.usda.gov/topics/farm-economy/farm-commodity-policy.aspx***

O'Donoghue, E., A. Hungerford, J. Cooper, T. Worth, and M. Ash. 2016. The 2014 Farm Act Agriculture Risk Coverage, Price Loss Coverage, and Supplemental Coverage Option Programs' Effects on Crop Revenue. Economic Research Report No. (ERR-204).

Hartung, F., and J. Schiemann. 2014. Precise Plant Breeding Using New Genome Editing Techniques: Opportunities, Safety, and Regulation in the EU. *The Plant Journal* 78(5): 742–52.

Hatfield, J., G. Takle, R. Grotjahn, P. Holden, R.C. Izaurralde, T. Mader, E. Marshall, and D. Liverman. 2014. Agriculture. In Climate Change Impacts in the United States: The Third National Climate Assessment, J.M. Melillo, T.C. Richmond, and G.W. Yohe, eds. *US Global Change Research Program* 150–174.

Klümper, W., and M. Qaim. 2014. A Meta-Analysis of the Impacts of Genetically Modified Crops. *PLoS ONE* 2014(9): e111629. ***http://journals.plos.org/plosone/article?id=10.1371/journal.pone.0111629***

Mercier, S. 2012. Farm Program History and Policy Considerations. AGree. ***http://www.foodandagpolicy.org/sites/default/files/AGreeFarmProgramHistoryandPolicyConsiderations_SMercier_0.pdf***

NAS (National Academies of Sciences, Engineering, and Medicine). 2016. Genetically Engineered Crops: Experiences and Prospects. Washington DC: The National Academies Press.

Olen, B., and J. Wu. 2014. The 2014 Farm Bill: What Are the Major Reforms and How Do They Affect Western Agriculture? OreCal Issues Brief No 013. ***http://oregonstate.edu/caep/sites/default/files/PDF/PolicyBriefs/orecal_issues_brief_013_0.pdf***

Pardey, P.G., C. Chan-Kang, J. Beddow, and S. Dehmer. 2015. Long-Run and Global R&D Funding Trajectories: The US Farm Bill in a Changing Context. *American Journal of Agricultural Economics* 97(5): 1312–1323. ***http://ajae.oxfordjournals.org/content/early/2015/06/15/ajae.aav035***

Plume, K. 2014. Farm Groups, Ag Tech Companies Agree on Data Privacy Standards. Reuters. ***http://www.reuters.com/article/2014/11/13/us-usa-agriculture-data-idUSKCN0IX2NU20141113***

Ray, D.E. 2001. Invited Paper Prepared for Presentation at the Southern Agricultural Economics Association Annual Meeting, Ft. Worth, Texas, January 28–31, 2001. ***https://www.google.com/url?sa=t&rct=j&q=&esrc=s&source=web&cd=2&ved=0ahUKEwiFqva5sqHOAhVYz2MKHWtbBVgQFggkMAE&url=http%3A%2F%2Fwww.iatp.org%2Ffiles%2FImpacts_of_the_1996_Farm_Bill_Including_Ag_Hoc.doc&usg=AFQjCNHPkcbwlJ2yBB7Hb5MuXI4Xsg0w7Q&cad=rja***

Ray, D.K., N. Ramankutty, N.D. Mueller, P.C. West, and J.A. Foley. 2012. Recent Patterns of Crop Yield Growth and Stagnation. *Nature Communications* 3(1293). ***http://www.nature.com/ncomms/journal/v3/n12/suppinfo/ncomms2296_S1.html***

RMA (Risk Management Agency). n.d. State Profiles. ***http://www.rma.usda.gov/pubs/state-profiles.html***

Rosch, H. 2016. Why Genome Editing Offers a Targeted Way of Breeding Better Crops. ***http://phys.org/news/2016-04-genome-crops.html#jCp***

Shields, D.A. 2010. Federal Crop Insurance: Background and Issues. Report by the Congressional Research Service, R40532. December 13, 2010. ***http://adriansmith.house.gov/sites/adriansmith.house.gov/files/CRS%20-%20Crop%20Insurance.pdf***

UN (United Nations, Department of Economic and Social Affairs, Population Division). 2015. World Population Prospects: The 2015 Revision, Key Findings and Advance Tables. ESA/P/WP.241. ***https://esa.un.org/unpd/wpp/Publications/Files/Key_Findings_WPP_2015.pdf***

USDA (United States Department of Agriculture). 2014. USDA Strategic Plan FY 2014–2018. ***http://www.usda.gov/documents/usda-strategic-plan-fy-2014-2018.pdf***

USDA (United States Department of Agriculture). 2016. USDA Building Blocks for Climate-Smart Agriculture and Forestry: Implementation Plan and Progress Report. ***http://www.usda.gov/documents/building-blocks-implementation-plan-progress-report.pdf***

Womach, J. n.d. Agriculture: A Glossary of Terms, Programs, and Laws, 2005 Edition. Washington D.C. UNT Digital Library. ***http://digital.library.unt.edu/ark:/67531/metacrs7246/***

About the Authors

John Abatzoglou, PhD, is an associate professor of geography at the University of Idaho.

Elizabeth Allen, PhD, is a research associate in the Center for Sustaining Agriculture and Natural Resources at Washington State University.

John Antle, PhD, is a professor of applied economics at Oregon State University.

Rakesh Awale, PhD, is a post-doctoral researcher at the Columbia Basin Agricultural Research Center, Oregon State University.

Andy Bary is a senior scientific assistant of crop and soil science at Washington State University.

Edward Bechinski, PhD, is a professor of entomology, plant pathology, and nematology at the University of Idaho.

Prakriti Bista, PhD, is a post-doctoral researcher in crop and soil sciences at Oregon State University.

Kristy Borrelli, PhD, is an Extension educator in sustainable agriculture at Pennsylvania State University, formerly an Extension specialist at the University of Idaho.

Nilsa Bosque-Pérez, PhD, is a professor of entomology, plant pathology, and nematology at the University of Idaho.

Tabitha Brown, PhD, is a research associate of crop and soil science at Washington State University.

Ian Burke, PhD, is an associate professor of crop and soil science at Washington State University.

Susan Capalbo, PhD, is a professor of agricultural resource economics and Senior Vice Provost for Academic Affairs at Oregon State University.

Paul Carter, PhD, is a Columbia County Extension agronomist and soil scientist at Washington State University.

Xianming Chen, PhD, is a research plant pathologist for the USDA-ARS Wheat Health, Genetics, and Quality Research Unit and Washington State University.

David Crowder, PhD, is an assistant professor of entomology at Washington State University.

Sanford Eigenbrode, PhD, is a professor of entomology at the University of Idaho.

Rajan Ghimire, PhD, is an assistant professor of cropping systems at New Mexico State University, formerly a post-doctoral researcher at Oregon State University.

Haiying Tao, PhD, is an assistant professor of crop and soil science at Washington State University.

Laurie Houston is a senior faculty research assistant in applied economics at Oregon State University.

David Huggins, PhD, is a research soil scientist with the USDA-ARS Northwest Sustainable Agroecosystems Research Unit and Washington State University.

Kendall Kahl is a research assistant in soil science at the University of Idaho.

Elizabeth Kirby is a research associate in the Center for Sustaining Agriculture and Natural Resources at Washington State University.

Chad Kruger is the director of the Center for Sustaining Agriculture and Natural Resources at Washington State University.

Tai Maaz, PhD, is a post-doctoral researcher in crop and soil science at Washington State University.

Stephen Machado, PhD, is a professor and dryland cropping system agronomist at Oregon State University.

Timothy Murray, PhD, is a professor of plant pathology at Washington State University.

Kate Painter, PhD, is an assistant professor of agricultural economics and Extension educator at the University of Idaho.

Bill Pan, PhD, is a professor of crop and soil science at Washington State University.

Timothy Paulitz, PhD, is a research plant pathologist for the USDA-ARS Wheat Health, Genetics, and Quality Research Unit and Washington State University.

Kirti Rajagopalan, PhD, is a research associate in the Center for Sustaining Agriculture and Natural Resources at Washington State University.

Arash Rashed, PhD, is an assistant professor of ecological entomology at the University of Idaho.

Silvia Rondon, PhD, is an associate professor of crop and soil science at Oregon State University.

Kurtis Schroeder, PhD, is an assistant professor of cropping systems agronomy and plant pathology at the University of Idaho.

Clark Seavert is a professor of applied economics at Oregon State University.

Bradley Stokes is an Extension associate professor at the University of Idaho.

Nicole Tautges, PhD, is a post-doctoral researcher in crop and soil science at Washington State University.

Bertie Weddell, PhD, is a research associate in the Center for Sustaining Agriculture and Natural Resources at Washington State University.

Don Wysocki, PhD, is an associate professor of crop and soil science at Oregon State University.

Georgine Yorgey is the assistant director of the Center for Sustaining Agriculture and Natural Resources at Washington State University.

Frank Young, PhD, is a research agronomist for the USDA-ARS Northwest Sustainable Agroecosystems Research Unit and Washington State University.

Glossary

adaptation. In relation to global climate change, an effort made to reduce the negative consequences of a changing climate or take advantage of new opportunities, for example by adjusting practices, processes, or structures.

aggregate stability. The capacity of soil aggregates to resist disintegration when exposed to external destructive forces.

aggregate. A group of primary soil particles that are held together by organic and inorganic materials.

agroecological class (AEC). A classification system used to characterize the diversity of agricultural cropping patterns. The three classes in the inland Pacific Northwest's dryland cereal production region are (1) Annual Crop with < 10% fallow, (2) Annual Crop-Fallow Transition with 10–40% fallow, and (3) Grain-Fallow with > 40% fallow. The AEC classification system differs from other classification systems that have been applied in the region because distinctions among classes are based on actual land use rather than biophysical characteristics. For additional description of AECs, see Chapter 1: Climate Considerations.

biochar. A solid, carbon-rich, porous material that is generated when organic materials are heated (thermochemically decomposed) in an oxygen-limited environment.

biosolids. The materials produced by municipal wastewater treatment of organic solids, transformed through the treatment process into a product that is made up of living and dead wastewater treatment microorganisms, small inorganic particles, and insoluble compounds.

bulk density. The ratio of the mass of oven-dry soil to its bulk volume.

carbon dioxide (CO_2) fertilization. The principle that increased carbon dioxide in the atmosphere may enhance crop growth by supplying more carbon dioxide to plants.

carbon-to-nitrogen (C:N) ratio. The ratio of the mass of organic carbon to the mass of total nitrogen in soil or in organic material.

cation exchange capacity (CEC). The sum of exchangeable cations that a soil can hold at a specific pH. It is a measure of total negative charges on soil surfaces.

compaction. The physical consolidation of the soil when soil particles are pressed together, reducing porosity (pore space) between them.

conservation (tillage) systems. Management practices that reduce tillage of the soil to provide soil erosion control, soil organic matter retention or accumulation, and soil fertility improvement.

conservation tillage. Tillage systems which retain residues from the previous crop on the surface, resulting in at least 30% coverage of the soil surface after the planting of the next crop.

conventional tillage. Intensive tillage requiring four or more tillage passes a year for seedbed preparation, weed control during a fallow period, and fertilizer incorporation prior to fall planting of wheat or spring planting of other crops. This term can vary in meaning depending on geographical location and date of published information.

cover crop. A crop planted primarily to manage soil erosion, soil fertility, soil health, water, weeds, pests, or diseases in a crop rotation.

direct seeding. A form of conservation tillage where typically a 1–2 pass operation is used for seedbed preparation, planting, and fertilizing. Low-disturbance direct seed operations typically disturb less than

40% of the row width and retain maximum residue cover, whereas high-disturbance direct seed systems may disturb up to 65% of the row width, with moderate surface residue retention. Direct seed operations resulting in greater disturbance, including up to full-width, may be classified as mulch tillage. Low-disturbance direct seeding is often used synonymously with the term no-till.

dispersion. For soils, a process in which individual soil particles are kept separate from one another in a non-aggregated state.

dockage. Weed seeds, weed stems, chaff, straw, or grain other than wheat, which can be readily removed from the wheat by the use of appropriate sieves and cleaning devices; also, underdeveloped, shriveled, and small pieces of wheat kernels removed in properly separating, properly rescreening, or recleaning. The term also may be used to describe the amount of reduction in price taken because of a deficiency in quality.

electrical conductivity (EC). The ability of a soil solution to conduct electrical current; a measure of soluble salts in soil.

element toxicity. Suppressive effect of a nutrient to crop growth and health, which occurs when a nutrient is found in excess of crop needs.

facultative. Facultative small grains or oilseeds generally have reduced vernalization requirements (the period of exposure to cold temperatures required by some plants to enter their reproductive phase) compared to typical winter-type cultivars. Facultative types may be suitable as either a winter- or spring-sown crop. In wheat, facultative types generally have less cold hardiness, a shorter, distinct vernalization period, and initiate spring growth and flowering earlier than true winter wheats.

fallow. The part of a rotation in which cultivated land is not planted.

flex cropping. Cropping systems which provide flexibility in deciding whether the land is left fallow or planted to an alternative crop.

Decisions to grow spring or winter crops or to summer fallow would depend upon market prices, pre-planting soil water status, and weed and disease incidence.

flocculation. For soils, a process of soil aggregation.

green bridge. The presence of host plants for pests (insects and diseases) during times of the year when the crop is typically not present.

growing degree day (GDD). A measure of heat accumulation used to predict plant and pest development rates.

harvest index. The ratio of crop yield to the crop's total aboveground biomass.

herbicide resistance. The acquired ability of a weed population to survive an herbicide application that was previously known to control the population.

herbicide tolerance. The inherent ability of a species to survive and reproduce after herbicide treatment. There has been no selection acting on the tolerant weeds species, and there has been no change in the weed species' lack of response to the herbicide over time.

hydraulic conductivity. The soil's ability to transmit liquid such as water within soil.

immobilization. A reverse of the mineralization process, where inorganic ions are converted into organic forms that are not directly accessible by plants.

infiltration. Downward entry of water into the soil after precipitation or irrigation events.

inland Pacific Northwest (inland PNW). As used in this publication, the area extending from the Cascade Mountain Range in Washington and

Oregon eastward into parts of northern Idaho (see the introduction for a map showing this area). The major area of dryland and irrigated agricultural production includes three major land resource areas with distinctive geologic features and soils as defined by the USDA: the Columbia Basin, the Columbia Plateau, and the Palouse and Nez Perce Prairies, all of which are within the Northwestern Wheat and Range Region. The area also includes a small portion of dryland cropping area in the North Rocky Mountains major land resource area, adjacent to the eastern edge of the Palouse and Nez Perce Prairies.

instars. Larval growth stages of insects.

integrated pest management (IPM). An approach to pest management combining nonchemical approaches with judicious use of pesticides to achieve economically viable pest control. Integrated weed management (IWM) is a type of IPM.

leaching. For soils, downward movement of a dissolved substance with water percolating through soil.

matric potential. A component of soil water potential that results from the combined effects of adsorptive (adhesion) forces and capillarity (cohesion) within the soil matrix to influence water retention and movement.

mineralization. Microbial breakdown of organic substances into plant accessible (inorganic) forms.

mitigation. In relation to global climate change, an effort aimed at reducing the magnitude of climate change by decreasing the cause of that change, such as by reducing greenhouse gas emissions.

nitrification. Enzymatic oxidation of ammonium (NH_4) to nitrate (NO_3) by certain microorganisms in soil.

nitrifier. Microorganisms involved in the nitrification process, such as *Nitrosomonas*, *Nitrobacter*, etc.

nitrogen supply (N supply). The total amount of inorganic nitrogen in the rooting zone, including residual inorganic nitrogen at planting, fertilizer nitrogen inputs, and net nitrogen mineralization during the growing season.

nitrogen uptake efficiency. Ratio of crop nitrogen uptake to nitrogen supply, which characterizes how well the wheat recovers the supplied nitrogen.

nitrogen use efficiency (NUE). Grain yield per unit of nitrogen supply (e.g., wheat grain yield per lb of nitrogen) which characterizes how well the wheat uses the nitrogen supply to produce grain.

nitrogen utilization efficiency. Grain yield per unit of crop nitrogen uptake, which characterizes how well wheat utilizes its recovered nitrogen to produce grain.

no-till. A form of conservation tillage where a 1-pass operation is typically used for seedbed preparation, planting, and fertilizing, resulting in minimal soil disturbance (less than 25–35% of the row width) and maximum retention of surface residue. Often used synonymously with low-disturbance direct seeding.

PAMS. An integrated pest management (IPM) approach based on Prevention, Avoidance, Monitoring, and Suppression strategies outlined as part of the Natural Resources Conservation Service (NRCS) 595 Integrated Pest Management Practice Standard.

pest. In agricultural crops, an organism that causes damage or reduces yield. Weeds, insects, and diseases are types of pests.

pH. A measure of hydrogen ion (H^+) activity in a soil solution on a logarithmic scale (0 to 14) on which 7 is neutral, higher values (> 7) are basic or alkaline, and lower values (< 7) are acidic.

precision agriculture. The management of variable field patches to achieve explicit goals such as those relating to yield, percent protein, and nitrogen use efficiency (NUE). Precision agriculture applies data with high spatial and temporal resolution in order to guide management decisions that are relevant to the high degree of variability associated with agricultural production.

recrop. Planting after another crop without the typical fallow period (e.g., winter wheat after spring wheat harvest).

residue-to-grain (R:G) ratio. The ratio of dry residue yield to grain yield.

resistant. As applied to diseases and insects, the ability of a crop to prevent or suppress infection and reproduction of a pest, limiting the potential for yield loss.

soil health. The capacity of soil to function in a manner that sustains biological activity, maintains environmental quality, and promotes plant and animal health. The term is often used interchangeably with 'soil quality.'

soil microbial biomass (SMB). A measure of the mass of the living component of soil organic matter composed of microorganisms.

soil structure. The arrangement of soil particles into aggregates (groups of primary soil particles) of different shapes and sizes that regulate the amount of air and water present between them.

susceptible. As applied to diseases and insects, prone to infection by and reproduction of a pest, commonly resulting in damage or yield loss.

threshold. As applied to pest control in crops, a limit beyond which pest control measures are advised.

tolerant. As applied to diseases and insects, the ability of a crop to endure infection by and reproduction of a pest, without serious damage or yield loss.

unit nitrogen requirement (UNR). The amount of nitrogen needed to produce a unit of grain, which is the inverse of nitrogen use efficiency (NUE) at the economically optimal yield. For example, for wheat UNR is the amount of nitrogen required to produce one bushel of wheat, based on a grain protein goal in combination with plant-available soil moisture and nitrogen factors.

water holding capacity. The amount of water that a given soil can hold, which is primarily controlled by soil's texture and organic matter content.

water use efficiency (WUE). Grain yield per unit of available water (e.g., wheat grain yield per inch).

weed seedbank. The species composition and density of weed seed in soil.

Index

Page number entries in **bold** denote illustrations or tables.

B

D

H

I

S

U

V

W